HANDBOOK OF PRACTICAL ELECTRICAL DESIGN

Other McGraw-Hill Books of Interest

Anthony • ELECTRIC POWER SYSTEM PROTECTION AND COORDINATION
Cadick • ELECTRICAL SAFETY HANDBOOK
Croft, Summers • AMERICAN ELECTRICIANS' HANDBOOK
Hanselman • BRUSHLESS PERMANENT-MAGNET MOTOR DESIGN
Johnson • ELECTRICAL CONTRACTING BUSINESS HANDBOOK
Johnson • SUCCESSFUL BUSINESS OPERATIONS FOR ELECTRICAL CONTRACTORS
Kolstad • RAPID ELECTRICAL ESTIMATING AND PRICING
Kusko • EMERGENCY/STANDBY POWER SYSTEMS
Linden • HANDBOOK OF BATTERIES AND FUEL CELLS
Lundquist • ON-LINE ELECTRICAL TROUBLESHOOTING
Maybin • LOW VOLTAGE WIRING HANDBOOK
Meland • ELECTRICAL PROJECT MANAGEMENT
McPartland • MCGRAW-HILL'S HANDBOOK OF ELECTRICAL CONSTRUCTION CALCULATIONS
McPartland • MCGRAW-HILL'S NATIONAL ELECTRICAL CODE HANDBOOK
Pete • ELECTRIC POWER SYSTEMS MANUAL
Richter, Schwan • PRACTICAL ELECTRICAL WIRING
Smeaton • SWITCHGEAR AND CONTROL HANDBOOK
Traister • SECURITY/FIRE ALARM SYSTEM DESIGN, INSTALLATION, AND MAINTENANCE

HANDBOOK OF PRACTICAL ELECTRICAL DESIGN

J. F. McPartland
Electrical Consultant   Tenafly, New Jersey

Brian J. McPartland
*Editor   Electrical Contractors
Design and Installation Update
Tappan, New York*

Consulting Editors

Steven P. McPartland
*Instructor   New Jersey State Apprenticeship Training Program
Howell, New Jersey*

James L. McPartland
Electrical Consultant   Beacon, New York

Brendan A. McPartland
*Facilities Engineer   Grubb and Ellis
White Plains, New York*

Second Edition

McGRAW-HILL, INC.
New York San Francisco Washington, D.C. Auckland Bogotá
Caracas Lisbon London Madrid Mexico City Milan
Montreal New Delhi San Juan Singapore
Sydney Tokyo Toronto

Library of Congress Cataloging-in-Publication Data

McPartland, Joseph F.
   Handbook of practical electrical design / J.F. McPartland and
Brian J. McPartland. — 2nd ed.
   p.   cm.
   Includes index.
   1. Electrical wiring—Handbooks, manuals, etc.   I. McPartland,
Brian J.   II. Title.
TK3271.M35   1995
621.319′24—dc20   95-18379
                                                                CIP

Copyright © 1995, 1984 by McGraw-Hill, Inc. All rights reserved. Printed in the United States of America. Except as permitted under the United States Copyright Act of 1976, no part of this publication may be reproduced or distributed in any form or by any means, or stored in a data base or retrieval system, without the prior written permission of the publisher.

2 3 4 5 6 7 8 9   DOC/DOC   9 0 0 9 8 7 6

ISBN 0-07-045820-0

The sponsoring editor for this book was Harold B. Crawford, the editing supervisor was Paul R. Sobel, and the production supervisor was Donald F. Schmidt. It was set in Times Roman by North Market Street Graphics.

Printed and bound by R. R. Donnelley & Sons Company.

> Information contained in this work has been obtained by McGraw-Hill, Inc. from sources believed to be reliable. However, neither McGraw-Hill nor its authors guarantees the accuracy or completeness of any information published herein and neither McGraw-Hill nor its authors shall be responsible for any errors, omissions, or damages arising out of use of this information. This work is published with the understanding that McGraw-Hill and its authors are supplying information, but are not attempting to render engineering or other professional services. If such services are required, the assistance of an appropriate professional should be sought.

This book is printed on acid-free paper.

CONTENTS

Preface   vii

Chapter 1. Planning for Electrical Design   1

Chapter 2. Lighting and Appliance Branch Circuits   15

Chapter 3. Motor Branch Circuits and Control Circuits   173

Chapter 4. Feeders for Lighting and Power   247

Chapter 5. Motor Feeders   393

Chapter 6. Transformer Applications (to 600 v)   415

Chapter 7. Services   491

Chapter 8. Equipment Selection and Layout   585

Chapter 9. Design Reference Data   635

Index   705

PREFACE

"How to design electrical systems" could be the subtitle for this handbook. It is directed to electrical personnel and also to persons in mechanical disciplines—heating, ventilating, air conditioning, manufacturing, processing—who are involved with related electrical applications. This handbook has been prepared as a guide and reference for all designers of electrical systems. It addresses the everyday needs of electrical designers—consulting engineers, electrical and mechanical contractors, plant personnel, architects, electricians, and maintenance personnel. The book explains, in simple terms, the practical procedures involved in the design of electrical circuits and systems to supply lighting motors, appliances, machines, fans, air conditioning, heating, signals and controls in industrial, commercial, and residential applications.

Various phases of electrical design are covered in books, booklets, and magazine articles, but this handbook is a unified and complete treatment of the art of electrical design. It treats all facets of the design of electrical systems, from specific electrical loads and the circuits that serve them to connection to main power-supply lines, and relates the many parts to the whole. To satisfy the ever-increasing need for modern electrical design data, accepted practices are covered thoroughly, and design trends which have particular merit are emphasized. The presentation is based on sound engineering principles and intelligent conformity to the safety provisions of the **National Electrical Code,** and due recognition is given to substantial spare capacity as an essential element of modern electrical design.

The material is presented in the order in which design normally proceeds. First, general considerations relative to the contemplated system and its design are identified. Then, the detailed design of the system begins with provision of the branch circuits to supply the loads directly. From the nature and extent of the branch circuiting, the required system of feeders and subfeeders is developed. Next, the overall distribution scheme is planned. And, finally, the details of the energy supply to the entire system are developed.

Step-by-step details are presented on the sizing, selection, and application of conductors, raceways, switches, fuses, circuit breakers, motor starters, panel-boards, switchboards, motor control centers, transformers, contactors, relays, wiring devices, lighting fixtures, and other related system components. In addition, a very wide variety of practical pointers, illustrations, and examples are provided. All the design information is presented so as to be readily translated into a set of electrical diagrams and sketches—called "the plans"—which constitute the designer's instructions to the electrical installer. To complement these plans, a set of electrical specifications should be prepared. Together, the plans and specifications must fully and clearly instruct the installer in constructing the overall electrical system to achieve the design intent and implement the engineering concepts.

J. F. McPartland
Brian J. McPartland

HANDBOOK OF PRACTICAL ELECTRICAL DESIGN

CHAPTER 1
PLANNING FOR ELECTRICAL DESIGN

Every electrical design project should begin with careful planning based on four vital requirements. The designer must:

1. Ensure conformity to *applicable codes and standards*.
2. Study and establish the *electrical needs of the building*.
3. Determine the *characteristics of the energy supply* to the overall system.
4. Scale details of the overall electrical system to meet the *limitations of budgeted funds*.

The design of an electrical system for any building or outdoor area is basically a matter of providing an arrangement of conductors and equipment to safely and effectively transfer electric energy from a source of power to lamps, motors, and other functional devices which operate on electricity. This simple task is readily reduced to three basic steps which set the outline for the detailed design of any electrical system:

1. Select basic wiring concepts and configurations which will supply electric power of the required characteristics at each point of utilization.
2. Implement the electrical circuiting concepts with actual conductors, apparatus, and hardware, selecting types, sizes, models, characteristics, appearance, ratings, and other specifics of the required equipment.
3. Account for the installation of the overall electrical system, as determined in the first two steps, within the physical dimensions and structural makeup of a building, showing, as clearly as possible, the locations and details of equipment mountings, raceway runs, connections to main power-supply lines, and other elements that require special attention.

Of course, these three steps are interrelated, and particular decisions made within any one step will affect corresponding elements within either or both of the other steps.

As can be seen from the above three steps, the design of an electrical system is expressed in the form of electrical plans. All phases of a design, including the design of subsystems within the major system, should be reduced to a set of blueprints which present schematic wiring diagrams, single-line diagrams, full-wiring hookups, riser diagrams, isometric and other sketches, detail drawings, and equipment schedules—all as necessary to convey a clear picture of the system to the installer.

But before the actual design work begins, many factors must be considered and understood in relation to the contemplated design of an electrical system.

Approaching the Design Task

Successful electrical design for all types of modern buildings depends first of all upon the right approach. Electrical designers must be thoroughly familiar with all the background factors of design. They must, of course, have depth of engineering ability. But over and above this, they must fully understand the relation between the pure technology of an electrical system and such considerations as safe application, capacity for load growth, flexibility in the use of the system, and effective layout. Based on such understanding, the right approach to electrical design is inevitable.

Then the right approach must be correlated with modern standards. Because electrical design is a dynamic thing—the continually evolving product of years of accumulating technology—the electrical designer must learn new techniques and follow promising trends. Electrical design is not an automatic procedure that merely requires filling in formulas and adding parts together. It involves clear understanding of old, accepted techniques and the reasons why those techniques have survived the test of years of application. It requires combinations of old and new wiring techniques and the ability to devise original circuits and layouts for new or special equipment applications. In short, it demands that designers be as dynamic as the art and science they practice. Only with an all-around grasp of the subject can designers exercise judgment in the selection of circuit and feeder arrangements, equipment types and ratings, and installation methods.

Every electrical system must provide power and light without hazard to life or property, with sufficient extra capacity to meet foreseeable load growth, with ready adaptability to load modifications and revised layouts, and with provisions for necessary accessibility in the distribution arrangement.

Safety Is Basic

Compliance with the provisions of the **National Electrical Code** (**NEC**) can effectively minimize fire and accident hazards in any electrical design. The **NEC** sets forth requirements, recommendations, and suggestions and constitutes a minimum standard for the framework of electrical design. As is stated in its own introduction, the **NEC** is concerned with the "practical safeguarding of persons and property from hazards arising from the use of electricity" for light, heat, power, signaling, and other purposes.

Although the **NEC** assures minimum safety provisions, actual design work must include the constant consideration of safety as required by special types or conditions of electrical application. For example, the effective provision of automatic protective devices and the selection of control equipment for particular applications involve engineering skills beyond routine adherence to **NEC** requirements. Then, too, designers must know the physical characteristics—the application advantages and limitations—of the many materials they use for enclosing, supporting, insulating, isolating, and, in general, protecting electrical equipment. The task of safe application based on skill and experience is particularly important in designing for hazardous locations. Safety is not automatically made a characteristic of a system by simply observing codes. Safety must be designed into a system.

The **NEC** is recognized as a legal criterion of safe electrical design and installation. It is used in court litigation and by insurance companies as a basis for insuring buildings. Because the **NEC** is such an important instrument of safe design, it must be thoroughly understood by electrical designers. They must be familiar with all its sections and should know the accepted interpretations which have been given to many of its specific rulings. They should keep abreast of the official interpretations issued by the **NEC** committee. They should know the intent of **NEC** requirements i.e., the spirit as well as the letter of each provision. They should keep informed on interim amendments to the **NEC**. And, most important, they should keep the **NEC** text handy and study it often.

In addition to the **NEC** itself, the electrical designer should have on hand and be familiar with other standards and recommended practices made available in pamphlet form by the National Fire Protection Association. Designers may write to the Association at Batterymarch Park, Quincy, MA 02269 for any of the following:

70	National Electrical Code
70B	Recommended Practices for Electrical Equipment Maintenance
70A	Electrical Code for One- and Two-Family Dwellings
70L	Model State Electrical Law, Inspection of Electrical Installations
76A	Essential Electrical Systems for Health Care Facilities
79	Electrical Metalworking Machine Tools
71	Central Station Protective Signaling Systems
72A	Local Protective Signaling Systems
72B	Auxiliary Protective Signaling Systems
72C	Remote Station Protective Signaling Systems
72D	Proprietary Protective Signaling Systems
73	Public Fire Service Communications
74	Household Fire Warning Equipment
78	Lightning Protection Code
496	Purged Enclosures for Electrical Equipment
493	Standard on Intrinsically Safe Process Control Equipment
30	Flammable Combustible Liquids Code
33	Spray Application
101	Life Safety Code
501	Mobile Homes

Electrical designers must also be familiar with insurance company regulations and all local codes which affect particular installations. They should know in detail where and how local codes depart from the **NEC**. Again, they must understand the basis for special provisions to assure most effective conformity throughout a system. Often, a local condition which requires special code rulings has a significant effect upon the entire design.

OSHA and Third-Party Certification

All modern electrical work must also be related to Occupational Safety and Health Administration (OSHA) regulations and to the standards and application instructions of nationally recognized product testing and certifying organizations, such as Underwriters Laboratories Inc. (UL), Factory Mutual Engineering Corp., and Electrical Testing Laboratories, Inc. OSHA's relation to electrical work is clear. Subpart S, "Electrical," of the Occupational Safety and Health Standards sets forth the electrical regulations that OSHA makes mandatory, especially with respect to design safety standards for electrical systems.

OSHA regulations on electrical installations require third-party certification of the essential safety of the equipment and component products used to assemble an electrical installation and set forth the following:

Approval. The conductors and equipment required or permitted by this subpart shall be acceptable only if approved.

Approved. Acceptable to the authority enforcing this subpart. The authority enforcing this subpart is the Assistant Secretary of Labor for Occupational Safety and Health. The definition of "Acceptable" indicates what is acceptable to the Assistant Secretary of Labor, and therefore approved within the meaning of this subpart.

Acceptable. An installation or equipment is acceptable to the Assistant Secretary of Labor, and approved within the meaning of this Subpart S: (i) if it is accepted, or certified, or listed, or labeled, or otherwise determined to be safe by a nationally recognized testing laboratory, such as, but not limited to, Underwriters Laboratories Inc. and Factory Mutual Engineering Corp.; or (ii) with respect to an installation or equipment of a kind which no nationally recognized testing laboratory accepts, certifies, lists, labels, or determines to be safe, if it is inspected or tested by another federal agency, or by a state, municipal, or other local authority responsible for enforcing occupational safety provisions of the **National Electrical Code**, and found in compliance with the provisions of the **National Electrical Code** as applied in this subpart; or (iii) with respect to custom-made equipment or related installations which are designed, fabricated for, and intended for use by, a particular customer, if it is determined to be safe for its intended use by its manufacturer on the basis of test data which the employer keeps and makes available for inspection to the Assistant Secretary and his authorized representatives.

Other definitions contained in OSHA's electrical design standard are:

Listed. Equipment is "listed" if it is of a kind mentioned in a list which (a) is published by a nationally recognized laboratory which makes periodic inspection of the production of such equipment and (b) states such equipment meets nationally recognized standards or has been tested and found safe for use in a specified manner;

Labeled. Equipment is "labeled" if there is attached to it a label, symbol, or other identifying mark of a nationally recognized testing laboratory which (a) makes periodic inspections of the production of such equipment and (b) whose labeling indicates compliance with nationally recognized standards or tests to determine safe use in a specified manner;

Accepted. An installation is "accepted" if it has been inspected and found by a nationally recognized testing laboratory to conform to specified plans or to procedures of applicable codes;

Certified. Equipment is "certified" if it (a) has been tested and found by a nationally recognized testing laboratory to meet nationally recognized standards or to be safe for use in a specified manner, or (b) is of a kind whose production is periodically inspected by a nationally recognized testing laboratory, and (c) it bears a label, tag, or other record of certification;

Utilization equipment. Utilization equipment means equipment which utilizes electric energy for mechanical, chemical, heating, lighting, or similar useful purpose.

PLANNING FOR ELECTRICAL DESIGN

The use of custom-made equipment is covered in (iii) of the above-quoted OSHA rule. Every manufacturer of custom equipment must provide documentary safety-test data to the owner on whose work premises the custom equipment is installed. And it seems to be a reasonable conclusion from the whole rule itself that custom equipment assemblies must make maximum use of "listed," "labeled," or "certified" components.

Section 110-2 of the **NEC** regulates the use of electrical products and equipment in a rule similar to the OSHA regulation on acceptability of equipment, as described above. It states:

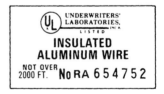

Third-party certification, such as listing by Underwriters Laboratories Inc., is an absolute must for products utilized in modern electrical design work. These typical labels, which are affixed to the products themselves, are visible evidence for the installer and electrical inspector that the products are certified to satisfy Section 110-2 of the **NEC**.

Approval. The conductors and equipment required or permitted by this Code shall be acceptable only if approved.

The **NEC** definition of "approved" is "acceptable to the authority having jurisdiction." This means that the electrical inspector having jurisdiction at any specific installation is the person who will decide what conductors and/or equipment are "approved." Although an inspector is not required to use "listing" or "labeling" by a national testing lab as the deciding factor in the approval of products, inspectors invariably base their acceptance of products on listings by testing labs. Certainly, the **NEC** exhibits almost the same insistence as OSHA that, whenever possible, acceptability must be based on some kind of listing or certification by a national lab. But on this matter, the OSHA law takes precedence—a "listed," "labeled," or otherwise "certified" product must always be used in preference to the same "kind" of product that is not recognized by a national testing lab.

NEC Section 110-3 gives general rules on the "examination, identification, installation and use of equipment." Part (b) of that section says:

> Listed or labeled equipment shall be used or installed in accordance with any instructions included in the listing or labeling.

That rule clearly and certainly says, for instance, that any and every product listed in the UL *Electrical Construction Materials Directory (Green Book)* must be used exactly as described in the application data given with the listing in the book. Because the *Electrical Construction Materials Directory* and the other UL books of product listings, such as the *Hazardous Location Equipment Directory (Red Book)* and the *Electrical Appliance and Utilization Equipment Directory (Orange Book)*, contain massive amounts of installation and application instructions, all those specific bits of application data become mandatory **NEC** regulations as a result of the rule in Section 110-3(b). The data given in the UL listings books supplement and expand upon rules given in the **NEC**. In fact, effective compliance with **NEC** regulations can only be assured by careful study and observance of the limitations and conditions spelled out in the application instructions given in the UL listings books or similar instructions provided by other national testing labs.

Capacity for Present-plus

In general, every electrical system should have sufficient capacity to serve the loads for which it is designed, plus spare capacity to meet anticipated growth in the load on the system. In particular, this means that conductors and raceways must be sized liberally for computed loads; substations, transformers, and switching and protective devices must have the needed capacity and ratings. And spare capacity throughout the branch circuiting should be reflected back through the entire electrical system to the point of power supply.

Allowance for load growth is probably the most neglected consideration in electrical design today. Although the **NEC** contains an almost obscure recommendation that electrical plans and specifications include allowances "for future increases in the use of electricity," experience with current electrical modernization practice indicates that lack of spare capacity plagues existing systems. In all types of buildings, conduit risers are filled to capacity, and the conductors in them are either loaded fully or overloaded. In most commercial and institutional buildings, the entire electrical system is at or near saturation—in branch circuits, feeders, and service. And in the majority of these buildings, modernization of the existing electrical system to handle increased load demands is impeded by an absence of space in which new risers and circuits may be run. It is obvious that the overall design of these electrical systems did not account sufficiently for future load growth.

Modern electrical design, therefore, must include careful planning for future increases in electrical utilization. Depending upon the particular conditions in any installation, the mains, switchgear, transformers, feeders, panelboards, and circuits should be sized to handle considerable load growth. Conductors should be selected on the basis of carrying capacity, voltage drop, and estimated future requirements. Conduit, wireways, troughs, and other raceways should be sized to allow future increases in occupancy. And the space used to house electrical equipment—electric closets, switchgear rooms, substation cages, riser and pipe shafts, etc.—should also be able to accommodate more equipment at a later date.

Every design must include:

Flexibility: Depending upon the type of building—industrial, commercial, or institutional—the electrical system must be designed to provide the required flexibility in distribution and circuiting. The layout and type of equipment should readily accommodate changes in the locations of motors and other utilization devices. Feeders, distribution panelboards, and circuits should be suited to a wide range of utilization patterns, allowing full and efficient use of power capacity for activities in the building's various areas.

Accessibility: Every electrical system should rate high in accessibility. In its final form, the design of the system must provide ease of access to equipment for maintenance and repair and for any possible extensions, modifications, or alterations to the system. The system of conductors, raceways, and equipment must allow for full use of its power-handling ability.

Reliability: Depending upon the nature of the activities within a building, the continuity of electrical supply and overall reliability of the wiring system itself can be a more or less important consideration. Where electric utility companies have a good record of supply continuity, and temporary loss of power would not be a direct life or property hazard, special provisions for a separate emergency power supply are usually not necessary. But in many industrial plants, hospitals, and buildings with essential electricity powered equipment, standby power plants or multiple services must be provided for absolute reliability of supply.

ANALYZING A SPECIFIC SYSTEM

Based on full recognition and appreciation of the foregoing system characteristics, the electrical design for a specific building begins with an analysis of the building type, its loads, and the source of supply. This analysis involves the careful determination of all usual and special electrical requirements for the type of building. The activities to be performed in the building and the nature of electrical usage by the occupants must be considered. Whether the building is a school, an office building, an industrial plant, or a hospital, the designer must know the history of electrical applications in that type of building and must be well informed on current trends and practices.

What Kind of Building?

The general design approach to any contemplated electrical system takes into account the characteristics of the building: Is it small or large; single-story or multi-level; an industrial plant, office building, apartment house, school, or hospital; and so on. These characteristics give insight to the types of electrical utilization, the need for flexibility, the accessibility of the system, and the duty cycles of various load devices. The many and varied requirements for hazardous locations also serve to give direction to the design approach. In such a general approach, the designer should select many design possibilities which suit the particular building and reject all methods and techniques which are immediately seen to be inapplicable.

In general, planning for the design of an electrical system should begin with determination and study of the size and nature of the total load to be served. This means approximation of lighting loads on a watts per square foot basis, analysis of the number and sizes of motors to be served in various areas of the building, and determination of the amounts of other utilization loads and their concentrations through the building. A full understanding of all loads—their sizes and points of application—is essential to the selection of the best type of distribution system.

Standardizing Equipment

Maximum standardization in equipment type and ratings should always be a design objective. The selection of standard supply voltage and standard values at all voltage levels results in significant economy since standard-rated transformers, switchgear, motors, and other equipment cost less than special equipment for nonstandard voltages.

Lack of standardization in an electrical system complicates maintenance because replacement parts are not easy to obtain, a large inventory of parts and equipment must be maintained, and the efficiency of maintenance personnel is reduced. And the use of special, nonstandard equipment and voltages may seriously impair expansion or alteration of the electrical system at a later date. In addition, use of proprietary equipment will limit the number of potential suppliers, which can result in little, or no, competition for a single supplier. Where the competition is limited, there will be little incentive for the "sole supplier" to offer competitive pricing. However, where special, nonstandard equipment is necessary for the particular function of a building, such equipment must be carefully selected and integrated into the system.

Providing the System Power Supply

Another preliminary consideration that fundamentally affects the design procedure is the nature of the power supply which will serve the building's electrical system. The power may be supplied by either the distribution system of an electric utility company or a private generating plant.

Utility company distribution is the most common type of supply to buildings and plants. The purchase of energy usually represents decided economy over the private generation of electric energy within the building or plant. Of course, in certain cases such as paper and pulp mills, the need for large amounts of process steam makes possible the use of excess steam to generate energy economically. In such cases, the generating plant may be operated independently of a utility company supply to the building or in parallel with it.

If a private generating plant operates independently, part of the total electrical load may be connected to it and part to the utility system. The use of a generating plant in parallel with a utility company supply must be checked and coordinated with utility company engineers.

If power is obtained from a utility company line, the characteristics of the supply must be matched to the requirements of the building. Depending upon the voltage and capacity of the supply, one particular type of distribution is often best suited to carrying the electrical loads in the building.

The purchase of power at utilization voltages indicates certain types of distribution within the building; higher-voltage services and primary supplies also indicate particular types of distribution systems. When several different supply voltages are available, each should be appraised in relation to the various distribution methods which might be used. Consultation with the utility company concerning the relation between the different services and interior distribution systems should precede any decisions.

CHECKLISTS OF MODERN DESIGN TECHNIQUES

Industrial Projects

Electrical load density varies with the particular type of industrial operation and can run well over 20 W/ft^2. Typical elements of modern industrial electrical systems include the following:

- Primary distribution to loadcenter substations
- Armored-cable feeders rated for 5-kV or 15-kV circuits
- Extensive use of 480-V power and 277-V fluorescent and/or metal-halide lighting with energy-efficient ballasts (electronic or other types)
- Busway and/or armored cable for secondary voltage feeders
- Cable troughs and/or racks carrying suitably protected cables, for maximum system flexibility and accessibility
- Rigid nonmetallic conduit in corrosive locations
- Underfloor power ducts for flexible circuiting to machine loads throughout widespread plant areas
- Plug-in busway for ready tapping to power and light loads
- Carefully designed hazardous-location installations
- Thoroughly engineered protection against system faults
- Adequate interrupting capacity for all devices that interrupt current flow—both control switching and protective equipment
- Coordinated use of short-circuit and overload protective devices to effectively isolate faulted sections of the system without interrupting service to other sections
- Modern, totally enclosed, dead-front switchgear for maximum safety, designed for easy maintenance
- High-power-factor operation with excellent regulation of voltage at all levels in the plant
- Variable speed motor drives
- Programmable logic controllers (PLCs) and other CPU-based controls

Office Buildings

Typical load densities for modern office buildings can run over 15 W/ft^2 where electric usage is at a maximum for light and power facilities. General area lighting may account for up to 8W/ft^2; power loads, 4 W/ft^2 or more; and small machines and incidental loads, up to 2 W/ft^2. Typical elements include:

- Primary supply to two or more unit substations strategically placed within the building
- Adequate interrupting capacities and coordinated protection
- Effective protection against ground-fault currents of low values
- Three-phase, 4-wire, 480/277-V distribution for power and general area fluorescent lighting
- Electronic ballasts and low-wattage fluorescent lamps
- Busway risers
- Air-handling luminaires to provide lighting and air conditioning from single units for module layouts
- Step-down local transformers for 120/208-V, 3-phase, 4-wire supply to incandescent lighting and convenience-receptacle circuits
- Underfloor raceway, through-floor fittings, or undercarpet flat-conductor cable systems for power, light, telephones, and signals in large office areas with shifting loads or layouts
- Effective alarms, signals, and communication facilities—fire alarms, burglar alarms, security-officer systems
- Remote-control switching

- Raised floors for underfloor circuiting to computers in data-processing centers
- Variable-speed motor drives
- CLPU-based fire, alarm, security, and building management systems
- Voice and data utilities between floors
- Power conditioning and/or UPS for data and communications equipment

Shopping Centers

Lighting-load densities vary widely among the different areas of shopping centers—from 3 to more than 10 W/ft^2—depending on the sizes and types of stores, the relative amounts of fluorescent and incandescent lighting, the light used for decoration, and the amount of outside lighting. Power-load densities depend upon the facilities selected. Typical elements include:

- Primary distribution to loadcenter substations
- Use of 480/277 V among and within buildings
- Contactors and split-bus panelboards for automatic control of large blocks of sign, display, and outdoor lighting
- Modern store lighting

- Music-distribution, intercom, and paging systems
- Modern parking-lot lighting: fluorescent, mercury-vapor, metal-halide, high-pressure sodium, and/or low-pressure sodium luminaires
- Branch circuits for computerized cash registers
- Data cables for cash-register interconnect

Schools

Lighting loads for modern schools vary with the type and layout of the building and the extent to which daylighting is utilized in classrooms. Typical load densities range from 3 to 7 W/ft^2. Power loads typically include air conditioning, electric heating, ventilation, compressors, and sometimes even elevators. Typical elements are:

- Distribution at 4160 V (or higher, where direct utility supply requires) to two or more loadcenters in large-area, multibuilding school layouts
- On-site electric generation for air-conditioned schools

- Use of 480/277-V distribution for power and lighting loads
- Electronic ballasts and low-wattage lamps
- "Showcase" wiring and equipment application in laboratories and craft shops, providing integration in the teaching of electricity and its use
- Modern gymnasium lighting, electric scoreboards, and public-address equipment
- Engineered lighting and power facilities for auditoriums, including programmed dimmer control of stage lighting and house lights
- Variable-speed drives for motor loads
- Intercom and paging systems
- Closed-circuit television systems for instructional purposes
- Clock-and-bell program system
- Modern fire-alarm system
- Vandal-alarm system to protect against vandals and other intruders during closed hours
- Data cabling and power conditioning/UPS protection for computer stations and centers

Hospitals

Typical electrical demand for modern hospital facilities can range up to 3000 W per bed where maximum use is made of electrical equipment. Typical elements include:

- Loadcenter distribution where possible
- Careful provision for electric supply continuity in case of failure of normal supply: two separate services, a special emergency service, an emergency standby power generator, or a standby supply for essential lighting and other essential facilities
- Electronic ballasts and low-wattage lamps
- Engineered application of modern signal and communication equipment: call-and-register systems, paging systems, intercom, closed-circuit television, radio-program distribution, data systems, and building management systems
- Modern wiring of laboratory facilities and x-ray equipment
- Careful compliance with all **NEC** requirements on wiring in operating rooms (ungrounded circuits and ground-detector-and-alarm system) and in hazardous (anesthetizing) areas

CHAPTER 2
LIGHTING AND APPLIANCE BRANCH CIRCUITS

A *branch circuit* is any segment of a wiring system extending beyond the final automatic overcurrent protective device that is approved for use as branch-circuit protection and designated by the **NEC** as the branch-circuit protective device (Fig. 2.1). Branch circuits generally originate in panelboards, but individual branch circuits to motors commonly originate at individual fused switches or circuit breakers tapped from busways or nippled out of auxiliary gutters.

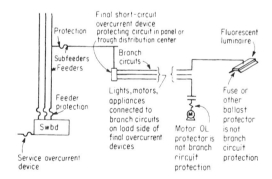

FIGURE 2.1 Definition of a branch circuit.

In an electrical system, the branch circuits are the circuits of lowest capacity and current rating. They are the circuits to which load devices—lights, motors, etc.—are connected. Of course, it is possible (and not at all uncommon) to have branch circuits of very high rating—for instance, a branch circuit with conductors rated at 450 A to supply a 150-hp, 230-V, 3-phase motor. But the vast majority of branch circuits are rated not over 30 A.

Thermal cutouts or motor overload devices are not branch-circuit protection. Neither are fuses in luminaires or in plug connections, where they are used for ballast protection or individual fixture protection. Such "supplementary overcurrent protection" is the type of individual fixture or appliance protection that is connected on the load side of branch-circuit protection. **NEC** Sections 424-22(c) to (e) cover the use of the supplementary overcurrent protection required for electric-resistance duct heaters.

Figure 2.2 shows a common situation for bridge and highway circuits, in which a number of large lamp loads are fed by a feeder with individual fusing for each fixture; this does result in the tap circuits becoming the branch circuits of the system.

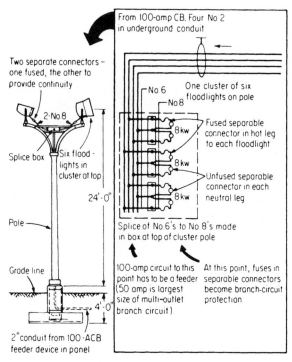

FIGURE 2.2 Common variation from standard branch-circuit layout.

In its simplest form, a branch circuit consists of two wires which carry current at a particular voltage from protective device to utilization device. Although the branch circuit represents the last step in the transfer of power from the service or source of energy to the utilization devices, it is the starting point for modern design procedures. First, the loads are circuited. Then, the circuits are lumped on the feeders. Finally, the distribution system is connected to one or more sources of power.

Each and every branch circuit—whether for a power or lighting load in a commercial, industrial, or residential building—should be sized for its load, with spare capacity added where possible load growth is indicated, where necessary to reduce heating in continuously operating circuits, and/or for voltage stability. The design should also provide for the economical addition of circuits to handle future loads. Each circuit must provide the required power capacity at full utilization voltage at every outlet.

Accessibility and flexibility are two important characteristics of effective branch circuiting. In commercial and industrial buildings, the shifting of load devices is common, as a result of tenant changes, rearrangement of office layouts, changes in production schedules, or relocation of production lines. Circuits in such areas must readily accommodate these shifts and must allow extensions.

Selecting Branch-Circuit Voltage

In general, modern branch-circuit design has developed from the trend toward high distribution voltages, incorporating the favorable economic and operating characteristics of loadcenter layout. As a result, branch circuits are short runs from strategically located panelboards or transformer loadcenters, instead of long circuits from a central distribution panelboard. The high voltage is delivered to the loadcenter, where it is reduced to utilization levels. Or, in a one-voltage system, the feeder is carried to the approximate physical center of a load concentration area before it is subdivided into branch circuits. The use of loadcenters and the resulting short branch circuits provide better-regulated voltage supplies to utilization devices and minimum circuit disturbances as a result of load changes.

Modern electrical systems use four basic voltage configurations (or systems) for branch circuits supplying utilization voltages at light and power outlets. These are derived from the common secondary distribution systems, as follows.

Single-Phase, 3-Wire System. This is a commonly used system in individual residences and small apartment and commercial buildings. Both lighting and single-phase motor loads can be served. Two-wire, 240-V branch circuits can be used for power loads; 3-wire, 120/240-V circuits for lighting outlets, split-wired duplex receptacles, and some power devices such as electric ranges; and 2-wire, 120-V circuits for lighting and receptacle outlets.

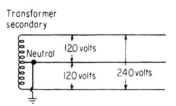

Three-Phase, 4-Wire Wye System. This is the most widely used 3-phase secondary distribution system. The most common use of this configuration is the 120/208-V, 3-phase, 4-wire system with the neutral grounded. With this system, a variety of circuits is available: 4-wire, 120/208-V circuits; 3-wire, 120/208-V circuits; 3-wire, 208-V circuits; 2-wire 208-V circuits; and 2-wire, 120-V circuits. Such a system can serve a combination of power and lighting loads; it offers flexibility for circuit layout and the application of required utilization equipment.

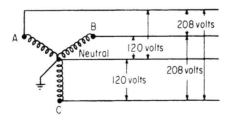

Another 3-phase, 4-wire, wye-connected system which has gained rapidly in acceptance and application is the 480/277-V wye-connected system (also called the

460/265-V system, owing to the voltage spread). Under certain conditions, this system offers more advantages and economy in commercial building applications than the 120/208-V system. The system makes available three types of branch circuits: 480-V, 3-phase circuits for motor loads; 277-V, single-phase circuits for fluorescent or mercury-vapor lighting: and 120-V, 240-V, or 120/208-V circuits from step-down transformers for receptacle circuits and miscellaneous loads.

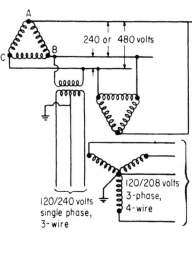

Three-Phase, 3-Wire System. This is the common delta-connected secondary system with phase-to-phase voltage of 240, 480, or 600 V between each pair of phase conductors. This system is used where the motor load represents a large part of the total load. In some such systems, 3-phase, 3-wire circuits at 480 V feed individual motor loads, with single- or 3-phase step-down transformers, stepping from 480 to 120 V used to supply lighting and receptacle circuits.

Three-Phase, 4-Wire Delta System. This is a variation on the 3-phase, 3-wire delta system above; it is often called a "red-leg" delta system because the phase leg with higher voltage to ground than the other two phase legs is commonly painted red to differentiate it. One of the transformer secondary windings supplying the system is center-tapped to derive a grounded neutral conductor to the two phase legs between which it is connected. Motors are supplied at 240 V, either 3-phase or single-phase, with single-phase 120-V branch circuits taken from the grounded center-tapped conductor and its associated phases.

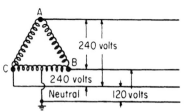

In some areas, electrical systems are supplied from two coexisting services—a 3-phase, 3-wire system for power loads and a separate 120/240-V, single-phase or 120/208-V, 3-phase grounded system for lighting and appliance loads.

Other system configurations are commonly used to supply branch circuits, but they are derived from these basic systems. Particular local conditions and/or requirements for load balance throughout an overall system frequently dictate the use of less common derivations of branch circuits. Such arrangements include 120/208-V, 3-wire feeds to branch-circuit panelboards, open-delta 3-phase supply to light and power branch circuits, and other special transformer hookups supplying branch-circuit loads. Motor branch circuits may be served directly from high-voltage supplies.

Basic Circuit Concepts

The design of a lighting layout for any commercial, institutional, or industrial area includes determination of the required intensity of lighting and selection of the general type of lighting (incandescent, fluorescent, mercury-vapor, metal-halide, high-pressure sodium, low-pressure sodium, or some combination of these), number of lamps per luminaire, number of luminaires, types of luminaires, mounting details, wiring methods, and other specifics. The lighting design for a particular area provides a value for the required circuit power (wattage) capacity. This value is used in determining the number and type of lighting circuits needed.

As shown in Fig. 2.3, **NEC** Article 100 gives definitions covering three types of branch circuits. The fundamentals of safety in lighting and appliance branch-circuit design are given in **NEC** Articles 200, 210, and 220. As set forth in Section 210-1, Article 210 covers branch circuits supplying lighting or appliance loads or combinations of such loads. Where motors or motor-operated appliances are connected to any branch circuit that also supplies lighting or other appliance loads, the provisions of both Article 210 and Article 430 apply. Article 430 applies where branch circuits supply motor loads only.

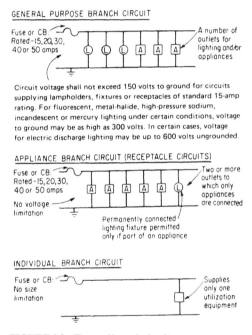

FIGURE 2.3 Types of branch circuits.

A "branch circuit" as covered by Article 210 may be a 2-wire circuit or a "multiwire" branch circuit. A multiwire branch circuit consists of two or more ungrounded conductors having a potential difference between them and an identified grounded conductor connected to the neutral conductor of the system and having equal potential difference between it and each of the ungrounded conductors. Thus, a 3-wire circuit consisting of two opposite-polarity ungrounded conductors and a neutral derived from a 3-wire, single-phase system or a 4-wire circuit consisting of three different phase conductors and a neutral of a 3-phase, 4-wire system is a *single multiwire branch circuit,* not three circuits (Fig. 2.4). It is only one circuit even though it may be protected by two or three single-pole protective devices or a multipole protective device in the panelboard. This is important, because other sections of the **NEC** refer to conditions involving "one branch circuit" or "a two-wire or multiwire branch circuit." (See Sections 250-24 and 410-31.)

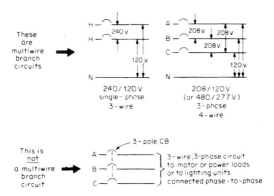

FIGURE 2.4 "Multiwire" branch circuit must include a neutral conductor.

A branch circuit is rated according to the setting or rating of the overcurrent device used to protect the circuit. Any branch circuit with more than one outlet, other than an "individual branch circuit" (see definition in Fig. 2.3), must be rated at 15, 20, 30, 40, or 50 A. That is, the protective device must have one of those ratings for multioutlet circuits, and the conductors must meet the other size requirements of Article 210. However, in industrial plants a circuit rated over 50 A is permitted to have two or more receptacle outlets to provide for easy relocation and connection of a single load such as a welder (Fig. 2.5).

Under the definition for "receptacle" in **NEC** Article 100, it is indicated that a duplex convenience outlet (a duplex receptacle) is considered to be two receptacles and not one—even though there is only one box (Fig. 2.6). Thus, a circuit that supplies only one duplex receptacle is not an "individual branch circuit" as defined in Article 100 because two loads (appliances, tools, etc.) could be connected to it. Even though there is only one box and, technically, only one "receptacle outlet," the circuit must conform to the rules for circuits with "more than one outlet."

It is very important to remember that the size of the overcurrent device determines the rating of any circuit covered by Article 210, even when the conductors used for the branch circuit have a current rating higher than that of the protective device. In a typical case, for example, a 20-A circuit breaker in a panelboard might be used to protect a branch circuit in which No. 10 conductors are used as the circuit wires. Although the load on the circuit does not exceed 20 A, and No. 12 conductors would have sufficient current-carrying capacity to be used in the circuit, the No. 10 conductors with their rating of 30 A were probably selected to reduce the voltage drop in a long homerun. But the rating of the circuit is 20 A because that is the size of the overcurrent device. The current rating of the wire must generally be not lower than the rating of the protective device, but it does not enter into the current rating of the circuit.

LIGHTING AND APPLIANCE BRANCH CIRCUITS

BASIC RULE —

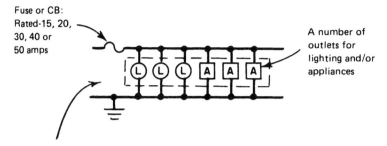

Circuit voltage shall not exceed 150 volts to ground for circuits supplying lampholders, fixtures or receptacles of standard 15-amp rating. For incandescent or electric-discharge lighting under certain conditions, voltage to ground may be as high as 300 volts. In certain cases, voltage for electric discharge lighting may be up to 500 volts "between conductors" and may be an ungrounded circuit.

EXCEPTION —

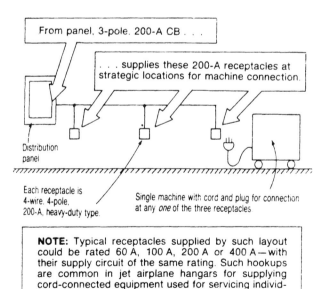

NOTE: Typical receptacles supplied by such layout could be rated 60 A, 100 A, 200 A or 400 A — with their supply circuit of the same rating. Such hookups are common in jet airplane hangars for supplying cord-connected equipment used for servicing individual planes in their hangar bays.

FIGURE 2.5 A multioutlet branch circuit must usually have a rating (of its overcurrent protective device) at one of the five values.

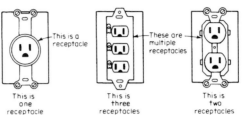

FIGURE 2.6 Each point at which a plug cap may be inserted is a receptacle.

Although multioutlet branch circuits are limited in rating to 15, 20, 30, 40, and 50 A, a branch circuit to a single-load outlet (for instance, a branch circuit to one machine or to one receptacle outlet) may have any current rating. For instance, there could be a 200-A branch circuit to a special receptacle outlet, or a 1000-A branch circuit to a single machine.

Number of Circuits

The number of branch circuits required to handle the general lighting load depends on the total load to be served, the layout of the lighting system and outlets, the amount of load to be placed on each circuit, and the capacity of the circuits to be used. According to the **NEC**, the number of circuits shall be not less than that determined from "the total computed load" and the capacity of the circuits to be used. The "total computed load" for general lighting is the load determined from **NEC** Section 220-3(b) and represents the minimum load for which branch-circuit capacity must be provided—even if the actual connected load for general lighting is lower in kilowatts or kilovoltamperes.

Throughout any building or any given area of a building, the "total computed load" for general lighting must be taken as not less than the value obtained by multiplying the total number of square feet of floor area covered by the general lighting by the minimum unit load in watts per square foot as specified in **NEC** Table 220-3(b) for the particular type of occupancy served by that general lighting. Note that the minimum-unit load values given in the table apply to the type of area that the general lighting serves. In a particular building, the entire area may be of the same type—all store areas or all office areas—or a number of different areas may have to be considered. For instance, for any and every area of a building that is office space, the unit load of 3½ W/ft^2 must be multiplied by the number of square feet of such area to determine the minimum "branch-circuit load" or "total computed load." For an area of the building that is a restaurant, a load of 2 W/ft^2 must be used. A courtroom in the same building has a minimum branch-circuit load ("total computed load") of 2 W/ft^2. Each store area must be taken at 3 W/ft^2.

In all cases, calculation of the "total computed load" establishes the minimum general lighting load for which branch circuits must be provided—regardless of the fact that very efficient modern light sources and fixtures can readily provide a high-quality high-footcandle lighting environment at unit loads of much less than those shown in the table and would require less branch-circuit capacity than that needed for the "total computed load." However, in any case where the actual connected load

LIGHTING AND APPLIANCE BRANCH CIRCUITS

of general lighting equipment exceeds the watts-per-square-foot load shown in the table, the required number of branch circuits (and the capacity of the branch circuits) must be increased to serve the connected loads adequately in accordance with all applicable **NEC** rules. Section 220-4(a) says, "In all installations, the number of circuits shall be sufficient to supply the load served."

WARNING: Although Section 220-4(a) does say that "the minimum number of branch circuits shall be determined from the total computed load and the size or rating of the circuits used," Section 220-4(d) makes an important exception to that requirement. Although Section 220-4 is headed "Branch Circuits Required," part (d) goes beyond branch circuits and makes clear that a feeder to a branch-circuit panelboard and the main busbars in the panelboard must have a minimum ampacity sufficient to serve the calculated total load of lighting, appliances, motors, and other loads supplied. And the amount of feeder and panel ampacity required for the general lighting load must not be less than the ampere value determined from the circuit voltage and the total power (watts), where the latter is obtained by multiplying the minimum unit load from Table 220-3(b) (watts per square foot) by the area of the occupancy supplied by the feeder—even if the actual connected load is less than the calculated load determined on the watts-per-square-foot basis. (Of course, as noted above, if the connected load is greater than that calculated on the watts-per-square-foot basis, the greater load value must be used in determining the number of branch circuits, the panelboard capacity, and the feeder capacity.)

It should be carefully noted that the first sentence of Section 220-4(d) states, "Where the load is computed on a volt-ampere-per-square-foot basis, the wiring system up to and including the branch-circuit panelboard(s) shall be provided to serve not less than the calculated load." The phrase "wiring system up to and including" requires that a feeder must have sufficient capacity for the total minimum branch-circuit load, determined as the area (square feet) times the minimum unit load [watts per square foot, from Table 220-3(b)]. And the phrase clearly requires that amount of capacity in every part of the distribution system supplying the load. The required capacity would, for instance, be required in a subfeeder to the panel, in the main feeder from which the subfeeder is tapped, and in the service conductors supplying the whole system.

Actually, reference to "wiring system" in the wording of Section 220-4(d) presents a requirement that goes beyond the heading "Branch Circuits Required" of Section 220-4 and, in fact, constitutes a requirement on feeder capacity that supplements the rule of the second sentence of Section 220-10(a). This requires a feeder to be sized to have enough capacity for the "computed load" as determined by Part (a) of this article (which means, as computed in accordance with Section 220-3).

A second part of Section 220-4(d) affects the required minimum number of branch circuits. Although the feeder and panelboard must have a minimum ampacity equivalent to the "computed load," it is only necessary to install the number of branch-circuit overcurrent devices and circuits required to handle a connected load that is less than the calculated load. The last sentence of Section 220-4(d) is clearly an exception to the basic rule of the first sentence of Section 220-4(a), which says that the minimum number of branch circuits "shall be determined from the total computed load." Instead of supplying that minimum number of branch circuits, it is necessary to have only the number of branch circuits required for the actual total "connected load."

For general lighting in an office area, Fig. 2.7 shows a typical comparison between the actual load and the minimum "total computed load." The layout of 277-V lighting fixtures is the result of a lighting design that provides a high-quality illumination level using efficient fluorescent lamps. The total load of all the lighting fixtures comes to only 92,000 W. However, the rule of Section 220-4 says that branch circuits for lighting "shall be provided to supply the loads computed in accordance with Section 220-3." Because the "total computed load" must be taken as 3.5 W/ft^2 × 40,000 ft^2 of floor area, or 140,000 W, the first sentence of Section 220-4 and the first sentence of part (a) of that section would appear to require a total branch-circuit capacity of 140,000 W. If 20-A, 265-V circuits are used, and the load on each circuit must be limited to 16 A because the lighting is "continuous" (operates for 3 hours or more), that basic rule would seem to call for 140,000 ÷ (16 A × 265 V) or 33 branch-circuit protective devices in the panel serving the office area. But, the last sentence of Section 220-4(d) says, "Branch-circuit overcurrent devices and circuits need only be installed to serve the connected load." As a result, the minimum required number of branch-circuit poles then becomes 92,000 W ÷ (16 A × 265 V), or 22 branch-circuit protective devices. However, as discussed later, the panelboard busbars, the feeder to the panel, and the service-entrance conductors for the building must have their load-current ratings based on the "total computed load" of 140,000 W and not on the 92,000-W actual load.

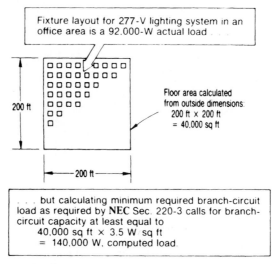

FIGURE 2.7 Minimum required branch-circuit capacity that must be provided for general lighting is based on actual load and "computed load."

There seems to be a great deal of confusion about utilization voltages—that is, the voltages we should expect to see on equipment nameplates and use in our calculations. This is understandable to some extent because there are two industry-recognized documents that cover this topic: the 1993 edition of the **NEC** and American National Standards Institute (ANSI) Standard C-84.1 "Voltage Ratings for Electric Power Systems and Equipment (60 Hz)." And these two documents do not seem to fully agree as to what voltage value is to be used when sizing branch circuit, feeder, and service conductors.

Article 220 of the **NEC**, "Branch-Circuit And Feeder Calculations," specifies the voltage values that are to be used for branch-circuit and feeder calculations. Those values are given in Section 220-2, which reads as follows: "Unless other voltages are specified, for purposes of computing branch-circuit and feeder loads, nominal system voltages of 120, 120/240, 208Y/120, 240, 480Y/277, 480, and 600 volts shall be used."

The ANSI standard is clear on nominal, no-load system voltages and agrees with the **NEC** values. The ANSI-defined nominal system voltages are shown in Table 2.1.

TABLE 2.1

Three-Wire	Four-Wire
120/240	208Y/120
480	240/120
4,160	480Y/277
13,800	12,470Y/7200
	13,200Y/7620
	29,940Y/14,400
	34,500Y/19,920

The most helpful tables covering equipment utilization voltages appear in App. C of this ANSI Standard. There are three tables in this section that apply to all motor and motor control equipment and all heating, refrigeration, and air conditioning equipment. Such equipment should be rated as shown in Table 2.2.

TABLE 2.2

Nominal system voltage	Nameplate (utilization) voltage	
	Three phase	Single phase
120	—	115
208	200	—
240	230	230
480	460	—
600	575	—
2,400	2,300	—
4,160	4,000	—
4,800	4,600	—
6,900	6,600	—
13,800	13,200	—

The standard goes on to say that all other residential and commercial equipment—such as lighting fixtures, appliances, water heaters, etc.—should have a nameplate utilization voltage the same as the system nominal voltage, with three exceptions. That is, food waste disposers, clothes washers, and dishwashers should be marked with a nameplate utilization voltage of 115 V.

The Problem

At this point some may be thinking, So what? The point is that where the size of branch circuit, feeder, and service conductors is calculated from the tabulated kilovoltampere load, use of a *lower* voltage value will result in a *higher* current. And that will result in larger conductors with greater capacity.

In service and feeder conductors, additional capacity is always desirable because invariably additional loads will have to be supplied. Although this reality is often accommodated by adding a "fudge" factor (some additional percentage of total) to the tabulated kilovoltamperes, use of the *lower* utilization voltages—even for equipment marked with nominal system voltages—will also serve to assure adequate spare capacity for more additional loading. For branch circuits, use of the lower (utilization) voltages will effectively limit the number of devices that may be supplied and will reduce the amount of load current that the conductors carry. As a result, the conductors will operate at lower temperatures, which should reduce the likelihood of conductor overload and insulation failure, as well as maximize the service life of the conductors.

While the use of the ANSI utilization voltages shown in Table 2.2 to determine conductor size from the total kilovoltamperes essentially provides for a safer, more adequate system, there is some question regarding acceptability of such practice.

As shown above, the wording of Section 220-2 includes the mandatory word "shall." Use of that word *requires* that those voltages be used when determining minimum circuit capacity in accordance with the **NEC**. In very nearly all areas of the country, the **NEC** has the force of law, and must be satisfied, while the ANSI standard is viewed as a guideline. That means the **NEC** takes precedence over the ANSI standard. With that in mind, the question is, "How can I use the lower voltage values and still satisfy **NEC** Section 220-2?"

The answer to that question lies in the basic task presented by the **NEC** when it comes to conductor sizing. That is, when calculating the size of branch-circuit, feeder, and service conductors in accordance with the **NEC**, we are determining the *minimum* acceptable size. Because the use of the ANSI utilization voltages instead of the **NEC** nominal system voltages will result in a greater current value and, therefore, a *larger* conductor or *less* loading, such application will exceed the **NEC** required *minimum* size.

The rule of Section 220-2 is essentially aimed at standardizing calculations for the purpose of uniform application and enforcement. This is also covered in Part (b) of Chapter 9, which specifies nominal system voltages for use when calculating minimum conductor size. But, as indicated, *all* branch-circuit, feeder, and service conductor calculations made using utilization voltages do satisfy **NEC** minimum requirements on conductor sizing.

Let's take an example. How many receptacles can be installed on a single-phase 15-A circuit in a nonresidential occupancy fed from a 208Y/120-V ac system?

As covered in Section 220-3(c)(6), each receptacle (strap) is to be counted as 180 VA. Using the nominal voltage shown in Sec. 220-2, we determine the maximum voltamperes that the circuit can provide, as follows:

$$VA = V \times A$$

$$VA = 120 \text{ V} \times 15 \text{ A}$$

$$VA = 1800$$

Next, divide the total voltampere capacity (1800 VA) by the voltampere value of a single device (180 VA), as follows:

$$1800 \text{ VA}/180 \text{ VA} = 10$$

Therefore, it would be permissible to use 10 receptacles on this single-phase branch circuit. If, however, I use the utilization voltage of 115 V, the calculations would be as follows:

$$VA = 115\ V \times 15\ A$$

$$VA = 1725$$

Then, divide by 180 VA:

$$1725\ VA/180\ VA = 9.5$$

which is rounded down to 9.

A similar comparison for a 20 A circuit is shown in Fig. 2.8.

20A, 115V CIRCUIT—Maximum of 12 receptacle outlets

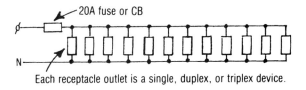

Each receptacle outlet is a single, duplex, or triplex device.

20A × 115V = 2300VA

2300VA ÷ 180VA = 12.7, or 12 receptacles

20A, 120V CIRCUIT—Maximum of 13 receptacle outlets

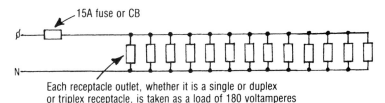

Each receptacle outlet, whether it is a single or duplex
or triplex receptacle, is taken as a load of 180 voltamperes

20A × 120V = 2400VA

2400VA ÷ 180VA = 13.3, or 13 receptacles

FIGURE 2.8 Number of receptacles per 20-A circuit, in nonresidential occupancies. As can be seen, use of 115 V instead of 120 V results in less loading of the circuit conductors. Use of lower voltages (115, 230, 440, etc.) as denominators in calculations would *not* be an **NEC** violation because the higher current values that result would assure **NEC** compliance because of greater capacity of circuit wires and equipment.

It is worth noting that in some places the **NEC** adopts 115 V as the basic operating voltage of equipment designed for operation at 110 to 125 V. That is indicated in **NEC** Tables 430-148 to 430-151. References are made to "rated motor voltages" of 115, 230, 460, 575, and 2300 V—all values over 115 are integral multiples of 115 V. The last notes to Tables 430-149 and 430-150 indicate that motors of those voltages are applicable systems rated 110-120, 220-240, 440-480, and 550-600 V. Although motors can operate satisfactorily within those ranges, it is better to design circuits to deliver rated voltage. These **NEC** voltage designations for motors are consistent with the trend over the years for manufacturers to rate equipment for values of voltage that correspond to the voltage that is likely to be available at the equipment.

Another point about performing calculations in accordance with the **NEC** relates to fractions of amperes. Where calculations result in values involving fractions of an ampere the fraction may be dropped if it is less than 0.5 A. A value such as 20.7 A should be continued to be used as 20.7, or it should be rounded up to the next higher whole number, in this case 21. Again, this is to be on the safe side. There are occasions, however, when current values must be added together. In such cases, it is on the safe side to retain fractions less than 0.5 A, because several fractions added together can result in the next whole ampere.

A Caution

It must be emphasized that the preceding discussion was intended for designers and installers on real-life applications. For those taking tests covering **NEC** calculations, *always* use the **NEC**-recommended nominal system voltages when determining conductor size and permissible loading. Failure to do so will result in "wrong" answers because these tests are measuring your knowledge of the **NEC**, part of which should include knowing that use of nominal system voltages is called for by Section 220-2. In a similar manner, retaining fractions less than 0.5 A is also not called for by the **NEC** and should *not* be used for calculations on contractor tests, master/journeyman electrician tests, etc.

Load Limitation for Continuous Operation

Where a branch-circuit load will be in "continuous" operation—such as for general illumination in large office areas, private offices, schools, hospitals, industrial plants, and the like—the total load on each branch circuit may not exceed 80 percent of the rating of the circuit, and the "rating of the circuit" is established as the ampere rating of the branch-circuit protective device. The **NEC** defines a continuous load as a load "where the maximum current is expected to continue for three hours or more."

This 80 percent limitation on circuit loading is set forth in three different sections of the **NEC**: Sections 210-22(c), 220-3(a), and 384-16(c) (Fig. 2.9).

Section 210-22(c) says that, for loads other than motor loads, the total load on a branch circuit must not exceed 80 percent of the circuit rating when the load will constitute a continuous load—such as store lighting.

Because branch circuits for multioutlet circuits, such as lighting circuits, are rated in accordance with the rating or setting of the overcurrent device, the maximum permitted load is 80 percent of the rating of the protective device. Although this limitation applies only to loads other than motor loads, Section 384-16(c) says that "the total load on any overcurrent device located in a panelboard shall not exceed 80 percent of its rating where in normal operation the load will continue for three hours or more."

<u>NOTE</u>: In both the above cases, the 80 percent load limitation does not apply "where the assembly including the overcurrent device is approved for continuous duty at 100 percent of its rating," which means where it is listed by UL or another qualified electrical testing laboratory. But Section 384-16(c) requires any circuit breaker (CB) or fuse in a panelboard to have its load limited to 80 percent; and only one exception is made for such protective devices in a panel: A continuous load of 100 percent is permitted only when the protective-device assembly (fuse in switch or

LIGHTING AND APPLIANCE BRANCH CIRCUITS

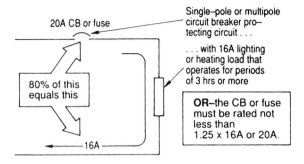

FIGURE 2.9 Branch-circuit protective device must be rated not less than 125 percent of the continuous load current. [Sec. 210-22(c).] Unless "listed" for continuous operation at 100 percent of its rating, a branch-circuit overcurrent device must not be loaded over 80 percent of its rating.

CB) is approved for 100 percent continuous duty. (And there are no such devices rated less than 225 A. UL has a hard-and-fast rule that any CB not marked for continuous load must have its load limited to 80 percent of its rating.)

For panelboard applications, the use of conductors that have had their ampacity derated because more than three are located in a raceway (Note 8 to **NEC** Tables 310-16 through 310-19) does not eliminate the 80 percent limit on the continuous load on the circuit protective device. An example would be six No. 12 TW copper wires in a conduit, with the ampacity of each derated from the ampacity of 25 A shown in **NEC** Table 310-16 to 20 A (0.80 × 25 A) and then protected at 20 A. Such circuits may not be continuously loaded up to the 20-A value. Section 384-16(c) would require that a continuous load be not over 80 percent of the 20-A rating of the protective device, or 16 A in this example (Fig. 2.10).

Circuit Voltage Limits

Voltage limitations for branch circuits are presented in Sec. 210-6. In general, branch circuits serving lampholders, fixtures, or receptacles of the standard 15-A or less rating are limited to operation at a maximum voltage rating of 120 V.

Part (a) applies specifically to dwelling units—one-family houses, apartment units in multifamily dwellings, and condominium and co-op units—and to guest rooms in hotels and motels and similar residential occupancies. In such occupancies, any lighting fixture or any receptacle for plug-connected loads rated up to 1,440 VA or for motor loads of less than ¼ hp must be supplied at not over 120 V between conductors.

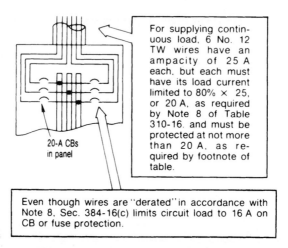

FIGURE 2.10 Overcurrent-device load limitation of 80 percent is required even for circuit conductors derated in accordance with Note 8 of **NEC** Table 310-16.

NOTE: The 120-V supply to the above-type loads may be derived from (1) a 120-V, 2-wire branch circuit, (2) a 240/120-V, 3-wire branch circuit, or (3) a 208/120-V, 3-phase, 4-wire branch circuit. Appliances rated 1440 VA or more (ranges, dryers, water heaters, etc.) may be supplied by 240/120-V circuits in accordance with Section 210-6(c)(5).

Part (b) permits a circuit with not over 120 V between conductors to supply medium-base screw-shell lampholders, ballasts for fluorescent or high-intensity discharge (HID) lighting fixtures, and plug-connected or hard-wired appliances—in any type of building or on any premises (Fig. 2.11).

Part (c) applies to circuits with over 120 V between conductors (208, 240, 277, or 480 V) but not over 277 V (nominal) to ground. This is shown in Fig. 2.12, where all of the circuits are "circuits exceeding 120 V, nominal, between conductors and not exceeding 277 V, nominal, to ground." Circuits of any of those voltages are permitted to supply incandescent lighting fixtures with mogul-base screw-shell lampholders, ballasts for electric-discharge lighting fixtures or plug-connected or hard-wired appliances, or other utilization equipment. It is important to note that this section no longer contains the requirement for a minimum 8-ft mounting height for incandescent or electric-discharge fixtures with mogul-base screw-shell lampholders used on 480/277-V systems.

A UL-listed electric-discharge lighting fixture rated at 277 V, nominal, may be equipped with a medium-base screw-shell lampholder and does not require a mogul-base screw-shell. The use of the medium-base lampholder, however, is limited to "listed electric-discharge fixtures," for 277-V incandescent fixtures. Section 210-6(c)(3) continues the requirement that such fixtures be equipped with "mogul-based screw-shell lampholders."

Fluorescent, mercury-vapor, metal-halide, high-pressure sodium, low-pressure sodium, and/or incandescent fixtures may be supplied by 480/277-V, grounded-wye circuits—with loads connected phase-to-neutral and/or phase-to-phase. Such circuits operate at 277 V to ground even, say, when 480-V ballasts are connected phase-to-phase on such circuits. Or lighting could be supplied by 240-V delta systems—either ungrounded or with one of the phase legs grounded, because such systems operate at not more than 277 V to ground.

LIGHTING AND APPLIANCE BRANCH CIRCUITS　　　　　　　　31

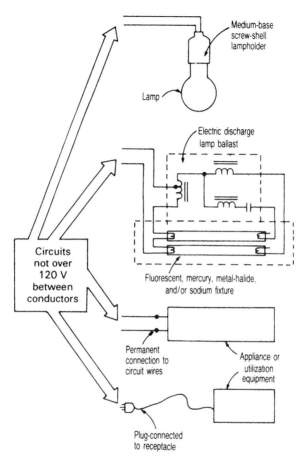

FIGURE 2.11 In any occupancy, 120-V circuits may supply these loads.

Use of incandescent lighting at over 150 V to ground is accepted by the **NEC** in commercial, institutional, and industrial buildings and premises.

Although 277-V incandescent lamps are available with medium screw bases, and fixtures for them are available with medium screw-shell lampholders, use of such equipment violates the **NEC** rule in Section 210-6(c)(1). And Section 210-6(b)(1) limits medium-based screw-shell lampholders to use only where connected to circuit wires with not more than 120 V between the wires—such as 2-wire 120-V circuits.

Section 210-6(c) permits installations on 480/277-V, 3-phase, 4-wire wye systems—with equipment connected from phase-to-phase (480-V circuits) or connected phase-to-neutral (277-V circuits). In either case, the voltage to ground is only 277 V. In any such application, it is important that the neutral point of the 480/277-V wye be grounded to limit the voltage aboveground to 277 V. If the neutral were not grounded and the system operated ungrounded, the voltage to ground, according to the **NEC**, would be 480 V (see definition "voltage to ground," Art. 100), and lighting equipment could be used on such circuits only for outdoor applications as specified under Section 210-6(d) (discussed later).

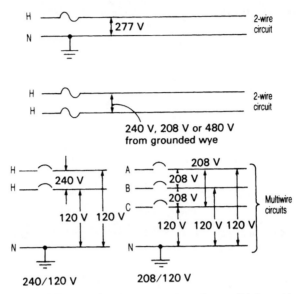

FIGURE 2.12 These circuits may supply incandescent lighting with mogul-base screw-shell lampholders for over 120 V between conductors, electric-discharge ballasts, and cord-connected or permanently wired appliances or utilization equipment.

On a neutral-grounded 480/277-V system, incandescent, fluorescent, mercury-vapor, metal-halide, high-pressure sodium, and low-pressure sodium equipment can be connected from phase-to-neutral on the 277-V circuits. If fluorescent or mercury-vapor fixtures are to be connected phase-to-phase, some **NEC** authorities contend that autotransformer-type ballasts cannot be used when these ballasts raise the voltage to more than 300 V, because, they contend, the **NEC** calls for connection to a circuit made up of a grounded wire and a hot wire. (See Section 410-78.) On phase-to-phase connection these ballasts would require use of 2-winding, electrically isolating ballast transformers. The wording of Section 410-78 does, however, lend itself to interpretation that it is only necessary for the *supply system* to the ballast to be *grounded*—thus permitting the two hot legs of a 480-V circuit to supply an autotransformer because the hot legs are derived from a neutral-grounded "system." But Section 210-9 can become a complicating factor. Use of a 2-winding (isolating) ballast is clearly acceptable and avoids all confusion.

Section 210-6(d) of the **NEC** permits fluorescent and/or mercury-vapor units to be installed on circuits rated over 277 V, nominal, to ground and up to 600 V between conductors—but only where the lamps are mounted in permanently installed fixtures on poles or similar structures for the illumination of areas such as highways, bridges, athletic fields, parking lots, at a height not less than 22 ft. or on other structures such as tunnels at a height not less than 18 ft. (See Fig. 2.13.) Part (d) covers use of lighting fixtures on 480-V ungrounded circuits—such as fed from a 480-V delta-connected or wye-connected ungrounded transformer secondary.

This permission for use of fluorescent and mercury units under the conditions described is based on phase-to-phase voltage rather than on phase-to-ground voltage. This rule has the effect of permitting the use of 240- or 480-V ungrounded circuits

LIGHTING AND APPLIANCE BRANCH CIRCUITS 33

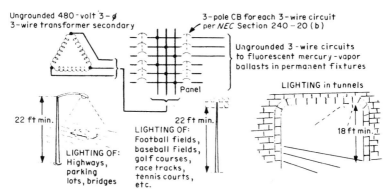

FIGURE 2.13 Ungrounded circuits, at up to 600 V between conductors, may supply lighting only as shown.

for the lighting applications described. But as described above, autotransformer-type ballasts may not be permitted on an ungrounded system if they raise the voltage to more than 300 V (Section 410-78). In such cases, ballasts with 2-winding transformation would have to be used.

Certain electric railway applications utilize higher circuit voltages. Infrared lamp industrial heating applications may be used on higher circuit voltages as allowed in Section 422-15(c) of the **NEC**.

CAUTION: The concept of maximum voltage not over "120 V . . . *between* conductors," as stated in Section 210-6(a), has caused considerable discussion and controversy in the past when applied to split-wired receptacles and duplex receptacles of two voltage levels. It can be argued that split-wired general-purpose duplex receptacles are not acceptable in dwelling units and in hotel and motel guest rooms because they are supplied by conductors with more than 120 V between them—that is, 240 V on the 3-wire, single-phase, 120/240-V circuit so commonly used in residences. The two hot legs connect to the brass-colored terminals on the receptacle, with the shorting tab broken off, and the voltage between those conductors *does* exceed 120 V. The same condition applies when a 120/240-V duplex receptacle is used—the 240-V receptacle is fed by conductors with more than 150 V between them. But, the **NEC**-making panel ruled that the use of duplex receptacles connected on 240/120-V, 3-wire, single-phase circuits in dwelling units and guest rooms in hotels, motels, dormitories, etc., is *not* prohibited. See Fig. 2.14.

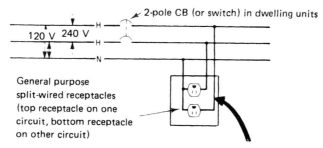

FIGURE 2.14 Split-wired receptacles are permitted in residential occupancies ("dwelling units") and in all other types of occupancies (commercial, institutional, industrial, etc.).

Objection to split-wired receptacles in dwelling units has a sound basis: Two appliances connected by 2-wire cords to a split-wired receptacle in a kitchen do present a real potential hazard. With, say, a coffee maker and a toaster plugged into a split-wired duplex, if the hot wire in each appliance should contact the metal enclosure of the appliance, there would be 240 V between the two appliance enclosures. The user would be exposed to the extremely dangerous chance of touching each appliance with a different hand—putting 240 V across the person, from hand to hand, through the heart path (Fig. 2.15). Use of nonsplit-wired receptacles on the usual spacing of up to 4 ft does tend to separate appliances on different hot legs.

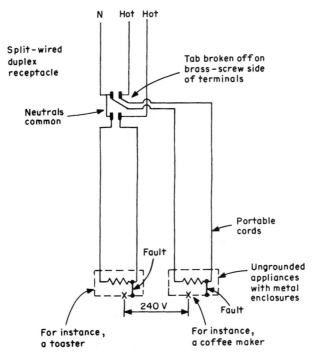

FIGURE 2.15 This can be dangerous.

Use of split-wired receptacles and other receptacles with more than 120 V between terminals is, of course, completely acceptable in any commercial, institutional, or industrial location. And it is also perfectly acceptable at any time to use a split-wired receptacle for switch control of one plug-in point leaving the other hot all the time (Fig. 2.16).

Section 210-6(c)(6) clearly permits "permanently connected utilization equipment" to be supplied by a circuit with voltage between conductors in excess of 120 V, and permission *is* intended for the use of 277-V heaters in dwelling units, as used in high-rise apartment buildings and similar large buildings that may be served at 480/277 V.

Acceptable Circuit Loads

A single branch circuit to one outlet (for a single machine, appliance, or other individual utilization device) may serve any load and is unrestricted as to ampere rating.

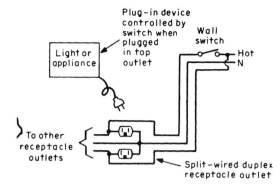

FIGURE 2.16 Split-wiring of receptacles to control one of the receptacles may be done from the same hot leg of a 2-wire circuit or with separate hot legs of a 3-wire, 240/120-V circuit.

Circuits with more than one outlet (even a circuit that supplies a single duplex receptacle—which is two receptacles and may supply two load devices) are subject to **NEC** load limitations. The **NEC** limitations on the use of branch circuits with two or more outlets are as follows:

1. Branch circuits rated 15 and 20 A may serve lighting units and/or appliances. The rating of any one-cord and plug-connected appliance must not exceed 80 percent of the branch-circuit rating. Appliances fastened in place may be connected to a circuit serving lighting units and/or plug-connected appliances, provided the total rating of the fixed appliances fastened in place does not exceed 50 percent of the circuit rating (Fig. 2.17). However, modern design provides separate circuits for individual fixed appliances. In commercial and industrial buildings, separate circuits should be provided for lighting and separate circuits for receptacles.

2. Branch circuits rated 30 A may serve fixed lighting units (with heavy-duty-type lampholders) in other than dwelling units or appliances in any occupancy. An individual cord-and-plug-connected appliance that draws more than 24 A may not be connected to this type of circuit (Fig. 2.18).

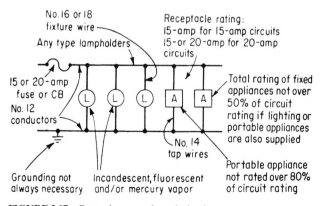

FIGURE 2.17 General-purpose branch circuits.

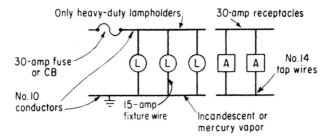

FIGURE 2.18 Multioutlet 30-A circuits.

3. Branch circuits rated 40 and 50 A may serve fixed lighting units (with heavy-duty lampholders) or infrared heating devices in other than dwelling units, or cooking appliances in any occupancy (Fig. 2.19). It should be noted that a 40- or 50-A circuit may be used to supply any kind of load equipment—such as a dryer or water heater—where the circuit is an individual circuit to a single appliance.

Application of the above rules—and other **NEC** rules that refer to dwelling units—must take into consideration the **NEC** definition of that phrase: A dwelling unit is defined as "one or more rooms" used "as a housekeeping unit" and must contain space or areas specifically dedicated to "eating, living, and sleeping" and must have "permanent provisions for cooking and sanitation" (Fig. 2.20). A one-family house is a "dwelling unit." So is an apartment in an apartment house or a condominium unit. But a guest room in a hotel or motel or a dormitory room or unit is not a "dwelling unit" if it does not contain "permanent provisions for cooking"—which must mean a built-in range or counter-mounted cooking unit (with or without an oven). Nevertheless, the rule of Section 210-8, which requires ground-fault circuit interrupter (GFCI) protection of receptacles in bathrooms of "dwelling units," applies to bathroom receptacles in standard hotel or motel units, even though such units do not contain "permanent provisions for cooking."

It should be noted that the requirement for heavy-duty-type lampholders in lighting units on 30-, 40-, and 50-A multioutlet branch circuits excludes the use of fluorescent lighting on these circuits because fluorescent lampholders are not rated "heavy duty" in accordance with Section 210-21(a). Other electric-discharge lighting fixtures (mercury-vapor, metal-halide, sodium) with mogul lampholders may be used on these circuits provided tap-conductor requirements are satisfied.

As indicated, multioutlet branch circuits for lighting are limited to a maximum loading of 50 A. Individual branch circuits may supply any loads. Excepting motors, this means that an individual piece of equipment may be supplied by a branch circuit that has sufficient carrying capacity in its conductors, is protected against current in excess of the capacity of the conductors, and supplies only the single outlet for the load device.

Fixed outdoor electric snow-melting and deicing installations may be supplied by any of the above-described branch circuits or by individual branch circuits of higher rating, provided the circuit conductors and overcurrent protection are rated not less than 125 percent of the total heater load current. (See Section 426-4 in Article 426, "Fixed Outdoor Electric De-Icing and Snow Melting Equipment.")

LIGHTING AND APPLIANCE BRANCH CIRCUITS

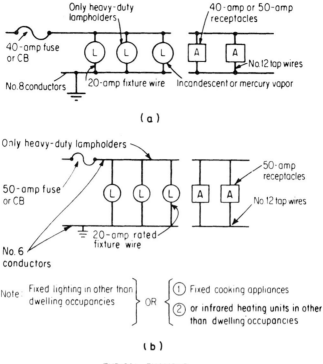

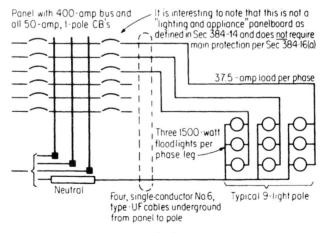

FIGURE 2.19 (*a*) Multioutlet 40-A circuits. (*b*) Multioutlet 50-A circuits. (*c*) Typical example.

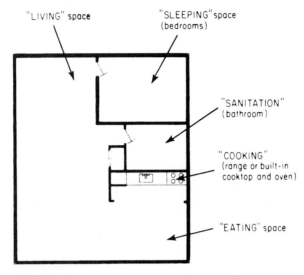

FIGURE 2.20 Specific conditions identify a "dwelling unit." Eating, living, and sleeping space could be one individual area, as in an efficiency apartment. But the unit must contain a "bathroom," defined in Section 210-8(b) as "an area including a basin with one or more of the following: a toilet, a tub, or a shower." And the unit must contain permanent cooking equipment.

Lighting-Circuit Details

A number of specific **NEC** rules affect the design and installation of lighting equipment:

1. Branch-circuit wires that come within 3 in of a ballast within a luminaire (that is, wires running through end-to-end-connected fluorescent fixtures) must be rated at 90°C, as are RHH, THHN, XHHW, and THW wires. THW wire is normally rated at 75°C, but it has a 90°C rating for use in fluorescent fixtures (Fig. 2.21). As explained later, these 75 and 90°C wires must be used at the ampacities of 60°C wires (such as TW) of corresponding sizes for circuits up to 100 A, unless equipment is marked to permit the use of wires rated higher than 60°C.

2. For supplying recessed incandescent luminaires, ⅜-in flexible metal conduit, liquid-tight flexible metal conduit, or flexible metallic tubing is acceptable for the 4- to 6-ft length that carries the 150°C fixture wires from the branch-circuit junction box to the luminaire to satisfy Section 410-67(c) (Fig. 2.22). **NEC** Table 350-3 shows that ⅜-in flexible metal conduit may contain up to four No. 14 AF wires and up to seven No. 18 wires. For a typical recessed incandescent lighting fixture, two AF wires may be used, without a grounding conductor, in a 4- to 6-ft length of ⅜-in flex in accordance with Exception No. 1b of Section 210-19(c). Two No. 16 AF wires (rated at 8 A) or two No. 18 wires (rated at 6 A) may be used on circuits rated not over 20 A, where the current rating of the fixture does not exceed the AF wire rating. Fixture wire used in the flex from the box to the fixture may be protected and sized according to Exception No. 2 of Section 210-19(c) and Section 240-4, which specifically permits the use of fixture wires for taps as covered in Exception No. 1 of Section 210-19(c) and states that No. 16 and No. 18 fixture wires are considered protected by a 20-A CB or fuse.

LIGHTING AND APPLIANCE BRANCH CIRCUITS

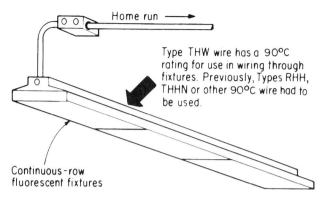

FIGURE 2.21 Type THW wire has a 90°C rating for through-fixture use, within 3 in of a ballast.

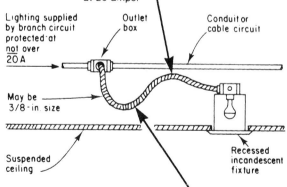

IMPORTANT! Flexible metal conduit, liquidtight flex, or flex tubing is equipment grounding conductor because AF wires in flex are tapped from circuit protected at not over 20 amps, as permitted in Section 250-91(b), Exception No 1 or Exception No. 2.

FIGURE 2.22 Flexible raceways and fixture wires must be carefully applied to recessed fixtures.

3. Although flexible metal conduit, flex tubing, and liquid-tight flex are not approved as equipment grounding means in long lengths, Exceptions No. 1 and No. 2 of Section 250-91(b) do recognize the three types of flex as a grounding means if the total length of flex is not over 6 ft. Standard metal flex (Greenfield) may be used as an equipment ground only where the wires within the flex are protected at not over 20 A (Fig. 2.23). However, an equipment grounding conductor *must* be used (either inside or outside the flex) in every case where metal flex supplies equipment that is not fixed in one place or location. In such cases, where the flex is subject to moving and flexing, an equipment grounding conductor must be used, regardless of length or conductor protection.

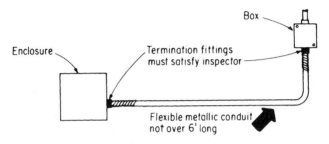

FLEX NOT OVER 6 FT LONG is suitable as a grounding means (without a separate ground wire) if the conductors in it are protected by OC devices rated not more than 20 amps.

FIGURE 2.23 Flex not over 6 ft long may serve as grounding means.

4. An electric-discharge lighting fixture, when supported independently of the outlet box, may be supplied by cord-and-plug connection to a receptacle installed above the fixture, provided it is an open installation so that the cord is continuously visible for its entire length and is not subject to strain or physical damage. Figure 2.24 shows the hookups covered by **NEC** Section 410-30.

5. Lampholders must not have a rating lower than the load to be served; and lampholders connected to circuits rated over 20 A must be of the heavy-duty type (that is, rated at least 660 W for an "admedium" type, and at least 750 W for other types). Because fluorescent lampholders are not of the heavy-duty type, this excludes the use of fluorescent luminaires on 30-, 40-, and 50-A circuits. The intent is to limit the rating of lighting branch circuits supplying fluorescent fixtures to 20 A. The ballast (not the lamp) is directly connected to the branch circuit, but by controlling the lampholder rating, a 20-A limit is established for the ballast circuit.

Most lampholders manufactured and intended for use with electric-discharge lighting for illumination purposes are rated less than 750 W and are not classified as heavy-duty lampholders. If the luminaires are individually protected, such as by a fuse in the cord plug of a luminaire cord connected to, say, a 50-A trolley or plug-in busway, some inspectors have permitted use of fluorescent luminaires on 30-, 40-, and 50-A circuits. But such protection in the cord plug or in the luminaire is supplementary (Section 240-10), and branch-circuit protection rated at 30, 40, and 50 A would still exclude the use of fluorescent fixtures according to Section 210-21(a).

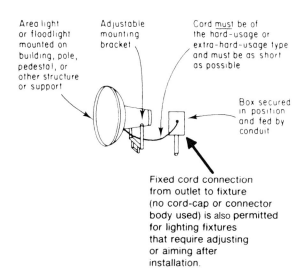

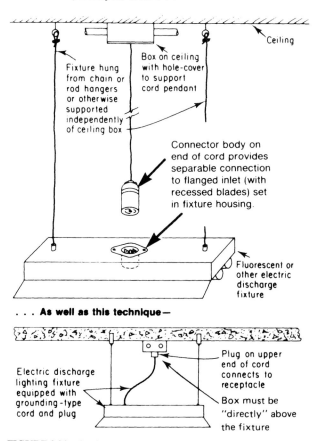

FIGURE 2.24 Cord connection of lighting fixtures.

Modular Wiring

So-called modular wiring systems are highly engineered, ultraflexible, plug-in branch-circuit systems for supplying and controlling lighting fixtures in suspended ceilings. These systems are UL-listed and made of UL-listed raceway, cable, and connector components.

NEC Article 604 recognizes the various types of modular wiring systems that provide plug-in connections to lighting fixtures, switches, and receptacles in all kinds of commercial and institutional interiors that use suspended ceilings made of lift-out ceiling tiles. Except for those fixtures that are UL-listed for cord-and-plug connection as part of listed modular wiring systems for use in suspended-ceiling spaces, cord connection may not be used for fixtures installed in lift-out ceilings.

These manufactured wiring systems are a logical solution to a variety of needs in electrical systems for commercial and institutional occupancies. In the interest of giving the public a better way at a better price, a number of manufacturers have developed basic wiring systems to provide plug-and-receptacle-type interconnection of branch-circuit wires to lighting fixtures in suspended-ceiling spaces. Such systems afford ready connection between the hard-wired circuit homerun and the cables and/or ducts that form a grid- or treelike layout of circuiting to supply incandescent, fluorescent, or HID luminaires in the ceiling.

The acknowledged advantages of modular wiring systems are numerous and significant:

- Factory-prewired raceways and cables provide highly flexible and accessible plug-in connection to multicircuit runs of 120- and/or 227-V conductors.
- Drastic reductions can be made in requirements for conventional pipe-and-wire hookups of individual circuits, which are costly and inflexible.
- Plug receptacles afford a multiplicity of connection points for fixtures to satisfy needs for specific types and locations of lighting units. They permit the location of fixtures to serve any initial layout of desks or workstations and still offer unlimited, easy, and extremely economical modification of layouts and the addition of fixtures for any future rearrangements of office landscaping or activities.
- These systems may include switches and/or convenience receptacles in walls, partitions, or floor-to-ceiling poles, with readily altered switching provisions to provide energy conservation through effective on-off control of any revised lighting layout.
- Work on the systems has been covered by agreement between the IBEW and associated trades.
- Such systems have the potential for tax advantage via accelerated depreciation as office equipment rather than real estate.

Modular systems may be used in air-handling ceilings. The equipment may be used in the specific applications and environments for which it is listed by UL.

Residential Circuits

The design of general-purpose lighting and appliance branch circuits for dwelling units (single-family houses and apartments) should be based on a minimum provision of one 20-A, 2-wire, 120-V circuit for each 800 ft^2 of floor area or one 15-A, 2-wire, 120-V circuit for each 600 ft^2 of floor area. This is an **NEC** minimum that works out to be a capacity of 3 W/ft^2. For hotel or motel rooms, at least 2 W/ft^2 should be used. The number of circuits determined in this way will handle general illumination and convenience receptacles spaced. As indicated, receptacle outlets on fixed spacing must be installed in every room of a dwelling unit except the bathroom. The **NEC** rule lists the specific rooms that are covered by the rule requiring receptacles spaced no greater than 12 ft apart in any continuous length of wall.

In Section 210-52(a), the required receptacles must be spaced around the designated rooms and any "similar room or area of dwelling units." The wording of this section assures that receptacles are provided—the correct number with the indicated spacing—in those unidentified areas so commonly used today in residential architectural design, such as *greatrooms* and other big areas that combine living, dining, and/or recreation areas.

As shown in Fig. 2.25, general-purpose convenience receptacles, usually of the duplex type, must be laid out around the perimeters of living room, bedrooms, and all the other rooms. Spacing of receptacle outlets should be such that no point along the floor line of an unbroken wall is more than 6 ft from a receptacle outlet. Care should be taken to provide receptacle outlets in smaller sections of wall space segregated by doors, fireplaces, bookcases, or windows. Although Section 210-52(h) calls for one receptacle outlet for each dwelling-unit hallway that is 10 ft or more in length, this section does not specify location or require more than a single receptacle outlet. However, good design practice would dictate that a convenience receptacle should be provided for each 10 ft of hall length. And they should be located as close as possible to the middle of the hall.

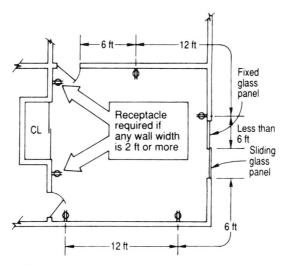

FIGURE 2.25 From any point along wall, at floor line, a receptacle must be not more than 6 ft away. Required receptacle spacing considers a fixed glass panel as wall space and a sliding panel as a doorway.

In determining the location of a receptacle outlet, the measurement is to be made along the floor line of the wall and is to continue around corners of the room but is not to extend across doorways, archways, fireplaces, passageways, or other space unsuitable for having a flexible cord extended across it. The location of outlets for special appliances within 6 ft of the appliance [Sec. 210-50(c)] does not affect the spacing of general-use convenience outlets but merely adds a requirement for special-use outlets.

Figure 2.26 shows two wall sections 9 and 3 ft wide extending from the same corner of the room. The receptacle shown located in the wider section of the wall will permit the plugging in of a lamp or appliance located within 6 ft of either side of the receptacle. The same rule would apply to the other wall shown.

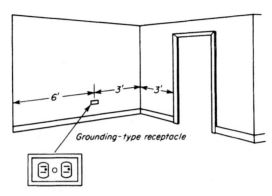

FIGURE 2.26 Location of the receptacle as shown will permit the plugging in of a lamp or appliance located 6 ft on either side of the receptacle.

Receptacle outlets shall be provided for all wall space within the room except individual isolated sections which are less than 2 ft in width. For example, a wall space 23 in wide and located between two doors would not need a receptacle outlet.

In measuring receptacle spacing for exterior walls of rooms, the fixed section of a sliding glass door assembly is considered to be "wall space" and the sliding glass panel is considered to be a doorway. In previous **NEC** editions the entire width of a sliding glass door assembly—both the fixed and movable panels—was required to be treated as wall space in laying out receptacles "so that no point along the floor line in any wall space is more 6 feet" from a receptacle outlet. The wording takes any fixed glass panel to be a continuation of the wall space adjoining it, but the sliding glass panel is taken to be the same as any other doorway (such as with hinged doors) (Fig. 2.25).

The last sentence of the first paragraph of Part (a) requires fixed room dividers and railings to be considered in spacing receptacles. This is illustrated by the sketch of Fig. 2.27. In effect, the two side faces of the room divider provide additional wall space, and a table lamp placed as shown would be more than 6 ft from both receptacles A and B. Also, even though no place on the wall is more than 6 ft from either A

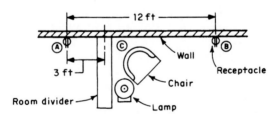

FIGURE 2.27 Fixed room dividers must be counted as wall space requiring receptacles.

or B, a lamp or other appliance placed at a point such as C would be more than 6 ft from B and out of reach from A because of the divider. This rule would ensure placement of a receptacle in the wall on both sides of the divider or in the divider itself if its construction so permitted.

As noted in the next-to-last paragraph of Section 210-52(a), any receptacle that is an integral part of a lighting fixture or an appliance or a cabinet may not be used to satisfy the specific receptacle requirements of the section. For instance, a receptacle in a medicine cabinet or lighting fixture may not serve as the required bathroom receptacle. And a receptacle in a post light may not serve as the required outdoor receptacle for a one-family dwelling.

In spacing receptacle outlets so that no floor point along the wall space of the rooms designated by Section 210-52(a) is more than 6 ft from a receptacle, a receptacle that is part of an appliance must not generally be counted as one of the required spaced receptacles. However, the Exception at the end of Part (a) states that a receptacle that is "factory installed" in a "permanently installed electric baseboard heater" (not a portable heater) may be counted as one of the required spaced receptacles for the wall space occupied by the heater. Or a receptacle "provided as a separate assembly by the manufacturer" may also be counted as a required spaced receptacle. But, such receptacles must not be connected to the circuit that supplies the electric heater. Such a receptacle must be connected to another circuit.

Because of the increasing popularity of low-density electric baseboard heaters, their lengths are frequently so long (up to 14 ft) that required maximum spacing of receptacles places receptacles above heaters and produces the undesirable and dangerous condition where cord sets to lamps, radios, TVs, etc., will droop over the heater and might droop into the heated-air outlet. And UL rules prohibit use of receptacles above certain electric baseboard heaters for that reason. Receptacles in heaters can afford the required spaced receptacle units without mounting any above heater units. They satisfy the UL concern and also the preceding note near the end of Section 210-52(a) that calls for the need "to minimize the use of cords across doorways, fireplaces, and similar openings"—and the heated-air outlet along a baseboard heater is a "similar opening" that must be guarded (Fig. 2.28).

A fine-print note at the end of Part (a) points out that the UL instructions for baseboard heaters (marked on the heater) may prohibit the use of receptacles above the heater because cords plugged into the receptacle are exposed to heat damage if they drape into the convection channel of the heater and contact the energized heating element.

In kitchens and dining areas, the rules of Section 210-52(c) require a receptacle at each counter space wider than 12 in. They also define counter spaces and disqualify as "required outlets" any receptacles rendered inaccessible by the installation of appliances that are either fastened in place or positioned in a space that is "dedicated"—i.e., assigned for permanent positioning of an appliance. Refrigerators and freezers would be typical of appliances "occupying dedicated space" (Fig. 2.29).

The rule of Part (c) further requires that, at any countertop space, "no point along the wall line" of the countertop is permitted to be more than 24 in from a receptacle outlet, measured horizontally along the wall line. The same section also calls for similar installation of receptacle outlets for any "island" or "peninsula" countertop with a long dimension of 24 in or greater and a short dimension of 12 in wide or wider. Such receptacles would have to be installed in the vertical surfaces of those kinds of counter constructions—but they must be within 12 in of the countertop surface to satisfy the rule of Part (c).

At least one wall receptacle outlet must be installed adjacent to each wash basin location in bathrooms of dwelling units—and Section 210-60 requires the same receptacle in bathrooms of hotel and motel guest rooms (Fig. 2.30).

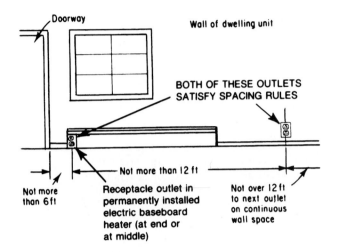

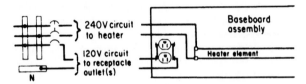

FIGURE 2.28 Receptacles in baseboard heaters may serve as "required" receptacles.

At least one outdoor receptacle must be installed at the front and back for every one-family house ("a one-family dwelling") and grade-level accessible unit in a two-family dwelling. The definition of "one-family dwelling" (Article 100) makes clear that an outdoor receptacle is not required for outdoor balconies of apartment units, motels, hotels, or other units in multiple-occupancy buildings.

It is also required that one-family dwellings be provided with one GFCI-protected outdoor receptacle outlet at the front of each dwelling and one at the rear of each dwelling. This rule is aimed at providing adequacy in the availability of outdoor receptacles for one-family dwelling units (Fig. 2.31)—especially those town-house-type units.

The use of outdoor appliances at two-family houses has been judged to be as common as at one-family houses, and the need for outdoor receptacles to eliminate use of extension cords from within the house is recognized by the rule calling for outdoor receptacles for two-family houses.

For a two-family dwelling, the rule requires separate outdoor receptacle outlets for each dwelling unit in a two-family house where each dwelling unit is an upstairs-and-downstairs unit—that is, each unit has living space (i.e., kitchen and living room) located "at grade level." And, as with a one-family dwelling, the receptacle outlets could contain a single, duplex, or triplex receptacle—installed on the outside wall or fed underground. [Note that a receptacle in a post light does not qualify as the required outdoor receptacle, because a receptacle "that is part of any lighting fixture" is excluded by the last paragraph of Section 210-52(a).] The clear intent of the

Any point along the wall line of each length of counter top must *not* be over 24 in., measured horizontally, from a receptacle outlet.

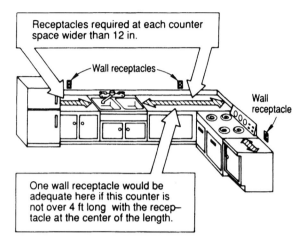

COUNTER SPACES in kitchen and dining rooms such as shown by arrows (above) must be supplied with receptacles if they are over 12 in. wide. Appliances are frequently used even on narrow counter widths; this requirement is designed to remove the dangerous practice of stretching cords across sinks, behind ranges, etc., to feed such appliances.

Inaccessible receptacles.

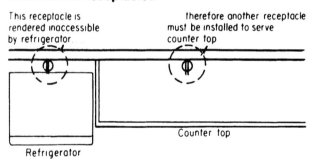

RECEPTACLE LOCATED behind an appliance, making the receptacle inaccessible, does not count as one of the required "counter-top" receptacles. (Neither does it count as one of the appliance-circuit receptacles required to be located every 4 ft.)

FIGURE 2.29 Countertop receptacles are needed and must be accessible.

rule, however, is *not* to require outdoor receptacles for a dwelling unit that is totally on the second floor of a two-family house, with only its entrance door on the first floor, providing access to the stairway.

In a multiple-occupancy building—such as adjacent up-and-down duplex units in "town houses"—if adjacent units are separated by fire-rated walls, each unit is con-

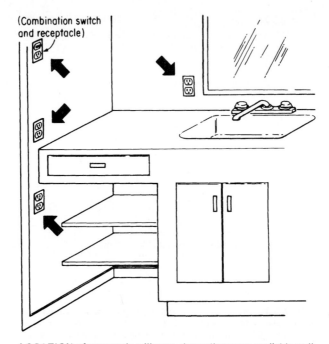

LOCATION of receptacle will vary, depending upon available wall space. Arrows show several possibilities. A receptacle in a medicine cabinet or in the bathroom lighting fixture does not satisfy this rule.

FIGURE 2.30 Receptacle required adjacent to wash basin in residence.

sidered to be a separate building and each is, by **NEC** definition, a "one-family dwelling," even though the appearance of a continuous structure might make it seem like a multifamily dwelling or apartment house. Each such unit is, therefore, required to have at least two GFCI-protected outdoor receptacles.

At least one receptacle—single or duplex or triplex—must be installed for the laundry of a dwelling unit. Such a receptacle and any other receptacles for special appliances must be placed within 6 ft of the intended location of the appliance. And a receptacle outlet is also required in a basement in addition to any receptacle outlet(s) that may be provided as the required receptacle(s) to serve a laundry area in the basement. One receptacle in the basement at the laundry area located there may *not* serve as *both* the required "laundry" receptacle and the required "basement" receptacle. A separate receptacle has to be provided for each requirement to satisfy the **NEC**.

Section 210-52(g) requires that at least one receptacle (other than for the laundry) must be installed in the basement of a one-family dwelling, one in an attached garage, and one in a detached garage *if* power is run to the detached garage. This rule calls for at least one receptacle outlet in the basement of a one-family house, in addition to any required for a basement laundry (Fig. 2.32). It calls for at least one receptacle in an attached garage of a one-family house. But for a detached garage of a one-family house, the rule simply requires that one receptacle outlet must be installed in the detached garage *if*—for some reason other than the **NEC**—electric

LIGHTING AND APPLIANCE BRANCH CIRCUITS 49

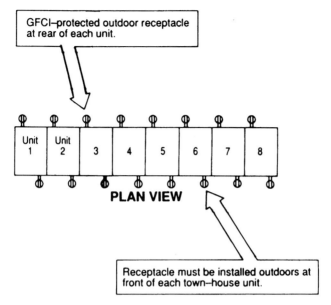

FIGURE 2.31 Front- and rear-receptacle outlets are required outdoors for town-house-type one-family dwellings.

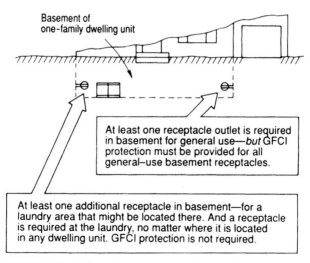

FIGURE 2.32 Only one basement receptacle is required (in addition to any for the laundry), but *all* general-purpose receptacles in *unfinished* basements must be GFCI protected.

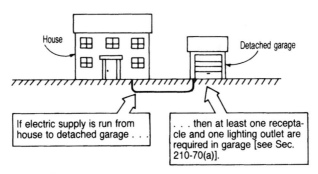

FIGURE 2.33 Detached garage may be required to have a receptacle and lighting outlet.

power is run to the garage, such as where the owner might desire it or some local code might require it (Fig. 2.33). The rule itself does *not* require that electric power be run to a detached garage to supply a receptacle there.

If the required "basement" receptacle is installed in an "unfinished" basement—that is a basement that has *not* been converted to, or constructed as, a recreation room, bedroom, den, etc.—such a receptacle would be required to be provided with GFCI protection [Section 210-8(a)(4)]. And, that same rule requires that any additional receptacles in an unfinished basement be GFCI-protected. In addition, *all* receptacles installed in a dwelling-unit garage (attached or detached) must have GFCI protection, as required by Section 210-8(a)(2).

With the wording of the rules of Sections 210-52(f) and 210-8(a)(4), it would be acceptable for a one-family dwelling to have one basement receptacle with GFCI protection if the basement is unfinished, but any other receptacles that are optionally installed must also have GFCI protection. The one or more receptacles provided for a laundry area in the basement are *excluded* from need for GFCI protection by Exception No. 2 of Section 210-8(a)(4).

As previously indicated, a receptacle outlet is required in any dwelling-unit hallway that is 10 ft or more in length. This provides for connection of plug-in appliances that are commonly used in halls—lamps, vacuum cleaners, etc. The length of a hall is measured along its centerline. Figure 2.34 shows required receptacles for a one-family dwelling.

The number of receptacles in a guest room of a hotel or motel must be determined by the every-12-ft rule of Section 210-52(a), but they *may* be located where convenient for the furniture layout, exempted from the rule that "no point along the floor line in any wall space is more than 6 ft . . . from an outlet." The intent of the rule is that the *number* of receptacles must satisfy Section 210-52(a) but *spacing* of the receptacles is exempted from the every-12-ft rule. In such cases, the spacing requirements of not more than 12 ft between receptacles, etc., do not have to be observed.

The rule here calls for one receptacle in a show window for each 12 ft of length (measured horizontally) to accommodate portable window signs and other electrified displays (Fig. 2.35).

A general-purpose 125-V receptacle outlet must be installed within 25 ft of heating, air-conditioning, and refrigeration equipment on rooftops *and* in attics and crawl spaces (Fig. 2.36).

This rule provides a readily accessible outlet for connecting 120-V tools and/or test equipment that might be required for the maintenance or servicing of mechanical equipment in attics and crawl spaces as well as rooftops. Each such receptacle must be

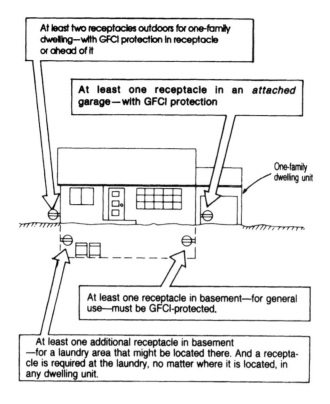

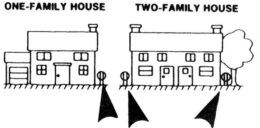

FIGURE 2.34 These specific receptacles are required for dwelling occupancies.

on the same level and within 25 ft of the heating, refrigeration, and air-conditioning equipment. This receptacle must not be fed from the load side of the disconnecting means for the mechanical equipment. And it must be GFCI-protected to satisfy Section 210-8(b)(2). Only rooftop units on one- and two-family dwelling units are excluded from this requirement.

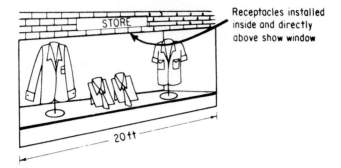

For a 20-ft-long store show window,
a minimum of two receptacles must be installed, one for
each 12 linear ft or major fraction thereof of show window
length.

FIGURE 2.35 Receptacles are required for show windows in stores or other buildings.

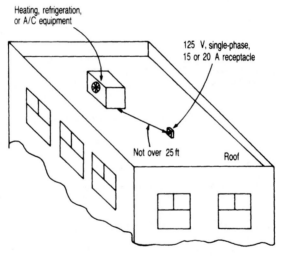

FIGURE 2.36 Maintenance receptacle outlet required for rooftop mechanical equipment as well as for such equipment in attics and crawl spaces.

The **NEC** requires that the two or more 20-A branch circuits required by Section 220-4(b) supply all the receptacle outlets in the kitchen, pantry, dining room, breakfast room, and any similar area of any dwelling unit—one-family houses, apartments, and motel and hotel suites with cooking facilities or serving pantries.

The two or more small appliance circuits serving the kitchen and other specified rooms must not have outlets in any other rooms (Fig. 2.37). Part (b)(2) requires that at least two such circuits must supply countertop receptacle outlets in the kitchen itself. The two circuits feeding outlets in the kitchen may also feed outlets in the

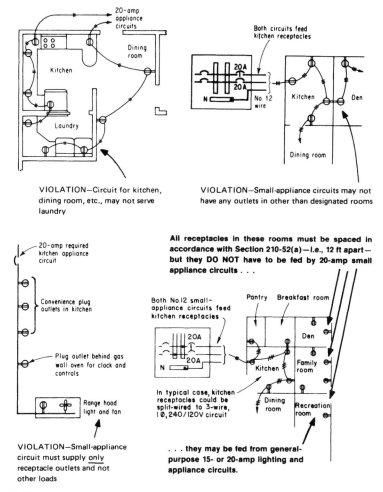

FIGURE 2.37 Small-appliance circuits for eating areas must not have outlets for other rooms or uses.

other areas above (dining room, pantry, etc.). What the **NEC** prohibits is, say, one circuit feeding the kitchen outlets and the other circuit or circuits feeding the outlets in the other prescribed areas. And it would not be acceptable for one 2-wire circuit to supply all of the countertop kitchen receptacles and the other circuit to supply kitchen wall outlets that are not located above countertop area. Of course, use of a 3-wire, 240/120-V, 20-A circuit feeding all of the kitchen receptacles with each receptacle split-wired to the different circuit hot legs would satisfy the rule, because each countertop receptacle would be fed by two circuits (a 3-wire, 240/120-V circuit is equivalent to two 120-V circuits) (Fig. 2.38).

It should be noted that the wording of Part (b)(1) requires that the "two or more" small appliance circuits must supply the receptacle outlets for any "refrigeration equipment" in the designated rooms. Therefore, receptacles for refrigerators

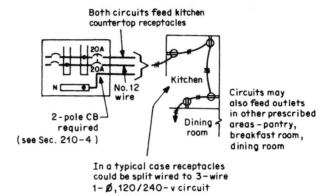

METHOD 1 – A 3-wire circuit to all outlets in prescribed areas.

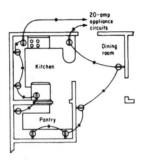

METHOD 2 – Two 2-wire circuits, each with at least one kitchen countertop receptacle outlet.

FIGURE 2.38 Two appliance circuits must have outlets in area specified in Section 210-52(b).

and freezers in those rooms must be connected on the 20-A small-appliance circuits. Because such appliances are often high-amperage appliances, some inspectors require a single (not duplex) receptacle fed by an individual 20-A branch circuit. And in such cases, the inspector usually requires that there must be two 20-A small-appliance circuits, in addition to the separate circuit for the refrigerator or freezer, to supply the general-purpose receptacle outlets spaced around the kitchen and other designated rooms. Such requirements are defended on the basis that the rule of the **NEC** calls for two "or more" circuits, that it is not acceptable to allow a refrigerator or freezer to consume the major capacity of one small-appliance circuit when only two such circuits are provided, and that it is the intent of the **NEC** to require at least two small-appliance circuits in addition to any individual appliance circuits.

NOTE: Part (b)(1) requires that the two or more 20-A small-appliance circuits must supply "all" receptacle outlets in the kitchen, dining room, etc., *and must also* supply receptacle outlets in any "similar area of a dwelling unit." That phrase suggests that all the rules on small-appliance circuits and receptacles *might* apply to *family rooms, recreation rooms, greatrooms, dens,* and any other room that would be used for preparing and/or serving food.

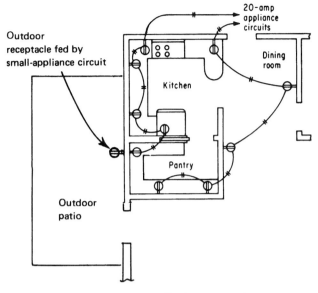

3-wire grounding. Outlet recessed for plug cap, allows clock to hang flush with wall. Clock hook furnished with each device.

Clock-hanger receptacle

FIGURE 2.39 Certain outlets may be fed by small appliance circuit or other circuit.

As noted in Exceptions to Part (b)(1), electric clock-hanger outlets and/or outdoor receptacle outlets may be connected on *either* a small appliance circuit or a general-purpose circuit (Fig. 2.39).

The intent of Section (b)(1), Exception No. 3, is that a wall switch-controlled receptacle(s) may be used to supply a plug-in lamp for lighting in a dining room, breakfast room, or pantry of a dwelling unit, *but* the switched receptacle(s) must not be connected on the one or more required 20-A small-appliance circuit(s) supplying those rooms (Fig. 2.40). This Exception correlates the rule of Section 210-52(b)(1) with Exception No. 1 of Section 210-70(a), as follows:

- The basic rule of Section 210-70(a) requires a wall switch-controlled lighting outlet in each habitable room of a dwelling unit—which includes the kitchen, dining room, breakfast room, and pantry.

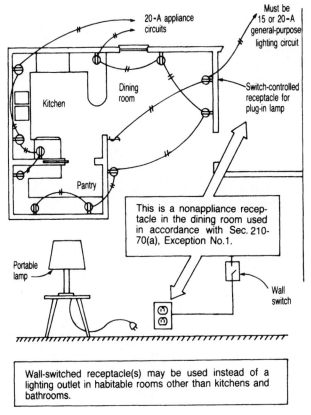

FIGURE 2.40 This type of circulating correlates Section 210-52(b)(1) with Section 210-70(a), Exception No. 1.

- Exception No. 1 of Section 210-70(a) says that a wall switch-controlled receptacle may be used instead of a switch-controlled outlet to a lighting fixture, with a floor or table lamp plugged into the switched receptacle to provide lighting in a room. But this may be done only in rooms other than the kitchen or bathroom, which must always have a switched lighting outlet for a ceiling or wall-mounted fixture.
- If a wall switch-controlled receptacle is used to supply a plug-in lamp for lighting in a dining room, breakfast room, or pantry—as permitted by Exception No. 1 of Section 210-70(a)—that poses a conflict with the rule of Section 210-52(b)(1), which says that two or more 20-A small-appliance branch circuits must be used to supply *all* receptacle outlets in the kitchen, dining room, breakfast room, and pantry and that those circuits must not have outlets other than for plug-in appliances—meaning no outlets supplying lighting.
- Section 210-52(a) requires the small-appliance receptacle outlets in the dining room, breakfast room, and/or pantry to be installed so that "no point along the floor line in any wall space is more than 6 ft, measured horizontally, from an outlet in that space."

- Exception No. 3 of Section 210-52(b)(1) says that, *in addition* to the number of small-appliance receptacle outlets required to satisfy the above rules, it is permissible to install one or more switched receptacle outlets for plug-in lamps *provided* that such receptacles are supplied from a general-purpose 15- or 20-A branch circuit and *not* from one of the required 20-A small-appliance circuits.

In a dwelling unit, any number of receptacle outlets may be connected on a circuit, whether or not that circuit also supplies lighting outlets. Recognizing that receptacle outlets have great diversity of usage, that loads are intermittent and/or light-duty cord-connected appliances, and that the number of such receptacles simply provides convenience of connection, the **NEC** does not require each receptacle to be taken as a fixed 180-VA load—as is required in nonresidential occupancies.

In occupancies other than dwelling units, circuit capacity must be provided for all receptacle outlets and outlets serving loads separate from general illumination [Section 220-2(c)], as covered later under "Receptacle Circuits."

Figures 2.41 through 2.43 present recommended design characteristics for residential appliance circuits.

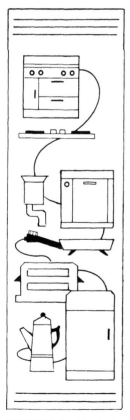

	Typical Load in Watts	Volts	Wires	Circuit Breaker or Fuse	Number of Outlets	Notes
RANGE	12000	120/240	3- #6	50-60A.	1	Use of more than one outlet is permitted, but not recommended.
OVEN (Built in)	4500	120/240	3- #10	30A.	1	Appliance may be direct connected.
RANGE TOP	6000	120/240	3- #10	30A.	1	Appliance may be direct connected.
RANGE TOP	3300	120/240	3- #12	20A.	1 or more	
DISHWASHER	1200	120	2 #12	20A.	1	These appliances may be direct connected on a single circuit. Grounded receptacles required, otherwise.
WASTE DISPOSER	300	120	2 #12	20A.	1	
BROILER	1500	120	2 #12	At Least Two Kitchen Appliance Circuits		Heavy duty appliances regularly used at one location should have a separate circuit. Only one such unit should be attached to a single circuit at a time.
FRYER	1300	120			20A. 2 or more	
COFFEEMAKER	1000	120				
REFRIGERATOR	300	120	2 #12	20A.	2	Separate circuit serving only refrigerator and freezer is recommended.
FREEZER	350	120	2 #12	20A.	2	

FIGURE 2.41 Circuits for kitchen appliances.

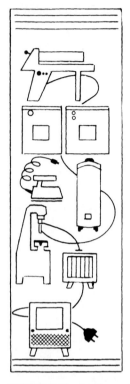

Load Devices	Typical Load in Watts	Volts	Wires	Circuit Breaker or Fuse	Number of Outlets	Notes
Circuits for Laundry Areas						
IRONER	1650	120	2 #12	20A.	1	Grounding type receptacle required.
WASHING MACHINE	1200	120	2 #12	20A.	1	Grounding type receptacle required.
DRYER	5000	120/240	3 #10	30A.	1	Appliance may be direct connected — must be grounded.
Circuits for Other Loads						
HAND IRON	1000	120	2 #12	20A.	2 or more	
WATER HEATER	3000					Consult Utility Co. for load requirements
WORKSHOP	1500	120	2 #12	20A.	2 or more	Separate circuit recommended.
PORTABLE HEATER	1300	120	2 #12	20A.	1	Should not be connected to circuit serving other heavy duty loads.
TELEVISION	300	120	2 #12	20A.	2 or more	Should not be connected to circuit serving appliances.

FIGURE 2.42 Appliance circuits.

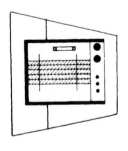

Size of Air Conditioner	Average Wattage	Circuits Required	Size of Each Circ.	Number of Outlets	Remarks
Air Conditioner ¾ hp	1200	Separate Circuit	2#12 120-v	1	Use of 3-wire, 120/240-volt circuits to unit conditioners offers circuit flexibility for 120 or 240 volts
Air Conditioner 1½ hp	2400	Separate Circuit	3#12 120/240-v	1	

FIGURE 2.43 Circuits for unit air conditioners.

Residential Lighting Outlets

Certain lighting outlets are mandatory for all dwelling-type occupancies—one-family houses, apartment houses, hotels, motels, dormitories, etc. Section 210-70 requires "at least one wall switch-controlled lighting outlet" in rooms, halls, stairways, attached garages, "detached garages with electric power," and at outdoor entrances. This rule requires a wall switch-controlled lighting outlet in every *attached* garage of a dwelling unit (such as a one-family house). But, for a *detached* garage of a dwelling unit, a switch-controlled lighting outlet is required *only* if the garage is provided with electric power—whether the provision of power is done as an optional choice or is required by a local code. Note that the **NEC** rule here does not itself require running power to the detached garage for the lighting outlet, such as a light outlet at any garage door that is provided as a vehicle entrance, because the lights of the car provide adequate illumination when such a door is being used during darkness. But the wording of this note does suggest that a rear or side door that is provided for personnel entry to an attached garage would be "considered as an outdoor entrance" because the note excludes only "vehicle" doors. Such personnel entrances from outdoors to the garage would seem to require a wall-switched lighting outlet.

At least one lighting outlet must be installed in every attic, underfloor space, utility room, and basement if it is used for storage or if it contains equipment requiring servicing. In such cases, the lighting outlet must be controlled by a wall switch at the entrance to the space. A lamp socket controlled by a pull chain or a canopy switch cannot be used. And each such required lighting outlet must be installed "at or near the equipment requiring servicing."

The last paragraph of Part (a) just before the Exceptions, states that lighting outlets for indoor stairways are required and must be controlled by a wall switch at each floor level connected by a stairway of six or more steps. This rule has the effect of requiring three-way switching for control of the lighting outlet illuminating such stairways.

Two exceptions are given to the basic requirements. Exception No. 1 notes that in rooms other than kitchens and bathrooms, a wall switch-controlled receptacle outlet may be used instead of a wall switch-controlled lighting outlet. The receptacle outlet can serve to supply a portable lamp, which would give the necessary lighting for the room. Exception No. 2 states that "in hallways, stairways, and at outdoor entrances remote, central, or automatic control of lighting shall be permitted." This latter recognition appears to accept remote, central, or automatic control as an alternative to the wall switch control mentioned in the basic rules.

But note carefully that every kitchen and every bathroom must have at least one wall switch-controlled lighting outlet (Fig. 2.44).

Section 210-70(b) notes that "at least one wall switch controlled lighting outlet or wall switch controlled receptacle shall be installed in guest rooms in hotels, motels, or similar occupancies."

Part (c) requires that a wall switch-controlled lighting outlet must be provided in attics or underfloor spaces housing heating, air conditioning, and/or refrigeration equipment—*in other than dwelling units*. The lighting outlet must be located at or near the equipment to provide effective illumination. And the control wall switch must be installed at the point of entry to the space (Fig. 2.45).

Outlet Loads Other than General Lighting

In all types of occupancies, branch-circuit capacity must be provided for specific outlets. For lighting other than general lighting, and for appliances other than motor-operated types, circuit capacity for each outlet must be provided as follows:

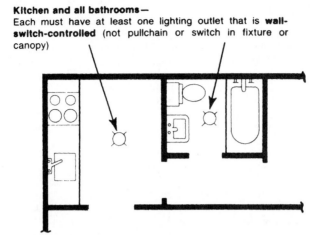

FIGURE 2.44 Switch-controlled lighting outlet in kitchen and bathroom.

1. For each outlet supplying a specific appliance or load device, circuit capacity equal to the current rating of the appliance or device must be provided.
2. In computing the minimum required branch-circuit capacity for outlets not used for general illumination, the actual voltampere rating of a recessed lighting fixture must be taken as the amount of load that must be included in the branch-circuit capacity. Local and/or decorative recessed lighting fixtures must be taken at their actual load values rather than as a load allowance of "180 VA per outlet"—even if each fixture is lamped at, say, 25 W. (In the case where a recessed fixture contained a 300-W lamp, an allowance of only 180 VA would be inadequate.) The inclusion of excessive or inadequate load in determining branch-circuit and feeder capacity must be carefully avoided.
3. For each outlet supplying a heavy-duty lampholder, 600 VA of circuit capacity must be allowed.
4. For each other outlet, such as each general-purpose convenience receptacle outlet in other than one-family and multifamily dwellings and guest rooms of hotels and motels, at least 180 VA of circuit capacity must be provided. This is fully discussed later.
5. When a circuit supplies only motor-operated devices, the provisions of **NEC** Article 430 or 440 must be taken into consideration.

These minimum load allowances for receptacle outlets and outlets for local lighting and appliances are modified as follows: **NEC** Table 220-19 establishes the basis for computing the required branch-circuit capacity for electric cooking appliances. For example, the branch circuit for a 16.6-kW household electric range that operates at 230/115 V would be sized as follows (refer to **NEC** Table 220-19):

1. Column A applies to ranges rated not over 12 kW, but the range in this example is rated 16.6 kW.
2. Note 1 below the table explains how to use the table for ranges over 12 kW and through 27 kW. For such ranges, the maximum demand in column A must be

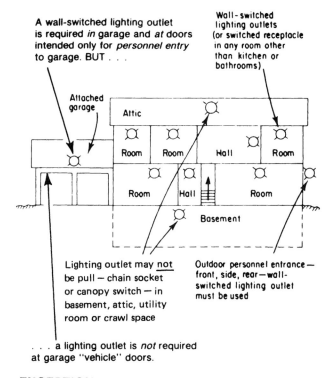

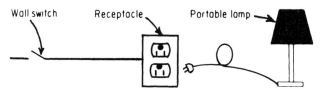

Wall-switched receptacle(s) may be used instead of a lighting outlet in habitable rooms other than kitchens and bathrooms.

FIGURE 2.45 Lighting outlets required in dwelling units.

increased by 5 percent for each additional kilowatt of rating (or major fraction thereof) above 12 kW.

3. The 16.6-kW range exceeds 12 kW by 4.6 kW (16.6 − 12).
4. In this example there is only one range. The maximum demand in column A for one range is 8000 W. Five percent of this value is 400 W (8000 × 0.05).
5. The maximum demand for this 16.6-kW range must be increased above 8000 W by 2000 W (400 W × 5). (The 5 is derived as 4 kW above 12 kW plus 1 for the remaining 0.6 kW.)
6. The required branch circuit must be sized, therefore, for a total demand load of 10,000 W (8000 W + 2000 W).

7. The required size of the branch circuit is, then,

$$\frac{10{,}000 \text{ W}}{230 \text{ V}} = 43 \text{ A}$$

The minimum branch-circuit size is 45 A.

For a circuit to the three cooking appliances shown in Fig. 2.46, the sizing of branch-circuit wires must be based on the following load calculation:

1. Note 4 of Table 220-19 says that the branch-circuit load for a counter-mounted cooking unit and not more than two wall-mounted ovens, all supplied from a single branch circuit and located in the same room, shall be computed by adding the nameplate rating of the individual appliances and treating this total as being equivalent to one range.
2. Therefore, for purposes of computing the branch-circuit load, the three appliances shown may be considered to be a single range with an 18-kW rating (6 kW + 6 kW + 6 kW).
3. From Note 1 of Table 220-19, since an 18-kW range exceeds 12 kW by 6 kW (18 − 12), the 8-kW demand of column A for a single range must be increased by 400 W (5 percent of 8000 W) for each of the six additional kilowatts above 12 kW.
4. Thus, the branch-circuit load is

$$8000 \text{ W} + (6 \times 400 \text{ W}) = 10{,}400 \text{ W}$$

For show-window lighting, instead of calculating circuit capacity at the specified load per outlet, 200 W of load may be allowed for each linear foot of show window, measured along its base (Fig. 2.47). And it may be safely assumed that the total load determined from such a calculation is based on the reality that such lighting is a continuous load, so the 125 percent multiplier is not needed.

A capacity of 1.5 A (or 180 VA) must be allowed for each 5 ft or fraction thereof of each separate and continuous length of fixed multioutlet assembly (surface raceway with spaced receptacles in the cover). Each such 5-ft length is considered as one outlet of 1.5-A capacity. In those cases where it is likely that a number of appliances will be used simultaneously, each 1 ft or fraction thereof is considered as an outlet and requires a load allowance of 1.5 A (Fig. 2.48). No branch-circuit capacity has to be

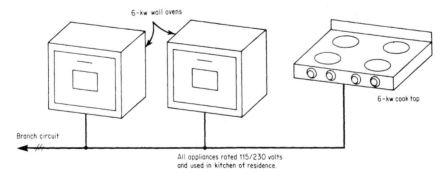

FIGURE 2.46 For branch-circuit sizing, these three cooking units are considered to be an 18-kW range.

LIGHTING AND APPLIANCE BRANCH CIRCUITS

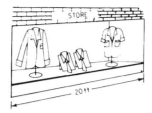

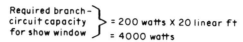

Required branch-circuit capacity for show window = 200 watts × 20 linear ft = 4000 watts

FIGURE 2.47 Design capacity for branch-circuit supply to show-window lighting is usually added without knowledge of the actual loads.

included for multioutlet assemblies in dwellings or guest rooms in motels and hotels. In such occupancies, the multioutlet assembly is considered part of the general lighting load—just as standard receptacle outlets are—and the **NEC** imposes no restriction on the number of receptacles that may be connected to any single branch circuit.

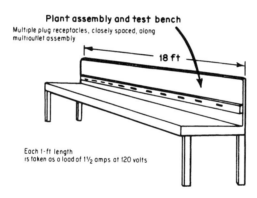

Load allowed for this bench = 18 × 1½ = **27 A**
At 120 volts, this represents a load of **3240 VA**

THEREFORE, 3240 VOLTAMPERES OF CIRCUIT CAPACITY MUST BE PROVIDED IN THIS EXAMPLE

FIGURE 2.48 Adequate capacity must be allowed for heavy use of cord-and-plug–connected loads.

Branch-Circuit Conductors

Although the **NEC** establishes No. 14 wire as the minimum size to be used for branch-circuit wiring, good design practice dictates the use of No. 12 as the minimum size, protected at either 15 or 20 A. For electric ranges of 8¾ kW or more, No. 8 copper or No. 6 aluminum is the **NEC** minimum size, but No. 6 copper or No. 4 aluminum is the recommended size. In a 3-wire, single-phase branch circuit to a household electric range, the neutral may be sized on the basis of 70 percent of the current-carrying capacity of one of the ungrounded conductors, although it may not be smaller than No. 10 wire.

The load current to be used in selecting the correct sizes of conductors for branch circuits can be calculated readily with the simple formulas given in Fig. 2.49. The selection of insulated conductors for branch-circuit wiring should be based on all conditions involved in any given application. The size should be sufficient to minimize voltage drop, to compensate for reasonable energy losses, and to contain short-circuit heating. The size also is influenced by metallic composition (copper, aluminum, or copper-clad aluminum), construction (stranded or solid), and the duty cycles of loads (continuous or intermittent operation), as follows:

2-WIRE CIRCUITS:

$$\text{Line current} = \frac{\text{connected load va (or watts at unity pf)}}{\text{line voltage}}$$

SINGLE-PHASE, 3-WIRE CIRCUITS

Apply above formula, considering each line to neutral separately and using line-to-neutral voltage. Result gives current in line conductors.

THREE-PHASE, 3-WIRE CIRCUITS:

$$\text{Line current} = \frac{\text{balanced 3-phase load va}}{\text{line voltage} \times 1.732}$$

FIGURE 2.49 Lighting and appliance branch-circuit calculations at any power factor.

- The type of insulation on conductors must be suited to operation at the ambient temperature of the area in which they are used.
- Where the ambient temperature exceeds 86°F (or 30°C, the room-temperature maximum for which insulated conductors are rated in **NEC** Tables 310-16 through 310-19), the ampacities of conductors should be derated according to correction factors given at the bottom of each of the ampacity tables (Tables 310-16 through 310-19).
- Where more than three conductors are used in a raceway or cable, the conductor ampacities must be derated from the ampacity values given in Table 310-16 or 310-18 to compensate for the increased heating effect due to reduced ventilation of an enclosed group of closely spaced conductors. (See Note 8 of Tables 310-16 through 310-19.)
- The type of insulation must be suited to the moisture content of its surroundings in the given application. Advantages and limitations related to the dryness or wetness of the area in which conductors are to be used must be considered. **NEC** regulations regarding the moisture content of the place of installation conform to the following definitions:

Damp location: A location subject to a moderate degree of moisture, such as some basements, some barns, some cold-storage warehouses, under canopies, marquees, roofed open porches, and the like.

Dry location: A location not normally subject to dampness or wetness. A location classified as dry may be temporarily subject to dampness or wetness, as in the case of a building under construction.

Wet location: A location subject to saturation with water or other liquids, such as locations exposed to weather, washrooms in garages, and like locations. Installations underground or in concrete slabs or masonry in direct contact with the earth shall be considered as wet locations.

- **NEC** Section 310-8 says that insulated conductors used underground, in concrete slabs or other masonry in direct contact with earth, in wet locations, or where condensation or accumulation of moisture within the raceway is likely to occur shall be (1) lead-covered, (2) of Type RHW, TW, THW, THHW, THWN, or XHHW, or (3) of a type listed by a testing laboratory as suitable for use in such locations. Cables of one or more conductors used in wet locations must be listed for use in wet locations. Such conductors are not suitable for direct burial in the earth unless they are of a type specifically listed for direct-earth burial. MI cable [Section 330-3, item (10)] and ALS cable (Sections 334-3 and 334-4) are recognized for direct-earth burial by the **NEC**. Type USE, UF, or any other type of cable listed and approved for such use may also be directly buried in the ground. Section 300-5 covers the installation of underground circuits.

Conductor insulations are designed to operate reliably at the anticipated operating temperature (ambient plus temperature rise due to load-induced heat); to resist environmental conditions (moisture, fumes, chemicals, oil and grease, acids, alkalies); to promote installation ease; to keep the costs of installation, maintenance, and replacement within reasonable limits; and to ensure efficient, reliable service by providing proper physical and electrical properties.

Most general-purpose 600-V building wire contains materials that are members of the rubber and thermoplastic families. Rubber, which also includes various butyl and mineral bases, is available in many grades and thicknesses to comply with cost, application, and overloading conditions. It also is applicable, when provided with auxiliary jacketing, in locations subject to oil or heat.

Thermoplastic insulations can resist moisture, oils, and chemicals without additional protection. This makes them smaller in diameter and lighter in weight than comparable rubber-based insulations. They are durable, tough, and lubricated with silicone and therefore easy to handle and install. Temperature ratings range from 60 to 90°C.

For higher-temperature applications, silicone asbestos insulation can be selected. Types are available for continuous operation at 125°C and for intermittent duty at 200°C.

Type FEP is a small-diameter wire insulated with fluorinated ethylene propylene. Type FEPB has a thinner layer of insulation but has an outer glass braid in sizes 14 to 8 and an asbestos braid in sizes 6 to 2. These types have a 90°C rating when used in dry locations for general wiring and a 200°C maximum rating for use in special applications.

Wiring in Air-Handling Ceilings. A common application which raises questions about the effects of temperature and moisture on wire insulation is that of wiring above suspended ceilings when such spaces are used for distributing air. **NEC** Section 300-22 says:

> **(c) Other Space Used for Environmental Air.** Section 300-22(c) applies to space used for environmental air-handling purposes other than ducts and plenums as specified in Sections 300-22(a) and 300-22(b). Only totally enclosed nonventilated insulated busway having no provisions for plug-in connections and wiring methods consisting of Type MI cable, Type MC cable without an overall nonmetallic covering, Type AC cable, or other factory-assembled multiconductor control or power cable that is specifically listed for the use shall be installed in such other space.

Other type cables and conductors shall be installed in electrical metallic tubing, flexible metallic tubing, intermediate metal conduit, rigid metal conduit, flexible metal conduit, or, where accessible, surface metal raceway or wireway with metal covers or solid bottom metal cable tray with solid metal covers.

Electric equipment with a metal enclosure or with a nonmetallic enclosure listed for the use and having adequate fire-resistant and low-smoke-producing characteristics, and associated wiring material suitable for the ambient temperature, shall be permitted to be installed in such other space unless prohibited elsewhere in this Code.

(FPN): The space over a hung ceiling used for environmental air-handling purposes is an example of the type of other space to which Section 300-22(c) applies.

Exception No. 1: Liquidtight flexible metal conduit in single lengths not exceeding 6 feet (1.83 m).

Exception No. 2: Integral fan systems specifically identified for such use.

Exception No. 3: This section does not include habitable rooms or areas of buildings, the prime purpose of which is not air handling.

Exception No. 4: Listed prefabricated cable assemblies of metallic manufactured wiring systems without nonmetallic sheath shall be permitted where listed for this use.

Exception No. 5: This section does not include the joist or stud spaces in dwelling units where wiring passes through such spaces perpendicular to the long dimension of such spaces.

(d) Data Processing Systems. Electric wiring in air-handling areas beneath raised floors for data processing systems shall comply with Article 645.

Section 300-22(b) covers wiring methods and equipment within "ducts or plenums," which are channels or chambers intended and used only for the supply or return of conditioned air. Such ducts or plenums are sheet metal or other types of enclosures which are provided expressly for air handling; they must be distinguished from "other space used for environmental air," such as the space between a suspended ceiling and the floor slab above it or the space between a raised floor (as used for data-processing wiring) and the slab below the raised floor, which may or may not be intended and used for air handling. Space of that type is covered by Section 300-22(c), although a raised floor used for data-processing circuits must also comply with Section 645-2(d). NFPA Standard 90A defines a duct system as "a continuous passageway for the transmission of air which, in addition to ducts, may include duct fittings, dampers, plenums, fans, and accessory air handling equipment." The word "duct" is not defined, but a plenum is defined as "an air compartment or chamber to which one or more ducts are connected and which forms part of an air distribution system."

NEC Sections 300-22(b) and (c) clearly limit acceptable wiring methods *only* to the ones described (Fig. 2.50). Part (c) permits use of totally enclosed, nonventilated, insulated busway in an air-handling ceiling space *provided* it is a non-plug-in-type busway that cannot accommodate plug-in switches or breakers. This one specific busway wiring method was added for hung-ceiling space used for environmental air. Surface metal raceway or wireway with metal covers or solid-bottom metal cable tray with solid metal covers may be used in air-handling ceiling space provided that the raceway is accessible, such as above lift-out panels. The air-handling space under a raised floor in a data-processing location is covered by Article 645 on electronic computer and data processing equipment.

The **NEC** panel has made clear that they generally oppose nonmetallic wiring methods in ducts and plenums and in air-handling ceilings, except for nonmetallic cable assemblies that are specifically listed for such use. It is also the intent of the **NEC**

LIGHTING AND APPLIANCE BRANCH CIRCUITS

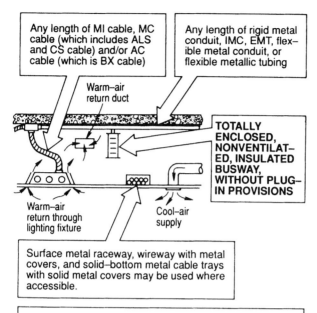

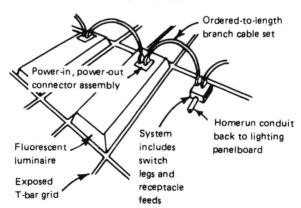

FIGURE 2.50 In air-handling ceilings, branch-circuit design may utilize *only* these wiring methods.

that cables with an outer nonmetallic jacket should not be permitted in ducts or plenums. Although the jacket material, usually PVC, would not propagate a fire, it would contribute to the smoke and provide additional flammable material in the air duct. The last paragraph of Part (c) permits use of nonmetallic equipment enclosures and wiring that are specifically listed or classified for use in air-handling ceiling spaces.

In effect, the rules in Part (b) exclude from use in all air-handling spaces any wiring that is not metal-jacketed or metal-enclosed, to minimize the creation of toxic fumes due to burning plastic under fire conditions. Section 800-53(a) basically requires telephone, intercom, and other communications circuits to be wired with type CMP cable or other types installed in compliance with Section 300-22 when such circuits are used in ducts or plenums or air-handling ceilings.

Wiring in air-handling space under raised floors in computer centers must use the wiring methods described in Section 645-5(d). Ventilation in the raised-floor space must be used only for the data-processing area and the data-processing equipment.

Although the rules of Section 300-22(c) apply only to air-handling spaces above suspended ceilings and beneath raised floors in other than computer rooms, such spaces are also subject to the general rules that apply to non-air-handling spaces. For a thorough understanding of this complex matter, refer to these definitions in Article 100: "accessible," "concealed," "exposed," and "readily accessible."

Because all those words or phrases are used in the **NEC** and are critically important to applications of wiring methods and equipment, their definitions must be carefully studied and cross-referenced with each other, as well as related to **NEC** rules using those words or phrases. Many common controversies about **NEC** rules revolve around those words and phrases and interpretation of the definitions. Refer to the discussion on "suspended ceilings" given under the definition for "accessible" in Article 100 of this handbook. In addition to that information, other rules relate to use above a suspended ceiling as follows:

1. All switches and CBs must be located so they may be operated from a readily accessible place, and the distance from the floor or platform up to the center of the handle in its highest position must not be over 6 ft 6 in (Section 380-8). Exception No. 2 of that rule does permit switches to be installed at high locations that are not readily accessible, even above suspended ceilings, *but only unfused switches,* because use of a fused switch would violate Section 240-24 on ready accessibility of overcurrent devices (the fuses in the switch). However, Section 430-102 requires a motor disconnect switch to be in sight from the motor controller location. And Section 430-107 says one disconnecting means shall be readily accessible. That means *not* above a suspended ceiling, where it would *not* be readily accessible.

2. Section 430-102(b) permits a motor to be out of sight from the location of its controller, and there is no rule requiring that motor controllers be readily accessible. Motor controllers may be installed above suspended ceilings.

3. Section 450-13 requires transformers to be installed so they *are* readily accessible, but certain exceptions are made. Exception No. 1 permits dry-type transformers rated 600 V or less to be located "in the *open* on walls, columns, or structures"— without the need to be readily accessible. And Exception No. 2 permits dry-type transformers up to 600 V, 50 kVa, to be installed in "fire-resistant hollow spaces of buildings not permanently closed in by structure," provided the transformer is designed to have adequate ventilation for such installation.

Air-Handling Ceilings. All the foregoing rules also apply to wiring and equipment installed above suspended ceilings in space used for air-conditioning purposes. But, in addition to those rules, the broad and detailed rules of Section 300-22(c) cover electrical installations in spaces above suspended ceilings when the space is used to handle environmental air. This section makes two basic determinations:

1. It lists all the wiring methods that are permitted in air-handling ceilings (which also may be used in non-air-handling ceiling spaces) and gives conditions and limitations for such use. This is a straightforward materials list which needs little or no interpretation.
2. Section 300-22(c) further comments on other "electric equipment" that is permitted in such spaces. That refers to switches, starters, motors, etc. The basic condition that must be satisfied is that the wiring materials and other construction of the equipment must be suitable for the expected ambient temperature to which they will be subjected.

Application of the **NEC** permission on use of "equipment" calls for substantial interpretation. The designer and/or installer must check carefully with equipment manufacturers and with inspection agencies to determine what is acceptable in air-handling space above a suspended ceiling. Practice in the field varies widely on this rule, and **NEC** intepretation has proved difficult.

Exception No. 2 of this section recognizes the installation of motors and control equipment in air-handling ducts where such equipment has been specifically approved for the purpose. Equipment of this type is listed by Underwriters Laboratories Inc. and may be found in the *Electrical Appliance and Utilization Equipment List* under the heading "Heating and Ventilating Equipment."

Exception No. 3 is intended to exclude from the requirements those areas which may be occupied by people. Hallways and habitable rooms are being used today as portions of air-return systems, and while they have air of a heating or cooling system passing through them, the prime purpose of these spaces is obviously not air handling.

Exception No. 4 permits modular wiring systems to be used in air-handling spaces *provided* that the wiring system consists of metallic-jacketed cable assemblies and there is *not* a plastic outside sheath over the metal.

Exception No. 5 permits Type NM cable to "pass through" a closed-in joist or stud space that is used for cold-air return. This is allowed because NM cable is suitable to be used under the temperature and moisture conditions in such spaces, as used in "dwelling units," to which the Exception is limited.

Overcurrent Protection

According to the basic **NEC** rule, the rating or setting of an overcurrent device in any branch circuit must not exceed the current-carrying capacity of the circuit conductor. Figure 2.51 shows the basic rules that apply to overcurrent protection for branch circuits.

When a branch circuit supplies a load for which loss of power would create a hazard, Section 240-3(a) states it is not necessary to provide "overload protection" for its conductors, *but* "short-circuit protection" must be provided. By "overload protection," this section means protection at the conductors' ampacity—that is, protection that would prevent overload by opening the circuit at any value of load current *over* the ampacity of the wires.

Several points should be noted about this section, which is shown at the bottom of Fig. 2.51:

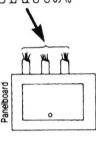

Panelboard

Circuit conductors protected by fuse or CB poles in panelboard—with amp rating of fuse or CB not in excess of conductor ampacity, per Sec. 210-19. Conductor ampacity from Table 310-16.

NOTE THIS

Each fuse or CB in a panel must not carry a continuous (3 hrs. or more) load current greater than 80% of fuse or CB amp rating—

BUT, THIS IS AN IMPORTANT EXCEPTION! If "overload protection" creates a hazard, it may be eliminated.

Protective device **may** have rating higher than ampacity of circuit conductors

Loss of power to magnet would present hazard of falling weight to personnel

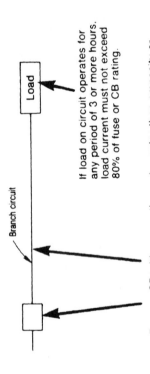

Branch circuit

Load

If load on circuit operates for any period of 3 or more hours, load current must not exceed 80% of fuse or CB rating.

Fuse or CB with amp rating not greater than ampacity or load-current rating of circuit conductor. See **NEC** Sec. 210-19 and Tables 310-16 through -19 with notes.

NOTE: Although the rule of NEC Sec. 210-19(a) requires branch-circuit conductors to have an ampacity of not less than the rating of the branch circuit (which is the rating of the branch-circuit overcurrent device) for a multi-outlet branch circuit that supplies and readily accessible receptacles, conductors of a branch circuit supplying lighting outlets only is required to have ampacity adequate only for the load-current being supplied. The latter circuit may have an overcurrent device that is the next standard rating of device above the conductor ampacity where there is not a standard rating of a device that is exactly equal to conductor ampacity — such as where conductors have their ampacity derated because more than three conductors are installed in a conduit or EMT run.

Power circuit to lifting magnet of electromagnetic material-handling rig

FIGURE 2.51 Basic design rules for branch circuits to other than motor loads.

1. Use of this exception is reserved only for applications where circuit opening on "overload" would be more objectionable than the overload itself, "such as in a material handling magnet circuit." In that example, mentioned in the exception, the loss of power to such a magnet while it is lifting a heavy load of steel would cause the steel to fall and would certainly be a serious hazard to personnel working below or near the lifting magnet. To minimize the hazard created by such power loss, the circuit to the magnet need not be protected at the conductor ampacity. A higher value of protection may be used to let the circuit sustain an overload rather than open on overload and drop the steel. Because such lifting operations are usually short-duration, intermittent tasks, an occasional overload is far less of a safety concern than the dropping of the magnet's load.
2. The permission to eliminate "overload protection" is not limited to a lifting-magnet circuit, which is mentioned simply as an example. Other electrical applications that present similar "hazards" would be equally open to the use of this exception.
3. Although the exception permits the elimination of overload protection and requires short-circuit protection, it gives no guidance on the selection of the actual rating of the protection that must be used. For such circuits, fuses or a CB rated, say, at 200 to 400 percent of the full-load operating current would give freedom from overload opening. Of course, the protective device ought to be selected with as low a rating as would be compatible with the operating characteristics of the electrical load. And it must have sufficient interrupting capacity for the circuit's available short-circuit current.
4. Finally, this exception is not a mandatory rule but a permissible application. It says "overload protection shall not be required"; it does *not* say that overload protection shall not be used. Overload protection may be used, or it may be eliminated. Obviously, careful study should always precede the use of this exception.

A number of general and specific **NEC** rules apply to overcurrent protection for branch circuits (and the same rules apply to feeder circuits). These rules apply to both fuses and circuit breakers, as follows:

1. As shown in Fig. 2.52, an overcurrent device (fuse or CB trip unit) must be placed at the supply end of each ungrounded conductor of the circuit to be protected (except where the conductors are permitted to be used as tap conductors without overcurrent protection at the supply end—such as the 10-, 25-, and 100-ft and other tap conductors recognized by **NEC** Section 240-21).
2. Where the device protecting a conductor is of a rating or setting that also provides protection for smaller conductors tapped from the larger conductor, there is no need to provide protection at the point where the smaller conductors are tapped from the larger conductor. The literal permission to provide such installation was eliminated in the 1993 **NEC** for no apparent reason. Clearly such application is adequate and many inspectors recognize the use of smaller conductors tapped from larger conductors as shown at the bottom of Fig. 2.53.
3. An overcurrent device must not be placed in any permanently grounded conductor, except (*a*) where the device simultaneously opens all conductors of the circuit, as a multipole CB could or (*b*) where the device is used for motor-running overload protection and is required by Sections 430-36 and 430-37 for the grounded conductor of a 3-phase, 3-wire circuit from a delta supply with a corner (one of the phase legs) grounded (Fig. 2.54).

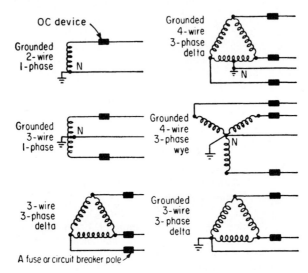

FIGURE 2.52 An overcurrent device must be placed in each ungrounded conductor.

BASIC RULE

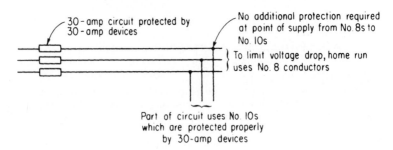

FIGURE 2.53 An exception deleted from the Code in the '93 edition permitted one set of protective devices to protect different sizes of wire. Many inspectors still recognize such applications.

LIGHTING AND APPLIANCE BRANCH CIRCUITS

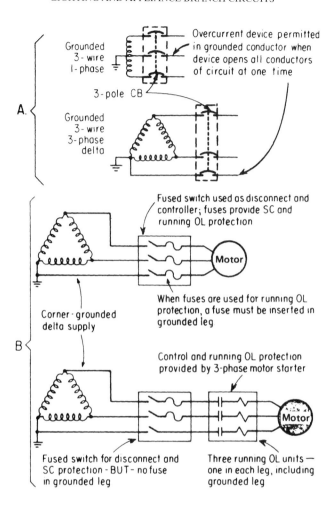

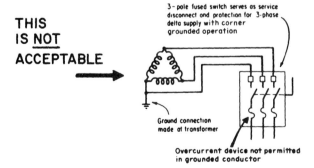

FIGURE 2.54 Two cases where an overcurrent device is permitted in a grounded conductor.

4. Branch-circuit taps (the smaller tap conductors used to supply individual lighting outlets or appliances, as covered in Sections 210-19 and 210-20) are considered protected by the branch-circuit overcurrent devices (Fig. 2.55). For recessed incandescent fixtures that are wired with No. 16 or No. 18 fixture wire in flex "whips" that are between 4 and 6 ft long, the branch-circuit protective device, rated at 15 or 20 A is adequate protection for the fixture wires.

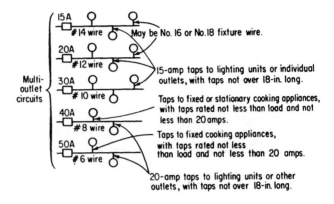

EXAMPLE:

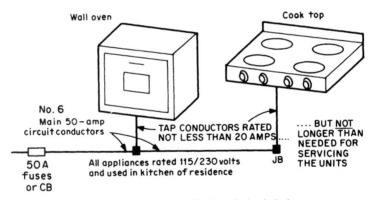

FIGURE 2.55 Sizes of tap wires protected by branch-circuit devices.

5. Overcurrent devices must be located so they are readily accessible, with the following exceptions: overcurrent devices for circuits tapped from a busway, which must be in the device for tapping the busway, in the cord plug of a fixed or semi-fixed luminaire supplied from a trolley busway or mounted on a luminaire plugged into a busway; overcurrent devices that are supplementary overcurrent protection for subdivided electric-resistance heating elements used in ductwork above suspended ceilings, as required by Sections 424-22(b) and (c); and overcurrent devices that are supplementary to the protection required by the **NEC**, as recognized by Section 240-10 (Fig. 2.56). "Readily accessible" means "capable of being reached quickly, for operation, renewal, or inspections, without requir-

LIGHTING AND APPLIANCE BRANCH CIRCUITS

ing those to whom ready access is requisite to climb over or remove obstacles or to resort to portable ladders, chairs, etc."

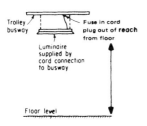

FIGURE 2.56 Overcurrent devices may not be readily accessible under these conditions.

6. Plug fuses must not be used in circuits of more than 125 V between conductors, but they may be used in grounded-neutral systems where the circuits have more than 125 V between ungrounded conductors but not more than 150 V between any ungrounded conductor and ground. That means plug fuses may be used on circuits rated 120 V, derived from 240/120-V or 208Y/120-V systems. But an exception in that the **NEC** rule does not permit the use of plug fuses for each hot leg of a 240-V, 2-wire or 3-wire circuit derived from a 240/120-V system: for each hot leg of a 208-V, 2-wire or 3-wire circuit derived from a 208Y/120-V system; or for each hot wire of a 208/120-V, 3-wire or 4-wire circuit derived from a 208Y/120-V system (Fig. 2.57). The screw shell of a plug fuseholder must be connected to the load side of the circuit, and the **NEC** does not require a disconnecting means on the supply side of a plug fuse.

7. Fuseholders for plug fuses must be of Type S to accommodate Type S plug fuses. Fuseholders must be designed or equipped with adapters to take either a 0- to 15-A Type S fuse, a 16- to 20-A Type S fuse, or a 21- to 30-A type S fuse. A 0- to 15-A fuseholder or adapter must not be able to accept a 20- or 30-A fuse and a 20-A fuseholder or adapter must not be able to take a 30-A fuse. (Fig. 2.58). The purpose of this rule is to prevent overfusing of 15- and 20-A circuits. Edison-base (non-Type S) plug fuses are recognized only for replacement use.

NEC Section 240-60 recognizes the use of "300-volt type" cartridge fuses and fuseholders in circuits rated not over 300 V between conductors and in circuits supplied by a system with a grounded neutral and rated not over 300 V to ground from any ungrounded conductor. Such cartridge fuses are, therefore, recognized for use on circuits derived from a 480Y/277-V system.

8. Cartridge fuses and fuseholders must be such that a fuse of any given class cannot be used in a fuseholder of a lower current or higher voltage rating.

9. Fuseholders for current-limiting fuses must not accept fuses that are not current-limiting. Low-voltage current-limiting cartridge fuses up to 600 A are avail-

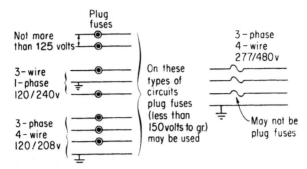

FIGURE 2.57 Circuit voltage determines the acceptability of the use of plug fuses.

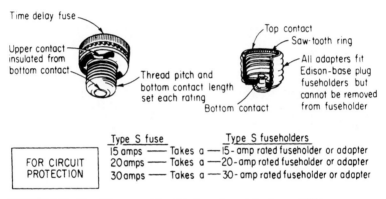

FIGURE 2.58 Type S fuses provide safety through noninterchangeability.

able with special ferrules or knife blades to provide installation in conformity with the requirements of Section 240-60(b) (Fig. 2.59). Such fuses can fit into current-limiting fuseholders and also fit standard **NEC** fuseholders. But standard **NEC** fuses (Class H) cannot be inserted into the current-limiting holders. This arrangement maintains safety in applications where, say, the bracing of busbars in a panel or switchboard is based only on the amounts of short-circuit letthrough current that a current-limiting fuse will pass but where the busbars could not take the higher current that would flow at that point in the event noncurrent-limiting fuses were used.

Class R fuses are UL-listed and meet the current **NEC.** They fit Class R fuseholders that accept only current-limiting fuses and reject fuses of low interrupting rating, such as Class H fuses. Class R fuses have the same overall physical dimensions as standard Class H (formerly **NEC**) fuses and will fit existing fused equipment, eliminating the need for duplicate stocking.

10. Fuses rated for 600 V may be used at any lower voltage.
11. A disconnecting means must be provided on the supply side of all fuses in circuits of more than 150 V to ground and of cartridge fuses in circuits of any voltage where the fuses would be accessible to other than qualified personnel (Fig. 2.60).

LIGHTING AND APPLIANCE BRANCH CIRCUITS

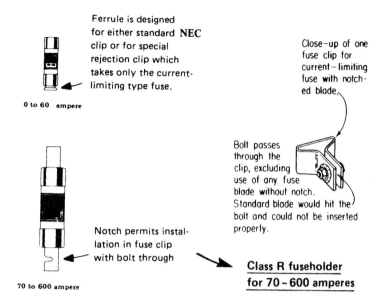

FIGURE 2.59 Class R fuses and Class R fuseholders provide for rejection of inadequately rated fuses.

12. Circuit breakers must open simultaneously all ungrounded conductors of the circuits they protect; i.e., they must be multipole CB units, except that individual single-pole CBs may be used for the protection of each ungrounded conductor of certain types of circuits, including ungrounded 2-wire circuits, 3-wire single-phase circuits, and lighting or appliance branch circuits connected to 4-wire, 3-phase systems provided such lighting or appliance circuits are supplied from a grounded-neutral system and the loads are connected line to neutral (Fig. 2.61).

Those are the rules of **NEC** Section 240-20(b). They coordinate with Section 210-10, which permits the use of 2-wire branch circuits tapped from the outside conductors of systems where the neutral is grounded on 3-wire dc or single-phase, 4-wire, 3-phase and 5-wire, 2-phase systems.

All poles of the disconnecting means used for branch circuits supplying permanently connected appliances over 300 VA or ⅛ hp must be operated at the same time. This requirement applies where the circuit is supplied through either circuit breakers or switches.

In the case of fuses and switches, when a fuse blows in one pole, the other pole may not necessarily open; and the requirement to manually switch together involves only the manual operation of the switch. Similarly, when two circuit breakers are connected with handle ties, an overload of one of the conductors with the return circuit through the neutral may open only one of the circuit breakers; but the manual operation of the pair when used as a disconnecting means will open both poles. The words "manually switch together" should be considered as meaning "operating at the same time," i.e., during the same operating interval; they apply to the equipment when used as a disconnecting means and not when used as an overcurrent protective device.

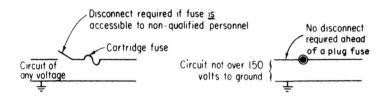

FIGURE 2.60 Cartridge fuses must generally be used in switches.

Circuit breakers with handle ties are, therefore, considered as providing the disconnection required by this section. The requirement to "manually switch together" can be achieved by a "master handle" or "handle tie," since the operation is intended to be effected manually. The intent is not to require a common trip for the switching device but to require that it have the ability to disconnect ungrounded conductors with a single movement of the hand.

The basic rule of Section 210-4 requires that multiwire branch circuits (such as 240/120-V, 3-wire, single-phase and 3-phase, 4-wire circuits at 208/120 V or 480/277 V) be used only with loads connected from a hot or phase leg to the neutral conductor. The intent of the **NEC** is that line-to-line connected loads may be used, other than in Exception No. 1, only where the poles of the circuit protective device operate together—simultaneously and automatically—in response to overcurrent. A multipole CB satisfies the rule, but a fused multipole switch does not (Fig. 2.62).

The use of multipole switching (that is, the disconnect and not the "overcurrent device" itself) on such circuits provides safety to persons working on or replacing lighting fixtures or receptacle outlets connected line to line in existing systems. When a receptacle is fed by more than one hot wire, someone attempting to work on

LIGHTING AND APPLIANCE BRANCH CIRCUITS

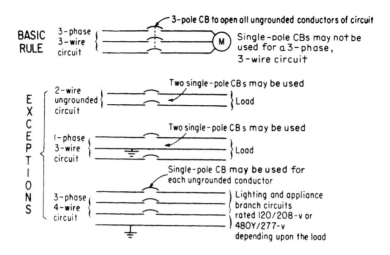

EXAMPLES:

Two 1-pole CBs may be used for a 240-volt or 208-volt circuit to an electric heating unit or air-conditioning unit if the CBs are not intended to provide the disconnect means required by Secs. 424-19(b)(1), 440-63, 426-20, 422-21(b), 430-85, or 430-103.

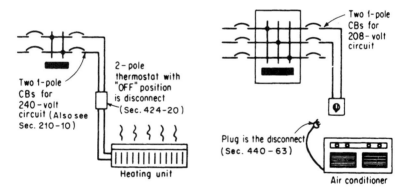

NOTE: If multipole breakers are used instead of the single-pole units shown, the breakers could serve as disconnect means because they would satisfy Section 210-10 requiring poles of a disconnect to switch together.

NOTE: A 3-pole CB must always be used for a 3-phase, 3-wire circuit supplying phase-to-phase loads fed from an ungrounded delta system, such as 480-volt outdoor lighting for a parking lot, etc., as permitted by Section 210-6(b).

FIGURE 2.61 Selecting between a multipole CB and multiple single-pole CBs.

the receptacle may shut off only one of the branch-circuit breakers or switching devices in the branch-circuit panelboard and believe that the circuit has been killed. However, the other leg to the receptacle will still be hot and will pose a shock hazard to that person. The same is true of lighting fixtures connected phase-to-phase at 240, 208, or 480 V.

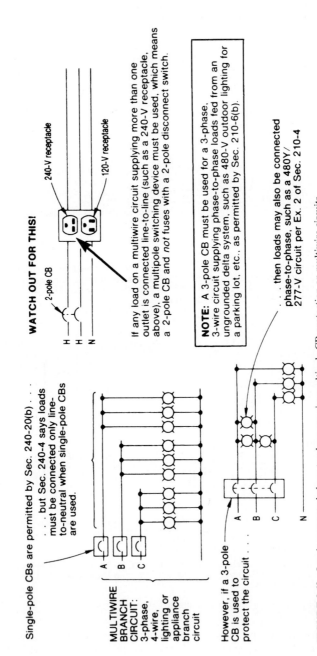

FIGURE 2.62 Loads connected phase to phase require multipole CB protection on multiwire circuits.

The last sentence in Section 210-4(b) requires that a two-pole (simultaneously operating) circuit breaker or switch be used at the panelboard (supply) end of a 240/120-V, single-phase, 3-wire circuit that supplies one or more split-wired duplex receptacles or duplex switches (two switches on a single mounting yoke) in a dwelling unit (a one-family house, an apartment in an apartment house, a condominium unit, or any other occupancy that conforms to the definition of "dwelling unit" given in Article 100). Figure 2.63 shows that a two-pole CB, two single-pole CBs with a handle tie that enables them to be used as a two-pole disconnect, or a two-pole switch ahead of branch-circuit fuse protection will satisfy the requirement that both hot legs must be interrupted when the disconnect means is opened to deenergize a multiwire circuit to a split-wired receptacle. This provides the greater safety of disconnecting both hot conductors simultaneously to prevent shock hazard.

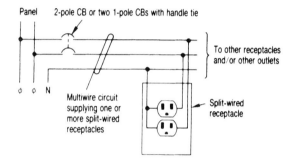

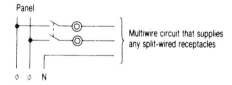

FIGURE 2.63 Two-pole disconnect must be used for switching residential split-wired receptacles.

A shock hazard may exist when a piece of electrical equipment is replaced or maintained after only one of two hot supply conductors has been opened. If only one of the two single-pole CBs protecting a 240/120-V 3-wire circuit to a split-wired receptacle is shut off, the load device, say a lamp, connected to one of the receptacles will go out. Repair personnel may presume that they will not contact any hot terminals within the receptacle box, but they will receive an electric shock from the other hot leg.

NOTE: Circuits supplying split-wired receptacles in commercial, industrial, and institutional occupancies (places that are not "dwelling units") may contain single-pole CBs, plug fuses without switches, or single-pole switches ahead of fuses.

It should also be noted that although a two-pole switch ahead of fuses may serve as the simultaneous disconnect required ahead of split-wired receptacles, such a switch does not satisfy the requirement of Exception No. 2 of Section 210-4 for a simultaneous multipole "branch-circuit protective device" when a multiwire circuit supplies any loads connected phase-to-phase. In such a case, a two-pole CB must be used because fuses are single-pole devices and do not assure simultaneous opening of all hot legs on overcurrent or ground fault.

Figure 2.64 points out two specific UL limitations on the use of molded-case CBs: A CB marked "120 V ac" may be used only for a load that is connected phase-to-neutral. Two such CBs are not suitable for protecting, say, a 2-wire, 240 V circuit with the load connected phase-to-phase. For such a circuit, UL regulations require single-pole CBs marked "120/240 V ac," to be equipped with handle ties when used for any phase-to-phase loads.

1. "Single-pole CBs rated 120 and 120/240 volts ac are suitable for use in a single-phase multiwire circuit where the neutral is connected to the load."

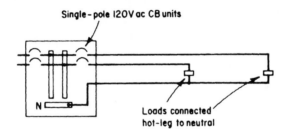

2. "Single-pole circuit breakers with handle ties rated 120/240 volts ac are suitable for use in a single-phase multiwire circuit *with or without* the neutral connected to the load."

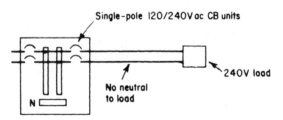

FIGURE 2.64 Single-pole CBs must be suited to their particular design function.

Conductor Temperature Rating

An important step in the design of branch circuits is the selection of the type of conductor to be used: TW, THW, THWN, RHH, THHN, XHHW, etc. The various types of conductors are covered in **NEC** Article 310, and the ampacities of conductors with different insulations and temperature ratings are given in Tables 310-16 through 310-19 for varying conditions of use—in raceways, in open air, at normal or higher-than-normal ambient temperatures. Conductors must be used in accordance with all the data in those tables and in the detailed notes given with them.

The selection of the best wire or cable for a proposed application has become increasingly complex because of the hundreds of different types of wire and cable available. Modern building wire (rated 600 V) can have thermoplastic insulation rated for use at various ambient temperatures, in wet or dry locations, or with a somewhat smaller overall diameter (THHN, THWN) to permit greater conduit fill or easier pulling. Or, a wire with cross-linked polyethylene insulation (XHHW) can be selected for tough applications. Cable assemblies range from nonmetallic sheathed cable (Type NM) and service-entrance cable (Type SE), which have been in common use for many years, to the new flat-conductor cable assemblies (Type FCC) for surface wiring on floors under carpet squares.

In selecting wire or cable, it is important to check the UL *Electrical Construction Materials Directory (Green Book)*. **NEC** Section 110-2 says, "The conductors and equipment required or permitted by this Code shall be acceptable only when approved." This wording is being widely interpreted to require the use of only UL-listed products if UL listing is given. This is consistent with the OSHA law, which requires the use of products listed by a nationally recognized testing lab.

In selecting the type and temperature rating of wire for branch circuits, consideration must be given to a very important UL qualification regarding the temperature ratings of equipment terminations. Although application data on temperature ratings of conductors connected to equipment terminals are not given in the **NEC,** they nevertheless become part of the mandatory regulations of the **NEC** because of Section 110-3(b). This section incorporates the instructions in UL and other listings books as part of the **NEC** itself. It reads as follows: "Listed or labeled equipment shall be used or installed in accordance with any instructions included in the listing or labeling."

UL treats different equipment in a slightly manner. For "Appliances and Utilization Equipment," the UL *Green Book* says:

> **Appliances and Utilization Equipment Terminations**—Except as noted in the information at the beginning of some product categories most terminals unless marked otherwise are for use only with copper wire. If aluminum or copper-clad aluminum wire can be used, marking to indicate this fact is provided. Such marking is required to be independent of any marking on terminal connectors, such as on a wiring diagram or other visible location. The marking may be in an abbreviated form, such as "AL-CU."
>
> Except as noted in the information at the beginning of some product categories, the termination provisions are based on the use of 60C insulated conductors in circuits rated 100 amperes or less and the use of 75C insulated conductors in higher rated circuits as specified in Table 310-16 of the National Electrical Code. If the termination provisions on equipment are based on the use of other conductors, the equipment is either marked with both the size and temperature rating of the conductors to be used or with only the temperature rating of the conductors to be used. If the equipment is only marked for use with conductors having a higher (75C or 90C) temperature rating (wire size not specified), the 60C ampacities (for circuits rated 100 amperes or less) and 75C ampacities (for circuits rated over 100 amperes) should be used to determine wire size.

Conductors having a temperature rating higher than specified may be used, though not required, if the size of the conductors is determined on the basis of the 60C ampacity (100 ampere or less circuits) or 75C ampacity (over 100 ampere circuits).

And for "Distribution and Control Equipment," it says:

Distribution and Control Equipment Terminations—Most terminals are suitable for use only with copper wire. Where aluminum or copper-clad aluminum wire can or shall be used, (some crimp terminals may be Listed only for aluminum wire) there is marking to indicate this. Such marking is required to be independent of any marking on terminal connectors, such as on a wiring diagram or other visible location. The marking may be in an abbreviated form such as "AL-CU."

Except as noted in the following paragraphs or in the information at the beginning of some product categories, the termination provisions are based on the use of 60C ampacities for wire size Nos. 14-1 AWG, and 75C ampacities for wire size Nos. 1/0 AWG and larger, as specified in Table 310-16 of the National Electrical Code.

Some distribution and control equipment is marked to indicate the required temperature rating of each field-installed conductor. If the equipment, normally intended for connection by wire sizes within the range 14-1 AWG, is marked "75C" or "60/75C", it is intended that 75C insulated wire may be used at full 75C ampacity. Where the connection is made to a circuit breaker or switch within the equipment, such a circuit breaker or switch must also be marked for the temperature rating of the conductor.

A 75C conductor temperature marking on a circuit breaker or switch normally intended for wire sizes 14-1 AWG does not in itself indicate that 75C insulated wire can be used unless (1) the circuit breaker or switch is used by itself, such as in a separate enclosure, or (2) the equipment in which the circuit breaker or switch is installed is also so marked.

A 75 or 90C temperature marking on a terminal (e.g. AL7, CU7AL, AL7CU, or AL9, CU9AL, AL9CU) does not in itself indicate that 75 or 90C insulated wire can be used unless the equipment in which the terminals are installed is marked for 75 or 90C.

Higher temperature rated conductors than specified may be used if the size is based on the above statements.

This temperature limitation on terminals applies to the terminals on all equipment—switches, motor starters, contactors—except where some other specific condition is recognized in the general information preceding the product category. Figure 2.65 illustrates this vitally important matter as it applies to CBs, which are treated in a completely different manner than either of the other "types" of equipment:

CIRCUIT BREAKERS (DHJR)

This listing covers circuit breakers which, unless otherwise noted, are of the manually operable, air-break type, providing automatic overcurrent protection.

Circuit breakers and circuit breaker enclosures as listed herein are for use with copper conductors unless marked to indicate which terminals are suitable for use with aluminum conductors. Such markings are independent of any marking on terminal connectors and are located on a wiring diagram or in an other readily visible location.

1. Circuit breaker enclosures are marked to indicate the temperature rating of all field installed conductors.
2. Circuit breakers with a current rating of 125 amperes or less are marked as being suitable for 60 C, 75 C only or 60/75 C rated conductors. It is acceptable to use conductors with a higher insulation rating, if the ampacity is based on the conductor temperature rating marked on the breaker.
3. Circuit breakers rated 125 amperes or less and marked suitable for use with 75 C rated conductors are intended for field use with 75 C rated conductors at full 75 C

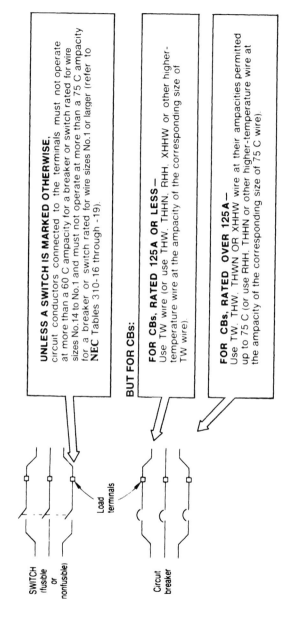

FIGURE 2.65 UL specifies maximum temperature ratings for conductors connected to equipment terminals.

ampacity only when the circuit breaker is installed in a circuit breaker enclosure or individually mounted in an industrial control panel with no other component next to it, unless the end use equipment (panelboard, switchboard, service equipment, power outlet, etc.) is also marked suitable for use with conductors rated 75 C.
4. A circuit breaker with a current rating of more than 125 amperes is suitable for use with conductors rated 75 C.
5. Circuit breakers intended for continuous operation at 100 percent of rated current may be marked to be connected with 90 C rated wire with the size based on 75 C ampacity.

A suitable marking is required in a circuit breaker enclosure, whether or not terminals are mounted therein, if it is intended that the breaker to be mounted therein is to be used with aluminum wire.

When terminals are tested for suitability at 60 or 75°C, the use of 90°C conductors operating at their higher current ratings poses a definite threat of heat damage to switches, breakers, etc. Many termination failures in equipment suggest overheating even where the load current did not exceed the current rating of the breaker, switch, or other equipment.

When a terminal that is rated at 60°C is fed by a conductor loaded to its 90°C ampacity and operating at 90°C, a substantial amount of heat will be conducted from the 90°C conductor metal to the 60°C terminal. Over a period of time, this can damage the termination even though the load current does not exceed the equipment current rating and does not exceed the ampacity of the 90°C conductor. Whenever two metallic parts at different operating temperatures are tightly connected together, the higher-temperature part (say 75 or 90°C wire) will transmit heat to the lower-temperature part (the 60°C terminal) and thereby raise its temperature.

The ampacity of any given size of conductor is established by the ability of the conductor insulation to withstand the I^2R heat produced by the current flowing through the conductor. But it must not be assumed that the equipment to which that conductor is connected also is capable of withstanding the heat that will be thermally conducted from the metal of the conductor to the metal of the terminal to which the conductor is tightly connected.

Although this limitation on the operating temperature of terminals in equipment might appear to eliminate any advantage that higher-temperature conductors have over lower-temperature conductors, there are still advantages to using the higher-temperature wires.

NEC Ampacities

The ampacity of a conductor is the amount of current, in amperes, that the conductor can carry continuously under specified conditions of use without developing a temperature in excess of the value that represents the maximum temperature that the conductor insulation can withstand. **NEC** Table 310-16, for instance, specifies ampacities for conductors where not more than three conductors are contained in a single raceway or cable or directly buried in the earth, provided that the ambient temperature is not in excess of 30°C (86°F). For higher ambient temperatures, the ampacity must be derated in accordance with the *correction factors* at the bottom of the table. For applications with more than three conductors in a conduit or cable, the ampacities of the conductors must be derated in accordance with the *correction factors* at the bottom of the table. For applications with more than three conductors in a conduit or cable, the ampacities of the conductors must be derated in accordance

LIGHTING AND APPLIANCE BRANCH CIRCUITS   87

with Note 8 to Tables 310-16 to 310-19. And where there are both an elevated ambient temperature and more than three conductors in a cable or conduit, both deratings must be made; that is, the conductors must be derated twice.

NEC Table 310-16 clearly indicates that No. 14, 12, and 10 conductors rated at 90°C do, in fact, have higher ampacities than those of the corresponding sizes of 60 and 75°C conductors. As shown in Fig. 2.66, No. 12 TW and No. 12 THW copper conductors are both assigned an ampacity of 25 A under the basic application conditions of the table. But, for a copper conductor, a No. 12 THHN, RHH, or XHHW (dry location) has an ampacity of 30 A. However, the footnote to Table 310-16 (shown in Fig. 2.67) requires that the "load-current rating" and "overcurrent protection" for No. 14, 12, and 10 copper conductors be taken as 15, 20, and 30 A, respectively, regardless of the type and temperature rating of the insulation on the conductors. When applied to the selection of branch-circuit wires in cases where conductor ampacity derating is required by Note 8 of Tables 310-16 through 310-19 for conduit fill (over three wires in a raceway), the footnote to Table 310-16 affords advantageous use of the 90°C wires for branch-circuit makeup. The reason is that, as stated in Note 8, the derating of ampacity is based on taking a percentage of the actual ampacity value shown in the table, and the ampacity values for 90°C conductors are higher than those for 60 and 75°C conductors.

Where not more than three conductors are used in a conduit, copper and aluminum conductors may supply load currents up to the ampacities given in **NEC** Table 310-16 for the various sizes and insulations of conductors with temperature ratings up

Table 310-16. Allowable Ampacities of Insulated Conductors Rated 0-2000 Volts, 60° to 90°C (140° to 194°F) Not More Than Three Conductors in Raceway or Cable or Earth (Directly Buried), Based on Ambient Temperature of 30°C (86°F)

Size AWG kcmil	Temperature Rating of Conductor. See Table 310-13.						Size AWG kcmil
	60°C (140°F)	75°C (167°F)	90°C (194°F)	60°C (140°F)	75°C (167°F)	90°C (194°F)	
	TYPES TW†, UF†	TYPES FEPW†, RH†, RHW†, THHW†, THW†, THWN†, XHHW†, USE†, ZW†	TYPES TA, TBS, SA, SIS, FEP†, FEPB†, MI, RHH†, RHW-2, THHN†, THHW†, THW-2, THWN-2, USE-2, XHH, XHHW†, XHHW-2, ZW-2	TYPES TW†, UF†	TYPES RH†, RHW†, THHW†, THW†, THWN†, XHHW†, USE†	TYPES TA, TBS, SA, SIS, THHN†, THHW†, THW-2, THWN-2, RHH†, RHW-2, USE-2, XHH, XHHW, XHHW-2, ZW-2	
	COPPER			ALUMINUM OR COPPER-CLAD ALUMINUM			
18	....	....	14	....	....	....	....
16	....	....	18	....	....	....	....
14	20†	20†	25†	....	....	....	....
12	25†	25†	30†	20†	20†	25†	12
10	30	35†	40†	25	30†	35†	10
8	40	50	55	30	40	45	8
6	55	65	75	40	50	60	6
4	70	85	95	55	65	75	4
3	85	100	110	65	75	85	3
2	95	115	130	75	90	100	2
1	110	130	150	85	100	115	1
1/0	125	150	170	100	120	135	1/0
2/0	145	175	195	115	135	150	2/0
3/0	165	200	225	130	155	175	3/0
4/0	195	230	260	150	180	205	4/0
250		255		170			250

FIGURE 2.66 NEC Table 310-16 shows higher ampacities for No. 14, No. 12, and No. 10 branch-circuit conductors.

> †The load current rating and the overcurrent protection for conductor types marked with an obelisk (†) shall not exceed 15 amperes for 14 AWG, 20 amperes for 12 AWG, and 30 amperes for 10 AWG copper; or 15 amperes for 12 AWG and 25 amperes for 10 AWG aluminum and copper-clad aluminum.

FIGURE 2.67 This note below **NEC** Table 310-16 radically alters conductor applications for No. 14, 12, and 10 circuit wires.

to 90°C, with the exceptions indicated in the footnote to Table 310-16. These ampacities apply when the ambient temperature does not exceed 30°C (86°F). If higher ambients exist, the ampacity must be derated in accordance with the table of correction factors given at the bottom of Table 310-16.

Note 8: Ampacity Derating. When there are more than three current-carrying conductors in a raceway or cable, their current-carrying capacities must be decreased to compensate for proximity heating effects and reduced heat dissipation due to reduced ventilation of the individual conductors, which are bunched or which form an enclosed group of closely placed conductors. In such cases, the ampacity of each conductor must be reduced as indicated in the table of Note 8, which appears after **NEC** Tables 310-16 to 310-19.

Note 8 to Table 310-16 says: "Where the number of current-carrying conductors in a raceway or cable exceeds three, the allowable ampacities shall be reduced as shown in the following table." If, for instance, four No. 8 THHN conductors are used in a conduit, the ampacity of each No. 8 is reduced from the 50-A value shown in the table to 80 percent of that value. Each No. 8 then has a new (reduced) ampacity of 0.8 × 50 A, or 40 A. Moreover, Section 240-3 of the **NEC** states, "Conductors other than flexible cords shall be protected against overcurrent in accordance with their ampacities as specified in Section 310-15. Thus, fuses or CB poles rated at 40 A are, as a general rule, required for overcurrent protection.

The application of these No. 8 conductors and their protection rating is based on the general concept behind the **NEC** tables of maximum allowable current-carrying capacities (called *ampacities*). The **NEC** tables of ampacities of insulated conductors installed in raceways or cables have always set the maximum continuous current that a given size of conductor can carry continuously (for 3 hours or longer) without exceeding the temperature limitation of the insulation on the conductor—that is, the current above which the insulation would be damaged.

This concept is verified in the FPN following Section 240-1 which says:

> Overcurrent protection for conductors and equipment is provided to open the circuit if the current reaches a value that will cause an excessive or dangerous temperature in conductors or conductor insulation.

NEC Table 310-16 gives ampacities under two conditions: that the raceway or cable containing the conductors is operating in an ambient temperature not over 30°C (86°F) and that there are not more than three current-carrying conductors in the raceway or cable. Under those conditions, the ampacities shown correspond to the thermal limit of each particular insulation. But in any case where either of the two conditions is exceeded, the ampacity of the conductors of a circuit must be reduced (and protection provided at the reduced ampacity) to ensure that the temperature limit of the insulation is not exceeded:

1. If the ambient temperature is above 30°C, the ampacity must be reduced in accordance with the correction factors given with Table 310-16.
2. If more than three current-carrying conductors are used in a single cable or raceway, the conductors tend to be bundled in such a way that their heat-dissipating capability is reduced, and excessive heating will occur at the ampacities shown in the table. As a result, Note 8 requires the reduction of ampacity, and conductors have to be protected at the reduced ampacity.

It should be clearly understood that any reduced ampacity, required because of a higher ambient and/or conductor bundling, has the same meaning as the value shown in the table: derated ampacity and normal (table) ampacity both represent a current value above which excessive heating would occur under the particular conditions. And if there are two conditions that lead to excessive heating, then more reduction of current is required than if only one such condition existed.

The application of Note 8 depends upon how many current-carrying conductors are in a raceway. A true neutral conductor (a neutral carrying current only under conditions of unbalanced loading on the phase conductors) is not counted as a current-carrying conductor. If a 208Y/120-V circuit or a 480Y/277-V circuit is made up of three phase legs and a true neutral in a conduit, the circuit is counted as consisting of only three conductors in the conduit, and it is not necessary to derate for conduit fill, as described in Note 8 of Tables 310-16 to 310-19. But neutrals for circuits with these voltage ratings must be counted as current-carrying conductors if the major portion of the load consists of electric-discharge lighting and/or electronic computer or data-processing equipment [Note 10(c) of the tables]. Thus, if the circuit supplies fluorescent, mercury, or metal-halide lamps, the neutral is counted as the fourth current-carrying conductor because it carries third-harmonic current, which approximates the phase-leg current under balanced loading. Any such 4-wire circuit, as well as such a circuit supplying other so-called nonlinear loads must have its ampacity derated to 80 percent of the ampacity given in Table 310-16, as required by Note 8 to the tables.

As shown in Fig. 2.68, the makeup of a branch circuit consists of first selecting the correct size of wire for the particular load current (based on the number of wires in the raceway, ambient temperature, and ampacity deratings) and then relating the rating of the overcurrent protective device to all the conditions.

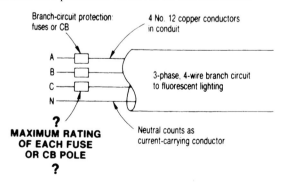

FIGURE 2.68 Loading and protection of branch-circuit wires must account for derating of conductor "ampacity."

In Table 310-16, which applies to conductors in raceways and in cables and covers most of the conductors used in electrical systems for power and light, the ampacities for No. 14, 12, and 10 conductors are particularly significant because copper conductors of those sizes are involved in most of the branch circuits in modern electrical systems. Number 14 wire has an ampacity of 20 A; No. 12 has an ampacity of 25 A; and No. 10 has an ampacity of 30 A. The typical impact on circuit makeup and loading is as follows.

1. Number 12 TW or THW copper is shown to have an ampacity of 25 A instead of 20 A; based on the general UL requirement that equipment terminals be limited to use with 60°C conductors up to a 100-A rating, No. 12 THHN or XHHW copper conductors must also be treated as having only a 25-A continuous rating and not a 30-A rating, as shown in Table 310-16. But the footnote to Table 310-16 limits all No. 12 copper wires to a maximum load of 20 A and requires that they be protected at not more than 20 A.

2. The ampacity of 25 A for No. 12 TW and THW copper wires interacts with Note 8 to Tables 310-16 to 310-19 where there are, say, six No. 12 TW current-carrying wires for the phase legs of two 3-phase, 4-wire branch circuits in one conduit supplying, say, receptacle loads. In such a case, the two neutrals of the branch circuits do not count in applying Note 8, and only the six phase legs are counted in determining the extent to which all circuit conductors must have their ampacities derated to the "percent of values in Tables" as given in the table in Note 8. In the case described here, that literally means that each No. 12 phase leg may be used at an ampacity of 0.8 × 25 A, or 20 A. And the footnote to Table 310-16 would require the use of a fuse or CB rated not over 20 A to protect each No. 12 phase leg. Each No. 12 would be protected at the current value that represents the maximum I^2R heat input that the conductor insulation can withstand. The only possible qualification is that Section 384-16(c) would require the load current on each of the phase legs to be further limited to no more than 80 percent of the 20-A rating of the overcurrent device—that is, to 16 A—if the load current is "continuous" (operates steadily for 3 hours or more), an unlikely condition for receptacle-fed loads. But the above-described application is different from what it would have been under previous **NEC**s.

3. If two 3-phase, 4-wire branch circuits of No. 12 TW or THW copper conductors were installed in a single conduit or EMT run to supply, say, fluorescent lighting or other nonlinear loads, the two neutrals of the branch circuits would carry harmonic current even under balanced conditions and would have to be counted, along with the six phase legs, as current-carrying wires for ampacity derating in accordance with Note 8. In such a case, as the table of Note 8 shows for 7 to 9 conductors, their ampacities would have to be derated to 70 percent of the 25-A value shown in Table 310-16, which gives each conductor an ampacity of 17.5 A (0.7 × 25 = 17.5). If 20-A overcurrent protection were used for each No. 12 phase leg and the load were limited to no more than 17.5 A (or 16 A, which is 0.8 × 20 A, for a continuous load), the application would satisfy Note 8 in the **NEC**. And Section 240-3(b) permits the use of 20-A protection as the next higher standard rating of protective device above the conductor ampacity of 17.5 A, provided that the circuit supplies only lighting outlets—and not readily accessible receptacle outlets that might allow the conductors to be overloaded above 17.5 A.

Use of the 20-A protection on conductors rated at 17.5 A is recognized only for fixed circuit loading (like lighting-fixtures outlets) and not for the variable loading of receptacle outlets, because any increase in load current over 17.5 A would produce excessive heat input to the eight bundled No. 12 conductors in the conduit, which would damage and ultimately break down conductor insulation. Of course, the question arises: Over the operating life of the electrical system, how can excessive current be prevented? The practical, realistic answer is: It can't! It would be better to use 15-A protection on the No. 12 conductors or use conductors rated at 90°C.

4. Number 12 THHN or XHHW conductors, with their 90°C rating and consequently greater resistance to thermal damage, could be used for the two 3-phase, 4-wire, 20-A circuits to the electric-discharge lighting load; they would satisfy all **NEC** rules and would not be subject to insulation damage. With eight current-carrying wires in the conduit, the ampacity derating to 70 percent as required by Note 8 would be applied to the 30-A value shown in Table 310-16 as the ampacity of No. 12 THHN, RHH, or XHHW (dry locations). Then, because 0.7×30 A $= 21$ A, the maximum of 20-A protection required by the footnote to Table 310-16 would ensure that the conductors could never be subjected to excessive current and its damaging heat. And if the original loading on the conductors is set at 16 A [the 80 percent load limitation of Section 384-16(c)] for continuous operation of the lighting, any subsequent increase in load—even up to the full 20-A capacity—would not reach the 21-A maximum ampacity set by Note 8 but would exceed the maximum permitted loading (continuous at 125 percent plus noncontinuous at 100 percent) by Sections 220-3(a), and 210-22(c).

Figure 2.69 summarizes the application described in items 3 and 4 above.

Advantage of 90°C Wires. If the four circuit wires in Fig. 2.68 are rated at 90°C, as are THHN, RHH, or XHHW conductors, then the loading and protection of the circuit must be related to the required ampacity derating as shown in Fig. 2.70. The application is based on these considerations:

1. As described in Figs. 2.66 and 2.67, each No. 12 THHN has an ampacity of 30 A from **NEC** Table 310-16, but the footnote to the table limits the load and overcurrent protection on any No. 12 THHN to not more than 20 A.
2. Because the neutral of the 3-phase, 4-wire circuit must be counted as a current-carrying wire, there are four conductors in the conduit; each conductor must therefore have its ampacity derated to 80 percent of its table ampacity value, as required by Note 8 of Tables 310-16 to 310-19. Each No. 12 then has a new (derated) ampacity of 0.8×30 A, or 24 A.
3. By using a 20-A, single-pole protective device (fuse, single-pole CB, or one pole of a three-pole CB), which is the maximum protection permitted by the footnote to Table 310-16, each No. 12 THHN is easily made to comply with the requirement of Section 210-19 that the branch-circuit wire have an ampacity "not less than" the maximum load current being supplied.
4. If the lighting load on the circuit is noncontinuous—that is, does not operate for any period of 3 hours or more—the circuit may be loaded up to its 20-A maximum rating.
5. If the load is continuous—full-load current flows for 3 hours or more—the load on the circuit must be limited to 80 percent of the rating of each 20-A fuse or CB pole, as required by Section 384-16(c). Then 16 A is the maximum load.

Figure 2.71 shows the use of two 3-phase, 4-wire circuits of THHN conductors in a single conduit. The 90°C wires offer distinct advantages (substantial economies)

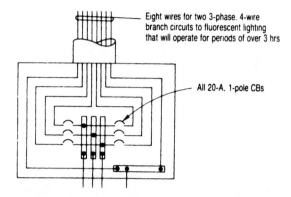

CASE 1
With TW or THW No. 12 copper conductors:
No. 12 ampacity = 25 A, from Table 310-16
From Note 8, max load = 0.7 × 25 = 17.5 A
From Sec. 384-16 (c), max continuous load = 0.8 × 20 = 16 A

CASE 2
With THHN, XHHW or RHH No. 12 copper conductors:
No. 12 ampacity = 30 A, from Table 310-16
From Note 8, max load = 0.7 × 30 = 21 A
From Sec. 384-16 (c), max continuous load = 0.8 × 20 = 16 A

IN CASE 1, CONDUCTORS ARE NOT PROTECTED IN ACCORDANCE WITH THE 17.5-A MAXIMUM ALLOWABLE LOAD CURRENT. IN CASE 2, THEY *ARE* PROTECTED AGAINST EXCESSIVE LOAD CURRENT.

FIGURE 2.69 Conductors with 90°C insulation eliminate the chance of conductor damage due to overload.

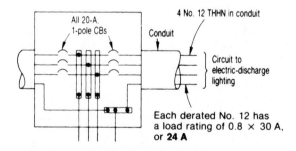

NOTE: TW or THW wires would have to be derated from 25A to 20 A.

FIGURE 2.70 Derating of 90°C branch-circuit wires (No. 14, 12, and 10) is based on higher ampacities.

over the use of either 60°C (TW) or 75°C (THW) wires for the same application, as follows.

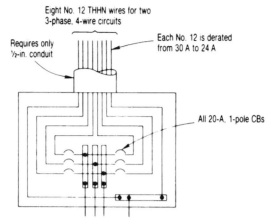

FIGURE 2.71 90°C conductors can take derating without losing the full circuit load-current rating.

Resistive Load. If the circuit shown feeds only incandescent lighting or other resistive loads (or electric-discharge lighting does not make up "a major portion of the load"), then Note 10(c) of Tables 310-16 to 310-19 does not require the neutral conductor to be counted as a current-carrying conductor. In such cases, circuit makeup and loading could include these considerations:

1. With the six phase legs as current-carrying wires in the conduit, Note 8 requires that the ampacity of each No. 12 be derated from its basic table value of 30 A to 80 percent of that value, or 24 A.
2. Then each No. 12 is properly protected by a 20-A CB or fuse, satisfying both Section 210-19 and the footnote to Table 310-16.
3. If the circuit load is not continuous, each phase leg may be loaded to 20 A.
4. If the load is continuous, a maximum of 16 A (80 percent) must be observed to satisfy Section 384-16(c).

Electric-discharge Load. If the two circuits of Fig. 2.71 supply electric-discharge lighting (fluorescent, mercury-vapor, metal-halide, high-pressure sodium, or low-pressure sodium) or computers or other nonlinear types of loads, the makeup and loading must be as shown in case 2 of Fig. 2.69. If that same makeup of circuits supplies noncontinuous loads, the circuit conductors may be loaded up to 20 A per pole.

In case 2 of Fig. 2.69, the only difference between such circuit makeups using XHHW or RHH conductors and those using THHNs is the need for ¾-in conduit instead of ½-in conduit because of the larger cross-sectional area of RHH and XHHW (see Tables 3A and 3B in **NEC** Chapter 9).

Color Coding of Circuit Conductors

NEC rules on the color coding of conductors (Section 210-5) apply only to branch-circuit conductors and do not directly require the color coding of feeder conductors.

However, Section 384-3(f) does require identification of the different phase legs of feeders to panelboards, switchboards, etc.—and that requires some technique for marking the phase legs. Many design engineers insist on color coding for feeder conductors, to afford effective balancing of loads on the different phase legs.

The color coding of branch-circuit conductors can be divided into three categories; grounded, hot, and grounding.

Grounded Conductor. The grounded conductor of a branch circuit (the neutral of a wye system or a grounded phase of a delta) must be identified by a continuous white or natural-gray color for the entire length of conductors of size No. 6 or smaller. Where wires of different systems (such as 208/120 and 480/277 V) are installed in the same raceway, box, or other enclosure, the neutral or grounded wire of one system must be white or gray; the neutral of the other system must be white with a color trace (stripe) along its insulation to distinguish it from the white neutral (Fig. 2.72). If there are three or more systems in the same raceway or enclosure, the additional neutrals must be white with color traces other than green. The point is that neutrals of different systems must be distinguished from each other when they are in the same enclosure.

Exceptions to Section 200-6 modify the basic rule that requires the use of continuous white or natural-gray color along the entire length of any insulated grounded conductor (such as a grounded neutral) in sizes No. 6 or smaller. One exception permits the use of other colors (black, purple, yellow, etc.) for a grounded conductor in a multiconductor cable under certain conditions:

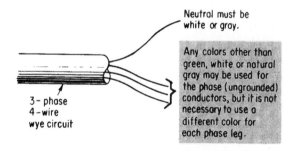

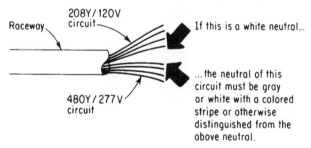

FIGURE 2.72 Neutrals of different systems must be distinguished where circuits are in the same raceway or enclosure.

1. Such a conductor is used only where qualified persons supervise and perform service or maintenance on the cable—such as in industrial, institutional, commercial, and mining applications.
2. Every grounded conductor of color other than white or gray is effectively and permanently identified at all terminations by distinctive white marking or other effective means applied at the time of installation.

This permission is intended to allow the practice in commercial and industrial facilities where multiconductor cables are commonly used.

Hot Conductors. The **NEC** requires that individual hot conductors of a multiwire circuit be identified where a building has more than one nominal voltage system (Fig. 2.73). [See Section 210-4(d).]

NEC Section 215-8 does require positive identification of the "high" leg (the hot conductor with higher voltage to ground than the other hot legs) of a 240/120-V, 3-phase, 4-wire delta system, as shown in Fig. 2.74. The wording of this section recognizes orange as the preferred color for the high leg of a 4-wire delta supply without disturbing current practices in various local areas, where other colors (such as red, yellow, or blue) or other means of identification are required by electric-utility regulations or local code.

Note that identification of the phase leg with 208 V to ground is required only at those points in the system where the neutral is present, such as in panelboards, motor-control centers, and other enclosures where circuits are connected. The purpose is to provide a warning that 208 V, and not 120 V, exist from the high leg to the neutral. Such a warning minimizes the chances that a 208-V circuit might be accidentally or unwittingly connected to a 120-V load, such as lamps or appliances or 120-V operating coils in motor starters. The connection would burn out the 120-V equipment and present a hazard to personnel.

Grounding Conductor. An equipment grounding conductor of a branch circuit (if one is used) must be color-coded green or green with one or more yellow stripes; or, the conductor may be bare, as covered in Section 210-5(b). However, Exception No. 1 of Section 210-5(b) refers to Section 250-57(b), which says that an equipment grounding conductor of size larger than No. 16 may be other than a green insulated conductor or green with yellow stripe(s). Section 250-57 permits an equipment grounding conductor with insulation that is black, blue, or any other color—provided that one of the three techniques specified in Section 250-57 is used to identify this conductor as an equipment grounding conductor.

The first technique consists in stripping the insulation from the insulated conductor for the entire length of the conductor appearing within a junction box, panel enclosure, switch enclosure, or any other enclosure. With the insulation stripped, the conductor appears as a bare conductor, which is recognized by the **NEC** for the purpose.

The second technique that is acceptable is to paint the exposed insulation green for its entire length within the enclosure. If, say, a black insulated conductor is used in a conduit coming into a panelboard, the length of the black conductor within the panelboard can be painted green to identify it as an equipment grounding conductor.

The third acceptable method is to mark the exposed insulation with green-colored tape or green-colored adhesive labels. Green-colored conductors must not be used for any purpose other than equipment grounding.

Figure 2.75 summarizes these rules regarding the identification of branch-circuit equipment grounding conductors.

WHEN BUILDING CONTAINS ONLY ONE SYSTEM VOLTAGE FOR CIRCUITS:

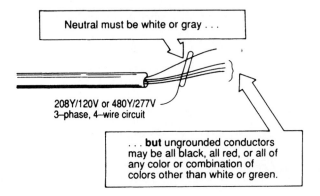

IF THERE ARE TWO SYSTEM VOLTAGES:

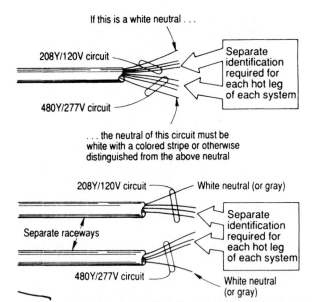

FIGURE 2.73 Separate identification of ungrounded conductors is required only if a building utilizes more than one nominal voltage system. Neutrals must be color-distinguished if circuits of two voltage systems are used in the same raceway (Section 210-5) but not if different voltage systems are run in separate raceways.

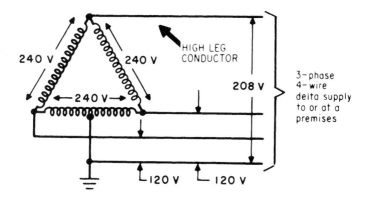

HIGH-LEG CONDUCTOR may be orange in color or may be some other color—such as red or yellow—as long as the color or tagging or other identification clearly distinguishes this as the one with higher voltage to ground at any connection point where the neutral is present.

FIGURE 2.74 High-leg conductor (heavy arrow) is the only hot leg required by the **NEC** to be identified.

The standardization effected by coding according to a set of simple, specific color designations will ensure all the safety and operating advantages of color coding to all electrical systems. Particularly today, with electrical systems being subjected to an unprecedented amount of alterations and additions because of the development and expansion of electric usage, conductor identification for safety is required over the entire life of each system.

Of course, there are alternatives to color identification throughout the lengths of conductors. Color differentiation is almost worthless for color-blind electricians. And it can be argued that color differentiation of conductors poses problems because electrical work is commonly performed in darkened areas where color perception is reduced, even for those with good eyesight. The **NEC** already recognizes the use of white tape or paint over conductor insulation ends at terminals, to identify neutrals where the conductor size is larger than No. 6 (Section 200-6). Number markings (1, 2, 3, etc.) spaced along the length of a conductor, on the insulation—say, white numerals on black insulation—might prove very effective for differentiating conductors. Or the letters A, B, and C could be used to designate specific phases. Or a combination of color and such markings could be used. But some kind of conductor identification is essential to the safe, effective hookup of the ever-expanding array of conductors used throughout buildings and systems today.

Receptacle Circuits

The design of branch circuits for plug receptacles requires careful determination of particular requirements. The type and size of occupancy and the nature of the work performed there will indicate the best manner of handling plug-connected loads. For known appliances, individual or multioutlet branch circuits should be used, depending on the size of the load.

Automatic appliances should always be provided with separate circuits to isolate them from the effects of faults or other disturbances in other load devices. Fans or

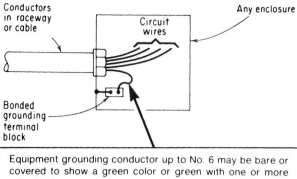

FIGURE 2.75 Equipment grounding conductor for a branch circuit must be identified.

heaters might be individually fed by branch circuits, grouped on their own branch circuits, or connected to receptacles on general lighting circuits.

Of course, the number of plug-in appliances in a particular area will greatly affect the circuiting. In general, plug-in devices should be supplied from receptacles on circuits other than general-lighting circuits. In this way, the loading of circuits can be kept under control, and spare-capacity provisions can be much more realistic.

The number of plug outlets connected on a single circuit should be related to the amount of load likely to be connected to any one receptacle in a particular occupancy. When plug-connected loads are known, determination of the proper number of plug outlets is relatively easy. And the number of circuits required in such cases follows directly from the loads and circuit capacities.

When plug-connected loads are not known, the type of occupancy will indicate the possible appliances to be provided for. If the possible appliances are relatively heavy-current devices, two or three outlets per circuit might be the maximum number to allow for efficient and convenient use of the circuits. If the possible appliances are low-current devices, up to 10 or even more plug outlets may be connected on the circuit without likelihood of overload.

NEC limitations regarding the number of plug outlets on any one branch circuit should be carefully observed in selecting the number of receptacle circuits, as follows.

Nonresidential Occupancies. NEC Section 210-52 specifies where and when receptacle outlets are required on branch circuits. Note that there are no specific requirements for receptacle outlets in commercial, industrial, and institutional installations other than for store windows, hotel guest rooms, and rooftops. There is the general rule that receptacles do have to be installed where flexible cords are used. In nonresidential buildings, if flexible cords are not used, there is no requirement for receptacle outlets. They have to be installed only where they are needed, and the number and spacing of receptacles are completely up to the designer. The NEC takes the position that receptacles in nonresidential buildings have to be installed only where needed for connection of specific flexible cords and caps and demands that where such receptacles are installed, each must be taken as a load of 180 VA.

Section 220-3(c)(6) requires that every general-purpose single, duplex, or triplex convenience receptacle outlet be taken as a load of 180 VA and that amount of circuit capacity must be provided for each such outlet, as shown in Fig. 2.76. If a single 15-A, 115-V circuit is used to supply only receptacle outlets, then the maximum number of general-purpose receptacle outlets that may be fed by that circuit is

$$15 \text{ A} \times 115 \text{ V} \div 180 \text{ VA, or 9 outlets}$$

For a 20-A, 115-V circuit, the maximum number of general-purpose receptacle outlets is

$$20 \text{ A} \times 115 \text{ V} \div 180 \text{ VA, or 12 outlets}$$

NOTE: In these calculations, the actual results work out to be 9.58 receptacles on a 15-A circuit, and 12.77 on a 20-A circuit. Some inspectors round off these values to the nearest integer and permit 10 receptacle outlets on a 15-A circuit and 13 on a 20-A circuit.

The value of 115 V is used in these calculations, although, in a reference to "voltage" covering the sample calculations in its Chapter 9, the NEC does specify that for uniform application of the provisions of Articles 210, 215 and 220, a voltage of 120, 120/240, 240, or 208Y/120 V be used in computing the ampere load on conductors.

Using 115 and 230 V instead of 120 and 240 V does produce higher current values when volts are divided into watts or voltamperes (for example, 6000 W ÷ 120 V = 50 A, and 6000 W ÷ 115 V = 52 A). Using higher current values for sizing conductors and other equipment gives greater assurance of adequate capacity and completely satisfies all NEC rules.

Based on experience and very clear evidence that electrical systems quickly tend to be overloaded, and because load growth is a certainty in this all-electric age, many designers resolve all questions of circuit loading and permitted capacity in favor of greater capacity. The use of 9 instead of 10 as the maximum number of receptacles on a circuit calls for more circuits and gives higher capacity to each receptacle outlet. "When in doubt, design up, not down."

For office areas, the NEC permits either of two methods for determining the required minimum branch-circuit capacity for receptacle outlets. The first is the method just described, in which a load of 180 VA is taken for each single or duplex receptacle outlet and sufficient branch-circuit capacity is provided for the total load (180 VA times the number of receptacle outlets). The second, or alternative, approach is described in the last footnote to NEC Table 220-3(b) (the double-asterisk note).

Based on extensive analysis of load densities for general lighting in office buildings, Table 220-3(b) now requires a minimum unit load of only 3.5 W/ft^2—rather than

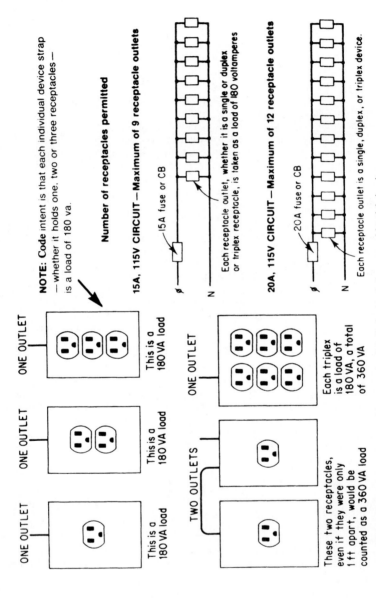

FIGURE 2.76 Number of receptacles on a circuit is limited only for nonresidential circuits.

the previous unit load of 5 W/ft^2—for general illumination in "office buildings" and for "banks." And branch-circuit capacity for general lighting in such occupancies must not be less than 3.5 times the area (in square feet) of the space being supplied with the general lighting. But the double-asterisk note at the bottom of the table requires the addition of another 1 W/ft^2 to the 3.5-W/ft^2 value to cover the loading added by general-purpose receptacles in those cases where the actual number of receptacles is not known at the time feeder and branch-circuit capacities are being calculated. In such cases, a unit load of 4.5 W/ft^2 must be used; the calculation based on that figure will yield the minimum branch-circuit capacity for both general lighting and for all general-purpose receptacles that may later be installed. Whatever number of receptacles may later be installed, they may be divided among lighting circuits or separate receptacle circuits, with the loads "evenly proportioned" among the circuits. The allowance of 180 VA for each receptacle outlet may be disregarded.

But the rule does require that, where the actual number of general-purpose receptacles is known, the general lighting load be taken at 3.5 W/ft^2 for branch-circuit and feeder capacity, and each receptacle be taken as a load of 180 VA, in determining the total required branch-circuit capacity. The demand factors of Table 220-13 may be applied to get the minimum required feeder capacity for receptacle loads.

As noted previously, Exception No. 1 of Section 220-3(c) requires branch-circuit capacity to be calculated for multioutlet assemblies (prewired surface metal raceways with plug outlets spaced along their length). Exception No. 1 says that each 1-ft length of such strip must be taken as a load of 1.5 A (180 VA) when the strip is located where a number of appliances are likely to be used simultaneously. For instance, in industrial applications, on assembly lines involving frequent simultaneous use of plugged-in tools, a loading of 1.5 A/ft must be used. (A loading of 1.5 A for each 5-ft section may be used in commercial or institutional applications of multioutlet assemblies when the use of plug-in tools or appliances is not heavy.)

From the standpoint of design, the determination of the number of plug outlets required for various nonresidential occupancies is not a matter of easy or standard calculation. The following are suggestions for receptacle layouts in various areas:

1. In office buildings, the growing use of business machines dictates heavier load allowances for plug outlets. In separate offices less than 400 ft^2 in area, at least one plug outlet should be allowed for each 10 linear feet of wall space. In each office over 400 ft^2 in area, eight plug outlets should be allowed for the first 400 ft^2 of floor area and three plug outlets for each additional 400 ft^2 of floor area or fraction thereof. The number of outlets obtained in this way should be evenly distributed throughout the area.

2. In schoolrooms, common practice has been to provide at least one plug outlet on each wall of the room. Figure 2.77 shows a school panelboard with wide variation in the number of receptacle outlets connected to 120-V circuits. The layout and circuiting of receptacles were based on a study of likely loads, possible simultaneous use of appliances, and habits of use of electrical appliances. In a home-economics classroom, for example, heavy appliance loads and sewing-machine desks dictated the use of only two receptacles per No. 12 2-wire circuit.

3. In stores, at least one plug outlet should be provided for each 400 ft^2 of floor area or major fraction thereof.

4. In industrial areas, plug outlets should be provided on the basis of particular conditions. Figure 2.78 shows the circuit layout and supply to receptacles operating at 60 and 400 Hz in a modern airplane hangar.

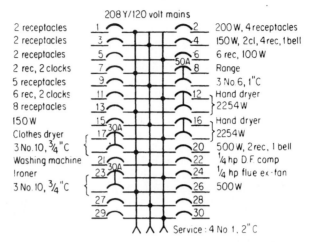

FIGURE 2.77 Variation in the number of receptacles per circuit is dictated by different applications.

Of course, the possible number of appliances to be used in any area will also affect the design of receptacle circuits. When special requirements for receptacle outlets arise after tenants move into an area, the spare circuits in the panelboard, and extra capacity in the existing system of raceways or underfloor system will offer a solution to these requirements.

In addition to conventional receptacle circuits, modern buildings increasingly require new and often complex types of receptacle circuits. In particular, data-processing equipment and other special machines used in today's commercial and industrial buildings call for multiwire receptacle circuits of varying voltage, phase, and frequency characteristics.

Figure 2.79 shows a nine-pole receptacle used for dozens of outlets in a computer room. The receptacles were installed in a wireway under a raised floor and provided for flexibility in the movement and connection of various machines, each of which required single-phase, 120-V power and 3-phase, 4-wire, 120/208-V, 400-cycle power. Each machine is equipped with a nine-conductor cord and plug for ready connection to one of the receptacles. And an expanded color code was used for the circuit wiring.

Figure 2.80 shows the amperage loading schedule for the nine-pole receptacles—the single-phase, 60-Hz and the 3-phase, , 400-Hz. There is a receptacle for each entry under the heading "Machine No." Note that the single-phase loads of the receptacles are balanced among the three phases of the 60-Hz, 3-phase, 4-wire branch circuits. And note the substantial spare capacity, enough for over 100 percent load growth in most circuits. The multioutlet branch circuits used here—50- and 70-A—are special and do not conform to **NEC** regulations regarding conventional receptacle circuits.

Residential Occupancies. The **NEC** covers the maximum permitted number of receptacle outlets on a circuit in commercial, industrial, institutional, and other non-residential installations, but there are no such limitations on the number of receptacle outlets on residential branch circuits. Instead, the **NEC** simply assumes that cord-connected appliances will always be used in all residential buildings, and it

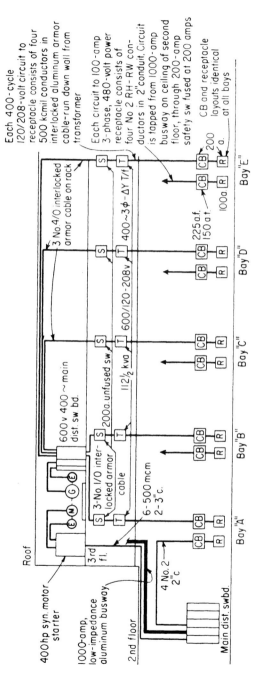

FIGURE 2.78 Large-capacity receptacle circuits are common in industrial locations.

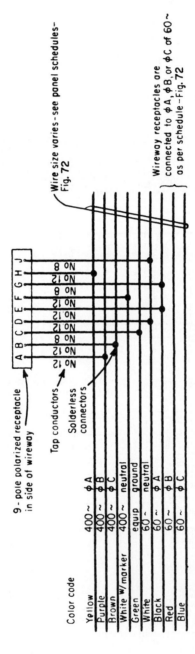

FIGURE 2.79 Branch-circuit wiring to special receptacles used for supplying computer equipment.

LIGHTING AND APPLIANCE BRANCH CIRCUITS

PANEL SCHEDULES

60 CYCLES—PANEL "B" / 400 CYCLES—PANEL "A"

Circuit No.	Circuit Breaker Frame	Poles	Trip	Machine No.	60~ Kva	60~ Amps Per φ φA	φB	φC	400~ Kva	400~ Amps Per φ φA	φB	φC	Machine No.	Circuit Breaker Frame	Poles	Trip	Circuit No.
1. 4/8 RHW	100A	3	50A	Console 304 Future	0.6 0.5 0.5	5.0	4.2	4.2	0.5 6.9 0.5	1.4 19.0 1.4	1.4 19.0 1.4	1.4 19.0 1.4	Console 304 Future	100A	3	50A	1. 4/8 RHW
				Total	1.6	5.0	4.2	4.2	7.9	21.8	21.8	21.8					
2. 4/6 RHW	100A	3	70A	332 332 332 332 332 332 332 332 332 330 Future " " "	1.3 1.3 1.3 1.3 1.3 1.3 1.3 1.3 1.3 0.3 0.3 1.3 1.3 1.3	11.0 .... 11.0 .... 11.0 .... 11.0 .... 2.8 2.8 11.0	.... 11.0 .... 11.0 .... 11.0 .... 11.0 .... 2.8 .... 11.0	.... 11.0 .... 11.0 .... 11.0 .... 11.0 .... 11.0	0.3 0.3 0.3 0.3 0.3 0.3 0.3 0.3 0.3 3.9 3.9 0.3 0.3 0.3	0.9 0.9 0.9 0.9 0.9 0.9 0.9 0.9 0.9 10.8 10.8 0.9 0.9 0.9	0.9 0.9 0.9 0.9 0.9 0.9 0.9 0.9 0.9 10.8 10.8 0.9 0.9 0.9	0.9 0.9 0.9 0.9 0.9 0.9 0.9 0.9 0.9 10.8 10.8 0.9 0.9 0.9	332 332 332 332 332 332 332 332 332 330 Future " " "	100A	3	70A	2. 4/6 RHW
				Total	16.2	46.8	46.8	44.0	11.4	32.4	32.4	32.4					
3. 4/6 RHW	100A	3	70A	340B 340T 320 332 Maint. Rm. Future " " "	0.2 5.8 0.3 1.3 2.0 0.2 5.8 0.1 1.3	1.7 .... 11.0 16.7 1.7 .... 11.0	.... 48.0 .... 0.9	.... 2.5 .... 48.0	3.9 0.4 4.4 0.3 1.0 3.9 0.4 2.1 0.3	10.8 1.1 12.2 0.9 2.8 10.8 1.1 5.9 0.9	10.8 1.1 12.2 0.9 2.8 10.8 1.1 5.9 0.9	10.8 1.1 12.2 0.9 2.8 10.8 1.1 5.9 0.9	340B 340T 320 332 Maint. Rm. Future " " "	100A	3	70A	3. 4/6 RHW
				Total	17.0	42.1	48.9	50.5	16.7	46.5	46.5	46.5					
4. 4/6 RHW	100A	3	70A	360 370 380 Future	2.5 1.3 5.0 1.3	21.0	.... 11.0 .... 11.0	.... 42.0	1.0 0.7 1.0 1.0	2.8 1.9 2.8 2.8	2.8 1.9 2.8 2.8	2.8 1.9 2.8 2.8	360 370 380 Future	100A	3	50A	4. 4/8 RHW
				Total	10.1	21.0	22.0	42.0	3.7	10.3	10.3	10.3					

FIGURE 2.80 Load schedules for special receptacles.

requires general-purpose receptacle outlets of the number and spacing indicated in Sections 210-52(a) and (c). These rules cover one-family houses, apartments in multifamily houses, guest rooms in hotels and motels, living quarters in dormitories, etc. But because so many receptacle outlets are required in such occupancies and because the use of plug-connected loads is intermittent and has great diversity of load values and operating cycles, the **NEC** notes at the bottom of Table 220-3(b) that the loads connected to such receptacles are adequately served by the branch-circuit capacity required by Section 220-4; no additional load calculations are required for such outlets.

In dwelling occupancies, it is first necessary to calculate the total "general lighting load" from Section 220-3(b) and Table 220-3(b) (at 3 W/ft^2 for dwellings or 2 W/ft^2 for hotels and motels, including apartment houses without provisions for cooking by tenants). Then the minimum required number and rating of 15 and/or 20-A general-purpose branch circuits must be provided to handle that load, as covered in Section 220-4(a). As long as the basic circuit capacity is provided, any number of lighting outlets may be connected to any general-purpose branch circuit, up to the rating of the branch circuit if loads are known. The lighting outlets should be evenly distributed among all the circuits. Although residential lamp wattages cannot be anticipated, the **NEC** method covers fairly heavy loading.

When the above rules on circuits and outlets for general lighting are satisfied, general-purpose convenience receptacle outlets may be connected on circuits supplying lighting outlets, receptacles only may be connected on one or more of the required branch circuits, or additional circuits (over and above the required minimum) may be used to supply the receptacles. But no matter how general-purpose receptacle outlets are circuited, any number of general-purpose receptacle outlets may be connected on a residential branch circuit—with or without lighting outlets on the same circuit.

And, when small-appliance branch circuits (in kitchen, dining room, pantry, etc.) are provided in accordance with the requirements of Section 220-4(b), any number of small-appliance receptacle outlets may be connected on the 20-A small-appliance circuits. However, only receptacle outlets may be connected to these circuits, and only in the specified rooms.

Section 210-52(a) applies to the spacing of receptacles connected on the 20-A small-appliance circuits, as well as the spacing of general-purpose receptacle outlets. That section, therefore, establishes the minimum number of receptacles that must be installed. Of course, a greater number of receptacles may be installed for greater convenience of use.

Applying Plug Receptacles

The first rule on the use of receptacles is that any receptacle must have an ampere rating at least equal to the load it supplies. In addition, on circuits having two or more outlets, receptacles must be rated as follows:

On 15-A circuits	Not over 15-A rating
On 20-A circuits	15- or 20-A rating
On 30-A circuits	30-A rating
On 40-A circuits	40- or 50-A rating
On 50-A circuits	50-A rating

The **NEC** requires that only grounding-type receptacles be installed on 15- and 20-A branch circuits. This means that all general-purpose, 120-V plug outlets must be of the three-pole type. The rule applies to all types of buildings and areas and to receptacles at any voltage level. The grounding terminals on such receptacle outlets must be connected to ground either by a specific grounding conductor run with the circuit conductors or by bonding (with a jumper or self-grounding screws) to the metal circuit outlet box which is, itself, grounded by the metallic jacketing on cable, by an equipment grounding conductor in nonmetallic sheathed cable, or by a metallic raceway system.

With a properly grounded metallic system (conduit or armored cable) run to a surface-mounted metal outlet box, receptacles are grounded through the mounting bracket that is metallically connected to the metal box—which, in turn, is grounded through the metal raceway or cable armor supplying it. At recessed boxes set into the wall, a separate jumper must be used from the green hex-head screw on the receptacle to a solid connection on the grounded metal box (Fig. 2.81), unless the receptacle is a self-grounding type; that is, unless it has a yoke especially designed to provide a sure grounding path of low resistance through the screw to the threaded hole on the box. A grounding wire run with the circuit conductors to provide for equipment grounding is mandatory in nonmetallic wiring systems.

Although the general rule states that a flush-type box, installed in a wall for a receptacle outlet, does require a bonding jumper from a grounded box to the receptacle grounding terminal, Exception No. 1 pertains to surface-mounted boxes and eliminates the need for a separate bonding jumper between a surface-mounted box

LIGHTING AND APPLIANCE BRANCH CIRCUITS

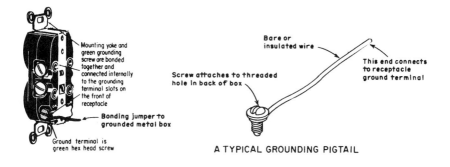

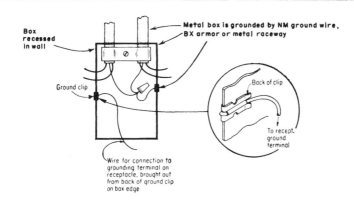

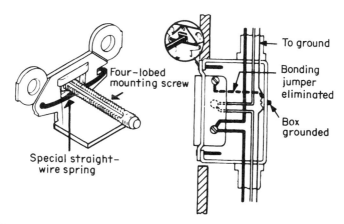

FIGURE 2.81 Bonding jumper connects receptacle ground to grounded box. Self-grounding screws ground receptacle in recessed box without bonding jumper.

and the receptacle grounding terminal under the conditions described. Although the Exception generally exempts surface-mounted boxes from the need for a bonding jumper from the box to the ground terminal of a receptacle installed in the box—because there is solid contact between the receptacles grounded mounting yoke and the ears on the box when installed—that is not applicable to a receptacle mounted in a raised box cover. A receptacle mounted in a raised box cover is connected to the cover by only a single screw, and that has been judged inadequate for grounding. A bonding jumper must be used on such a receptacle (Fig. 2.82).

The bottom of Fig. 2.82 illustrates a grounding device which is intended to provide the electrical grounding continuity between the receptacle yoke and the box on which it is mounted and serves the dual purpose of both a mounting screw and a means of providing electrical grounding continuity in lieu of the required bonding jumper. As shown in the sketch, special wire springs and four-lobed machine screws are part of a receptacle design for use without a bonding jumper to box. This complies with Section 250-74, Exception No. 2.

Exception No. 3 permits non-self-grounding receptacles without an equipment grounding jumper to be used in floor boxes which are designed for and listed as providing proper continuity between the box and the receptacle mounting yoke.

NEC Section 410-56(b) requires that when aluminum conductors are connected directly to the terminals of any receptacle rated 20 A or less, the device must be marked "CO/ALR" (Fig. 2.83).

Section 210-21(b) contains two paragraphs of importance. The second paragraph reads: "A single receptacle installed on an individual branch circuit shall have an ampere rating of not less than that of the branch circuit." Since the branch-circuit overcurrent device determines the branch-circuit rating (or classification), a single receptacle (not a duplex receptacle) supplied by an individual branch circuit cannot have a rating below that of the branch-circuit overcurrent device (Fig. 2.84).

Poke-Through Receptacles. Poke-through wiring—the technique in which floor outlets in commercial buildings are wired through holes in concrete-slab floors—is an acceptable wiring method if use is made of UL-listed poke-through fittings that have been tested and found to preserve the fire rating of the concrete floor (Fig. 2.85). Poke-through wiring continues to be a popular and very effective method of wiring floor outlets in office areas and other commercial and industrial locations, as an alternative to underfloor raceway or wiring in cellular floor construction. Holes are cut or drilled in concrete floors at the desired locations of floor outlets, and floor-box assemblies are installed and wired from the ceiling space of the floor below. The method permits installation of each and every floor box at the precise location that best serves the layout of desks and other office equipment. But a serious question is posed about how the use of the poke-through wiring technique can be properly reconciled with **NEC** Section 300-21.

Each floor outlet at a poke-through location may be wired basically in either of two ways: by some job-fabricated assembly of pipe nipples and boxes or by means of a manufactured through-floor assembly made expressly for the purpose and tested and listed by a nationally recognized testing lab such as UL. Violation of the **NEC** has been indicated where floor receptacles are fed by using job-fabricated assemblies through holes in the floor, thereby destroying the fire rating of the floor. Figure 2.86 shows a typical objectionable method. Section 300-21 requires that electrical installations be made so as to substantially protect the integrity of rated fire walls and fire-resistant or fire-stopped walls, partitions, ceilings, and floors. There are UL-listed and UL-approved through-floor fittings that permit poke-through wiring that satisfies Section 300-21.

LIGHTING AND APPLIANCE BRANCH CIRCUITS

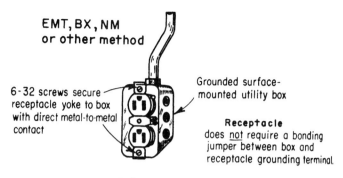

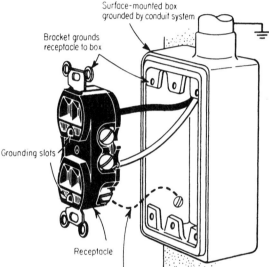

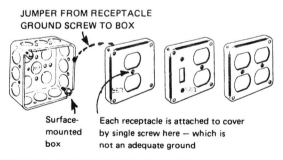

RECEPTACLE IN RAISED COVER REQUIRES BONDING JUMPER (cast covers for FS and FD boxes and other covers may contain a receptacle without a bonding jumper *if they* are "listed" as suitable for grounding.)

FIGURE 2.82 Typical applications where a surface box does and does not need a receptacle bonding jumper.

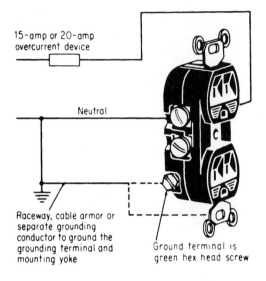

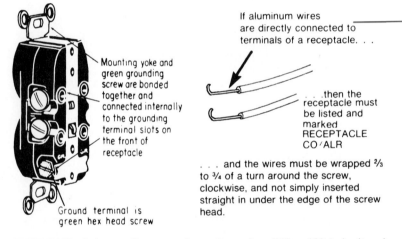

FIGURE 2.83 Only grounding receptacles may be used on all 15- and 20-A circuits and only CO/ALR receptacles with aluminum wire may be used.

Designers must give careful consideration to the liabilities incurred through less-than-strict compliance with the letter and intent of **NEC** and OSHA regulations. Given the nature of today's electrical work—with much broader, more intensified enforcement of national codes and standards—the use of new types of products, such as fire-rated floor outlets, must be thoroughly evaluated. This also applies to cable and/or conduit penetrations of fire-rated walls, floors, and ceilings without altering the fire rating of the structural surface.

FCC Wiring. Flat-conductor cable (FCC) is a surface-mounted cable for use on floors, under carpet squares not over 36 in square; it is the basis for a complete wiring

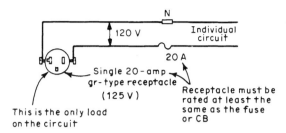

FIGURE 2.84 A single receptacle on an individual circuit must have an ampere rating equal to the circuit rating.

system for supplying power, signal, and communications floor outlets. This alternative to underfloor raceway, cellular-floor wiring, and conventional pipe-and-wire and poke-through receptacles is a UL-listed, **NEC**-recognized (Article 328) branch-circuit wiring system that may be used to supply floor outlets in office areas and other commercial and institutional interiors. The system may be used for new buildings or for modernization or expansion in existing interiors. FCC wiring may be used on any hard, sound, smooth floor surface—concrete, wood, ceramic, etc. The great flexibility and ease of installation of this surface-mounted wiring system are given added importance by the fact that the average floor power outlet in an office area is relocated every 2 years (Fig. 2.87).

Undercarpet wiring to floor outlets eliminates any need for core drilling of concrete floors, avoiding noise, dripping water, falling debris, and the disruption of normal activities in an office area. Alterations or additions to Type FCC circuit runs are neat, clean, and simple and may be made during office working hours; this eliminates the overtime labor rates incurred by floor drilling, which must be done at night or on weekends. The FCC method also eliminates the use of conduit or cable and the need to fish conductors.

Type FCC wiring offers versatile supply to floor outlets for power and communication at any location on the floor. The flat cable is inconspicuous under the carpet squares. And, the elimination of floor penetrations maintains the fire integrity of the floor, as required by Section 300-21.

Residential GFCI

NEC Section 210-8(a) is headed "Dwelling Units." The very clear and detailed definition of those words, as given in **NEC** Article 100, indicates that all the ground-fault circuit interruption rules apply to:

- All one-family houses
- Each dwelling unit in a two-family house
- Each apartment in an apartment house
- Each dwelling unit in a condominium

GFCI protection is required [Part (a)(1)] for all 125-V, single-phase, 15- and 20-A receptacles installed in *bathrooms* of dwelling units [Part (a)(1)] and hotel or motel units [Part (b)] and in garages of dwelling units only (Fig. 2.88). The requirement for GFCI protection in "garages" is included because home owners do use outdoor appliances (lawn mowers, hedge trimmers, etc.) plugged into garage receptacles.

THIS IS ACCEPTABLE—UL LISTED

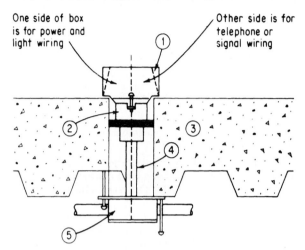

Numbered components include (1) combination floor service box, (2) fire-rated center coupling, (3) concrete slab, (4) barriered extension, and (5) barriered junction box.

...OR THIS FITTING

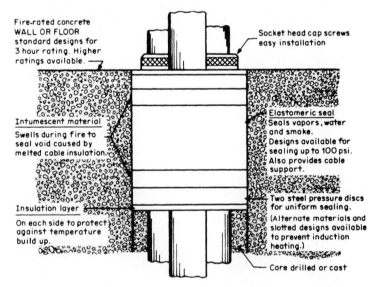

FIGURE 2.85 Poke-through wiring of floor outlets must be carefully designed.

LIGHTING AND APPLIANCE BRANCH CIRCUITS   113

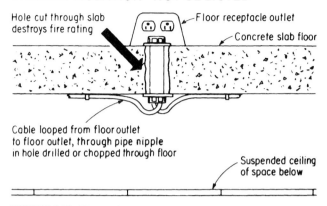

FIGURE 2.86 Untested floor penetrations must not be used for poke-through wiring.

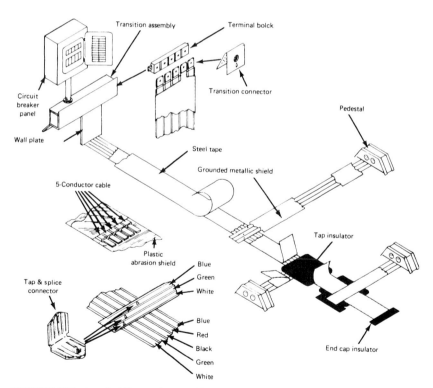

FIGURE 2.87 A typical flat-conductor-cable system includes the cable assembly and all accessory devices to make a complete layout.

Such receptacles require GFCI protection for the same reason as "outdoor" receptacles. In either place, GFCI protection may be provided by a GFCI circuit breaker that protects the whole circuit and any receptacles connected to it, or the receptacle may be a GFCI type that incorporates the components that give it the necessary tripping capability on low-level ground faults.

A lot of the controversy that was generated by the question "What is a bathroom?" is eliminated because a definition of the word "bathroom" is given at the

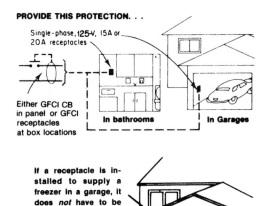

FIGURE 2.88 GFCI protection is required for receptacles in garages as well as in bathrooms.

LIGHTING AND APPLIANCE BRANCH CIRCUITS 115

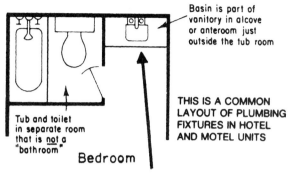

Guest rooms in hotels and motels are required by Section 210-60 to have the same receptacle outlets required by Section 210-52 for "dwelling units." The requirement for a wall receptacle outlet at the "basin location" applies to bathrooms; and the anteroom area with only a basin is, by definition and intent, part of the "bathroom." A receptacle at the basin would, therefore, be required. Section 210-8(b) applies to bathrooms, in hotels and motels — and GFCI protection is required for this receptacle.

FIGURE 2.89 GFCI is needed in bathrooms of hotel and motel "guest rooms."

end of Section 210-8(b). A "bathroom" is "an area" (which means it could be a room or a room plus another area) that contains first a "basin" (sometimes called a "sink") and then at least one more plumbing fixture—a toilet, a tub, and/or a shower. A small room with only a "basin" (a washroom) is not a "bathroom." Neither is a room that contains only a toilet and/or a tub or shower (Fig. 2.90).

As noted above, GFCI protection is required by Section 210-8(b) in bathrooms of all occupancies. This includes commercial office buildings, industrial facilities, schools, dormitories, theaters—bathrooms in *all* nondwelling occupancies. Figure 2.89 shows application in hotel and motel bathrooms. The rule here extends the same protection of GFCI breakers and receptacles to bathrooms in all nondwelling-type occupancies as for receptacles in bathrooms of dwelling units.

The rule of Section 210-8(a)(2) requiring GFCI protection in garages applies to both attached and detached (or separate) garages associated with "dwelling units"—such as one-family or multifamily houses where each unit has its own garage. Section 210-52 requires at least one receptacle in an attached garage and in a detached garage if electric power is run out to the garage.

Although the basic rule of Part (a)(2) requires all 125-V, single-phase, 15- or 20-A receptacles installed in garages to have ground-fault circuit-interrupter protection. Exception No. 1 excludes a ceiling-mounted receptacle that is used solely to supply a cord-connected garage-door operator. And, as worded, the Exception excludes *any* receptacles that "are not readily accessible"—that is, any receptacle that requires use of a portable ladder or a chair to get up to it.

From Exception No. 2 to Part (a)(2), garage receptacles for "dedicated appliances"—those that are put in place and not normally moved because of their weight and size—are excluded from the need for GFCI because they will not normally be

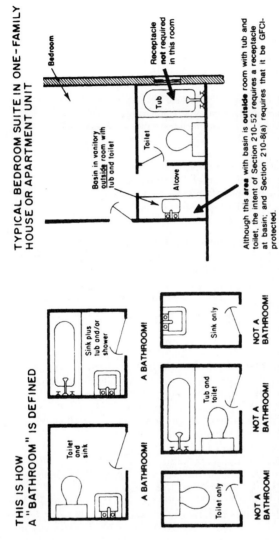

FIGURE 2.90 For GFCI use, a bathroom may be a "room" or an area that includes one or more "rooms."

used to supply hand-held cord-connected appliances (lawn mowers, hedge trimmers, etc.) that are used outdoors.

NOTE:Any receptacle in a garage that is excluded from the requirement for GFCI protection, as noted in the two Exceptions of this section, may not be considered as the receptacle that is required by Section 210-52(g) to be installed in an attached or detached garage with electric power. Thus if a non-GFCI-protected receptacle is installed in a garage at the ceiling for connection of a door operator or if a non-GFCI-protected receptacle is installed in a garage for a freezer, at least one additional receptacle must be installed in the garage to satisfy Section 210-52(g) for general use of cord-connected appliances, and such a receptacle *must* be installed not over 5½ ft above the floor and *must* have GFCI protection (either in the panel CB or in the receptacle itself).

Application of the two Exceptions of Section 210-8(a)(2) may prove troublesome, because receptacle outlets are most commonly installed in a garage during construction of a house and before it is known what appliances might be used. Under such conditions, GFCI receptacles would be required because "dedicated space" appliances are not in place. Then if, say, a freezer is later installed in the garage, the GFCI receptacle could be replaced with a non-GFCI device. But such replacement would not be acceptable for a receptacle that is the only one in the garage, because such a receptacle is required by Section 210-52(g) and is not subject to Exception No. 2 of Section 210-8(a).

Part (a)(3) of Section 210-8, on outdoor receptacles, requires GFCI protection of "all 125-V, single-phase, 15- and 20-A receptacles installed outdoors" at dwelling units. Because hotels, motels, and dormitories are not "dwelling units" in the meaning of the **NEC** definition, outdoor receptacles at such buildings do not require GFCI protection. The rule specifies that such protection of outdoor receptacles is required only "where there is direct grade level access to the dwelling unit and to the receptacles . . ." (Fig. 2.91). The phrase "direct grade level access" is defined in Part (a)(3) and means not over 6½ ft above grade and readily accessible. Because there is no "grade level access" to apartment units constructed above ground level, there would be no need for GFCI protection of receptacles installed outdoors on balconies for such apartments or condominium units. Likewise, GFCI protection would not be required for any outdoor receptacle installed on a porch or other raised part of even a one-family house provided that there was not grade-level access to the receptacle, as in the examples of Fig. 2.91.

The definition of "direct grade level access" in the 1987 **NEC** said that the receptacle requiring GFCI protection had to be "readily accessible without entering or passing through a dwelling unit." That phrase was interpreted to exclude from the need for GFCI protection any outdoor receptacles behind "town houses" (contiguous single-family dwelling units) because such receptacles are readily accessible only by passing through the unit to get to its outdoor back yard. The same is true of "atrium" yards, which are outdoor areas totally surrounded by the structure of a dwelling unit. To make it clear that any outdoor receptacle mounted not over 6½ ft above the ground and readily accessible must be GFCI protected, even though you have to go through the house to reach it, the phrase "without entering or passing through a dwelling unit" was removed from this rule for the 1990 **NEC**.

In any case of an outdoor receptacle at a dwelling unit, GFCI protection is not required if there is no access to the receptacle from the grade level of ground around the building. This important rule on GFCI protection of outdoor outlets reflects the **NEC**-making panel's conviction that such protection is not needed where the receptacle cannot readily be used by someone plugging in a tool or appliance and making contact with earth or any masonry walk or other surface or grade.

118 CHAPTER TWO

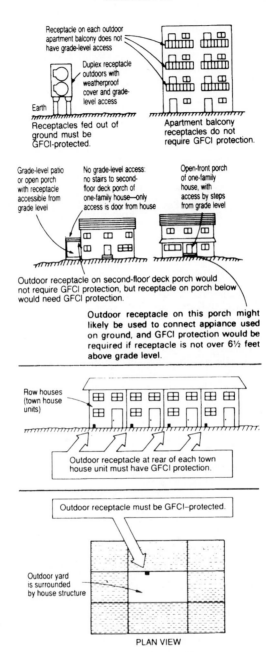

FIGURE 2.91 For dwelling units, only outdoor receptacles with "direct grade level access" require GFCI protection.

According to the rule of Section 210-8(a)(4), all 125-V, single-phase, 15- and 20-A receptacles installed in crawl spaces at or below grade and/or in unfinished basements must be GFCI-protected. This is intended to apply only to those basements or portions thereof that are unfinished (not habitable). Section 210-52(g) requires that at least one receptacle outlet must be installed in the basement of a one-family dwelling, in addition to any installed for laundry equipment. This rule applies to basements of all one-family houses but not to apartment houses, hotels, motels, dormitories, and the like.

Exception No. 1 to Section 210-8(a)(4) eliminates the need for GFCI protection of any single receptacle (not duplex or triplex) that is fed by a "dedicated" (an individual) branch circuit that is "located and identified" for use by a plug-in refrigerator or freezer.

Exception No. 2 exempts a laundry circuit from the need for GFCI protection of its receptacle outlets.

Exception No. 3 eliminates the need for GFCI protection of a single receptacle supplying a permanently installed sump pump.

According to Part (a)(5), GFCI protection is required for all 125-V, single-phase, 15- or 20-A receptacles installed within 6 ft of any kitchen sink or wet bar sink "to serve the counter-top surfaces"—whether such receptacles are above or below the countertop. This will provide GFCI protected receptacles for appliances used on countertops in kitchens or bars in dwelling units. This would include any receptacles installed in the vertical surfaces of a kitchen "island" that includes countertop surfaces with or without additional hardware such as a range, grill, or even a sink. Because so many kitchen appliances are equipped with only 2-wire cords (toasters, coffee makers, electric fry pans, etc.), their metal frames are not grounded and are subject to being energized by internal insulation failure, making them shock and electrocution hazards. Use of such appliances close to a sink creates the strong possibility that a person might touch the energized frame of such an appliance and at the same time make contact with a faucet or other grounded part of the sink—thereby exposing the person to shock hazard. Use of GFCI receptacles within arm's reach of the sink (6 ft to either side of the sink) will protect personnel by opening the circuit under conditions of dangerous fault current flow through the person's body (Fig. 2.92).

Part A(6) requires that all general-use receptacles in a boathouse of a dwelling unit must have GFCI protection. This rule is in recognition of the potential shock hazard due to damp or wet conditions in a boathouse.

Branch-Circuit Controls

Branch-circuit design must include provision for control of the circuits. This may take the form of control over individual outlets or groups of outlets or may consist of control of the entire circuit.

Local control of one or more lighting outlets (less than the entire circuit) is generally effected with manual, direct-acting, wiring-device switches. Alternating-current snap switches feature a wide variety of modern designs. Such switches permit full loads for incandescent and fluorescent lighting and 80 percent loads for motors. The switches are available with exposed screw terminals or screwless pressure-locking terminals. Mercury switches have ac-dc snap-switch ratings (Fig. 2.93). Local control is not generally needed or required for energizing and deenergizing receptacle outlets, although such control is sometimes desirable and recommended. A commonly used technique for receptacle-outlet control is to split-wire a duplex receptacle (by breaking off the metal tab between the brass-colored screws on the

120 CHAPTER TWO

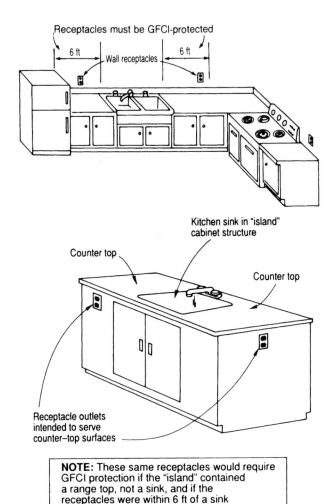

FIGURE 2.92 GFCI protection must be provided for receptacles within 6 ft of kitchen sink, whether above or below the countertop. Receptacles in the face of an island cabinet structure in a kitchen must be GFCI-protected if "within 6 ft of kitchen sink and intended to serve countertop area."

ungrounded side of the device) so the top outlet is controlled by a wall switch and the bottom outlet is always hot (Fig. 2.94).

Although no **NEC** rules require switch control of general-purpose lighting or receptacle outlets in commercial, institutional, or industrial electrical systems, the **NEC** does make certain switched lighting outlets mandatory for all dwelling units (one-family houses, apartment units, etc.). Section 210-70 requires that at least one wall-switch-controlled lighting outlet be installed in every habitable room, in hallways, stairways, and attached garages, and at outdoor entrances. The rule does not

LIGHTING AND APPLIANCE BRANCH CIRCUITS   121

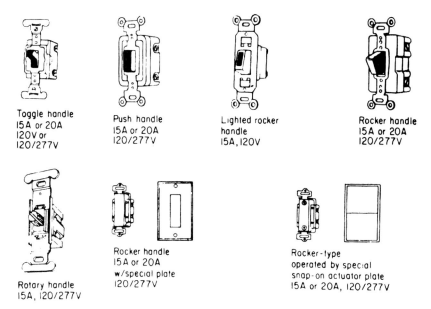

FIGURE 2.93 Wide variety of wiring-device switches offers great flexibility for design of branch-circuit control.

stipulate that these must be ceiling lighting outlets; any or all of them may be wall-mounted lighting outlets (Fig. 2.95).

A vehicle door in an attached garage is not considered as an outdoor entrance. The **NEC** does not require a lighting outlet at any garage door that is provided as a vehicle entrance because the lights of the car provide adequate illumination when such a door is being used in darkness. But the wording of the **NEC** does suggest that a rear or side door that is provided for personnel entry to an attached garage would be "considered as an outdoor entrance" because the note excludes only "vehicle"

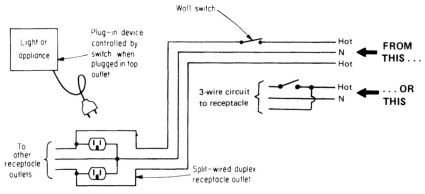

FIGURE 2.94 Split wiring of a receptacle provides control of the receptacle from either a 2-wire or 3-wire circuit.

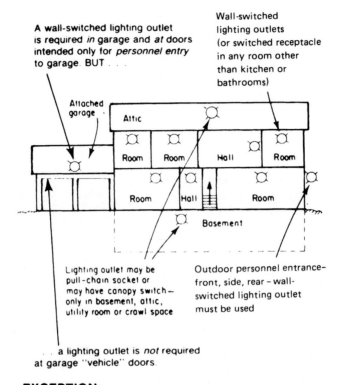

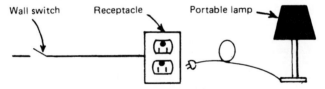

Wall-switched receptacle(s) may be used instead of a lighting outlet in habitable rooms other than kitchens and bathrooms

FIGURE 2.95 In dwelling units only, specific switch-controlled lighting outlets are required.

doors. Each such personnel entrance from outdoors to the garage would require a wall-switched lighting outlet.

In addition, at least one lighting outlet must be installed in every attic, underfloor space, utility room, and basement if it is used for storage or if it contains equipment requiring servicing. In these latter cases, the lighting outlet does not have to be controlled by a wall switch, although it may be. A lamp socket controlled by a pull chain or a canopy switch could be used.

Two exceptions are given to the basic requirements. Exception No. 1 notes that in habitable rooms other than kitchens or bathrooms, a wall-switch-controlled receptacle outlet may be used instead of a wall-switch-controlled lighting outlet. The receptacle outlet can serve to supply a portable lamp, which would give the necessary lighting for the room. Exception No. 2 states that in hallways, stairways, and at outdoor entrances, remote, central, or automatic control of lighting shall be permitted. This latter recognition accepts remote, central, or automatic control as an alternative to the wall-switch control mentioned in the basic rules.

But note carefully that every kitchen and every bathroom must have at least one wall-switch-controlled lighting outlet.

Section 210-70(b) states that at least one wall-switch-controlled lighting outlet or wall-switch-controlled receptacle must be installed in guest rooms of hotels, motels, and similar occupancies.

An important rule on the use of wiring-device switches is contained in **NEC** Section 380-8(b), as follows:

> Snap switches shall not be grouped or ganged in enclosures unless they can be so arranged that the voltage between adjacent switches does not exceed 300, or unless they are installed in enclosures equipped with permanently installed barriers between adjacent switches.

This applies to switches with either screw-type connections or screwless connections (push-in type). It applies where 277-V switches, mounted in a common box (such as two- or three-gang switches), control 277-V loads, with the voltage between exposed line terminals of adjacent switches in the common box being 480 V. If the adjacent switches have exposed live terminals, anyone changing one of the switches without disconnecting the circuit at the panel could contact 480 V, as shown in Fig. 2.96. The rule requires permanent barriers between adjacent switches located in the same box when the voltage between such switches exceeds 300 V and terminals are exposed.

Screwless terminal switches (with no exposed live parts) would seem to satisfy the intent (and literal text) of Section 380-8 because the switches could be considered to be "arranged" to prevent exposure to 480 V. Of course, the hookup shown would be acceptable if a separate single-gang box and plate were used for each switch, or a common wire from only one phase (*A, B,* or *C*) supplied all three switches in the three-gang box. The inspector should be consulted concerning questionable cases.

Full-Circuit Control. When local switching is not necessary, the entire circuit can be controlled by the branch-circuit breaker or switch in the panelboard. Although panelboard control is a common and satisfactory practice in many occupancies—particularly industrial and large commercial areas—most areas should be provided with the convenience and economy of operation afforded by local switching. Switching at the panelboard requires a separate circuit for each group of outlets to be controlled together.

Such factors as distance from panelboard to lighting fixtures, partitioning of the area of lighting coverage fed from the panelboard, and need for three- or four-way switch control will determine the amount of local switching, the amount of panelboard switching, and the number of circuits needed.

In commercial and industrial electrical systems, the breakers in the lighting panel are commonly used for on-off control of lighting; this eliminates the need for local wiring-device switches. The UL states that "circuit breakers marked SWD are suitable for switching 120-V fluorescent lighting on a regular basis." Such listing indicates that any CB used for regular switching of 120-V fluorescents must be marked SWD (switching duty) and that breakers not so marked are not suitable for panel

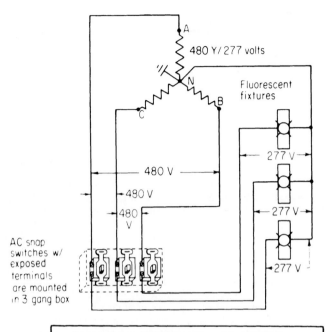

FIGURE 2.96 Snap switches on different phase legs of a circuit fed into a multigang box must not present a personnel hazard.

switching of lighting. **NEC** Section 240-83(d) requires SWD-marked circuit breakers for switching 120- and 277-V fluorescent lighting (Fig. 2.97). Both 120- and 277-V CBs are designed to be suitable for regular switching of lighting of any kind; these are marked SWD. Only those breakers bearing the designation SWD should be used as snap switches for lighting control. Type SWD breakers have been tested and found suitable for the greater frequency of on-off operations required for switching duty, as compared with strictly overcurrent protection, where the breaker is used only for infrequent disconnecting for circuit repair or maintenance.

Remote and/or Automatic Control. Automatic direct switching of lighting circuits is commonly accomplished by means of time switches, which may be used to control all or some of the loads fed by branch circuits. A time switch may provide direct control, with its contacts making and breaking the current to the load, as in the case of the double-pole switch at the left of Fig. 2.98; or, the time switch may be used to control the operating coil of a magnetic contactor, as shown at the right in that diagram. And there is a growing trend toward the convenience and flexibility of low-voltage

LIGHTING AND APPLIANCE BRANCH CIRCUITS

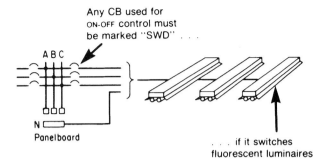

FIGURE 2.97 Circuit breakers used for on-off switching of full-circuit lighting load must be suitable.

relay switching and the use of various types of remote-control magnetic switching through relays and contactors. Such applications are particularly effective for controlling individual lighting fixtures or groups of fixtures when the ability to turn off lights not in use is desired for energy conservation and reduced electric bills.

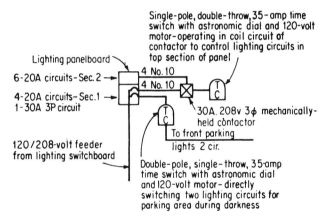

FIGURE 2.98 Time switches can provide automatic control of lighting and other loads—either directly or through a contactor.

Low-Voltage Relay Switching. Low-voltage relay switching is used where remote control or control from a number of spread-out points is required for each of a number of small 120- or 277-V lighting or heating loads. In this type of control, contacts operated by low-voltage relay coils are used to open and close the hot conductor supplying the one or more luminaries or load devices controlled by the relay. The relay is generally of the 3-wire, mechanically held, on-off type, energized from a step-down control transformer.

In a typical installation, all the relays may be mounted in an enclosure near the panelboard supplying the branch circuits which the relays switch, with a single transformer mounted there to supply the low voltage (Fig. 2.99). Where a single panelboard serves a large number of lighting branch circuits in a very large area, as in large office areas in commercial buildings, several relays associated with each section of the overall area may be group mounted in an enclosure in that section.

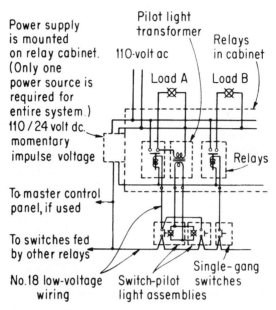

FIGURE 2.99 Typical low-voltage relay switching system provides great flexibility and convenience of control.

Figure 2.100 shows 24-V control of 277-V fixtures, with constantly illuminated switchplates alongside doorways to define interior exit routes and practical hollow-partition clip-in switch boxes in interior labs of a medical research center. Control relays are in compact boxes atop luminaires. Relays are connected to switches and to 50-VA continuous-duty 120/24-V transformers by multiwire Class 2 remote-control circuits (**NEC** Article 725) routed through overhead plenums and supported by insulator rings attached to fixture hangers by spring clips. No switching circuit controls more than four relays. Switching circuits and exit lights are served from an emergency power source, with general-lighting fixtures served through normal 480/277-V feeders. Interior-area route-indicating lights and general-lighting switches are mounted together in thin boxes set in partitions.

Figure 2.101 shows the schematic and physical hookup of the components of another low-voltage relay control system.

Contactor Control. One of the most popular types of modern branch-circuit control is contactor switching. Magnetic contactors—either electrically or mechanically held, depending upon the stability of the control voltage and such factors as coil hum and coil power drain—may be used in individual enclosures for control of circuit or load devices. The basic difference between electrically (or magnetically) held contactors and mechanically held contactors is shown in Fig. 2.102.

Magnetic Contactors. The basic magnetic contactor is magnetically operated by energizing its operating coil to close its contacts. It is magnetically held in position by

LIGHTING AND APPLIANCE BRANCH CIRCUITS

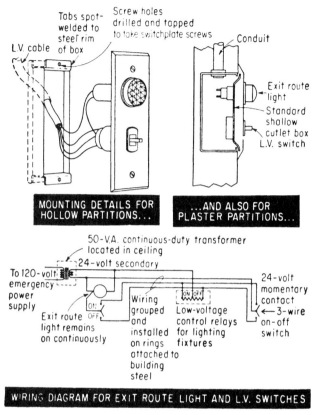

FIGURE 2.100 Details of a low-voltage relay system.

maintaining current flow through the coil. If the voltage to the coil is removed, the contactor opens. The voltage supply to the coil is generally taken from the line side of the contactor itself. Magnetic contactors are made in a wide range of ratings up to about 1000 A, with one to four poles, and to 600 V.

Magnetically held contactors provide effective remote switching control of power and lighting loads where stable circuit voltage will prevent objectionable low-voltage dropout and where the inherent ac hum of continually energized coils is not objectionable. They are particularly applicable to controlling motor loads where running-overload protection is not required or is provided elsewhere. In such use, the magnetically held contactor can be arranged for low-voltage protection or low-voltage release for the motor load.

Mechanically Held Contactors. A variation of the basic magnetic contactor is the mechanically held contactor, in which the operating coil is momentarily energized to close its contacts and momentarily energized to open its contacts.

Mechanically held contactors are best suited for use as remotely controlled safety switches. They are used for switching feeders, branch circuits, and individual loads—lighting, motor, or other power loads. They offer a particular advantage where a single contactor is to be controlled from a number of different remote locations. The

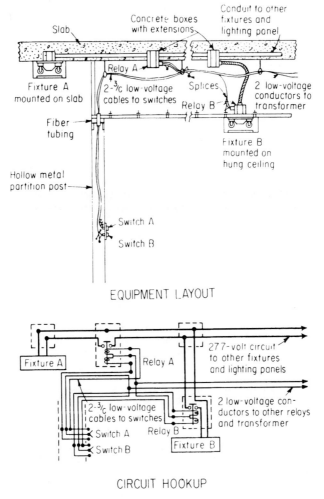

FIGURE 2.101 Low-voltage relay may be located at a fixture or remote from a fixture.

mechanically held design is also highly desirable where continuity of circuit connection must be maintained during voltage dips and power outages and where severe voltage fluctuations would drop out a magnetically held contactor.

A second important advantage of the mechanically held contactor is the absence of ac hum. Because the coil is only momentarily energized for opening and for closing, the coil does not constantly carry current as long as the contacts are in the closed position. This suits the mechanically held contactor to application in such quiet areas as schools, hospitals, and libraries.

In keeping with present energy-conservation efforts in most facilities, contactors are being applied increasingly to control lighting panels, heating and ventilating equipment, and production lines furnished with continuously running electrical equipment. Mechanically held contactors are commonly chosen for this type of control.

LIGHTING AND APPLIANCE BRANCH CIRCUITS

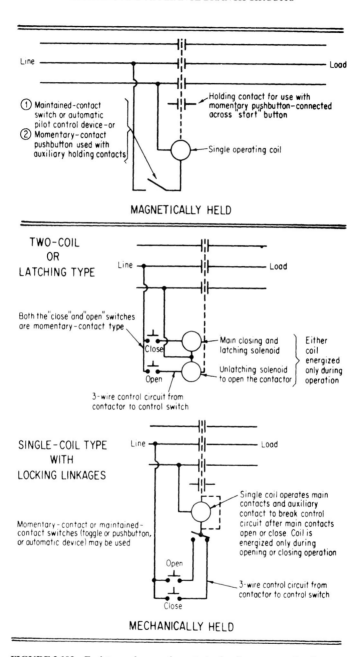

FIGURE 2.102 Each type of magnetic contactor has its own operating features.

The control voltage of the contactor operating coil must be carefully selected. Standard control voltages are 24, 120, 208, 240, 277, 480, and 550 V. Coils can be provided for any voltage, regardless of the voltage of the switched circuit. If the control-circuit voltage is the same as the phase-to-phase or phase-to-neutral voltage of the circuit being controlled by the contactor, the control circuit can be tapped from the line-side terminals of the contactor. Safety considerations frequently dictate the use of a lower control-circuit voltage when the voltage of the controlled circuit is 440 V or higher. In this event, a control transformer can be added to the contactor to obtain the desired voltage, or the control voltage may be derived from a separate ac or dc source and not from the controlled circuit.

Both the magnetic and the mechanically held contactor are operated via a control circuit, the operating action being initiated by a pilot device such as a pushbutton or time clock. The contactor is placed in the circuit to be switched. This may be in the middle of a split-bus panelboard, where the switch controls one section of bus (Fig. 2.103). It also may be adjacent to a panel, where the switch controls a branch circuit or feeder from the panel. Industrial applications of remote-control switching can follow a basic hookup, with the individual mechanically held contactors mounted in enclosures high up on columns in plant areas. Each contactor can be located at the approximate center of its lighting load, to keep circuit wiring as short as possible for minimum voltage drop, and can be tapped from the nearest point on the busway. Control circuits can then be carried to one or more control panels, as shown in Fig. 2.104. The operating coil of a contactor (or *remote-control switch*, as it also is known) may be switched either manually (using pushbuttons or momentary-contact switches) or automatically by a time switch or photoelectric device.

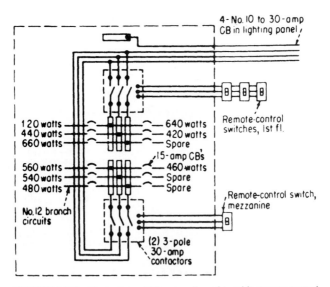

FIGURE 2.103 Contactors within a panelboard provide remote control of loads connected to the two sections of the split-bus panelboard.

A typical contactor application for full-panel control might involve locating the contactors at widely spaced lighting panelboards supplying outdoor lighting, with all

LIGHTING AND APPLIANCE BRANCH CIRCUITS 131

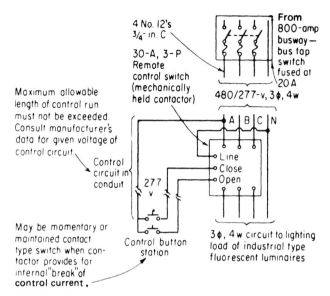

FIGURE 2.104 Remote-control switching for large-block control of lighting.

the control circuits brought to pilot switches at a common point of control. For the control of individual circuits supplying lighting loads, contactors may be located near the panelboard or near the loads, with the control circuits arranged—for maximum convenience of operation—at a central control point or at any number of points, depending upon the job requirements (Fig. 2.105). Where incandescent lighting is to be switched, the contactor must be tungsten-rated; where motors are involved, the contactor must be able to break the inductive load current.

Figure 2.106 shows a magnetic contactor used for the control of heating circuits in a night-setback hookup. During the day, the electric heaters are controlled by room thermostats at their locations; the overcall key switch is in the closed position, thereby keeping the contactor coil energized and its contacts closed. At night, the overcall switch is opened and the magnetic contactor is controlled by the zone thermostat shown, which operates to maintain the temperature at the night setting of 50°F. The local thermostats are all closed at that temperature.

Typical coil-circuit control devices for contactors include maintained-contact and momentary-contact pushbuttons, selector switches, control- and master-type switches, pressure switches, float switches, limit switches, time switches, thermostats, plugging switches, and contacts of control relays. Which of these devices is used depends upon the nature and layout of the load; more than one type may be used. Contactors may be used for the control of branch circuits or feeders and for supplying lighting loads, heating loads, and/or motor or other power loads.

The design and layout of magnetic-contactor control circuits and their components are regulated by **NEC** rules. Article 725 covers remote control. As shown in Fig. 2.107, protection of the coil-circuit conductors is set forth in Section 725-12.

Exception No. 3 to Section 725-12 covers protection of the remote-control circuit that energizes the operating coil of a magnetic contactor, as distinguished from a magnetic motor starter. Although it is true that a magnetic starter is a magnetic con-

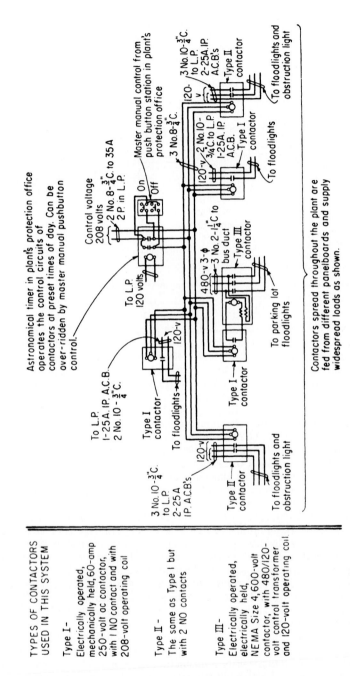

FIGURE 2.105 Central control of widespread lighting with contactors.

LIGHTING AND APPLIANCE BRANCH CIRCUITS

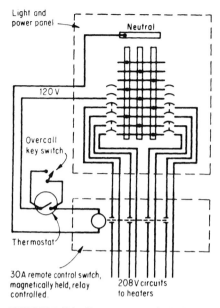

FIGURE 2.106 Contactor control of electric-heating circuits.

tactor with the addition of running-overload relays, Exception No. 3 covers only the coil circuit of a magnetic contactor. And this exception states that contactor remote-control conductors can be satisfactorily protected by overcurrent devices rated at not more than 300 percent of the carrying capacity of the control-circuit conductors. That applies to control wires for magnetic contactors used for the control of lighting and heating loads, but not motor loads. Section 430-72 covers requirements for control circuits of magnetic motor starters.

It should be noted that the overcurrent protection is required for the control conductors and not for the operating coil. Because of this, control conductors can be sized for use without separate overcurrent protection.

In Fig. 2.108, office lighting made up of continuous-row fluorescent luminaires is circuited from a remote-control contactor panelboard which is outside the office area, on the wall of the adjoining manufacturing area. The special panelboard is fed by conductors tapped from a busway plug switch, as indicated. The panel is equipped with a main contactor which switches the supply to the panel mains. Each branch circuit consists of a branch CB and a remote-control contactor. All control circuits are carried in conduits to a control panel on the wall at one end of the office area.

Branch-Circuit Panelboards

Lighting and appliance branch circuits originate in panelboards in which the overcurrent protective devices for the circuits are mounted. A panelboard usually contains a means for switching each circuit. Typical panelboard configurations are shown in Fig. 2.109. Some panelboards have a switch and a fuse for each circuit; most panelboards have circuit breakers that provide both protection and switching for the circuits.

The term "*loadcenter*" is commonly used to describe the light-duty type of panelboard used for residential and light commercial and industrial systems. However, both the UL and the **NEC** refer only to "panelboards" and include both loadcenters and the heavier-duty panelboards within this single category.

The selection of a panelboard is based first on the number of circuits it must serve. Then the designer must ensure that the busbars in the panelboard for any application have sufficient current rating for the demand load of all of the branch circuits.

Standards of UL and NEMA base the ratings of main switches, overcurrent devices, and busbars feeding lighting and/or appliance branch devices within the panel on an assumed average load current of not more than 10 A per branch circuit. Where the continuous loading will be in excess of an average of 10 A per branch circuit, busbars, integral main switches, and overcurrent devices of greater capacity are required.

All panelboards—for lighting and power—are required by **NEC** Section 384-13 to have a rating (the current capacity of the busbars) not less than the **NEC** mini-

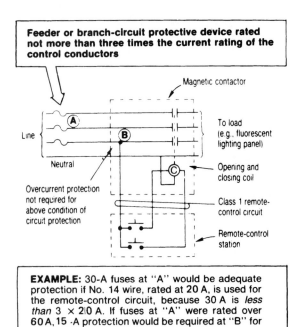

FIGURE 2.107 Design of contactor control for branch circuits includes sizing of coil-circuit wires and provision of overcurrent protection for such wires.

mum feeder-conductor capacity for the entire load served by the panel. That is, the panel busbars must have a nameplate ampere rating at least equal to the required ampere capacity of the conductors feeding the panel (Fig. 2.110). A panel may have a busbar current rating greater than the current rating of its feeder but must never have a current rating lower than that required for its feeder.

In addition to the rule of Section 384-13, the rule of Section 220-4(d) applies to panelboard ampacity ratings, it requires any panelboard to have busbar ampacity at least equal to the minimum calculated branch-circuit loads from Section 220-3. For a panelboard supplying general lighting, the rule of Section 220-4(d) requires busbars to have an ampere rating of "not less than the calculated load," as determined from the minimum watts-per-square-foot value for the branch-circuit load and the area served—even if the result is less than the actual connected branch-circuit loading.

A panelboard must have a minimum ampacity sufficient to serve the calculated total load of lighting, plus appliances, motors, and other loads supplied. Of course, if the connected lighting load is greater than that calculated on the watts-per-square-foot basis, the greater value of load must be used in determining the number of branch circuits, the panelboard capacity, and the feeder capacity.

Example. For a 200- × 200-ft office area, a 3-phase, 4-wire, 460/265-V feeder and branch-circuit panelboard must be selected to supply 265-V HID lighting that is a 92,000-W connected load and will operate continuously (3 hours or more). What are the minimum size of feeder conductors and minimum panelboard rating that must be used to satisfy Section 220-4?

1. Size of feeder from bus plug switch to main contactor is based on total volt-amperes of fluorescent-lighting load at 100% demand.

2. Additional feeder capacity is included for load growth.

3. Further upsizing of feeder conductors keeps voltage drop under 1% from service to branch circuit panelboard and reduces the heating effect of continuous load operation.

4. Main contactor is sized to match feeder capacity.

5. Lighting panelboard is rated for the total feeder capacity.

6. Branch circuit protection consists of 20-amp, 3-pole CB's for multi-wire branch circuits to 277-volt fluorescent ballast.

7. Individual branch circuit contactors are 3-pole, 30-amp remote-control switches connected in the lighting panelboard on the load side of the branch CB's.

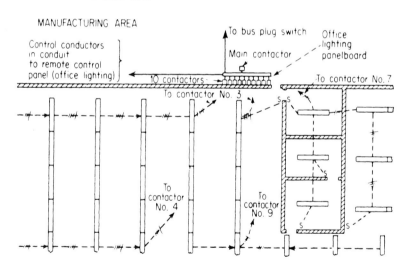

8. Each branch circuit is loaded to 50% of its capacity (80% loading is the NEC maximum for continuously operating circuits).

FIGURE 2.108 Major considerations in circuit design for contactor control.

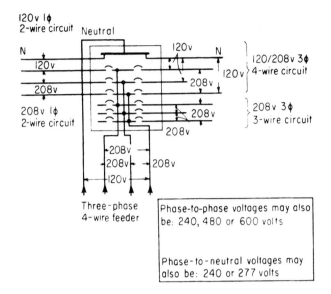

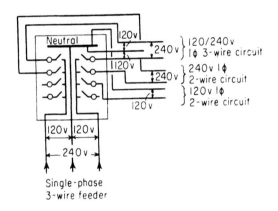

FIGURE 2.109 Typical circuit makeups in panelboards (rarely seen).

Solution. For the branch-circuit 92,000-W connected load, the panelboard would be sized for the total calculated load of 140,000 W as shown in Fig. 2.111.

The actual connected lighting load for the area, calculated from the lighting design, is 92,000 W ÷ (460 V × 1.732) = 116 A per phase. Sizing of the feeder and panelboard, however, must be based on 176 A, not 116 A, to satisfy Section 220-4(d). But then, other **NEC** rules require adjustment of that figure of 176 A to get the usable capacity for a 176-A continuous load.

LIGHTING AND APPLIANCE BRANCH CIRCUITS 137

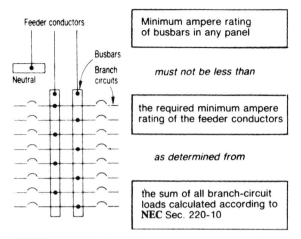

FIGURE 2.110 Panelboard bus must have a rating equal to the calculated minimum feeder ampacity.

The next step is to correlate the rules of Sections 220-4(a) and (d) with those of Section 220-10(a). The latter requires a feeder to be sized for the "computed load" as determined according to Part (a) [Section 220-3(b)]. The feeder protective device for the continuous calculated load of 176 A must have an ampacity at least equal to 125 percent of the calculated load; and, the feeder can be sized on the same basis:

$$176 \text{ A} \times 1.25 = 220 \text{ A} \text{ [Section 220-10(b)]}$$

which means a 225-A protective device. And a 225-A protective device would be used at the supply end of the feeder, say, No. 4/0 THW copper conductors.

Section 384-13 requires this panelboard to have a rating not lower than the minimum required ampacity of the feeder conductors—which, in this case, means the panel must have a busbar rating not lower than 220 A. A 225-A panelboard is therefore required, even though a 125-A panel would be adequate for the actual load current of 116 A.

As noted previously, Section 220-4(d) specifically eliminates the need to install enough panelboard protective devices and circuit wires to handle the calculated load of 140,000 W. The number of branch-circuit protective devices required in the panel (the number of branch circuits) is based on the size of the branch circuits used and their capacity as related to the connected load. If, say, all circuits are to be 20-A, 265-V, phase-to-neutral circuits, then each pole may be loaded at no more than 16 A, because Section 220-3(a) requires the continuous circuit load to be limited to 80 percent of the 20-A protection rating. With the 116 A of connected load per phase, a single-circuit load of 16 A calls for a minimum of 116 A ÷ 16 A, or 8 poles per phase leg. Thus a 225-A panelboard with 24 breaker poles would satisfy the rule of Section 220-4(d). But, because the panelboard must have capacity for the calculated continuous load of 176 A per busbar, the required panel must have space for a total number of poles (spare pole spaces plus the eight active poles per phase) at least equal to 176 A ÷ 16 A (per 20-A pole), or 11 poles per phase leg. Thus the required panel would have to contain at least 33 pole-places (3 × 11), although only 8 pole-places would have to be used per phase leg. Section 220-4(a) also requires that the mini-

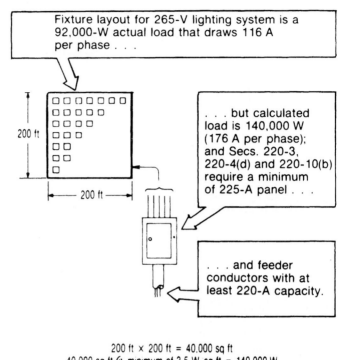

FIGURE 2.111 Panelboard busbar capacity must be rated for the "calculated load."

mum number of branch-circuit poles be based on the "total computed load" and not on a lesser value of connected load.

Panel Main Protection. Rules in **NEC** Section 384-16 concern the selection of lighting and appliance branch-circuit panelboards. Any panelboard that meets the **NEC** definition (Section 384-14) of a lighting and appliance branch-circuit panel must be individually protected on its supply side by not more than two main circuit breakers or two sets of fuses having a combined rating not greater than that of the panelboard (Fig. 2.112). In the vast majority of cases, a single main protective device (CB or set of fuses), rated at not more than the panel's rated busbar ampacity, is used to provide panelboard main protection. Individual protection is not required when a lighting and appliance branch-circuit panelboard is connected to a feeder that has overcurrent protection not greater than that of the panelboard.

NEC Section 384-14 defines a lighting and appliance branch-circuit panelboard as "one having more than 10 percent of its overcurrent devices rated 30 amperes or less, for which neutral connections are provided."

LIGHTING AND APPLIANCE BRANCH CIRCUITS   139

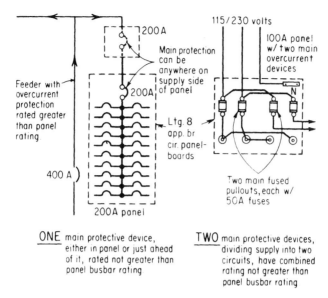

FIGURE 2.112 Lighting and appliance panelboard must have one or two main protective devices (that is, one or two CBs or sets of fuses).

Watch out for this definition! Many panels have makeups that appear to make them power panels or distribution panels, yet they are technically lighting and appliance panels in accordance with the above definition, and they must have protection for the busbars. Figure 2.113 shows an example of how it is determined whether a panelboard is or is not "a lighting and appliance branch-circuit panelboard." The determination is important because it indicates whether or not main protection is required for a particular panel. Figure 2.114 shows panels that do not need main protection because they are not lighting and appliance panels, which are the only panels required by Section 384-16 to have main protection.

Because of the wording of the definition in Section 384-14, it is important to evaluate a panel carefully to determine if protection is required. And just as it is strange, in Fig. 2.113, to identify a panel that supplies no lighting as a lighting panel, it is also strange that some panels that supply only lighting are technically not lighting panels. Because of the definition of Section 384-14, if the protective devices in a panel are all rated over 30 A, or if no neutral connections are provided in the panel, then the panel is not a lighting and appliance panel, and it does not require main overcurrent protection.

Figure 2.115 shows a panelboard which just meets **NEC** rules for classification as a lighting and appliance panel, as follows:

1. There is a total of 24 overcurrent devices in the panel.

2. Three of the overcurrent devices are rated 30 A or less, with neutral connections provided for them.

3. Ten percent of the number of overcurrent devices (24) is 2.4.

4. The panelboard therefore has more than 10 percent of its overcurrent devices rated 30 A or less, with neutral connections.

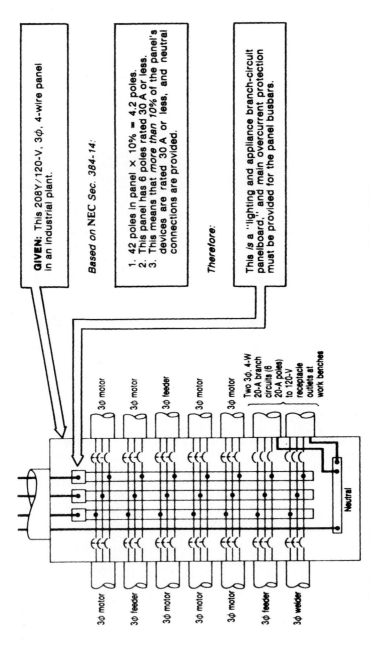

FIGURE 2.113 Main protection is required for any panel that meets the definition of Section 384-14.

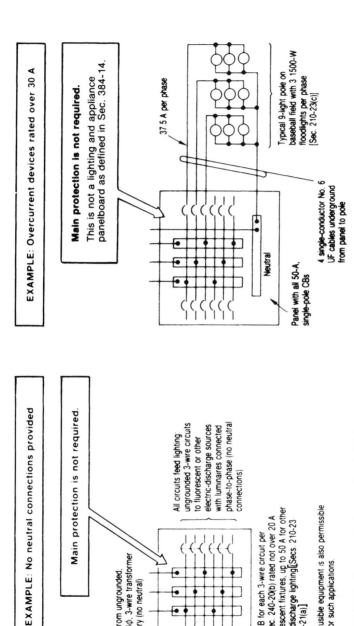

FIGURE 2.114 Some panels that supply only lighting are not lighting and appliance panels and do not require main protection.

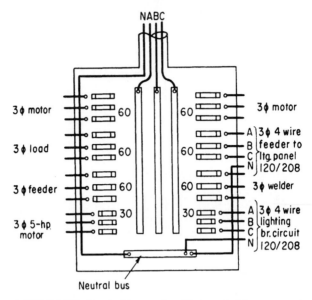

FIGURE 2.115 Panel with more than 10 percent of its overcurrent devices rated 30 A or less, for which neutral connections are provided, is a lighting and appliance panelboard.

5. Such a panelboard is designated by the **NEC** as a "lighting and appliance" panelboard, and it must be protected in accordance with **NEC** Section 384-16.

The overcurrent devices referred to in the foregoing determination may be fuses or CB poles. A two-pole CB is considered to be two overcurrent devices; a three-pole CB is three overcurrent devices. And the **NEC** says that not more than 42 overcurrent devices, "other than those provided for in the mains," may be installed in any one cabinet or cutout box for a lighting and appliance panelboard. Thus, a panel may contain 42 overcurrent poles plus main protection—such as a three-pole main CB ahead of the 42 devices.

Figure 2.116 shows the three locations at which panelboard main protection may be placed. Where a number of panels are tapped from a single feeder protected at a current rating higher than that of the busbars in any of the panels, the main protection may be installed as a separate device just ahead of the panel or as a device within the panel feeding the busbars (cases 2 and 3 in Fig. 2.116). The main protection would normally be a circuit breaker or fused switch with the number of poles corresponding to the number of busbars in the panel.

Figure 2.117 shows two panels that require main protection because they are both fed by the same feeder, which is protected at a current rating greater than the busbar rating of the panels. Figure 2.118 shows how feeder protection can protect panelboards.

Other Protection Rules. Section 384-16(b) says, "Panelboards equipped with snap switches rated at 30 amperes or less shall have overcurrent protection not in excess of 200 amperes." Thus any panel (lighting or power) which contains snap switches rated 30 A or less must have overcurrent protection not in excess of 200 A. Panels which are not lighting and appliance panels and do not contain snap switches rated

LIGHTING AND APPLIANCE BRANCH CIRCUITS

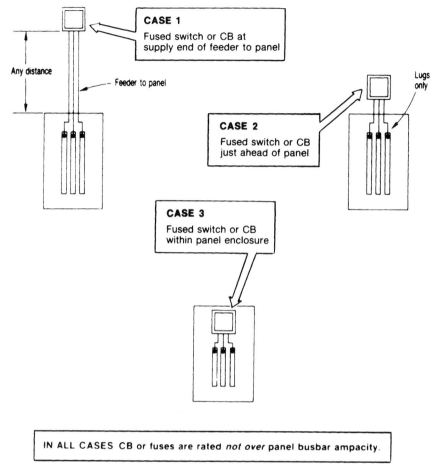

FIGURE 2.116 Main panel protection may be placed at any one of these locations.

30 A or less do not have to be equipped with main protection and may be tapped from any size of feeder. Figure 2.119 shows two examples of these overcurrent protection requirements for panelboards.

Section 384-16(d) applies to a panelboard fed from a transformer. The rule requires that overcurrent protection for such a panel, as required in parts (a) and (b) of the same section, must be located on the secondary side of the transformer.

But the primary-circuit protection is not acceptable as suitable protection for a panelboard on the secondary of a transformer with a 3-wire or 4-wire secondary—even if the panel busbars have an ampacity equal to the ampacity of the primary protection times the primary-to-secondary voltage ratio.

Primary protection of a panel on the secondary of a single-phase, 2-wire to 2-wire transformer is permitted if the primary overcurrent protection does not exceed the value determined by multiplying the panel busbar ampacity by the secondary-to-primary voltage ratio (Fig. 2.120).

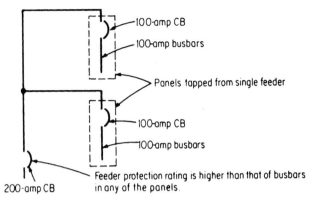

FIGURE 2.117 Single main protects each panel in accordance with the basic rule of Section 384-16(a).

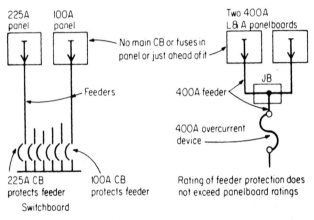

FIGURE 2.118 Panels do not require main protection because feeder protection protects panels [Exception No. 1 of Section 384-16(a)].

Residential Panel Mains. Main overcurrent protection is generally required for any lighting and appliance panelboard. However, Exception No. 2 of **NEC** Section 384-16(a) notes that a panelboard used as residential service equipment for an "existing" installation may have up to six main protective devices [as permitted by Exception No. 3 of Section 230-90(a)]. Such usage is limited to "individual residential occupancy," such as a private house or an apartment in a multifamily dwelling, where the panel is truly service equipment and not a subpanel fed from service equipment in the basement or from a meter bank on the load side of a building's service.

But it must be noted that the phrase "for existing installations" makes it a violation to use a service panelboard (loadcenter) with more than two main service disconnect-and-protection devices for a residential occupancy in a new building (Fig. 2.121). That applies to one-family houses and to "dwelling units" in apartment houses, condominiums, and the like. The use of, say, a split-bus panelboard with four

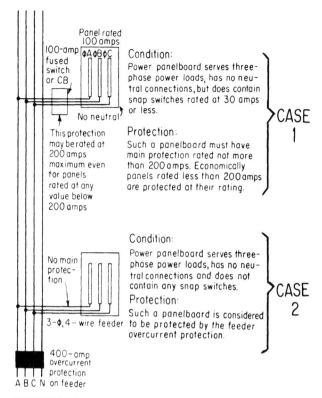

FIGURE 2.119 Power panels do not require main overcurrent protection except with snap switches rated 30 A or less.

or six main protective devices (such as two-pole CBs) in the busbar section that is fed by the service-entrance conductors is limited to a service panel in "an individual residential occupancy" in an *existing* building, for example, for service modernization (Fig. 2.122). For any new job, any panelboard—for service or otherwise, and in any type of occupancy—may have no more than two main protective devices if the panel is a lighting and appliance panelboard.

A lighting and appliance panelboard containing six single-pole breakers or six two-pole breakers (or even six three-pole breakers or fuses) of any rating may not be used as new residential service equipment without a main protective device ahead. For an existing service, where it is permissible to use a split-bus panel with six main devices in the section fed by the service conductors, it is permissible to use 15- and/or 20-A protective devices in that main section. Of course, 15- and/or 20-A protective devices may be used in the bottom section of the panel, which is protected by one of the mains in the top part of the panel.

Other Design Concerns. NEC Section 384-16(c) states that "the total load on any overcurrent device located in any panelboard shall not exceed 80 percent of its rating where in normal operation that load will continue for three hours or more." An exception to this would be an assembly, together with the overcurrent device, which has been found suitable for continuous duty at 100 percent of its rating and is UL-

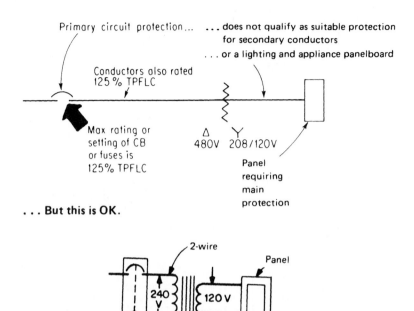

FIGURE 2.120 Main protection for a lighting and appliance panel is not generally permitted on the primary side of the supply transformer.

listed (or otherwise listed) for such application. The rule of Section 384-16(c) parallels that of Section 210-23(b), which has the same effect of limiting load on the protective device.

Panelboards may be equipped with a main switch or circuit breaker feeding the entire bus or a section of bus in the board. In cases where a lighting panel is supplied by a tap from a feeder with overcurrent protection greater than the main busbar rating of the panel, a main CB or fused switch required for the panel provides simultaneous disconnect of circuits from the feeder. In many cases, a remote-control switch, operated by a manual pilot switch or by some automatic pilot device like a time clock, is used to control the circuits connected to the bus in the panel by switching the feed to the bus, as shown in Fig. 2.123.

To supplement the spare capacity allowed during branch-circuit design, spare circuit capacity should be allowed in panelboards. One spare circuit should be allowed in the panel for each five active circuits. Inasmuch as panelboards are made in multiples of four switch-and-fuse units or circuit breakers, the spare circuits can often be provided without using a panelboard larger than is required for the active circuits.

When spare circuits are provided in flush-type panelboards, the required conduit capacity to wire these circuits also should be provided, to avoid the need to tear out walls in the future. This extra conduit capacity may consist of empty conduit runs to the ceiling and floor, terminated in covered outlet boxes, spare capacity in other raceways for branch circuits from panelboards, etc.

LIGHTING AND APPLIANCE BRANCH CIRCUITS 147

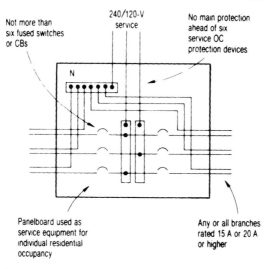

FIGURE 2.121 For new installations, main overcurrent protection is required for a lighting and appliance panel used as service equipment for a residence.

In addition to standard panelboard hookups, there are cases which call for careful design of special panel arrangements. A typical special panel application is commonly used in gas stations. **NEC** Section 514-5 (in Article 514, "Gasoline Dispensing and Service Stations") contains a specific requirement regarding "circuit disconnects." It states:

> Each circuit leading to or through dispensing equipment, including equipment for remote pumping systems, shall be provided with a switch or other acceptable means to disconnect simultaneously from the source of supply all conductors of the circuit, including the grounded conductor, if any.

Figure 2.124 shows how this can be accomplished using a gas-station-type panelboard, which has its bussing arranged to permit the hookup of standard solid-neutral circuits in addition to the required switched-neutral circuits. Another way of supplying such switched-neutral circuits is with circuit-breaker-type panelboards for which there are standard accessory breaker units, which have a trip element in the ungrounded conductor and only a switching mechanism in the other pole of the common-trip breaker, as shown in Fig. 2.125. Either two- or three-pole units may be used for 2-wire or 3-wire circuits, rated 15, 20, or 30 A. No electrical connection is made to the panel busbar by the plug-in grip on the neutral breaker unit. A wire lead connects the line side of the neutral breaker to the neutral block in the panel, or two clamp terminals are used for the neutral.

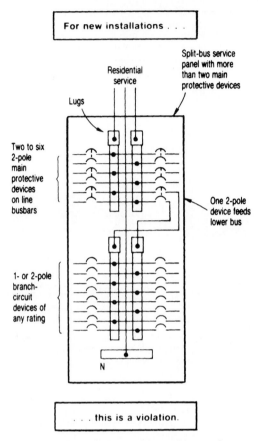

FIGURE 2.122 Split-bus loadcenter with more than two main overcurrent devices is permitted for residential service equipment only in "existing" installations.

Branch-Circuit Voltage Drop

In laying out circuits, the loading and lengths of homeruns and runs between outlets must be related to voltage drop and the need for spare capacity in the circuits for possible future increases in load. Each lamp, appliance, or other utilization device on the circuit is designed for best performance at a particular operating voltage. Although such devices will operate at voltages on either side of the design value, adverse effects will generally result from operation at voltages lower than the specified value.

A 1 percent drop in voltage to an incandescent lamp produces about a 3 percent decrease in light output; a 10 percent voltage drop will decrease the output about 30 percent. In heating devices of the resistance-element type, voltage drop has a similar effect on heat output. In motor-operated appliances, low voltage to the device will affect the starting and pullout torques, and the current drawn from the line will increase with drops in voltage. As a result, the heat rise in the motor windings will be above normal.

Voltage drop in the conductors is due to the resistance of the conductors plus, in ac circuits, reactance. And the heat developed by the dissipation of power (the I^2R

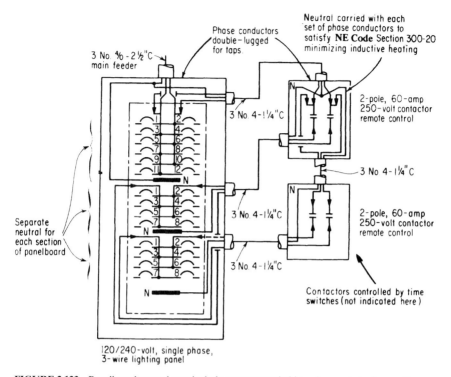

FIGURE 2.123 Panelboard control may include contactor switching of separate busbar sections.

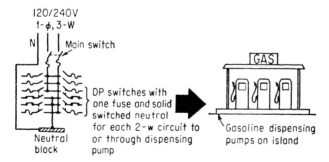

FIGURE 2.124 Branch circuits to gasoline dispensing pumps must have switched neutrals.

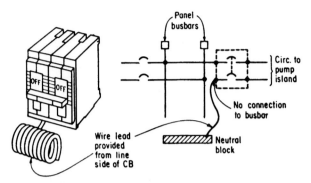

FIGURE 2.125 Two-pole, switched-neutral CB has a pigtail for connecting one pole to the neutral block.

losses) in the wiring, which itself costs money, deteriorates the conductor insulation. In good design, the voltage drop is held to no more than 1 percent.

Figure 2.126 shows the basic formulas used to calculate voltage drop, with examples of their application. Figure 2.127 shows how copper loss (the I^2R loss) is calculated for circuits with power factors of 100 percent and less than 100 percent.

In Section 210-19(a), the **NEC** comments (but does not make a mandatory rule) on branch-circuit voltage drop, as follows:

> Conductors for branch circuits as defined in Article 100, sized to prevent a voltage drop exceeding 3 percent at the farthest outlet of power, heating, and lighting loads, or combinations of such loads and where the maximum total voltage drop on both feeders and branch circuits to the farthest outlet does not exceed 5 percent, will provide reasonable efficiency of operation. See Section 215-2 for voltage drop on feeder conductors.

That wording is only a recommendation, and the voltage drop should be held to the above-recommended 1 percent in lighting branch-circuit conductors to minimize energy losses in the conductors and ensure optimal operation and efficiency for load devices. Figure 2.128 is a tabulation of circuit loads and the maximum circuit lengths from the panelboard to the first outlet (on a balanced 3-wire, single-phase circuit or a 4-wire, 3-phase circuit) that will hold the voltage drop to a 1 percent maximum. Figs. 2.129 and 2.130 contains a graph that may be used to directly select the wire size required to hold the voltage drop to 1 percent.

Although 50 percent loading of circuits is recommended for substantial protection against excessive voltage drop, the sizes of conductors for long runs should always be carefully selected to ensure that if spare capacity is put to use in the future, it will not increase the voltage drop beyond the limit. For this reason, when the design intent is to use 50 percent loading to provide spare capacity in the circuit, the conductors used in long runs should be sized for voltage drop on the basis of the maximum possible loading.

For lighting and appliance branch circuits, calculations of voltage drop should be made on a single-phase, 2-wire basis. That is, the voltage drop in the wires should be limited to some percentage of the voltage from the hot leg to the neutral—whether the circuit is 2- or 3-wire, single-phase or 4-wire, 3-phase. Since most branch-circuit loads are connected hot leg to neutral, the major concern should be with the voltage delivered to the loads rather than with the phase-to-phase voltage when that voltage is not supplied to any load device. Of course, 2-wire circuits made up of ungrounded

LIGHTING AND APPLIANCE BRANCH CIRCUITS

1 Two-wire, single-phase circuits (inductance negligible):

$$V = \frac{2k \times L \times I}{d^2} = 2R \times L \times I$$

$$d^2 = \frac{2k \times I \times L}{V}$$

V = drop in circuit voltage (volts)
R = resistance per ft of conductor (ohms/ft)
I = current in conductor (amps)
L = one-way length of circuit (ft)
d^2 = cross section area of conductor (circular mils)
k = resistivity of conductor metal (cir mil-ohms/ft) (k = 12 for circuits loaded to more than 50% of allowable carrying capacity; k = 11 for circuits loaded less than 50%; k = 18 for aluminum conductors) at 30C

EXAMPLE:
What is the voltage drop in this circuit?

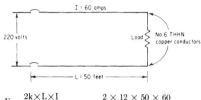

$$V = \frac{2k \times L \times I}{d^2} = \frac{2 \times 12 \times 50 \times 60}{26,240 \ (\text{NE Code Table 8, Ch. 9})}$$

$$V = \frac{72,000}{26,240} = 2.7 \text{ volts}$$

2 Three-wire, single-phase circuits (inductance negligible):

$$V = \frac{2k \times L \times I}{d^2}$$

V = drop between outside conductors (volts)
I = current in more heavily loaded outside conductor (amps)

EXAMPLE:

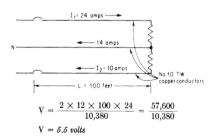

$$V = \frac{2 \times 12 \times 100 \times 24}{10,380} = \frac{57,600}{10,380}$$

$$V = 5.5 \text{ volts}$$

3 Three-wire, 3-phase circuits (inductance negligible):

$$V = \frac{2k \times I \times L}{d^2} \times 0.866$$

V = voltage drop of 3-phase circuit

EXAMPLE:

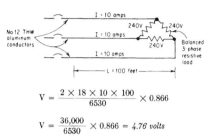

$$V = \frac{2 \times 18 \times 10 \times 100}{6530} \times 0.866$$

$$V = \frac{36,000}{6530} \times 0.866 = 4.76 \text{ volts}$$

4 Four-wire, 3-phase balanced circuits (inductance negligible):

For lighting loads: Voltage drop between one outside conductor and neutral equals one-half of drop calculated by formula for 2-wire circuits.

For motor loads: Voltage drop between any two outside conductors equals 0.866 times drop determined by formula for 2-wire circuits.

EXAMPLE:

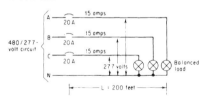

What size of copper wire must be used here to keep voltage drop to 1% under balanced conditions (i.e., 1% of 277 or 2.8 volts)?

For a 2-wire circuit, $d^2 = \dfrac{2k \times I \times L}{V}$

For a 3-ϕ, 4-wire circuit, $d^2 = \dfrac{2k \times I \times L}{V} \times \dfrac{1}{2}$

$$d^2 = \frac{2 \times 12 \times 15 \times 200}{2.8} \times \frac{1}{2} = \frac{72,000}{5.6}$$

= 12,857 circular mils

From Table 8, Ch. 9, NE Code —
No. 10 = 10,380 CM   No. 8 = 16,510 CM
No. 8 must be used.

FIGURE 2.126 Basic formulas for branch-circuit voltage drop.

legs of single-phase or 3-phase systems should be designed for a voltage drop that is some percentage of the voltage between them when they supply loads at other than phase-to-neutral connection.

The voltage drop of any 2-wire circuit (either phase-to-neutral or phase-to-phase) can be taken simply as the *IR* drop of the conductors, using the total dc resis-

With 100% load power factor and negligible conductor reactance

When load power factor is 100% and conductor reactance due to self-induction is negligible, calculation of voltage drop in the circuit and calculation of copper loss due to heating effect of current in the circuit conductors follows the standard relations between current and voltage. The supply voltage is equal to the arithmetic sum of the voltage drop in the conductors and the voltage across the load. Copper loss—energy wasted as heat produced in the circuit conductors—is a simple "I^2R" loss.

where E = circuit voltage (volts)
R_1 = total resistance of circuit conductors (ohms)
R_2 = resistance of noninductive load (ohms)
I = current flowing in circuit (amps)

$$I = \frac{E}{R_1 + R_2} = \frac{480}{1 + 15} = 30 \text{ amps}$$

Voltage drop in conductors = $I \times R_1 = 30 \times 1$
= *30 volts*

Copper loss in conductors = $I^2 \times R_1 = (30)^2 \times 1$
= *900 watts*

Percent voltage drop = $\frac{30}{480} \times 100 = 6\%$

With less than 100% load power factor and negligible conductor reactance

When the power factor of the load is less than 100% and the conductor reactance is negligible, the sum of the voltage drops across the load and the resistance of the circuit conductors is no longer equal to the circuit supply voltage. In the following example, note that although the voltage drop in the circuit conductors is 30.4 volts and might appear to leave only 449.6 volts for the load (480 − 30.4), there is actually 456 volts across the load (30.4 amps x 15 ohms of load impedance). The real loss in voltage to the load is, therefore, only 24 volts (480–456). The 30.4-volt drop in the conductors is only the voltage across the resistive conductor load, which differs in phase from the voltage across the reactive load device fed by the circuit. Although the arithmetic sum of the two voltages (across the conductor resistance and across the load) is equal to 486.4 (30.4 + 456), correct vector addition of the two voltages gives the 480-volt value of the supply. And note that the significant percent voltage drop is different from the apparent value. Copper loss, however, remains a simple "I^2R" value.

$Z_1 = \sqrt{(R_2)^2 + (X_L)^2}$ $(Z_1)^2 = (R_2)^2 + (X_L)^2$
then, $(15)^2 = (R_2)^2 + (X_L)^2$

but, POWER FACTOR = $0.8 = \frac{R_2}{Z_1} = \frac{R_2}{15}$

thus, $R_2 = 0.8 \times 15 = 12$ ohms
then from $(X_L)^2 = (Z_1)^2 - (R_2)^2$
= $(15)^2 - (12)^2$
= $225 - 144 = 81$
$X_L = \sqrt{81} = 9$ ohms

The total impedance of the circuit, Z_2, includes conductor resistance and must be calculated in order to find circuit current.

$Z_2 = \sqrt{(R_1 + R_2)^2 + (X_L)^2} = \sqrt{(1 + 12)^2 + (9)^2}$
= $\sqrt{169 + 81} = \sqrt{250} = 15.8$ ohms

Then $I = \frac{480}{15.8} = 30.4$ amps

Voltage drop in conductors = $I \times R_1 = 30.4 \times 1$
= *30.4 volts*

Copper loss in conductors = $I^2 \times R_1 = (30.4)^2 \times 1$
= *924 watts*

Apparent % voltage drop = $\frac{30.4}{480} \times 100 = 6.3\%$

Voltage delivered to load = $I \times Z_1 = 30.4 \times 15$
= *456 volts*

Significant % voltage drop

= $\frac{480 - 456}{480} \times 100 = \frac{24}{480} \times 100 = 5\%$

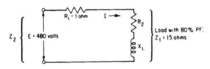

where X_L = inductive reactance component of load (ohms)
Z_1 = impedance of load (ohms)
Z_2 = total impedance of circuit (ohms)

FIGURE 2.127   Calculating voltage drop and copper loss.

tance of the two circuit wires. For conductors up to No. 3 AWG—which are used in the vast majority of branch circuits—the ac resistance of the wire is equal to the dc resistance; the latter is readily obtained from the right-hand columns of Table 8 in **NEC** Chapter 9. In such wire sizes, the inductive reactance of the circuit is negligible. Figure 2.131 shows an example of this approach.

LIGHTING AND APPLIANCE BRANCH CIRCUITS

Loads and lengths in feet for 1% drop on 3- and 4-wire, 115-volt circuits

Amp load	#10 wire	#12 wire	#14 wire
1	946	596	374
2	474	298	188
3	316	198	124
4	236	148	94
5	190	120	76
6	158	100	62
7	136	86	54
8	118	74	46
9	106	66	42
10	94	60	38
11	86	54	34
12	78	50	32
13	72	46	28
14	68	42	26
15	64	40	24
16	60	38	
17	56	36	
18	52	34	
19	50	32	
20	48	30	
21	46		
22	44		
23	42		
24	40		
25	38		
26	36		
27	36		
28	34		
29	32		
30	32		

Calculations based on copper resistance of 13 ohms per CM-ft at 60C (140F).

For a circuit made up of two phase wires and the neutral of a 3-phase, 4-wire wye system, multiply given lengths by 0.67.

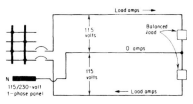

Balanced 3-wire single-phase circuit

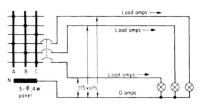

Balanced 4-wire 3-phase circuit

NOTE:
This table is based on a 1% voltage drop in each 115-volt circuit which is part of a 3-wire, single-phase circuit or part of a 4-wire, 3-phase circuit—provided the circuit is operating under balanced conditions. This is important because under balanced conditions, the neutral associated with each 115-volt circuit does not carry current. As a result, each 115-volt circuit is, in effect, fed by only one conductor. This produces half the voltage drop that would occur under unbalanced conditions, with both the phase leg and neutral carrying current. The distances given are, therefore, twice what they would be for the same voltage drop under unbalanced conditions.

EXAMPLE:

Q. In making up a 115/230-volt, 3-wire circuit to carry 15 amps on each hot leg for a total circuit length of 57 ft, what size wire must be used to keep the voltage drop to 1% with a balanced load?

A. Come down the left hand column, "Amp load," to the circuit load of 15 amps. Move across and note maximum distances for each wire size.

It can be seen that No. 12 wire reaches a drop of 1% (1.15 volts) at a length of 40 ft. No. 10 wire can carry a 15-amp circuit for up to 64 ft before reaching the voltage drop level.

Because the required circuit length of 57 ft is greater than 40, No. 12 cannot be used. But because it is less than 64 ft, No. 10 wire will keep voltage drop within limits.

FIGURE 2.128 Sizing branch-circuit wires for 1 percent voltage drop in the homerun (the wires between the panelboard and the first outlet).

The maximum tolerable voltage drop in a motor circuit usually depends upon the speed and torque requirements of the motor and how they would be affected by low voltage. In Fig. 2.132, for example, the characteristics of the dc motor and its load-torque requirements indicate that the motor needs at least 223 V for proper performance. For operation at normal ambient temperature, the branch-circuit conductors would be sized as follows to ensure at least the minimum allowable voltage at the motor.

The branch-circuit conductors must provide the voltage drop along with the required carrying capacity. **NEC** Section 430-22 requires branch-circuit conductors

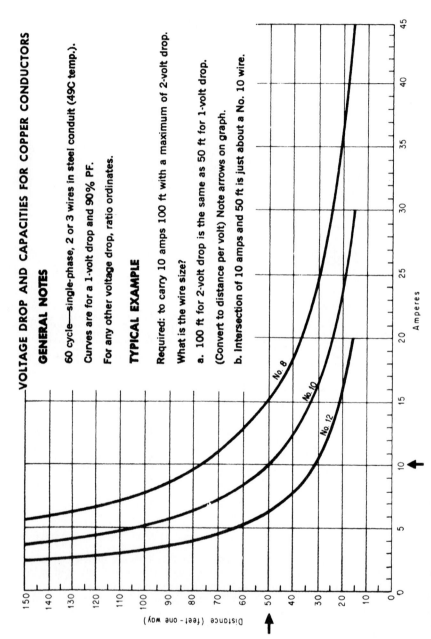

FIGURE 2.129 Voltage drop and capacities for copper conductors.

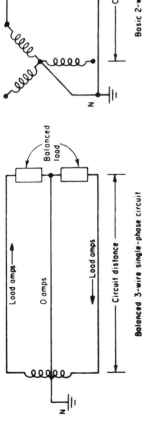

Balanced 3-wire single-phase circuit

Basic 2-wire single-phase circuit

NOTE:

In both of the types of circuits shown, there are two conductors carrying current—one out to the load and one back from the load. For the single-phase circuit derived from the 3-phase, 4-wire wye, it is best to select conductors for a 3-phase, 4-wire circuit on the basis of the voltage drop which would exist if only one phase was operating. This gives the worst condition for voltage drop. When such a circuit is truly balanced, there is no current in the neutral, and the phase-to-neutral connected load is being served by one conductor—which produces only half the voltage drop that will be present under conditions of maximum unbalance.

EXAMPLES:

1. Q. What wire size should be used to make up a 2-wire, 480-volt circuit to carry 15 amps a distance of 100 ft with a drop of 5 volts?

A. A drop of 5 volts over a 100-ft circuit is equal to a 1-volt drop for each 20 ft (100 ÷ 5). By converting to this value of distance per volt, we can use the table directly. On the horizontal scale, locate the load of *15 amps*. On the vertical scale, locate the distance of *20 ft*. The intersection of coordinates shows that *No. 12* conductors would be O.K.

2. Q. What wire size should be used for a 3-phase, 4-wire, 480/277-volt circuit that supplies a balanced lighting load of 16 amps per phase—to keep voltage drop at not more than 3 volts over the circuit length of 120 ft, even under conditions of maximum unbalance?

A. A drop of 3 volts over 120 ft is the same as 1 volt for 40 ft. Using the coordinates of 16 amps and 40 ft indicates that No. 10 conductors would do the job.

FIGURE 2.130 Graph shows branch-circuit voltage drop and minimum wire size for a given load current.

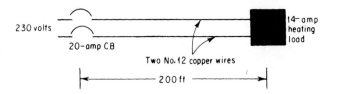

Total length of No. 12 conductor:
200 × 2 = 400 ft

From Table 8, Chapter 9, **NE Code**, No. 12 solid copper conductor has a dc resistance of 1.62 ohms per 1000 ft.

Then the 400-ft circuit length has a total resistance of

$$1.62 \times \frac{400}{1000} = 0.648 \ ohm$$

Voltage drop = I × R = 14 amps × 0.648 ohm
= *9 volts.*

In the preceding problem, the conductors were assumed to be exposed to an ambient temperature of 25C (77F). The resistance of the conductors, and thus the voltage drop, would increase under conditions of higher ambient temperature.

FIGURE 2.131   Calculating voltage drop from **NEC** table of wire resistances.

supplying a single motor to have an ampacity not less than 125 percent of the motor full-load current rating:

1.25 × 20 A = 25 A

From **NEC** Table 310-16, a No. 12 TW copper conductor has an ampacity of 25 A; this conductor would therefore satisfy Section 430-22 as regards the minimum ampacity for the motor circuit. And because the actual full-load current of the motor is 20 A, there is no conflict with the footnote to Table 310-16 requiring that the "overcurrent protection" of No. 12 copper must not exceed 20 A.

Although the same footnote sets 20 A as the maximum rating of overcurrent protection for No. 12 copper conductors, Section 240-3(f) exempts "motor-circuit conductors" from the basic requirement that "conductors . . . shall be protected against overcurrent in accordance with their ampacities" as specified in Tables 310-16 through 310-19 and all applicable notes to these tables. In addition this is one of those instances where the **NEC** "permits" a higher rating for the overcurrent protective device.

To ensure a voltage of at least 223 V at the motor, the maximum permitted voltage drop in the 250-ft circuit run from the panelboard to the motor must be

228 V − 223 V = 5 V

LIGHTING AND APPLIANCE BRANCH CIRCUITS

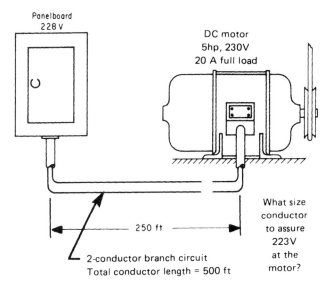

FIGURE 2.132 Voltage drop for dc motor branch circuit involves only resistance and current.

Table 8 in **NEC** Chapter 9 shows that a No. 12 copper conductor has a dc resistance of 1.62 Ω per 1000 ft at a temperature of 77°F. Because resistance is the only cause of voltage drop in a dc circuit (there is no inductive reactance as in ac circuits), the voltage drop of this circuit, using No. 12 copper wire, can be readily calculated.

With two conductors for the 250-ft-long circuit, the total length of No. 12 wire is 2 × 250 ft, or 500 ft. At 1.62 Ω per 1000 ft, the total resistance of the circuit would be half of 1.62 Ω, or 0.81 Ω.

At a full-load current of 20 A, the drop in voltage from the panel to the motor would be

$$V_d = I \times R = 20 \text{ A} \times 0.18 \text{ Ω} = 16.2 \text{ V}$$

That would produce a motor voltage of

$$228 \text{ V} - 16.2 \text{ V} = 211.8 \text{ V}$$

Obviously, this would not satisfy the requirement of 223 V at the motor. A larger conductor, with less resistance, must be used.

Because the drop of 16.2 V with No. 12 wire is over 3 times the maximum permitted drop of 5 V, the required conductor must have a resistance less than one-third that of No. 12 wire. Reference to the table of dc resistances shows that No. 8 wire has a resistance of 0.6404 Ω per 1000 ft, which is more than 1.62 Ω/3. But No. 6 wire, with a resistance of 0.410 Ω per 1000 ft, looks like it might do.

The calculation is repeated for the No. 6 wire: At a resistance of 0.410 Ω per 1000 ft, the required 500 ft of No. 6 wire for the 250-ft-long 2-wire circuit would have a total resistance equal to half of 0.410 Ω, or 0.205 Ω.

Then, at the 20-A load current, the voltage drop would be

$$V_d = I \times R = 20 \text{ A} \times 0.205 \text{ Ω} = 4.1 \text{ V}$$

The motor supply voltage then would be

$$228 \text{ V} - 4.1 \text{ V} = 223.9 \text{ V}$$

This satisfies the need for at least 223 V at the motor, and No. 6 copper conductor (TW, THW, RHH, THHN, or XHHW) must be used.

As shown in Table 9 in **NEC** Chapter 9, conductors smaller than No. 2 AWG have negligible self-inductance, and therefore there is no difference between the resistance of such conductors to direct current and their resistance to alternating current. In larger conductors, the "skin effect" (self-inductance forcing current to flow in the outer part of the conductor cross-sectional area) produces an ac resistance that is higher than the dc resistance. For most branch-circuit designs, the dc resistance of wire is entirely satisfactory for use in voltage-drop calculations.

In the smaller sizes of conductors, the inductive reactance produced under ac conditions is also generally negligible. However, it is important to distinguish between the increase in ac resistance due to self-inductance and that due to inductive reactance, which is also produced by the self-inductance of the conductor.

As noted above, ac flow in conductors is subject to the skin effect, which produces an apparent increase in resistance over the resistance value which would be obtained for dc flow. This is due to a reduction in the effective conductor cross section resulting from the tendency of alternating current to flow close to the surface (or "skin") of the conductor. Generally, this increase in resistance to alternating current is of little consequence in conductors smaller than No. 4/0 AWG.

Reactance in conductors carrying ac power depends upon the size of the conductor, the spacing between it and other conductors carrying current, the position of the conductor with respect to conductors close to it, the frequency of the alternating current, and the presence of magnetic materials close to the conductor. In an ac circuit, the reactance of the conductors may be reduced by placing the conductors close together and/or by placing them in nonmagnetic raceway instead of steel conduit or raceway. In many large-size or long ac circuits, the voltage drop due to impedance is often far greater than the drop due simply to the resistance of the conductors.

When current flows in a conductor in which the reactance due to self-induction is negligible, the voltage drop is equal to the product of the current in amperes and the total resistance of the conductor in ohms. But when the reactance of the conductor is not negligible, the voltage drop is equal to the product of the current in amperes and the total impedance of the conductor, which is determined from the formula

$$Z = \sqrt{R^2 + X^2}$$

where Z = total impedance, Ω
R = total ac resistance of conductor, Ω
X = reactance of conductor, Ω

The voltage drop in such a conductor is

$$V = \sqrt{I^2 + Z^2}$$

where V = voltage drop, Ω
I = current flowing in conductor, A
Z = total impedance of conductor, Ω

Resistance and reactance data on wires and cables are given in literature made available by the manufacturers. Tables and graphs are also available for quickly and easily computing the voltage drops in large, heavily loaded feeders operating at less than unity power factor and with considerable conductor reactance. But, for branch-circuit calculations of voltage drop, inductive reactance is not nearly as significant as it is for the larger conductors used for feeders.

SUMMARY ON CIRCUIT DESIGN

The requirements and recommendations of the foregoing discussion lead to a set of basic standards for modern branch-circuit design. Of course, in many cases, practical considerations such as size of area or type of load devices will require deviation from the letter of these standards. However, general standards for the majority of applications may be listed as follows:

- Separate branch circuits should be provided for general lighting, for automatic appliances, fixed appliances, and plug receptacles. Generally, each automatic or fixed appliance should be served by an individual circuit.
- Branch circuits with more than one outlet should not be loaded in excess of 50 percent of their carrying capacity.
- An individual branch circuit should have spare capacity to permit a 20 percent increase in load before reaching the level of maximum continuous load current permitted by the **NEC** for that circuit.
- At least one spare circuit should be allowed for each five circuits in use.
- The smallest wire size used in branch circuiting should be No. 12.
- The wire used in a branch-circuit homerun should be at least one size larger than that computed from the loading when the distance from the overcurrent protective device to the first outlet is over 50 ft.
- When the distance from the overcurrent protective device to the first plug outlet on a receptacle circuit is over 100 ft, the size of the circuit homerun should not be less than No. 10 for a 20-A branch circuit; it may be larger, depending upon the rating of the particular circuit, the actual distance, the voltage drop, and the load conditions.
- Homeruns on lighting circuits should be limited to a maximum of 100 ft, unless the load on the circuit is so small that the voltage drop between the overcurrent protective device and any outlet is under 1 percent. Careful layout of panelboard locations and use of sufficient number of panelboards will avoid this problem of long homeruns.

General design standards for the selection and layout of panelboards are as follows:

- No more than 42 branch-circuit phase conductors (for example, 42 single-pole protective devices) may originate from a single lighting and appliance panelboard. And, again, in determining the number of poles in a panel, a two-pole CB is considered to be two single poles, and a three-pole CB is considered to be three single poles.
- No branch circuit in a panelboard should run more than 100 ft to the first outlet of the circuit.
- All panelboards should be readily accessible.
- Each panelboard should be placed as near as possible to the center of the layout of load outlets that it handles.
- If circuit switching from the panelboard is desired, a switch-and-fuse or CB panelboard must be used, and such a panelboard should be used generally.

- Panelboard locations should be selected to conform as much as possible to the routing of feeders, ensuring the shortest possible feeder runs and a minimum number of bends.
- Every panelboard must have a rating that is not lower than the minimum feeder capacity required to serve the load, as determined from **NEC** Article 220.
- At least one lighting and appliance branch-circuit panelboard should be provided on each active floor of a building.

LAYOUT ON PLANS

The actual design procedure for lighting and appliance branch circuits involves working on a set of plans for the building. Such plans are usually available or can be made. The design work consists in using the foregoing information and whatever other data are known to lay out the outlets, circuit runs, and control legs. Of course, this step depends upon a decision having been made as to the general type of distribution and the characteristics and voltage of the feeders from which the branch circuits will be supplied. Figure 2.133 shows the kind of one-line diagram that can be included on electrical plans to show the overall relationship of branch circuits to the feeders supplying them.

The first step in circuit design is to indicate the locations of general-lighting outlets on floor plans. This layout of general-lighting outlets is part of overall lighting design. As a follow-up to this first part of the work, the wattage of each outlet is determined, the total load is determined, and the number of circuits is selected.

Design standards for lighting and appliance circuits are aimed at providing convenience, flexibility, operating efficiency, and reliability in using the available energy. To ensure all the advantages of sufficient circuit capacity plus spare capacity for load growth, modern design practice dictates the separation of loads into known, approximated, and unknown loads.

General illumination is a known load—whether derived from a detailed lighting layout or developed from a watts-per-square-foot calculation. The number, rating, and layout of outlets for general illumination can easily and accurately be apportioned among a number of branch circuits. Such circuits can be carefully loaded with due regard to voltage drop, operating cycle, possible future increases in lighting level, and required control.

Figure 2.134 shows branch-circuit loading schedules for two different installations. In Fig. 2.134*a*, the number of outlets and total load for each branch circuit in each panelboard are indicated. The relative balancing of loads on the two busbars is also shown. The top part of the panel in Fig. 2.134*b* shows the loading of 277-V fluorescent lighting on 20-A circuits, each of which has a load capacity of 5500 W. The schedule in Fig. 2.134*b* is actually for a loadcenter in a 480/277-V distribution system serving a modern school. Classroom and corridor lighting is supplied by 277-V fluorescent-lighting circuits from the 480/277-V panel (circuits 1 to 8 in the schedule). The 30-kVA transformer fed by circuit 20 in this panel supplies 120/208-V power to a general-purpose lighting and appliance panelboard. Note the spare circuits.

In the design of many commercial and industrial buildings, specific lighting requirements may not be known until after it is known what type of work will be performed and tenants decide upon their working layout. In such cases, the provision of sufficient capacity and circuits in the panelboard is the initial limit of circuit design work, pending determination of the loads.

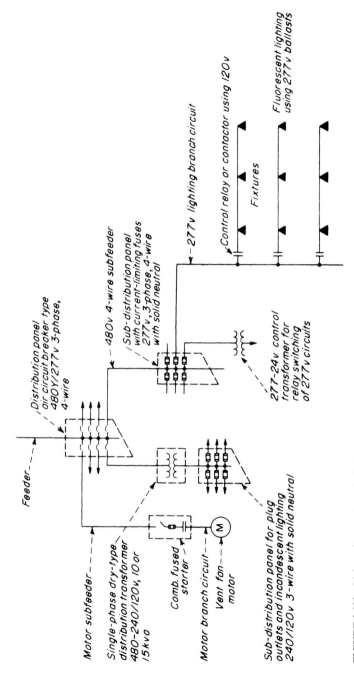

FIGURE 2.133 Light and power branch-circuit layout served from a 480Y/277-V system.

Circ. No.	G Lighting Panelboard			Circ. No.	K Lighting Panelboard		
	No. of Outlets	Wattage Bus A	Wattage Bus B		No. of Outlets	Wattage Bus A	Wattage Bus B
1	4	1000		1	4	480	
2	4		1000	2	3		1500
3	4	1000		3	3	1500	
4	4		1000	4	5		750
5	4	1000		5	5	750	
6	4		1000	6	2		800

(a)

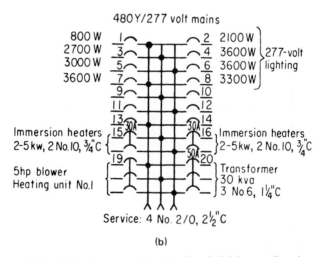

(b)

FIGURE 2.134 Examples of circuit loadings in lighting panelboards.

After lighting outlets are laid out, the next usual step is to indicate convenience and special receptacle outlets on the plans. Depending upon the occupancy, receptacle circuits may be run in conduit to wall- or column-mounted outlet boxes or fed up from the floor slab or down from the ceiling; the circuits may be run in underfloor raceway or a cellular metal floor system; the receptacles may be fed by a flat-conductor cable system under carpet squares, or they may be fed by poke-through floor fittings.

Figure 2.135 shows the standard symbols used to designate the elements of an electrical-system layout on a set of plans. Because electrical plans, along with a set of specifications, constitute the instructions to the electricians installing the system (that is, the persons executing the design concepts set forth in the plans), as much detail as possible should be included in the job drawings.

Graphical interpretation of electrical design involves indication of the types of electrical equipment on scaled representations of the building areas served by the

LIGHTING AND APPLIANCE BRANCH CIRCUITS

FIGURE 2.135 Electrical symbols for architectural plans.

equipment. Such indication includes the wire and raceway interconnections among the various elements and must convey all the design concepts to the mind of the installer who will work from the plans. Figure 2.136 shows the branch circuiting for a small school building with an outside eating area. A legend clarifies the symbols, and a schedule for the panelboard serving the circuits shows the various loads. For

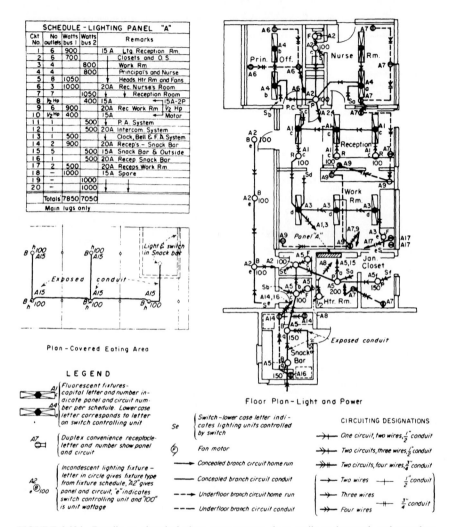

FIGURE 2.136 Details on electrical plans must convey clear, easily understood, and complete details of the branch-circuit design concepts.

the lighting circuits, these loads are known from the lighting design which established the number and ratings of the various lighting units. For the ½-hp motor on circuits 8 and 10, the load is known. For receptacle circuits, loads are determined either by allowing 1.5 A for each outlet or by using load figures for known or estimated ratings of appliances. Ratings of overcurrent protection devices for the branch circuits are given under "Remarks" in the schedule.

When lighting and appliance branch circuits, with their controls and panelboards, have been laid out, a panelboard schedule should be made up and included in the plans and specifications. This schedule should assign some code-letter designation to each panelboard (such as L1a for lighting panel *a* on the first floor, L2b for lighting

LIGHTING PANELBOARD SCHEDULE

Panel Designation	Location	Equip.	Branch Circuits					Mains			Service		Remarks
			No. Active	No. Spares	Total No.	Poles	Trip Calibr.	Type	Cap	Phase	Feed	Voltg.	
LP1A	1st Floor Stair No. 5	Cct Bkr	25	7	32	SP	20A	Cct. Bkr.	225/125	3	4-300 MCM	120/208	
LP1C	1st Floor Main Stair		27	7	34	SP	20A	"	225/150	3	4-4/0	"	
LP2A	2nd Floor Stair No. 6		29	7	36	SP	20A	"	225/150	3	4-300 MCM	"	
LP2B	2nd Floor Stair No. 1		22	8	2 / 30	2P / SP	20A / 20A	"	225/125	3	4-4/0	"	
LP2C	2nd Floor Main Stair		14	6	20	SP	20A	"	100/70	3	4-4/0	"	
LP3A	3rd Floor Stair No. 6		39	3	42	SP	20A	"	225/175	3	4-350 MCM	"	
LP3B	3rd Floor Stair No. 1		20	6	2 / 26	2P / SP	20A / 20A	"	100/100	3	4-4/0	"	
LP3C	3rd Floor Main Stair		12	4	16	SP	20A	"	100/100	1	—	"	Connect to Existing 3 Wire Feed
LP4A	4th Floor Stair No. 6		18	4	22	SP	20A	"	100/100	3	4-350 MCM	"	
LP4B	4th Floor Stair No. 1		13	5	18	2P / SP	20A / 20A	"	100/70	3	4-300 MCM	"	
LP5A	5th Floor Stair No. 6		11	9	20	SP	20A	"	100/100	3	4-400 MCM	"	
LP5B	5th Floor Stair No. 1		11	5	2 / 16	2P / SP	20A / 20A	"	100/70	3	4-300 MCM	"	
LP6A	6th Floor Stair No. 6		14	8	22	SP	20A	"	100/100	3	4-400 MCM	"	
LP6B	6th Floor Stair No. 1		15	5	2 / 20	2P / SP	20A / 20A	"	100/70	3	4-300 MCM	"	
LPPH-1A	Penthouse		3	3	6	SP	20A	Lugs Only	NEMA STD	1	3 No. 12	"	
LPBA	Basement Near Stair No. 1		9	3	12	SP	20A	Lugs Only	NEMA STD	3	4 No. 2	"	
LPBB	Basement Near Stair No. 1		11	3	14	SP	20A	Lugs Only	NEMA STD	3	4 No. 2	"	
LPBC	Basement Near Col. 32		5 / 0	5 / —	10 / —	SP / 2P	20A / 20A	Lugs Only	NEMA STD	3	4 No. 6	"	
LPBD	Basement Near Col. 8	↓	6 / 0	4 / —	10 / —	SP / 2P	20A / 20A	Lugs Only	NEMA STD	3	4 No. 8	"	

FIGURE 2.137 Typical detailed schedule of lighting panelboards gives selection and installation data for all panels.

panel *b* on the second floor, etc.) and should include a tabulation of panel locations, number and size of circuits in each panel, types and ratings of circuit protective devices, capacities of mains, size and type of main protection and disconnect, and any pertinent remarks about each panelboard which might clarify the design intent for the installer (Fig. 2.137).

Depending upon the type of occupancy and how the interior is to be laid out from a work standpoint, outlets for local and/or special lighting units represent approximated loads. Such outlets are included in the overall lighting design, either as fixed unit loads or estimated plug-in loads. Outlets—either lampholders or plug receptacles—for such lighting units may be connected to general-lighting circuits, provided with separate circuiting, or included in circuits allowed for plug receptacles.

Local or special lighting outlets include corridor lights, exit lights, entrance lights, washroom lights, emergency lights, and other known lighting-outlet requirements. Figure 2.138 shows a typical layout of special lighting outlets for the continuous operation of emergency lights which also serve as night lights in a hospital corridor. The circuiting arrangement for the two-lamp corridor lighting fixtures provides separate conduit and conductors for normal-circuit lighting with a 150-W lamp in each fixture and for emergency-circuit lighting with a 25-W night lamp in each fixture. In this diagram, D circuits 12 and 14 supply the 150-W lamps and may be switched off at night by means of the wall switch at the upper left. Panel D is fed from the normal (or nonessential) section of bus in the main switchboard. The 25-W night lamps are fed from panel 1 EM (circuits 3, 5, and 7 shown here) and operate day and night, with no local switching available. Panel 1 EM is fed from the section of main switchboard bus, which is fed from the normal supply and from the emergency generator when the normal supply fails.

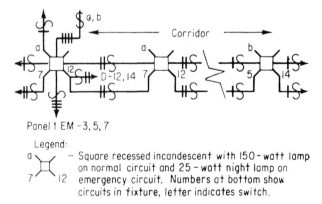

FIGURE 2.138 Diagrams on plans should show adequate details of special hookups, such as emergency lighting and night lighting.

In addition to one-line diagrams showing the basic power distribution and full wiring schematics, to detail the actual hookups and circuit wiring layouts, electrical plans should include isometric (three-dimensional) drawings to show the physical relationships between electrical equipment and building interior. Such drawings indicate the installed positions of the electrical equipment in relation to architectural and/or structural members of the building. They are particularly useful for the layout of wiring methods—which extend over very large areas of a building's floor

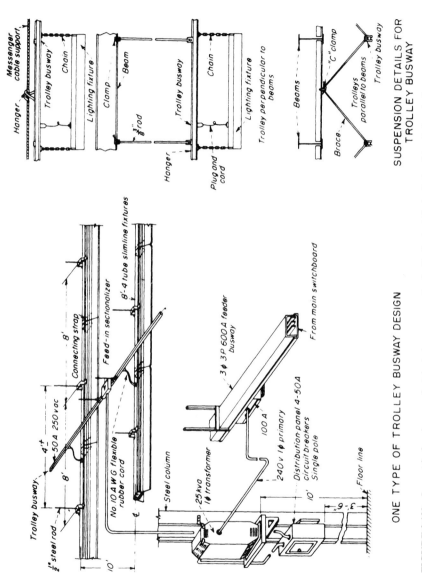

FIGURE 2.139 Isometric drawings on electrical plans can greatly clarify the installed relationship of electrical components.

or ceiling space. An example of this type of diagrammatic elaboration is presented in Fig. 2.139, which shows details from the plans of an industrial building in which fluorescent lighting is fed by cord-and-plug connection to trolley busway. As shown, the trolley busway serves as a lighting branch circuit and is fed from four 50-A CBs in a lighting panel. The panel is supplied from a transformer secondary, powered from a tap to a busway feeder or subfeeder. The lighting fixtures are suspended from the trolley busway, from beams, or from the ceiling. The trolley busway is supported by the ceiling, beams, messenger cable, or braces supported by the beams.

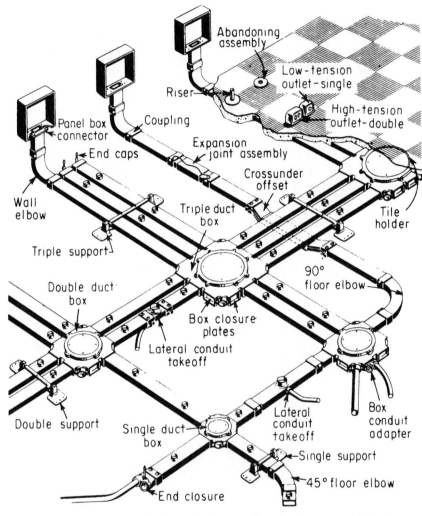

Underfloor Raceway for Use in Conventional Concrete Slab Floors

FIGURE 2.140 High-quality, detailed drawings on electrical plans assure fast and reliable communication of the designer's concepts to the installing electrician.

LIGHTING AND APPLIANCE BRANCH CIRCUITS

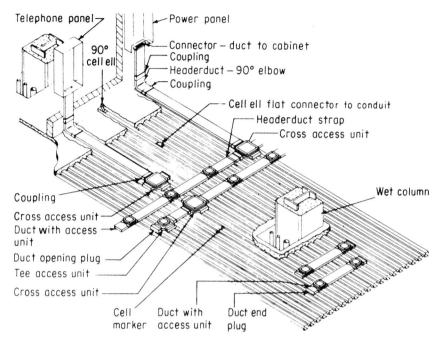

FIGURE 2.141 Layout of electrical components embedded in concrete must be fully detailed.

Figures 2.140 and 2.141 illustrate the advantage of isometric drawings in clarifying design concepts of electrical wiring methods with respect to floor construction.

In the design of many systems, the general construction of the building, including architectural features and the type of floor-slab construction, will affect the way the circuits are run and the locations of outlets. Detailed, effective drawings can greatly facilitate the task of making known to the installer exactly what is to be done. An example is the drawing in Fig. 2.142 for a data-processing center. The locations of the various types of receptacle outlets are laid out on the floor plan, and a legend is part of the plan. The numbers in the rectangles identify various types of IBM machines. Neoprene-jacketed cables with separable cord connectors run under the raised floor from the panels to the machines.

Special attention should be given to receptacle circuits serving special loads. Figure 2.143 shows the layout (on the electrical plans) for special receptacle outlets mounted in wireways to supply 9-wire circuits to computing machines in a data-processing center. The floor plan of the area shows the layout of the wireways and the locations of the receptacle outlets. The phase designation alongside each receptacle symbol (ϕA, ϕB, or ϕC) indicates which phase of the 60-Hz circuit is tapped to that particular receptacle. The arrow from each receptacle symbol points to the machine supplied by that receptacle. This grid layout of wireways provides for the installation of receptacles at any location, as necessary to serve machines anywhere in the area. The nine conductors for each circuit include three phases and a neutral for 60-Hz supply, three phases and a neutral for 400-Hz supply, and a separate ground wire (either No. 8 or No. 6).

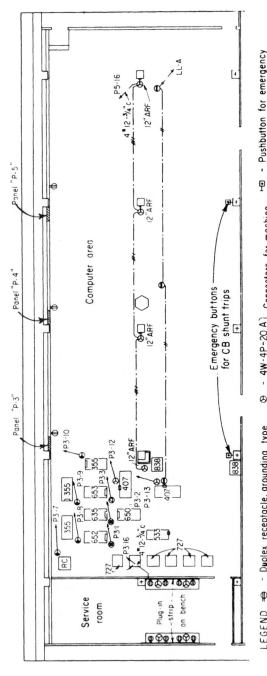

FIGURE 2.142 Floor-plan layout specifies installed location of a wide range of different receptacles for computers.

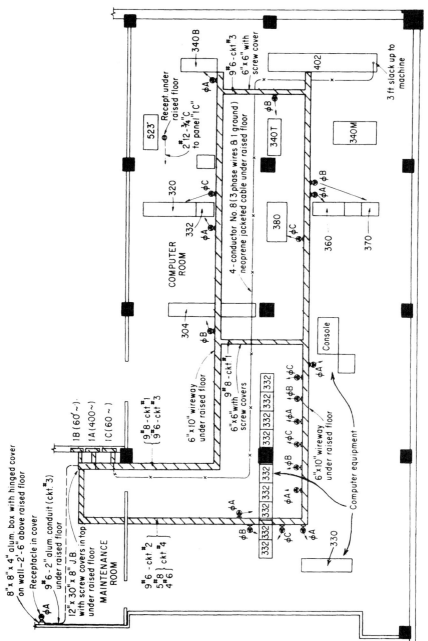

FIGURE 2.143 This kind of layout ensures that the complex 60- and 400-Hz system will be installed correctly.

CHAPTER 3
MOTOR BRANCH CIRCUITS AND CONTROL CIRCUITS

Motor loads vary widely in size and electrical characteristics, but all motor circuits require careful wiring and the protection of conductors and equipment to ensure safe and reliable operation. In any plant or building, of course, the problem of providing maximum safety and reliability must be solved along with other problems—minimizing voltage drop, avoiding excessive copper loss, providing sufficient flexibility for changes in the locations of equipment, designing for ease and economy of maintenance of the motors and equipment, and providing spare capacity for future load increases.

Although the effective application of motor controllers is based primarily on thorough engineering analysis, careful consideration should also be given to the **NEC**, which sets forth minimum safety provisions for the control of motors.

In no way is the **NEC** a substitute for the intelligent design of motor-control circuits suited to the particular characteristics of each individual application. However, because it does represent the accumulation of years of experience with motor circuits, it presents an excellent general outline of motor-circuit design. Within this basic framework, designers can add specific equipment features and circuit techniques to meet their own needs.

The minimum requirements that must be observed in designing branch circuits for motor loads are set out in two **NEC** articles which are directed specifically to motor applications:

1. Article 430 of the **NEC** covers the application and installation of motor circuits and motor-control hookups, including conductors, short-circuit and ground-fault protection, starters, disconnects, and running overload protection.

2. Article 440, "Air-Conditioning and Refrigerating Equipment," contains provisions for such motor-driven equipment and for branch circuits and controllers for this equipment, taking into account the special considerations involved with the use of sealed (hermetic-type) motor-compressors, in which the motor operates under the cooling effect of the refrigeration.

It must be noted that the rules of Article 440 apply in addition to or are amendments to the rules given in Article 430 for motors in general. The basic rules of Article 430 apply to air-conditioning and refrigerating equipment unless exceptions are indicated in Article 440. Article 440 further clarifies the application of **NEC** rules to air-conditioning and refrigeration equipment as follows:

1. Air-conditioning and refrigerating equipment which does not incorporate a sealed (hermetic-type) motor-compressor must satisfy the rules of Article 422 (appliances), Article 424 (space-heating equipment), or Article 430 (conventional motors), as they apply. For instance, where refrigeration compressors are driven by conventional motors, the motors and controls are subject to Article 430, not Article 440. Furnaces with air-conditioning evaporator coils installed must satisfy Article 424. Other equipment in which the motor is not a sealed compressor and which must be covered by Article 422, 424, or 430 includes fan-coil units, remote forced-air-cooled condensers, remote commercial refrigerators, and similar equipment.
2. Room air conditioners are covered in Part **G** of Article 440 (Sections 440-60 through 440-64 but must also comply with the rules of Article 422).
3. Household refrigerators and freezers, drinking-water coolers, and beverage dispensers are considered by the **NEC** to be appliances; their application must comply with Article 422 and must satisfy the rules of Article 440, because such devices contain sealed motor-compressors.

Figure 3.1 shows the six basic elements which the **NEC** requires the designer to account for in any motor circuit. Although these elements are shown separately here, in certain cases the **NEC** permits a single device to serve more than one function. For instance, in some cases, one switch can serve as both disconnecting means and controller. In other cases, short-circuit protection and overload protection can be combined in a single circuit breaker (CB) or set of fuses.

Throughout this discussion of motor and control circuits, it should be clearly understood that the **NEC** rules provide only the foundation for modern design. Many equipment applications, circuit layouts, and features must be incorporated in the design, over and above **NEC** rules, to provide convenience, flexibility, and fulfillment of the objectives of a specific installation.

Motor-Circuit Conductors

The basic **NEC** rule says that the conductors carrying load current to a single-speed motor used for continuous duty must have a current-carrying capacity of not less than 125 percent of the motor full-load current rating. In the case of a multispeed motor, the selection of branch-circuit conductors on the line side of the controller must be based on the highest of the full-load current ratings shown on the motor nameplate. The selection of branch-circuit conductors between the controller and the motor, which are energized for a particular speed, must be based on the current rating for that speed.

Figure 3.2 shows the sizing of branch-circuit conductors to four different motors fed from a panel. The full-load current for each motor is taken from **NEC** Table 430-150. Running overload protection is sized on the basis that nameplate values of motor full-load currents are the same as the values in Table 430-150. If the nameplate and table values are not the same, overload protection is *still* sized according to the nameplate value, *not* a table (Tables 430-147 through 430-150). The conductor sizes shown are for copper. The given current values and **NEC** Table 310-16 may be used to select the correct sizes of aluminum conductors. (Figure 3.2 also shows the

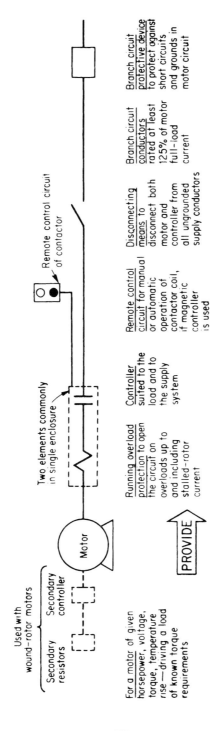

FIGURE 3.1 Design of any motor branch circuit must account for each of these elements.

sizing of branch-circuit protection and running overload protection, as discussed below. Refer to **NEC** Table 430-150 for motor full-load currents, and Table 430-152 for maximum ratings of protective devices.)

Conductors supplying two or more motors must have a current rating of not less than 125 percent of the full-load current rating of the largest motor supplied plus the sum of the full-load current ratings of all other motors supplied.

It is important to note that these are minimum conductor ratings based on temperature rise only; they do not take into account voltage drop or power loss in the conductors. Such considerations frequently require an increase in the size of the branch-circuit conductors.

Section 430-22(a) includes requirements for sizing individual branch-circuit wires serving motors used for short-time, intermittent, periodic, or other varying duty. For such types of duty, the frequency of starting and duration of operating cycles impose varying heat loads on conductors. Conductor sizing, therefore, varies with the application. But it should be noted that any motor is considered to be wired for continuous duty unless the nature of the apparatus that it drives is such that the motor cannot operate continuously with load under any condition of use.

Conductors connecting the secondary of a wound-rotor induction motor to the controller must have a carrying capacity at least equal to 125 percent of the motor's full-load secondary current if the motor is used for continuous duty. If the motor is used for less than continuous duty, the conductors must have a capacity not lower than the percentage of full-load secondary nameplate current given in **NEC** Table 430-22(a). Conductors from the controller of a wound-rotor induction motor to its starting resistors must have an ampacity in accordance with Table 430-23(c) (Fig. 3.3).

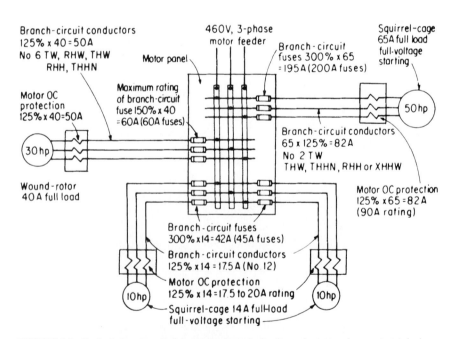

FIGURE 3.2 Basic design steps in sizing motor branch-circuit conductors and overcurrent devices.

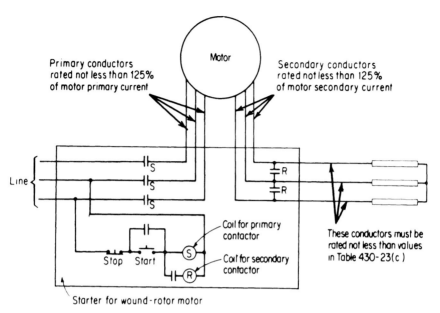

FIGURE 3.3 Circuit design for wound-rotor motors must include the sizing of both primary and secondary conductors.

What Current to Use? The **NEC** contains definite provisions for determining current-carrying capacities:

Section 430-6. For general motor applications (excluding applications of torque motors and sealed hermetic-type refrigeration motor-compressors), whenever the current rating of a motor is used to determine the current-carrying capacities of conductors, switches, fuses, or circuit breakers, the values given in Tables 430-147 through 430-150 must be used instead of the current rating marked on the motor nameplate. However, the selection of separate motor running overload protection must be based on the actual motor *nameplate current rating.*

For torque motors, shaded-pole motors, permanent split-capacitor motors, and ac adjustable-voltage motors, refer to Section 430-6.

Section 440-6. For sealed (hermetic-type) refrigeration motor-compressors, the actual nameplate full-load running current of the motor must be used in determining the current rating of the disconnecting means, controller, branch-circuit conductors, short-circuit and ground-fault protective devices, and motor running overload protection. When such equipment is marked with a branch-circuit selection current, it must be used instead of the rated-load current to determine the rating or ampacity of the disconnecting means, branch-circuit conductors, controller, and branch-circuit, short-circuit, and ground-fault protection. But the nameplate rated-load current must still be used for sizing separate running overload protection.

Example. The circuit components for a sealed hermetic motor-compressor with a nameplate current rating of 10 A are sized as follows:

1. The full-load current used for all calculations for a sealed (hermetic) motor-compressor is the current marked on the unit's nameplate [**NEC** Section 440-6(a)].
2. The branch-circuit conductors must have a rating at least equal to 125 percent of 10 A or 13 A (**NEC** Section 440-32).
3. Conductors for the branch circuit must, therefore, be not smaller than No. 14 copper or No. 12 aluminum (Type TW, THW, RHW, RHH, THHN or XHHW).
4. Running overload protection from overload relays in the starter must have a rating or setting not over 140 percent of the full-load (nameplate) current rating. This calls for an overload relay set to trip at 1.4×10 A, or 14 A [**NEC** Section 440-52(a)].
5. The maximum fuse rating for short-circuit protection for a hermetic motor-compressor is 175 percent of the motor nameplate full-load current rating, which is 1.75×10 A, or 17.5 A. Section 440-22 does not indicate that the next higher standard rating of fuse may be used, but it does permit up to 225 percent of the motor current where needed for starting of the motor. That would permit the use of 20-A fuses.

Short-Circuit Protection

The **NEC** requires that the branch-circuit protection for motor circuits must protect the circuit conductors, the control apparatus, and the motor itself against overcurrent due to short circuits or grounds (Sections 430-51 through 430-58).

The first, and obviously necessary, rule is that the branch-circuit protective device for an individual branch circuit to a motor must be capable of carrying the motor-starting current without opening the circuit. Given this condition, the **NEC** places maximum values on the ratings or settings of such overcurrent devices. It says that such devices must not be rated in excess of values determined from Table 430-152. If such values do not correspond to standard sizes or ratings of fuses, nonadjustable CBs, or thermal devices or to possible settings of adjustable CBs adequate to carry the load, the next higher size, rating, or setting may be used. Where absolutely necessary for motor starting, the device may be rated as follows:

1. The rating of a non-time-delay fuse not exceeding 600 A may be increased but may in no case exceed 400 percent of the full-load current.
2. The rating of a time-delay (dual-element) fuse may be increased but must never exceed 225 percent of the full-load current.
3. The setting of an instantaneous-trip circuit breaker (without time delay) may be increased, but never over 1300 percent of the motor full-load current.
4. The rating of an inverse-time (time-delay) circuit breaker may be increased but must not exceed 400 percent of a full-load current of 100 A or less and must not exceed 300 percent of a full-load current over 100 A.
5. The rating of a fuse rated 601 to 6000 A may be increased but must not exceed 300 percent of the full-load current.
6. Torque motors must be protected at the motor nameplate current rating; if a standard overcurrent device is not made in that rating, the next higher standard rating of protective device may be used but only up to 800 A. Or, if each winding is provided with individual overload protection, the rating or setting may be based on the largest winding, and a single device may be used as short-circuit and ground-fault protection.

For a multispeed motor, a single short-circuit and ground-fault protective device may be used for one or more windings of the motor, provided the rating of the protective device does not exceed the above applicable percentage of the nameplate rating of the smallest winding protected.

The **NEC** establishes maximum values for branch-circuit protection, setting the limit of safe application. However, the use of branch-circuit protective devices of smaller sizes is obviously permitted, and this does offer the opportunity for substantial economy in the selection of circuit breakers, fuses and the switches used with them, panelboards, etc. However, a branch-circuit device with a lower rating than the maximum permitted rating must have sufficient time delay in operation to permit the motor starting current to flow without opening the circuit. But a circuit breaker for branch-circuit protection must have a continuous current rating of not less than 115 percent of the motor full-load current. (See Fig. 3.2 for the sizing of overcurrent protection in accordance with Table 430-152.)

Where maximum ratings for the branch-circuit protection are shown in the manufacturer's heater table for use with a marked controller or are otherwise marked with the equipment, they must not be exceeded even though higher values are indicated in **NEC** Table 430-152 and in the other rules of this section. That requirement is in the last sentence of this **NEC** rule and is also specified in UL regulations which regulate the exposure of motor controllers to short-circuit currents to protect internal components, such as overload relays and contacts, from damage or destruction (Fig. 3.4). Those rules state:

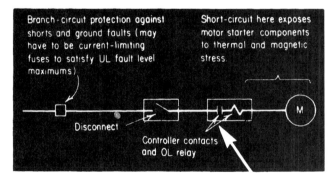

AVAILABLE SHORT-CIRCUIT CURRENT HERE MUST NOT EXCEED VALUES GIVEN BY UL OR MUST BE LIMITED TO THOSE VALUES

FIGURE 3.4 UL specifies maximum short-circuit withstand ratings for controllers.

Motor Controllers (NJOT).

This listing covers the following devices rated 600 V or less, and those rated 701–1500 V:

Auxiliary Devices
Combination Motor Controllers
Float- and Pressure-Operated Motor Controllers
Magnetic Motor Controllers

Manual Motor Controllers
Miscellaneous Motor Controllers
Power Conversion Equipment

Motor Controllers incorporating thermal cutouts, thermal overload relays, or other devices for motor-running overcurrent protection are considered to be suitably protected against overcurrent due to short circuits or grounds by motor branch-circuit, short-circuit, and ground-fault protective devices selected in accordance with the **NEC** and any additional information marked on the product. Motor Controllers may specify that protection is to be provided by fuses or by an inverse time circuit breaker. If there is no marking on protective device type, controllers are considered suitably protected by either type of device. Motor Controllers may specify a maximum rating of protective device. If not marked with a rating, the controllers are considered suitably protected by a protective device of the maximum rating permitted by the **NEC**.

Unless otherwise marked, Motor Controllers incorporating thermal cutouts or overload relays are considered suitable for use on circuits having available fault currents not greater than:

Ratings HP (600V Max)	Full Load Current, Amperes (701–1500V)	RMS Symmetrical Amperes
1 or less	—	1,000
1½ to 50	0–50	5,000
51 to 200	51–200	10,000
201 to 400	201–400	18,000
401 to 600	401–600	30,000
601 to 900	601–850	42,000
901 to 1600	851–1500	85,000

Motor Controllers which are marked "Suitable For Use On A Circuit Capable Of Delivering Not More Than ___ RMS Symmetrical Amperes, ___ Volts Maximum" have been investigated for the additional rating indicated.

Motor Controllers for group installations are marked with a maximum rating of fuse which is considered to suitably protect the controller for the group installation. Such fuse ratings may be in excess of the values given above.

Controllers for Electric Motor Drive Fire Pumps are listed in the Fire Protection Equipment List under the Pump Controller section.

Instantaneous-Trip CBs. The **NEC** recognizes the use of an instantaneous-trip CB (without time delay) for short-circuit protection of motor circuits. Such breakers— also called "magnetic-only" breakers—may be used only if they are adjustable and if combined with motor starters in combination assemblies. An instantaneous-trip CB or a motor short-circuit protector (MSCP) may be used *only* as a part of a "listed" (such as by UL) combination motor controller. A combination motor starter using an instantaneous trip breaker must have running overload protection in each conductor (Fig. 3.5). Such a combination starter offers use of a smaller CB than would be possible if a standard thermal-magnetic CB were used. And the smaller CB offers faster operation for greater protection against grounds and short circuits—in addition to offering greater economy.

A combination motor starter, as shown in Fig. 3.5, is based on the characteristics of the instantaneous-trip CB, which is covered by the third percent column from the left in the **NEC** Table 430-152. Molded-case CBs with only magnetic instantaneous-trip elements in them are available in almost all sizes. Use of such a device requires careful accounting for the absence of overload protection in the CB, up to the short-

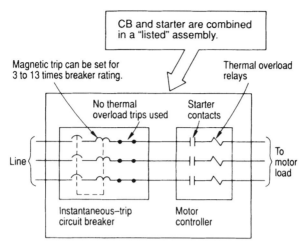

FIGURE 3.5 Section 430-52 accepts use of a magnetic-only circuit breaker if it is part of a "listed" assembly of a combination starter.

circuit trip setting. Such a CB is designed for use as shown in Fig. 3.5. The circuit conductors are sized for at least 125 percent of motor current. The thermal overload relays in the starter protect the entire circuit and all equipment against operating overloads up to and including stalled rotor current. They are commonly set at 125 percent of motor current. In such a circuit, a CB with an adjustable magnetic-trip element can be set to take over the interrupting task at currents above stalled rotor and up to the short-circuit duty of the supply system at that point of installation. The magnetic trip in a typical unit might be adjustable from 3 to 13 times the breaker current rating; i.e., a 100-A trip can be adjusted to trip anywhere between 300 and 1300 A. Thus the CB serves as motor-circuit disconnect and short-circuit protection.

Selection of such a listed assembly with an instantaneous-only CB is based on choosing a nominal CB size with a current rating at least equal to 115 percent of the motor full-load current to carry the motor current and to qualify under Sections 430-58 and 430-110(a) as a disconnect means. Then the adjustable magnetic trip is set to provide the short-circuit protection—the value of current at which instantaneous circuit opening takes place, which should be just above the starting current of the motor involved—using a multiplier of something like 1.5 on locked-rotor current to account for asymmetry in starting current. Asymmetry can occur when the circuit to the motor is closed at that point on the alternating voltage wave where the inrush starting current is going through the negative maximum value of its alternating wave. That is the same concept as asymmetry in the initiation of a short-circuit current.

Listed equipment using an instantaneous CB type is available with very simple instructions by the manufacturer to make proper selection and adjustment of the instantaneous-trip CB combination starter a quick, easy matter. The following describes the concept behind the application of combination starters with instantaneous-only CBs.

Example. A 30-hp, 230-V, 3-phase squirrel-cage motor is marked with the code letter *M*, indicating that the motor has a locked-rotor current of 10 to 11.19 kVA/hp. A full-voltage controller is also provided, with running overload protection in the controller to protect the motor within its heating-damage curve on overload. Select a circuit breaker which will provide short-circuit protection and will qualify as the motor-circuit disconnect means.

Solution. The motor has a full-load current of 80 A (Table 430-150). A CB suitable for use as the disconnect must have a current rating of at least 115 percent of 80 A. Table 430-152 permits the use of an inverse-time (the usual thermal-magnetic) CB rated not more than 250 percent of the motor full-load current (although a CB could be rated as high as 400 percent of full-load current if such size were necessary to pass motor starting current without opening). Based on 2.5 × 80 A or 200 A, a 225-A frame size with 200-A trip setting could be selected. This CB is large enough to generally take the starting current of the motor without tripping either the thermal element or the magnetic element in the CB. The starting current of the motor will initially be about 882 A [30 hp × 11.19 kVA/hp ÷ (220 V × 1.73)]. The instantaneous trip setting of the 200-A CB will be about 200 A × 10, or 2000 A. Such a CB will provide protection for grounds and shorts without interfering with the motor running overload protection.

But consider the use of a 100-A CB with thermal and adjustable magnetic trips. The instantaneous trip setting at 10 times the current rating would be 1000 A, which is above the 882-A locked-rotor current. But starting current would probably trip the thermal element and open the CB. This problem can be solved by removing the CB thermal element and leaving only the magnetic element in the CB. Then conditions of operating overload can be cleared by the running overload devices in the motor starter, right up to stalled-rotor current, with the magnetic trip adjusted to open the circuit instantaneously on currents above, say, 1300 A (882 A × 1.5). But because the value of 1300 A is greater than 1300 percent of the motor full-load current (80 A × 13 = 1040 A), the unnumbered exception in Section 430-52 would prohibit setting the CB at 1300 A. The maximum setting would be 1000 A or, rather, 1040 A.

Because a magnetic-only CB does not protect against low-level grounds and shorts in the circuit conductors on the line side of the running overload relays, this type of CB may be used only where the CB and starter are part of a listed combination starter in a single enclosure.

Motor Short-Circuit Protectors. An MSCP is a fuse-like device designed for use only in its own type of fusible-switch combination motor starter. The combination offers short-circuit protection, running overload protection, disconnect means, and motor control—all with assured coordination between the short-circuit interrupter (the motor short-circuit protector) and the running overload devices. It involves the simplest method of selecting the correct short-circuit protector for a given motor circuit. This packaged assembly is a third type of available combination motor starter, along with the conventional fusible-switch and circuit-breaker types.

The **NEC** recognizes the use of motor short-circuit protectors in Sections 430-40 and 430-52, provided the combination is especially approved for the purpose. Practically speaking, this means a combination starter equipped with motor short-circuit protectors should be listed by UL as a package called an MSCP starter.

One Circuit for Two or More Motors. A single branch circuit may be used to supply two or more motors as follows.

Small motors. Two or more motors, each rated not more than 1 hp and each drawing not over 6 A full-load current, may be used on a branch circuit protected at not more than 20 A at 125 V or less, or 15 A at 600 V or less. The rating of the branch-circuit protective device marked on any of the controllers must not be exceeded. [See Section 430-53(a).] Individual running overload protection is necessary in such circuits, unless the motor (1) is not permanently installed, (2) is manually started and is within sight from the controller location, (3) has sufficient winding impedance to prevent overheating due to stalled-rotor current, (4) is part of an approved assembly which does not subject the motor to overloads and which incorporates protection for the motor against stalled rotor, or (5) the motor cannot operate continuously under load.

Motors of any rating. Two or more motors of any rating, each having individual running overload protection, may be connected to a branch circuit which is protected by a short-circuit protective device selected in accordance with the maximum rating or setting of a device which could protect an individual circuit to the motor of the smallest rating. This may be done only where it can be determined that the branch-circuit device so selected will not open under the most severe normal conditions of service which might be encountered [Section 430-53(b)].

This permission of Section 430-53(b) offers wide application of more than one motor on a single circuit, particularly small, integral-horsepower motors installed on 440-V, 3-phase systems. The application primarily concerns the use of small, integral-horsepower 3-phase motors in 208-, 220-, and 440-V industrial and commercial systems. Only such 3-phase motors have full-load operating currents low enough to permit more than one motor on circuits fed from 15-A protective devices.

There are a number of ways of connecting several motors on a single branch circuit. Figure 3.6 covers the use of more than one motor on a branch circuit, where small integral-horsepower and/or fractional-horsepower motors are used in accordance with Section 430-53(b). Consider the following examples: In case I, with a three-pole CB used as the branch-circuit protective device, application is made in accordance with Section 430-53(b) as follows:

1. The full-load current for each motor is taken from **NEC** Table 430-150 [as required by Section 430-6(a)].

2. A circuit breaker is used instead of fuses for branch-circuit protection. The rating of the branch-circuit protective device, 15 A, does not exceed the maximum short-circuit protection required by Section 430-52 and Table 430-152 for the smallest motor of the group, which is the 1½-hp motor. Although 15 A is greater than the maximum value of 250 percent of the motor full-load current (2.5×2.6 A = 6.5 A) set by Table 430-152, the 15-A breaker is the "next higher size, rating or setting" for a standard circuit breaker as permitted by Section 430-52. A 15-A CB is the smallest standard rating recognized by the **NEC** in Section 240-6.

3. The total load of motor currents is

$$4.8 \text{ A} + 3.4 \text{ A} + 2.6 \text{ A} = 10.8 \text{ A}$$

This is well within the rating of the 15-A CB, which has sufficient time delay in its operation to permit starting of any one of these motors with the other two already operating. The torque characteristics of the loads on starting are not high, and it was determined that the CB will not open under the most severe normal service.

4. Each motor is provided with individual running overload protection in its starter.

5. The branch-circuit conductors are sized in accordance with Section 430-24:

$$4.8 \text{ A} + 3.4 \text{ A} + 2.6 \text{ A} + 0.25 \times 4.8 \text{ A} = 12 \text{ A}$$

CASE I—USING A CIRCUIT BREAKER FOR PROTECTION

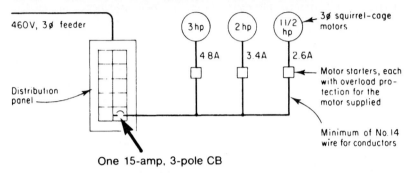

HERE IS THE KEY: A 15-amp, 3-pole CB is used, based on Section 430-52 and Table 430-152. This is the "next higher size" of standard protective device above 250% × 2.6 amps (the required rating for the smallest motor of the group). The 15-amp CB makes this application possible, because the 15-amp CB is the smallest standard rating of CB and is suitable as the branch-circuit protective device for the 1½-hp motor.

CASE II—USING A CIRCUIT BREAKER FOR PROTECTION

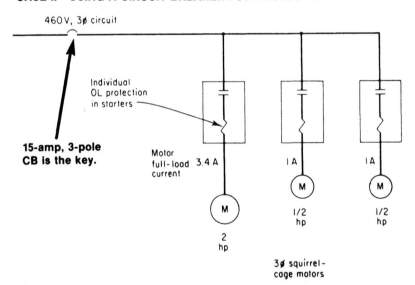

FIGURE 3.6 Branch circuit supplying two or more motors must be carefully designed.

The conductors must have an ampacity at least equal to 12 A, and No. 14 THW, TW, RHW, RHH, THHN, or XHHW conductors will fully satisfy this requirement.

In case II of Fig. 3.6, a similar hookup is used to supply three motors—also with a CB for branch-circuit protection:

1. Section 430-53(b) requires the branch-circuit protection rating to be not higher than the maximum current set by Section 430-52 for the lowest-rated motor of the group.
2. From Section 430-52 and Table 430-152, that maximum protection rating for a circuit breaker is 250 percent of 1 A (the rating of the lowest-rated motor), or 2.5 A. But, 2.5 A is not a "standard rating" of circuit breaker from Section 240-6; Exception No. 1 to Section 430-52 permits use of the next higher size, rating, or setting of standard protective device.
3. Because 15 A is the lowest standard rating of CB, it is the "next higher" device rating above 2.5 A and literally satisfies **NEC** rules on the rating of the branch-circuit protection.

The applications shown in cases I and II permit the use of several motors up to circuit capacity, based on Sections 430-24 and 430-53(b) and on the starting-torque characteristics, the operating duty cycles of the motors and their loads, and the time-delay of the CB. Such applications greatly reduce the number of CB poles, number of panels, and amount of wire used in the system. However, one limitation is placed on this practice in the next to the last sentence of Section 430-52, as follows:

> Where maximum branch-circuit short-circuit and ground-fault protective device ratings are shown in the manufacturer's overload relay table for use with a motor controller or are otherwise marked on the equipment, they shall not be exceeded even if higher values are allowed as shown above.

Therefore, where the starters or other equipment is marked, as is becoming more common, with a maximum rating for the overcurrent protective device and/or the type of device, that rating and/or type of device is required.

In case III, shown in Fig. 3.7, the three motors of case II would be hooked up differently to comply with the rules of Section 430-53(b) if fuses, instead of a circuit breaker, were used for branch-circuit protection:

1. To comply with Section 430-53(b), fuses used as branch-circuit protection must have a rating not in excess of the value permitted by Section 430-52 and Table 430-152 for the smallest motor of the group—here, one of the ½-hp motors.
2. Table 430-152 shows that the maximum permitted rating for non-time-delay fuses is 300 percent of the full-load current for 3-phase squirrel-cage motors. Applying that rule to one of the ½-hp motors gives a maximum fuse rating of 3.0×1 A = 3 A.
3. But there is no permission for the fuses to be rated higher than 3 A, because 3 A is a "standard" rating of fuse (but not a standard rating of circuit breaker). Section 240-6 considers fuses rated at 1, 3, 6, and 10 A to be "standard."
4. The maximum branch-circuit protection permitted by Section 430-53(b) for a ½-hp motor is therefore 3 A.
5. The two ½-hp motors may be fed from a single branch circuit with three 3-A fuses in a three-pole switch.
6. Following the same **NEC** rules, the 2-hp motor would require fuse protection rated not over 10 A (3.0×3.4 A = 10.2 A).

BUT, WATCH OUT!!!

CASE III—USING FUSES FOR CIRCUIT PROTECTION

> Interpretation of NE Code rules of Section 430-53(b) in conjunction with the "standard" ratings of fuses in Section 240-6 may require different circuit makeup when fuses are used to protect the branch circuit to several motors.

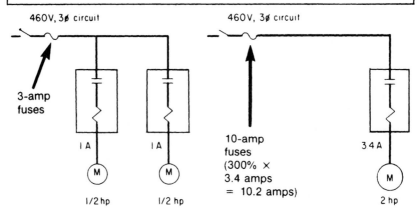

FIGURE 3.7 Fuses, instead of CBs, alter the design characteristics of a multimotor branch circuit.

NOTE: Because there are standard fuse ratings below 15 A, fuses have a different relationship to the applicable **NEC** rules than CBs; interpretation of the rules will be required to resolve the question of acceptable application in case II versus case III. Such interpretation will be necessary to determine if circuit breakers are excluded as circuit protection in those cases where the use of fuses, in accordance with the precise wording of the **NEC**, provides lower-rated protection than circuit breakers when the rule of the third paragraph of Section 430-52 is applied. And if the motors of case I are fed from a circuit protected by fuses, the literal effect of the rules is to require different circuiting for those motors.

Figure 3.8 shows one way of combining cases II and III to satisfy Sections 430-53(b), 430-52, and 240-6; however, the 15-A CB would then be feeder protection, because the fuses would be serving as the "branch-circuit protective devices" as required by Section 430-53(b). Those fuses might be acceptable in each starter, without a disconnect switch, in accordance with Section 240-40, which allows the use of cartridge fuses at any voltage without an individual disconnect for each set of fuses, provided only qualified persons have access to the fuses. But Section 430-112 would have to be satisfied if the single CB were to be used as a disconnect for the group of motors. And part *b* of the exception to that section recognizes one common disconnect in accordance with Section 430-53(a) but not with Section 430-53(b). Certainly, the use of a fusible-switch-type combination starter for each motor would fully satisfy all rules.

Figure 3.9 shows another hookup of several motors on one branch circuit—an actual job installation which was based on application of Section 430-53(b). The installation was studied as follows.

MOTOR BRANCH CIRCUITS AND CONTROL CIRCUITS 187

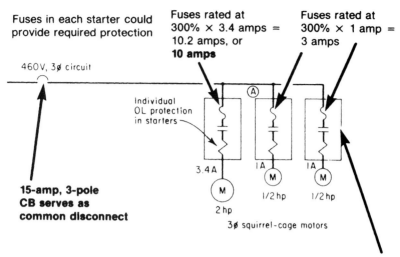

Fuses in each starter could provide required protection

460V, 3ø circuit

Individual OL protection in starters

Fuses rated at 300% × 3.4 amps = 10.2 amps, or **10 amps**

Fuses rated at 300% × 1 amp = 3 amps

3.4A

1A

1A

M 2 hp

M 1/2 hp

M 1/2 hp

3ø squirrel-cage motors

15-amp, 3-pole CB serves as common disconnect

Fuses without individual disconnects might be acceptable under Section 240-40; or a single disconnect switch, fused at 3 amps, could be installed at point "A," eliminating the need for fuses in the two starters for the ½-hp motors.

... AND A HOOKUP LIKE THIS MIGHT BE REQUIRED TO MAKE CASE 1 OF FIG. 3.6 COMPLY WITH NE CODE RULES

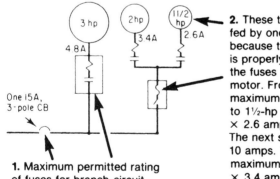

One 15A, 3-pole CB

3 hp 4.8A

2 hp 3.4A

(1½ hp) 2.6A

1. Maximum permitted rating of fuses for branch-circuit protection to this motor would be 300% × 4.8 amps = 14.4 amps, or 15-amp fuses. The 15-amp CB, therefore, satisfies.

2. These two motors may be fed by one branch circuit because the smaller motor is properly protected by the fuses sized for the 2-hp motor. From Section 430-53(b): maximum fuse rating for circuit to 1½-hp motor is 300% × 2.6 amps = 7.8 amps. The next standard size of fuse is 10 amps. That value is within the maximum rating of 300% × 3.4 amps, or 10.2 amps, for the 2-hp motor

FIGURE 3.8 Other designs for multimotor circuits.

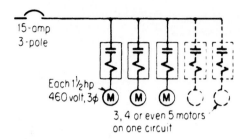

FIGURE 3.9 Economy often dictates the use of multiple fractional-horsepower or small integral-horsepower motors on a single circuit.

Example. A factory has one hundred 1½-hp, 3-phase motors with individual motor starters incorporating overcurrent protection, rated for 460 V. Provide circuits.

Solution. Prior to 1965, the **NEC** would not permit several motors on one branch circuit fed from a three-pole CB. Each of the 100 motors would have had to have its own individual 3-phase circuit fed from a 15-amp, three-pole CB in a panel. As a result, a total of 300 CB poles would have been required, calling for seven panels of 42 circuits each plus a smaller panel (or special panels of more than 42 poles per panel).

Under present rules, depending upon the starting-torque characteristics and operating duty of the motors and their loads, with each motor rated for 2.6 A, three or four motors could be connected on each 3-phase, 15-A circuit; this greatly reduces the number of panelboards and overcurrent devices and the amount of wire involved in the system. The time delay of the CB influences the number of motors on each circuit. However, an extremely important requirement is given in the paragraph above the FPN in Section 430-52, which says:

> Where maximum branch-circuit short-circuit and ground-fault protective device ratings are shown in the manufacturer's overload relay table for use with a motor controller or are otherwise marked on the equipment, they shall not be exceeded even if higher values are allowed as shown above.

Because Section 430-53(b) includes the condition that "the branch-circuit short-circuit and ground-fault protective device is selected not to exceed that allowed by Section 430-52," it is clear that all of Section 430-52 (including the sentence quoted above) must be fully satisfied (Fig. 3.10).

Motors for "Group Installation." Two or more motors of any rating may be connected to one branch circuit if each motor has running overload protection, if the overload devices and controllers are approved for group installation, and if the branch-circuit fuse or time-delay CB rating is in accordance with Section 430-52 for the largest motor plus the sum of the full-load current ratings of the other motors (Fig. 3.11). The branch-circuit fuses or circuit breaker must not be larger than the rating or setting of short-circuit protection permitted by Section 430-52 for the smallest motor of the group, unless the thermal device is approved for group installation with a given maximum size of fuse or time-delay CB for short-circuit protection. [See Sections 430-53(c) and 430-40.]

Tap Conductors. For installations of groups of motors as covered above, tap conductors run from the branch-circuit conductors to supply individual motors must be sized properly. Such tap conductors would, of course, be acceptable where they are the same size as the branch-circuit conductors themselves. However, tap con-

MOTOR BRANCH CIRCUITS AND CONTROL CIRCUITS **189**

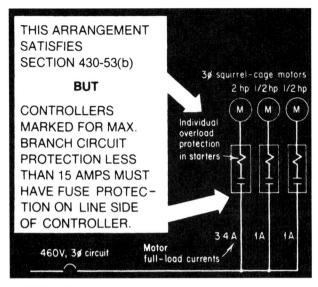

FIGURE 3.10 Branch-circuit protection must not exceed marked maximum value.

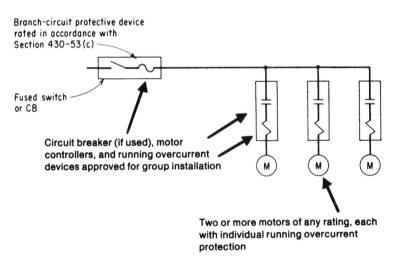

FIGURE 3.11 Motors of any horsepower rating require circuit equipment for group installation.

ductors to a single motor may be smaller than the main branch-circuit conductors provided that they have an ampacity at least ½ that of the branch-circuit conductors, their ampacity is not less than 125 percent of the motor full-load current, they are not over 25 ft long, and they are in raceway or are otherwise protected from physical damage (Fig. 3.12).

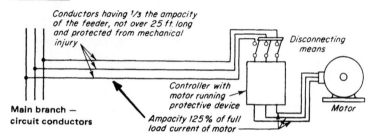

FIGURE 3.12 Overcurrent protection not required for taps to single motors of a group.

The principle applied here is that, since the conductors are short and protected from physical damage, it is unlikely that trouble will occur in the run between the mains and the motor protection which will cause the conductors to be overloaded, except some accident resulting in an actual short circuit. A short circuit will blow the fuses or trip the CB protecting the mains. An overload on the conductors caused by overloading the motor or trouble in the motor itself will cause the motor protective device to operate and so protect the conductors.

Air Conditioning and Refrigeration

NEC Section 440-22(a) covers the rating or setting of the branch-circuit short-circuit and ground-fault protective device for a circuit to an individually sealed hermetic motor-compressor. The rule says that the device "shall be capable of carrying the starting current of the motor." The required protection is considered to be obtained when the device has a rating or setting not exceeding 175 percent of the compressor rated-load current or branch-circuit selection current, whichever is greater (with a 15-A size minimum); however, where the specified protection is not sufficient for the starting current of the motor, it may be increased but may not exceed 225 percent of the motor rated-load current or branch-circuit selection current, whichever is greater.

Section 440-22(b) covers sizing of the short-circuit and ground-fault protective device for a branch circuit to equipment which incorporates more than one sealed hermetic motor-compressor or one sealed motor-compressor and other motors or other loads. This extensive coverage of branch-circuit protection for such complicated motor loads should be carefully studied to ensure effective compliance with the **NEC** on such work. In all such cases, where more than one motor is supplied by a single branch circuit, the rules of Section 430-53 ("Several Motors or Loads on One Branch Circuit") must be applied.

Section 440-62(a) points out that a room air conditioner must be treated as a single motor unit in determining its branch-circuit requirements when all the following conditions are met:

1. The unit is cord-and-plug connected.
2. Its total rating is not more than 40 A and 250 V, single phase.
3. The total rated-load current, rather than individual motor currents, is shown on the unit nameplate.
4. The rating of the branch-circuit, short-circuit, and ground-fault protective device does not exceed the ampacity of the branch-circuit conductors or the rating of the receptacle, whichever is less.

Section 440-60 describes a room air conditioner as an ac appliance of the air-cooled window, console, or in-wall type, with or without provisions for heating, installed in the conditioned room, and incorporating one or more hermetic refrigerant motor-compressors.

Figure 3.13 calls attention to the fact that branch-circuit protection must always be capable of interrupting the amount of short-circuit current which might flow through it. In the diagram, a short-circuit fault at C will draw current until the circuit is opened by the protection at B. The value of the short-circuit current available at C depends upon the apparent-power (kVA) rating of the supply transformer A, percentage reactance of the transformer, secondary voltage, and effective impedance of the current path from the transformer to the point of fault. In application, therefore, motor controllers must be carefully coordinated with the characteristics of the branch-circuit protective device, which must be able to safely interrupt the short-circuit current. And not only must the device be rated to interrupt the fault current, but it must also act quickly enough to open the circuit before let-through current can damage the controller. The speed in clearing the circuit must be compared to the abilities of the various circuit elements to withstand the damaging effects of short-circuit current flow during the time it takes the protective device to operate. Figure 3.14 shows a typical example of the problem that can arise when the operating speed of the branch-circuit overcurrent device does not protect other circuit elements from the effects of let-through current. The damage to the motor controller could have been avoided by use of a faster-opening CB or fast-acting fuses which would limit the fault current to a value that the controller could withstand.

Motor Controllers

As used in the **NEC**, the term "controller" includes any switch or device normally used to start and stop a motor, in addition to motor controllers as such. The basic requirements of Sections 430-81 and 430-83 regarding sizes and types of motor controllers are as follows (refer to Fig. 3.15, where controller types are numbered to correspond to the following list):

1. A *controller* must be capable of starting and stopping the motor which it controls, must be able to interrupt the stalled-rotor current of the motor, and must have a horsepower rating not lower than the rating of the motor.

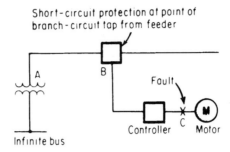

FIGURE 3.13 Branch-circuit protection must be properly rated for the maximum current drawn by a fault anywhere on its load side.

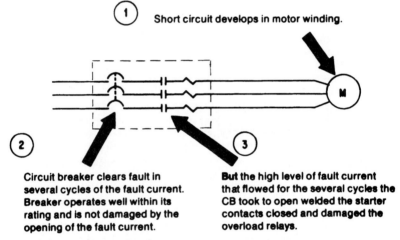

FIGURE 3.14 Effective design can prevent this kind of common fault.

2. The *branch-circuit protective device* may serve as the controller for a stationary motor rated ⅛ hp or less which is normally left running and is not subject to damage from overload or failure to start. (Clock motors are typical of this application.)
3. A *plug-and-receptacle connection* may serve as the controller for portable motors up to ⅓ hp.
4. A *general-use switch* rated at not less than twice the full-load motor current may be used as the controller for stationary motors up to 2 hp, rated 300 V or less. On ac circuits, a general-use snap switch suitable only for use on alternating current may be used to control a motor having a full-load current rating not over 80 percent of the ampere rating of the switch.
5. A *branch-circuit circuit breaker,* rated in amperes only, may be used as a controller. If the same circuit breaker is used both as a controller and to provide overload protection for the motor circuit, it must be rated accordingly.
6. Exception No. 3 of **NEC** Section 430-83 covers controllers for torque motors.

MOTOR BRANCH CIRCUITS AND CONTROL CIRCUITS

THIS IS THE BASIC RULE

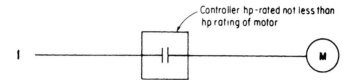

THESE ARE EXCEPTIONS TO THE BASIC RULE

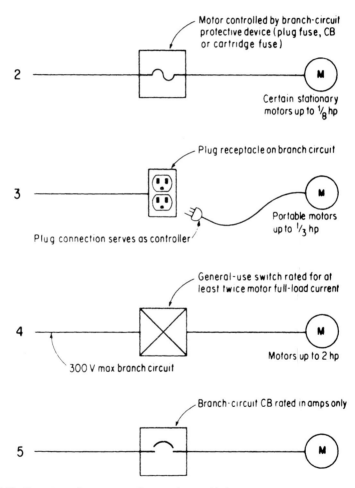

FIGURE 3.15 Some type of motor controller must be provided.

In the UL's *Electrical Construction Materials Directory,* data are presented on use of switches in motor circuits, as follows:

1. Enclosed switches with horsepower ratings in addition to current ratings may be used for motor circuits as well as for general-purpose circuits. Enclosed switches with ampere-only ratings are intended for general use but may also be used for motor circuits (as controllers and/or disconnects) as permitted by **NEC** Section 430-83 (Exception No. 1), Section 430-109 (Exceptions No. 2, 3, and 4), and Section 430-111.
2. A switch that is marked "MOTOR CIRCUIT SWITCH" is intended for use *only* in motor circuits.
3. For switches with dual-horsepower ratings, the higher horsepower rating is based on the use of time-delay fuses in the switch fuseholders to hold in on the inrush current of the higher-horsepower-rated motor.
4. Although Section 430-83 permits use of horsepower-rated switches as controllers and UL lists horsepower-rated switches up to 500 hp, UL does state in its Green Book that "enclosed switches rated higher than 100 hp are restricted to use as motor disconnect means and are not for use as motor controllers." But a horsepower-rated switch up to 100 hp may be used as both a controller and disconnect if it breaks all ungrounded legs to the motor, as covered in Section 430-111.

Figure 3.16 covers two of those points.

For sealed (hermetic-type) refrigeration motor-compressors, selection of the size of controller is slightly more involved than for standard applications. Because of their low-temperature operating conditions, hermetic motors can handle heavier loads than general-purpose motors of equivalent size and rotor-stator construction. And because the capabilities of such motors cannot be accurately defined in terms of horsepower, they are rated in terms of full-load current and locked-rotor current for polyphase motors and larger single-phase motors. Accordingly, the selection of controller size for a hermetic motor is different from that for a general-purpose motor, for which horsepower ratings must be matched.

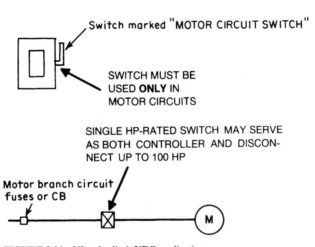

FIGURE 3.16 UL rules limit **NEC** applications.

NEC rules on controllers for motor-compressors are covered in Section 440-41. The controller must have a continuous-duty full-load current rating and a locked-rotor current rating that are not less than the full-load and locked-rotor currents of the motor.

For controllers rated in horsepower, the size required for a particular hermetic motor can be selected after the nameplate rated-load current or branch-circuit selection current (whichever is greater) and locked-rotor current of the motor have been converted to an equivalent horsepower rating. To get this equivalent horsepower rating, which is the required size of controller, the tables in Article 430 must be used. First, the nameplate full-load current of the motor is located in one of Tables 430-148 to 430-150, and the horsepower rating which corresponds to it is noted. Then the nameplate locked-rotor current of the motor is found in Table 430-151, and again the corresponding horsepower is noted. In all tables, if the exact value of current is not listed, the next higher value should be used. If the two horsepower ratings obtained in this way are not the same, the larger value is taken as the required size of controller.

Example. A 230-V, 3-phase, squirrel-cage induction motor in a compressor has a nameplate full-load current of 25.8 A and a nameplate locked-rotor current of 90 A. What size of horsepower-rated controller should be used?

Solution. From Table 430-150, 28 A is the next higher current above the nameplate current of 25.8 A, and the corresponding horsepower rating for a 230-V, 3-phase motor is 10 hp.

From Table 430-151, a locked-rotor current rating of 90 A for a 230-V, 3-phase motor requires a controller rated at 5 hp. The two values of horsepower obtained are not the same, so the higher rating is selected as acceptable for the given conditions. A 10-hp motor controller must be used.

Some controllers may be rated in full-load current and locked-rotor current rather than in horsepower. For use with a hermetic motor, such a controller must have current ratings equal to or greater than the nameplate full-load current and locked-rotor current of the motor.

Starter Poles. It is interesting to note that the **NEC** says that a controller need not open all conductors to a motor, except when the controller serves also as the required disconnecting means. For instance, a two-pole starter could be used for a 3-phase motor if running overload protection is provided in all three circuit legs by devices separate from the starter. The controller must interrupt only enough conductors to start and stop the motor.

However, when the controller is a manual (nonmagnetic) starter or is a manually operated switch or CB (as permitted by the **NEC**), the controller itself may also serve as the disconnect means if it opens all ungrounded conductors to the motor. This eliminates the need for another switch or CB to serve as the disconnecting means. But it should be noted that only a manually operated switch or circuit breaker may serve such a dual function. A magnetic starter cannot also serve as the disconnecting means, even if it does open all ungrounded conductors to the motor. These conditions are shown in Figs. 3.17 and 3.18.

The word "ungrounded" above refers to the condition that none of the circuit conductors is grounded. They may be the ungrounded conductors of grounded systems.

Installed Location. Basically, the **NEC** requires that the motor and its driven machinery be within sight from the controller for the motor. When the controller is out of sight (and the **NEC** considers a distance of 50 ft or more to be equivalent to

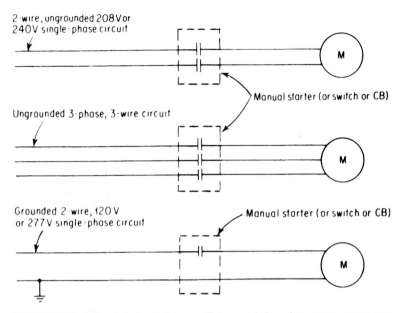

FIGURE 3.17 Manual starter that opens all ungrounded conductors to a motor may serve as *both* controller and disconnect.

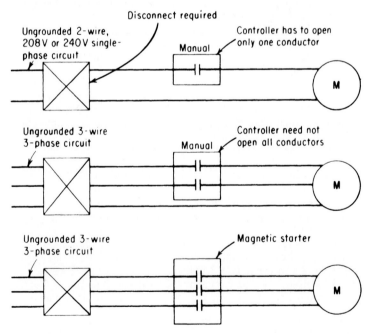

FIGURE 3.18 If a manual starter or switch does not open all ungrounded conductors, a separate disconnect is required.

MOTOR BRANCH CIRCUITS AND CONTROL CIRCUITS **197**

"out of sight," even though the motor and its load might actually be visible from the controller location), the controller must comply with one of the following conditions:

1. The controller disconnecting means must be capable of being locked in the open position, or
2. A manually operable switch, which will provide disconnection of the motor from its power-supply conductors, must be placed within sight from the motor location. And this switch may not be a switch in the control circuit of a magnetic starter.

These requirements are shown in Fig. 3.19. Specific layouts of the two conditions are shown in Fig. 3.20.

NOTE: The **NEC** provisions shown in these sketches are minimum safety requirements. The use of additional disconnects, with or without lock-open means, may be made necessary or desirable by job conditions. Additionally, compliance

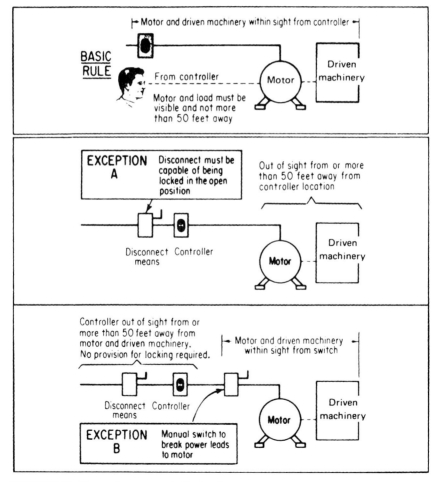

FIGURE 3.19 These rules cover controller location.

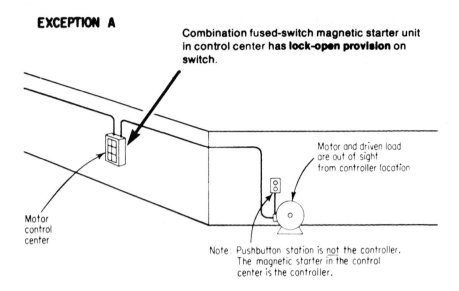

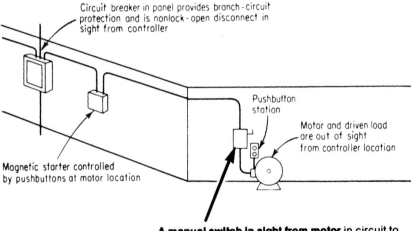

FIGURE 3.20 One of these design measures will permit a controller to be out of sight from its motor load.

with OSHA safe work practices may necessitate installation of lockable disconnects, even where **NEC** rules do *not* mandate them.

The **NEC** at one time permitted a controller to be installed out of sight of the motor and its load if a switch in the coil circuit of the starter was installed within sight of the motor. However, this is no longer acceptable; one of the two conditions shown in Fig. 3.20 must be met.

Generally, an individual motor controller is required for each motor. However, for motors rated not over 600 V, a single controller rated at not less than the sum of the horsepower ratings of all the motors in the group may be used with a group of motors in any one of the following cases:

1. Where a number of motors drive several parts of a single machine or piece of apparatus: metal- and wood-working machines, cranes, etc.

2. Where two or more motors are under the protection of one overcurrent device, as in the case of small motors supplied from a single branch circuit. This use of a single controller applies only to cases involving motors of 1 hp or less, as provided in Section 430-53(a).

3. Where a group of motors is located in one room and all are within sight from the controller location.

On the subject of motor controllers, the **NEC** further requires that speed-limiting devices be used with separately excited dc motors, with series motors, and with motor-generators and converters which can be driven at excessive speed from the direct-current end, as by a reversal of current or decrease in load. Certain exceptions are made in Section 430-89.

Adjustable-speed motors that are controlled by means of field regulation must be designed or equipped and connected so that they cannot be started under a condition of weakened field.

Types of Controllers. Controllers for ac motors can readily be divided into two types according to the condition of starting voltage: full-voltage and reduced-voltage controllers. A *full-voltage* or *across-the-line* controller is one which connects its controlled motor directly to the full value of the motor-circuit voltage. A *reduced-voltage* controller, as the name implies, initially connects the motor to a value of voltage less than that of the supply circuit and then increases the voltage gradually until full circuit voltage is impressed across the motor terminals.

Any polyphase induction motor can be started safely using a full-voltage controller, without doing damage to the motor. Under proper conditions, any size of motor on any voltage can be started at full voltage. However, when the full voltage is impressed, the initial or starting-current surge drawn from the line might be as much as 8 times the normal running current of the motor. The motor itself can handle the current and will start developing rotation; but the driven load may be damaged by the shock of the starting torque of the motor, and severe voltage disturbances may be set up in the distribution system supplying the motor. In such cases, it might be better (or necessary) to start the motor on reduced voltage.

In small and moderate-sized motor applications the torque, speed, and power requirements of the driven load generally permit full-voltage starting without objectionable results.

Across-the-line or full-voltage motor controllers can be divided into two categories according to the manner in which the contacts in the starter are closed or opened. In the manual starter, the contacts are operated by a mechanical linkage from the toggle handle or pushbuttons provided with the unit. In the magnetic starter, the contacts are operated by an electromagnetic coil which is controlled by switching energy to the coil.

The simplest type of manual starting switch is the one- or two-pole fractional-horsepower toggle switch used for infrequent starting and stopping of single-phase motors, up to a maximum of 1 hp at 120 or 240 V. This switch consists of a basic snap-action mechanism which connects the motor to the line in the on position and disconnects it in the off position. To provide running overload protection, the small assembly contains a thermal device to open the circuit on overload, as shown in Fig. 3.21.

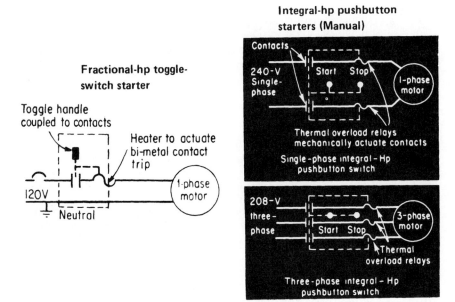

FIGURE 3.21 Manual starters are mechanical switches that include running overload trip devices.

Manual starting switches for use with single-phase and polyphase integral-horsepower motors are across-the-line starters containing electrical contact assemblies which are opened and closed by mechanical action. Manual operation may be effected either by pushbuttons or by a toggle handle, mechanically coupled to the contact assembly and protruding through the starter enclosure. Starters of this type may be used on motor circuits rated up to 600 V ac, for single-phase motors rated up to 5 hp, and for polyphase motors rated up to 7½ hp. Running overload protection is provided by thermal-relay assemblies similar to those used in fractional-horsepower starters.

Although manual starters are equipped with running overload protection, they do not have the form of protection known as *low-voltage* or *undervoltage* protection. As a result, the motor is not protected against the overheating of low-voltage operation, and a power failure or other loss of voltage to the motor circuit will stop the controlled motor but will not cause it to be disconnected from the supply circuit because the controller is mechanically fixed in the closed position. The starter contacts will remain closed, and the motor will restart immediately upon restoration of the circuit voltage. This can be hazardous when machine operators or maintenance personnel who are working on the motor are surprised by the return of power. To avoid such a hazard, manual starters must always be opened on power failure. As will be seen later, magnetic starters can provide "low-voltage protection" and "low-voltage release," which often dictate their use in cases where manual starters could otherwise be used.

The typical magnetic across-the-line starter is generally similar to the manual starter in the construction of the contact assembly; but, instead of requiring mechanical, hand-applied force to open and close the contacts, the magnetic starter uses electromagnetic energy to actuate the contacts. This starter connects the motor to its power-supply conductors at full-line voltage. And the unit is equipped with running overload protection and can provide protection against undervoltage. Magnetic across-the-line starters are made for single-phase and polyphase motors.

The most common type of control switch used for the coil in a magnetic starter is a pushbutton station—a set of start and stop pushbuttons. The pushbutton station is either in the front of the starter enclosure or installed separately from the starter at some remote location and wired to the starter, thereby providing remote control of the starter. Such remote control is one of the major advantages of magnetic starters.

When a magnetic starter is controlled by a remote pushbutton station, the coil circuit is called a *3-wire control circuit*. In such a hookup, three wires run from the starter assembly to the remote pushbutton station.

Other switching devices which may be used to provide remote and automatic control of the starter include float switches, pressure switches, limit switches, thermostats, and control relays. With any of these devices, when the switching contacts close, the coil in the starter is energized. Opening the pilot contacts breaks the circuit to the coil, and the starter disconnects the motor from its supply. Connection of these devices to the starter is made with two wires. Coil circuits which are controlled by on-off automatic pilot devices (float switches, limit switches, etc.) are called *2-wire control circuits*.

Motor running overload protection is a standard provision in magnetic across-the-line starters. Several different types of overload relays are used on the different types and sizes of controllers. In construction and operation, these devices are generally similar to the overload relays in manual starters.

Overload Protection

The **NEC** contains specific requirements regarding motor running overcurrent (overload protection); these are intended to protect the elements of the branch circuit—the motor itself, the motor-control apparatus, and the branch-circuit conductors—against excessive heating due to motor overloads. Such an overload is considered to be an operating overload up to and including stalled-rotor current. When the overload persists for a sufficient length of time, it will cause damage or dangerous overheating of the apparatus. This overload type does not include fault current due to shorts or grounds.

Typical **NEC** requirements for running overcurrent protection, shown in Fig. 3.22, are as follows:

1. Running overload protection must be provided for motors of more than 1 hp, if used for continuous duty. This protection may be an external overcurrent device actuated by the motor running current and set to open at not more than 125 percent of the motor full-load current for motors marked with a service factor of not

1. MOTORS RATED MORE THAN 1 HP

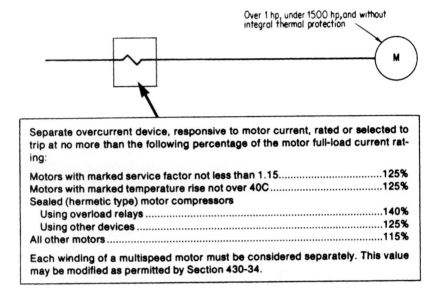

2. MOTORS RATED 1 HP OR LESS

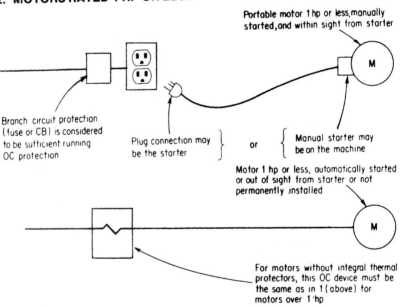

FIGURE 3.22 Circuit design must include some form of protection against overloads, up to locked-rotor current.

less than 1.15 and for motors with a temperature rise not over 40°C. Sealed (hermetic-type) refrigeration motor-compressors must be protected against overload and failures to start, as specified in Section 440-52, by one of the following:
   a. An overload relay set at up to 140 percent of the motor full-load current
   b. An approved, integral thermal protector
   c. A branch-circuit fuse or CB rated at not over 125 percent of the rated-load current
   d. A special protective system
   The overload device must be rated or set to trip at not more than 115 percent of the motor full-load current for all other motors, such as motors with a 1.0 service factor or a 55°C rise.
   Section 430-32 presents a thorough guide to the application of motor and branch-circuit overcurrent (overload) protection for conventional motors of the nonhermetic type. Section 440-52 gives detailed regulations on running overload protection for motor-compressors.
2. Motors of 1 hp or less which are not permanently installed and are manually started are considered protected against overload by the branch-circuit protection if the motors are within sight from the starter. Running overload devices are not required in such cases. A distance of over 50 ft is considered out of sight.
   Any motor of 1 hp or less which is not portable, is not manually started, and/or is not within sight from its starter location must have specific running overload protection. Automatically started motors of 1 hp or less must be protected against running overload in the same way as motors rated over 1 hp. That is, a separate or integral overload device must be used.

Basic **NEC** requirements are concerned with the rating or setting of overcurrent devices separate from motors. However, the rules permit the use of thermal protectors integral with motors, provided that such devices are approved for their particular applications and that they prevent dangerous overheating of the motors. Exceptions to the basic rules on providing specific running overload protective devices are as follows:

1. Where the values specified for motor running overload protection do not permit the motor to start or to carry the load, the next higher size of overload relay may be used, but not higher than the following percentages of motor full-load current rating:

   - Motors with a marked service factor not less than 1.5: 140 percent
   - Motors with a marked temperature rise not over 40°C: 140 percent
   - Sealed (hermetic-type) motor-compressors: 140 percent
   - All other motors: 130 percent

   Fuses or circuit breakers may be used for running overload protection but may not be rated or set up to those values. Fuses and CBs must have a maximum rating as listed above for motors of more than 1 hp. If the value determined as indicated there does not correspond to a standard rating of fuse or CB, the next smaller size must be used. A rating of 125 percent of the full-load current is the absolute maximum for fuses and circuit breakers.

2. Under certain conditions, no specific running overload protection need be used: The motor is considered to be properly protected if it is part of an approved assembly which does not normally subject the motor to overloads and which has controls to protect against stalled rotor. Or, if the impedance of the motor windings is sufficient to prevent overheating due to failure to start, the branch-circuit protection is considered adequate.

3. A motor used for service which is inherently short-time, intermittent, periodic, or varying duty (see **NEC** Table 430-22) is considered as protected against overcurrent by the branch-circuit overcurrent device. A motor is considered to be wired for continuous duty unless the motor cannot operate continuously with load under any condition of use.

Complete data on the required number and location of overcurrent devices are given in Table 430-37. Table 430-37 requires three running overload devices (trip coils, relays, thermal cutouts, etc.) for all 3-phase motors unless protected by other approved means, such as specially designed embedded detectors with or without supplementary external protective devices (Fig. 3.23).

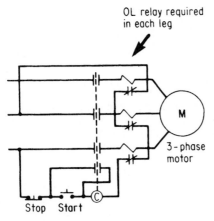

FIGURE 3.23 Three-phase starter must have three overload devices.

The usual acceptable rating of the overload protection in a motor starter is 125 percent of the motor full-load current (or as specifically required in **NEC** Sections 430-32 and 430-34). But when a power-factor capacitor is installed on the load side of the motor starter or at the motor itself, a correction must be made in the rating or setting of the overload device. The device must be rated to take into account the fact that the magnetizing current for the motor is being supplied by the capacitor, and the total current flowing from the supply circuit through the starter is lower than it would be without the power-factor capacitor.

Figure 3.24 explains the need for correcting the size of the running overload protection in motor controllers when power-factor capacitors are used on the load side of the controller, as required by Section 460-9 on the use of capacitors to correct power factor. And, as required by the **NEC**, the rating of the capacitors should not exceed the value required to raise the no-load power factor of the motor to unity. Capacitors of these maximum ratings usually result in a full-load power factor of 95 to 98 percent.

Although the **NEC** contains all those requirements on the running overload protection of motors, Section 430-44 does recognize that there are cases when automatic opening of a motor circuit due to overload may be objectionable from a safety standpoint. In recognition of the circumstances of many industrial applications, Section 430-44 permits alternatives to automatic opening of a circuit in the event of overload. This permission to eliminate overload protection is similar to the permission given in Section 240-12 to eliminate overload protection when automatic opening of the circuit on an overload would constitute a more serious hazard than the overload itself. As the rule notes, "if immediate automatic shutdown of a motor by a motor overload protective device(s) would introduce additional or increased hazard(s) to a person(s) and continued motor operation is necessary for safe shutdown of equipment or process" then automatic overload opening is not required. However, as shown in Fig. 3.25, the circuit must be provided with a motor overload sensing device conforming to the **NEC** requirement on overload protection, to indicate the presence of the overload by means of a supervised alarm. Such overload indication (instead of automatic opening) will alert personnel to the objectionable condi-

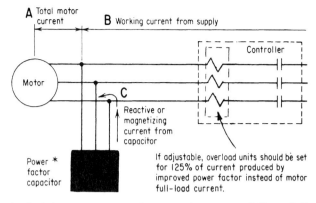

A (total motor current) = vector sum of B and C
where B = in-phase, working current
C = reactive, magnetizing current

EXAMPLE:

A motor with 70% power factor has a full-load current rating of 143 amps. Normally, the OL relay would be set for, say, 125% of 143 or 179 amps. *BUT*, because a PF capacitor is installed at the motor, the OL relay no longer will have 143 amps flowing through it at full load. If the capacitor corrects to 100% PF, the effect will be as follows:

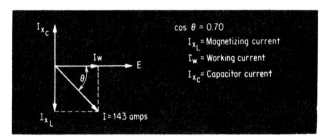

I_{x_c} cancels I_{x_L} at 100% PF, leaving only the working or in-phase current to be supplied from the circuit.

WORKING CURRENT =

TOTAL MOTOR CURRENT × POWER FACTOR

= 143 × 0.70 = 100 AMPS

THAT IS THE CURRENT THAT WILL BE FLOWING THROUGH THE OL RELAY AT FULL LOAD.

THE OL RELAY, THEREFORE, MUST BE SET AT 125% × 100 AMPS = 125 AMPS

FIGURE 3.24 Setting of overload protection must be corrected for reduced line current due to capacitors.

tion and will permit corrective action for an orderly shutdown, either immediately or at some more convenient time, to resolve the difficulty. But, as is required in Section 240-12, short-circuit protection on the motor branch circuit must be provided to protect against those high-level ground faults and short circuits that would be more hazardous than a simple overload.

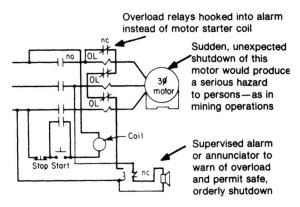

FIGURE 3.25 Overload protection may be eliminated under these conditions.

Motor Disconnect

The **NEC** specifically requires that a disconnecting means—basically, a motor-circuit switch rated in horsepower, or a CB—be provided in each motor circuit to disconnect the motor and its controller from all ungrounded supply conductors. In a motor branch circuit, every switch in the circuit in sight of the controller must satisfy the requirements regarding the type and rating of the disconnect means. And the disconnect switch or CB must be rated to carry at least 115 percent of the nameplate current rating of the motor for circuits up to 600 V. Figure 3.26 sets forth the basic requirements on types of disconnects.

The **NEC** includes a basic requirement that the disconnecting means for a motor and its controller be a motor-circuit switch rated in horsepower. For motors rated up to 500 hp, this rule is readily complied with, inasmuch as the UL lists motor-circuit switches up to 500 hp and manufacturers mark switches to conform. But for motors rated over 100 hp, the **NEC** does not require that the disconnect have a horsepower rating. An exception to the basic rule permits the use of ampere-rated switches or isolation switches, provided the switches have a carrying capacity of at least 115 percent of the nameplate current rating of the motor. The UL notes that a horsepower-rated switch that is rated over 100 hp must not be used as both a disconnect *and* a controller. But switches rated up to 100 hp may serve the dual function of disconnect and controller.

As noted above, although Exception No. 4 to Section 430-109 sets the maximum horsepower rating required for motor-circuit switches at 100 hp, higher-rated switches are now available, will provide additional safety, and should be used to assure adequate interrupting ability.

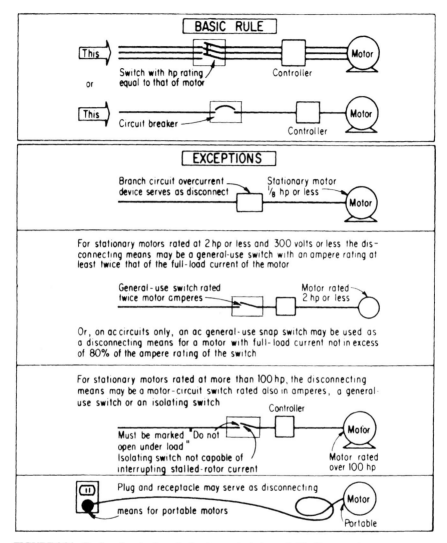

FIGURE 3.26 Design of motor branch circuit must include a suitable disconnecting means.

Isolation switches for motors over 100 hp, if they are not capable of interrupting stalled-rotor currents, must be plainly marked "Do not open under load" (Fig. 3.27).

Example. Provide a disconnect for a 125-hp, 3-phase, 460-V motor. Use a nonfusible switch, inasmuch as short-circuit protection is provided at the supply end of the branch circuit.

Solution. The full-load running current of the motor is 156 A. A suitable disconnect must have a continuous carrying capacity of 156 A × 1.15, or 179 A. This calls for

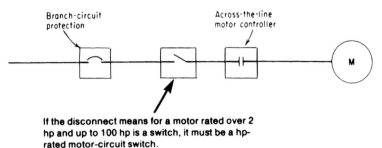

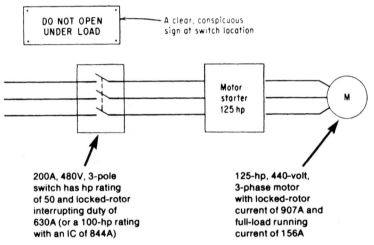

FIGURE 3.27 Motor disconnect selected on the basis of ampere rating must be carefully applied.

a 200-A, three-pole switch rated for 480 V. The switch may be a general-use switch, a current-and-horsepower marked motor-circuit switch, or an isolation switch. A motor-circuit switch with the required current and voltage rating for this case would be marked for 50 hp, but the horsepower rating is of no concern because the switch does not have to be horsepower-rated for motors larger than 100 hp.

If the 50-hp switch were of the heavy-duty type, it would have an interrupting rating of 10×65 A (the full-load current of a 440-V, 50-hp motor), or 650 A. But the locked-rotor current of the 125-hp motor might run as high as 900 A. In such a case, the switch should be marked "Do not open under load."

If a fusible switch had to be provided for this motor for disconnect and short-circuit protection, the size of the switch would be determined by the size and type of fuses used. For a fuse rating of 250 percent of the motor current (which does not exceed the 300 percent maximum in Table 430-152) for standard fuses, the application would call for 400-A fuses in a 400-A switch. This switch would certainly qualify as the motor disconnect. However, if time-delay fuses were used, a 200-A switch would be large enough to take the time-delay fuses and could be used as the disconnect (because it is rated at 115 percent of motor current).

In the foregoing, the 400-A switch might have an interrupting rating high enough to handle the locked-rotor current of the motor. Or the 200-A switch might be of the type that has an interrupting rating up to 12 times the rated load current of the switch itself. In either of these cases, there would probably be no need for marking the switch "Do not open under load."

Up to 100 hp, a switch which satisfies the **NEC** rule on rating for use as a motor controller may also provide the required disconnect means—the two functions being performed by the one switch—provided it opens all ungrounded conductors to the motor, is protected by an overcurrent device (which may be the branch-circuit protection or may be fuses in the switch itself), and is a manually operated air-break switch or an oil switch not rated over 600 V or 100 A.

As described under "Controllers," a manual starting switch capable of starting and stopping a given motor, capable of interrupting the stalled-rotor current of the motor, and having the same horsepower rating as the motor may serve the functions of controller and disconnecting means in many motor circuits, if the switch opens all ungrounded conductors to the motor. A single circuit breaker may also serve as controller and disconnect; this is permitted by Section 430-111. However, an autotransformer-type controller, even if manual, may not also serve as the disconnecting means. Such controllers must be provided with a separate means for disconnecting the controller and motor.

The acceptability of a single switch used as both the controller and disconnecting means is based on the single switch satisfying **NEC** requirements for a controller and for a disconnect. This "combination" finds application where general-use switches or horsepower-rated switches are used, as permitted by the **NEC**, in conjunction with time-delay fuses which are rated low enough to provide both running overload protection and branch-circuit (short-circuit) protection. In such cases, a single fused switch may serve a total of four functions: (1) controller, (2) disconnect, (3) branch-circuit protection, and (4) running overload protection. And it is also possible for a single circuit breaker to serve these four functions.

For sealed refrigeration compressors, Section 440-12 gives the procedure for determining the disconnect rating, based on the nameplate rated-load current or branch-circuit selection current (whichever is greater) and the locked-rotor current of the motor-compressor. As an example, suppose a 3-phase, 460-V hermetic motor rated at 11 A branch-circuit selection current and 60 A locked-rotor current is to be supplied with a disconnect switch rated in horsepower. The first step is to determine the equivalent horsepower rating of the motor by referring to **NEC** Table 430-150. This table lists $7\frac{1}{2}$ hp as the horsepower size of a 460-V, 11-A motor. To ensure adequate interrupting capacity, Table 430-151 is now used. This table shows $7\frac{1}{2}$ hp as the equivalent horsepower rating for any locked-rotor current over 45 A and up to 66 A for a 460-V motor. Both tables thus establish a $7\frac{1}{2}$-hp disconnect as adequate for the given motor. If the ratings obtained from the two tables had been different, the higher rating would have been chosen.

In general, each motor is provided with a separate disconnecting means. However, a single disconnect sometimes may serve a group of motors. Such a disconnect must have a rating sufficient to handle a single load equal to the sum of the horsepower ratings or current ratings of all the motors it serves. The single disconnect may

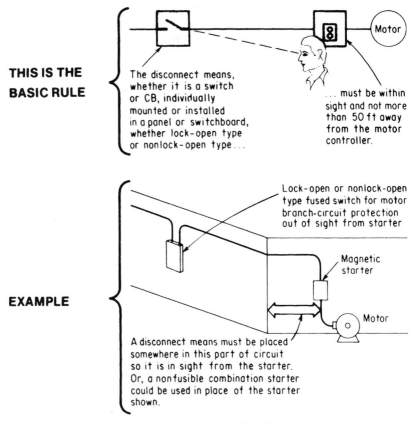

FIGURE 3.28 Basic motor-circuit design includes an "in-sight" disconnect.

be used for a group of motors driving different parts of a single piece of apparatus, for several motors on one branch circuit, or for a group of motors in a single room within sight from the disconnect location (Section 430-112).

An important rule in **NEC** Section 430-102 states simply and clearly that "a disconnecting means shall be located in sight from the controller location." This applies always, for all motor loads rated up to 600 V—even if an "out-of-sight" disconnect can be locked in the open position (Fig. 3.28). There are, however, two exceptions to this rule:

Exception No. 1 permits the disconnect for a high-voltage (over 600 V) motor to be out of sight from the controller location if the controller is marked with a warning label giving the location and identification of the disconnecting means to be locked in the open position.

Exception No. 2 is aimed at permitting practical, realistic disconnect means for large and complex industrial machinery utilizing a number of motors to power various interrelated machine parts. This exception recognizes that a single com-

MOTOR BRANCH CIRCUITS AND CONTROL CIRCUITS 211

mon disconnect for a number of controllers, as permitted by part a of the exception to Section 430-112, often cannot be installed "within sight" of all the controllers, even though the controllers are "adjacent one to the other." On much industrial process equipment, the components of the overall structure obstruct the view of many controllers. Exception No. 2 permits the single disconnect to be technically out of sight from some or even all of the controllers if the disconnect is simply "adjacent" to them—that is, nearby on the equipment structure.

Those two exceptions are shown in Fig. 3.29.

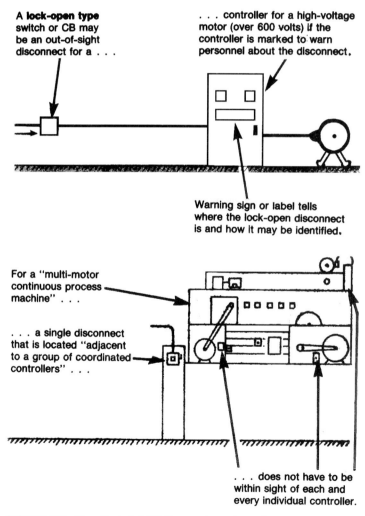

FIGURE 3.29 These "out-of-sight" disconnects may be used.

CONTROL CIRCUITS

A control circuit, as discussed here, is any circuit which has as its load device the operating coil of a magnetic motor starter, a magnetic contactor, or a relay. Strictly speaking, it is a circuit that exercises control over one or more other circuits. And these other circuits may themselves be control circuits, or they may be *load* circuits carrying utilization current to a lighting, heating, power, or signal device. Figure 3.30 clarifies the distinction between control circuits and load circuits.

The elements of a control circuit include all the equipment and devices concerned with the function of the circuit: conductors, raceway, contactor operating coil, source of energy supply to the circuit, overcurrent protective devices, and all switching devices that govern the energizing of the operating coil.

NEC rules on motor-starter control circuits must be evaluated against a number of very important background facts:

1. The **NEC** has no specific rule that covers the ampacity of control-circuit wires with respect to the amount of current that a starter operating coil draws. It simply presumes that the wires of a control circuit will have ampacity sufficient for the coil current.
2. **NEC** rules require the protection of control-circuit wires against current in excess of their ampacity or spell out very specific conditions under which overcurrent devices rated higher than wire ampacity may be used or under which the protection of control wires may be eliminated or must be eliminated.
3. The **NEC** rules on control-circuit overcurrent protection are not concerned with protecting the operating coil itself against overcurrent. Because of that, it is permissible to select the size of control-circuit wires solely in relation to the rating of branch-circuit protection when the use of larger control wires would eliminate the need for a control-circuit fuse block in the starter [i.e., control-wire size selected so that the branch-circuit fuses or CB are not more than 300 percent of the wire ampacity, as in Exception No. 2 to Section 430-72(b)].

The **NEC** covers the application of control circuits in Article 725 and in Sections 240-3 and 430-71 through 430-74.

Section 430-72 covers motor-control circuits that are derived within a motor starter from the branch-circuit wires that connect to the line terminals of the starter. When a control circuit derives its current supply from the same branch circuit that supplies power to the motor (either directly, by being tapped from the line terminals

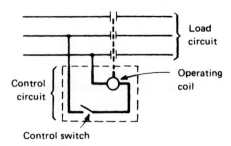

FIGURE 3.30 Control circuit provides on-off operation of load-circuit current.

within the starter, or indirectly by being fed from the secondary of a control transformer with its primary tapped from the starter line terminals and located within the starter enclosure), all the rules of Section 430-72 must be observed. The rules refer to such a control circuit as one "tapped from the load side" of the fuses or circuit breaker that provide branch-circuit protection for the conductors that supply the starter (Fig. 3.31). Motor-control circuits that are *not* tapped from the line terminals within a starter must be protected against overcurrent in accordance with Section 725-12 or 725-35. Such control circuits would be those that are derived from a panelboard or a control transformer—as where, say, 120-V circuits are derived from a source external to the starters and are typically run to provide lower-voltage control for 230-, 460-, or 575-V motors (Fig. 3.32).

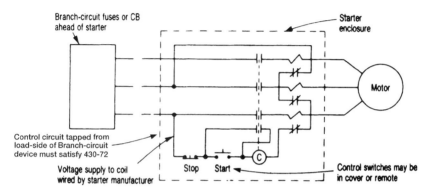

FIGURE 3.31 Control circuit derived within a motor controller must satisfy **NEC** Section 430-72 on overcurrent protection.

Section 725-2(e) states that the remote-control circuit of a motor starter must comply with the rules of Part **F** of Article 430 (Sections 430-71 through 430-74), as well as with any applicable rules of Article 725, in all cases where the control circuit is derived from the line terminals of a starter. When a control circuit for the operating coil of one or more magnetic motor starters is derived from a separate control transformer (one that is not within the starter housing and is not fed from a motor branch circuit), the control circuit or circuits and all components are covered by the rules of Article 725. In the same way, a control circuit that is taken from a panelboard for power supply to the operating coil of one or more motor starters is also covered by Article 725, and not by the rules of Section 430-72.

The design requirements for the coil circuit of a magnetic contactor—as distinguished from a motor starter—are based solely on the rules of Article 725.

The design and installation of control circuits are basically divided into three classes (in Article 725) according to the energy available in the circuit. Class 2 and Class 3 control circuits have low energy-handling capabilities; to qualify as a Class 2 or 3 control circuit, a circuit must have open-circuit voltage and overcurrent protection that is limited to the conditions given in Section 725-31.

Most control circuits for magnetic starters and contactors could not qualify as Class 2 or 3 control circuits because of the relatively high energy required for operating coils. And any control circuit rated over 150 V (such as 220- or 440-V coil circuits) can never qualify, regardless of energy.

CHAPTER THREE

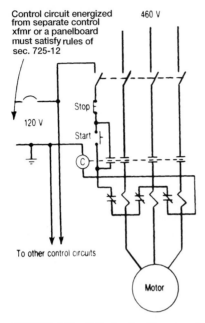

Control circuit energized from separate control xfmr or a panelboard must satisfy rules of sec. 725-12

FIGURE 3.32 Coil circuit with external voltage supply must satisfy Section 725-12 on overcurrent protection.

Class 1 control circuits include all operating-coil circuits for magnetic starters that do not meet the requirements for Class 2 or 3 circuits. Class 1 circuits must be wired in accordance with Sections 725-11 to 725-20. The rules may be summarized as follows:

1. In general, the wiring of Class 1 control circuits must be the same as general-purpose power and lighting wiring.
2. Conductor sizes generally are limited to a minimum of No. 14, but No. 16 or 18 fixture wires may be used if installed in a raceway or approved cable or flexible cord, loaded up to the ampacities shown in Section 402-5. Conductors of sizes No. 18 and 16 are considered as protected by overcurrent devices rated not over 7 and 10 A, respectively.
3. Wires larger than No. 16 must be of Type TW, THW, THHN, or some other approved type. Conductors of size No. 18 or 16 in a raceway or cable must be of Types RFH-2, FFH-2, etc. [See Section 725-16(b).] Other conductors may be used with specific approval for the particular purpose.
4. As specified in **NEC** Section 725-17, the number of Class 1 control-circuit conductors in a conduit must be determined from Tables 1 through 5 in **NEC** Chapter 9. There is a definite limit on the number of wires permitted in a raceway; if these wires carry continuous loads, the derating factors of Note 8 to Tables 310-16 through 310-19 must be applied to their current ratings. But Section 430-72(b)(1) exempts motor-starter coil-circuit wires from such derating.
5. When power-supply conductors and Class 1 remote-control conductors are used in a single conduit or EMT run (as permitted by Section 725-15), the derating factors of Note 8 must be applied as follows:
   a. Note 8 must be applied to all conductors in the conduit when the remote-control conductors carry continuous loads and the total number of conductors (remote-control and power wires) is more than three. In Fig. 3.33a, the conduit size must be selected according to the number and sizes of the wires. Because two of the control wires to the pushbutton and the power wires to the motor will carry a continuous load, a derating factor of 80 percent (from Note 8) must be applied to all the wires in the conduit.
   b. Note 8 must be applied only to the power wires when the remote-control wires do not carry a continuous load and when the number of power wires is more than three. In Fig. 3.33b, no derating at all is applied because the control wires do not carry continuous current (only for the instant of switching operation), and there are only three power wires.
   c. Conductors for two or more Class 1 control circuits (ac or dc or both) may be run in the same raceway if all conductors are insulated for the maximum voltage of any conductor in the raceway. Another permitted installation is shown in Fig. 3.33c.

Control Wires in Raceways

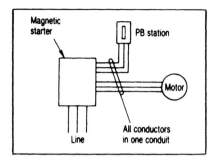

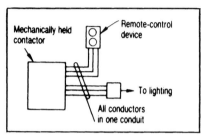

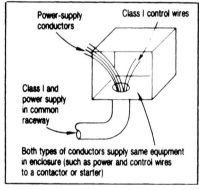

FIGURE 3.33 Design of every contactor or motor starter-coil circuit must satisfy certain basic rules.

Class 1 control wires (starter-coil circuit wires) may be run in raceways by themselves in accordance with the first sentences of Section 725-15. A given conduit, for instance, may carry one or several sets of Class 1 control-circuit wires. And Section 725-14 says that Class 1 circuit wires must conform to the same basic rules from **NEC** Chapter 3 that apply to standard power- and lighting-circuit wiring. Conduit fill must be made in accordance with Section 725-17, which cites the usual rules on number of wires in rigid metal conduit, in IMC, in EMT, etc. When more than three control wires are installed in a raceway, the ampacity derating factors of Note 8 to Tables 310-16 to 310-19 must be applied only if the conductors carry continuous loads (that is, carry their load current continuously for 3 hours or more), as required by Section 725-17. But that applies only to wires of control circuits that are derived external to the starter, as in those cases where the coil circuit is energized from a separate (external) control transformer or from a panelboard. Section 430-72(b), however, applies to coil circuits for starters in which control power is tapped from the starter line terminals; it requires basically that No. 14 wires and larger be protected at their ampacities as given in Tables 310-16 through 310-19, without derating factors.

Elimination of the requirement to derate conductor ampacities (or limit the load) according to Note 8 of Tables 310-16 to 310-19 and Section 725-17(a) has been a controversial point in the past. Although the derating required by Section 725-17 is still part of the 1993 **NEC**, that rule literally applies only to control circuits that are derived from an external source—such as a transformer or panelboard. The elimination of derating factors in Section 430-72(b) applies only to control circuits tapped within starters.

Two sections of the **NEC** cover the placement of Class 1 control-circuit conductors in the same raceway, cable, or enclosure that contains the circuit wires carrying power to the motor windings. Section 300-3 covers the general use of "conductors of different systems" in raceways as well as in cable assemblies and in equipment wiring enclosures, (that is, cabinets, housings, starter enclosures, junction boxes, etc.). But Class 1 control wires are also regulated by the rules of Article 725, covering "remote-control" wiring.

Figures 3.34 through 3.36 show three cases where Class 1 control wires are run in a raceway with the motor power conductors. Although the sketches show specific hookups of equipment, the rules described apply to any hookup or equipment layout. The cases shown are intended only to focus on the placement of control wires in a raceway with power wires—no matter how the hookup is made. The cases are as follows:

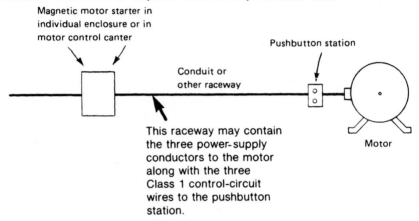

FIGURE 3.34 Control and power wires may always be in the same raceway for a single motor.

Case I. Locating the power and control wires for an individual motor in the same raceway, as here, is acceptable under Section 725-15.

Case II. The wording of Section 300-3(c)(1) recognizes the placement of power and control wires in a single raceway to supply more than one motor, but such usage must conform with the last sentence of Section 725-15. As explained in the accompanying sketches, the two **NEC** rules must be combined carefully.

A common raceway, as shown, may be used only where the two or more motors are required to be operated together to serve their load function. Many industrial and commercial installations have machines, manufacturing operations, or processes in which a number of motors operate in unison to perform the various parts of a task. In such cases, either all the motors operate or none do. The placement of all the control and power wires for such a group of motors in the same raceway would not produce a situation where a fault in one motor circuit could disable another circuit to a motor that might otherwise continue to operate.

But when a common raceway is used for power and control wires to separate, independent motors, a fault in one circuit could knock out all the others—and motors would be shut down unnecessarily. Because motor circuits are so closely associated with vital equipment such as elevators, fans, and pumps in modern buildings, it is a matter of safety to separate such circuits and minimize the chance of a large-scale outage due to a fault in a single circuit. For safety's sake, the **NEC** says, in effect, "Do not put all your eggs in one basket." But, again, the loss of more than one motor because of a single fault is not objectionable where all motors must be shut down when any one is stopped—as in multimotor machines and processes.

MOTOR BRANCH CIRCUITS AND CONTROL CIRCUITS 217

CASE II—SECTION 300-3(c)(1) PERMITS COMMON RACEWAY TO CONTAIN POWER AND CONTROL-CIRCUIT WIRES FOR TWO OR MORE MOTORS

This common raceway (conduit, EMT, wireway, or etc.) may contain **all** power conductors and **all** control conductors for two or more motors . . .

Motors operate together and are functionally associated as integral parts of a machine or process

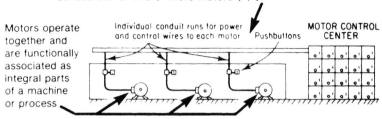

. . . **but**, the intent of this Code rule, along with that of Section 725-15, permits a common raceway **only** where the power and control conductors are for a number of motors that operate integrally—such as a number of motors powering different stages or sections of a multi-motor process or production machine. Such usage complies with Section 725-15 (last sentence), which permits power and control wires in the same raceway, cable, or other enclosure when the equipment powered is "functionally associated"—that is, the motors have to run together to perform their task.

FIGURE 3.35 Specific conditions permit all power and control wires for two or more motors to be installed in a single raceway.

CASE III—BUT THE LETTER AND INTENT OF SECTION 725-15 PROHIBITS INTERMIXING OF POWER AND CLASS 1 (STARTER COIL-CIRCUIT) WIRES WHEN MOTORS ARE NOT "FUNCTIONALLY ASSOCIATED"

Separate conduit or raceway is required for power and control wires to each motor . . .

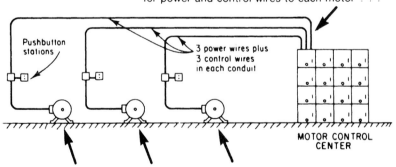

. . . When these are individual motors that do not operate together as parts of a machine or process—that is, each motor has a separate, independent load that may operate by itself.

FIGURE 3.36 Power and control wires must be segregated for motors that are not "associated."

Case III. In cases where each of several motors serves a separate, independent load (with no interconnection of their control circuits and no mechanical interlocking of their driven loads), the use of a separate raceway for each motor is required by the last sentence of Section 725-15—but only when control wires are carried in the raceways. It would be acceptable to run the power conductors for all three motors in a single raceway and all the control-circuit wires in another raceway. Such a hookup would not violate Section 725-15. However, the conductors would have to be derated, and there is a definite chance of losing more than one motor on a fault in only one of the circuits in either the power raceway or the control raceway.

Figure 3.37 shows the details of Section 300-3(c) as they apply to high-voltage motor starters and their control circuits.

SECTION 300-3(c)(2) PROHIBITS THIS—

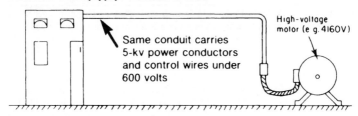

BUT, EXCEPTION NO. 3 PERMITS THIS—

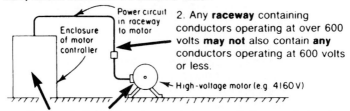

For any **individual** motor or starter—excitation, control, relay and/or ammeter conductors operating at 600 volts or less **may** occupy the same **starter or motor enclosure** as the conductors operating at over 600 volts.

FIGURE 3.37 Control wires for high-voltage starters may be used in the starter enclosure but not in raceway with power conductors.

Overcurrent Protection

In general, remote-control conductors must be protected against overcurrent. Section 430-72 covers overcurrent protection for remote-control conductors that are tapped from the line terminals within a motor starter. Section 725-12 covers overcurrent protection for remote-control conductors of motor starter-coil circuits that are derived from separate control transformers or from panelboards and overcurrent protection for the coil circuits of magnetic contactors (as used for the control of lighting and/or power). Exception No. 3 of Section 725-12 states that remote-control conductors other than those for motor control circuits can be considered to be satisfactorily protected by the branch-circuit overcurrent device if it is rated at not more than 300 percent of

MOTOR BRANCH CIRCUITS AND CONTROL CIRCUITS   219

the carrying capacity of the control-circuit conductors. That applies to control wires for magnetic contactors used for the control of lighting or heating loads, but not for motor loads. Section 430-72 includes a similar requirement for motor control circuits.

For motor starter-coil circuits tapped from line terminals within the starter enclosure, a number of points must be observed.

The basic rule of Section 430-72(b), on "Overcurrent Protection," requires coil-circuit conductors to have overcurrent protection rated in accordance with the maximum values given in column A of Table 430-72(b). That table shows 7 A as the maximum rating of protection for No. 18 copper wire and 10 A for No. 16 wire and refers to Table 310-16 for larger wires—15 A for No. 14 copper, 20 A for No. 12, etc. The Exceptions to the basic rule cover conditions under which other ratings of protection may be used, are as follows:

Exception No. 1 to Section 430-72(b) covers protection of control wires for magnetic starters that have their start-stop buttons in the cover of the starter enclosure.

In Exception No. 1, the value of branch-circuit protection must be compared to the ampacity of the control-circuit wires that are factory-installed in the starter and connected to the start-stop buttons in the cover. If the rating of the branch-circuit fuse or CB does not exceed the value of the current shown in column B of Table 43-72(b) for the particular size of either copper or aluminum wire used to wire the coil circuit within the starter, then other protection is not required to be installed within the starter (Fig. 3.38). If the rating of branch-circuit protection *does* exceed the value shown in column B for the size of coil-circuit wire, then separate protection must be provided within the starter, and it must be rated not greater than the value shown for that size of wire in column A of Table 430-72(b). For instance, if the internal coil circuit of a starter is wired with No. 16 copper wire and the branch-circuit device supplying the starter is rated over the 40-A value shown for No. 16 copper wire in column B of Table 430-72(b), then protection must be provided in the starter for the No. 16 wire, and the protective device(s) must be rated not over the 10-A value shown for No. 16 copper wire in column A of Table 430-72(b).

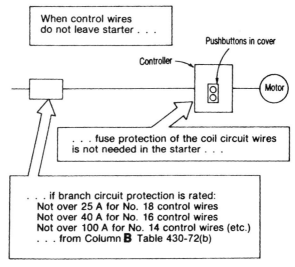

FIGURE 3.38   This is the rule of Exception No. 1 Section 430-72(b).

Because most starters are the smaller ones using No. 18 and No. 16 wires for their coil circuits, Exception No. 1 and its reference to column B are particularly applicable to those wire sizes. For No. 16 control wires, branch-circuit protection rated up to 40 A would eliminate any need for a separate control-circuit fuse in the starter. And for No. 18 control wires, separate coil-circuit protection is not needed for a starter with branch-circuit protection rated not over 25 A. For No. 14, No. 12, and No. 10 copper control wires, maximum protective-device ratings are given in column B as 100, 120, and 160 A, respectively. For conductors larger than No. 10, the protection may be rated up to 400 percent of (or 4 times) the free-air ampacity of the size of conductor from Table 310-17.

Exception No. 2 of Section 430-72(b) covers protection of control wires that run from a starter to a remote-control device (pushbutton station, float switch, limit switch, etc.). Such control wires may be protected by the branch-circuit protective device—without need for separate protection within the starter—if the branch-circuit device has a rating not over the value shown for the particular size of copper or aluminum control wire in column C of Table 430-72(b) (Fig. 3.39). Note that the maximum ratings of 7 A for No. 18 and 10 A for No. 16 require that fuse protection at those ratings must always be used to protect those sizes of control-circuit wires connected to motor starters supplied by CB branch-circuit protection, because 15 A is the lowest available standard rating of CB. But branch-circuit fuses of 7- or 10-A rating could eliminate the need for protection in the starter where No. 18 or No. 16 control wires are used. Figure 3.40 shows an application that was permitted for many years under previous wording of the **NEC** rule but is now contrary to the letter and intent of the rule.

For any size of control wire, if the branch-circuit protection ahead of the starter has a rating greater than the value shown in column C of Table 430-72(b), then the control wire must be protected by a device(s) rated not over the ampere value shown for that size of wire in column A of Table 430-72(b). For instance, if No. 14 copper wire is used for the control circuit from a starter to a remote pushbutton station and the branch-circuit protection ahead of the starter is rated at 40 A, then the branch-circuit device is not over the value of 45 A shown in column C, and separate control protection is not required within the starter. But if the branch-circuit protection were, say, 100 A, then No. 14 control wire would have to be protected at 15 A because column A shows that No. 14 must have maximum protection rating from Note 1—which refers to Table 310-16 where No. 14 wire in conduit is shown, by the footnote, to require protection at 15 A.

It should be noted that column A gives the values to be used for overcurrent protection placed within the starter to protect control-circuit wires in any case where the rating of branch-circuit protection exceeds the value shown in either column B (for starters with no external control wires) or column C (for control wires run from a starter to a remote pilot control device).

Exception No. 3 of Section 430-72(b) permits protection on the primary side of a control transformer to protect the transformer in accordance with Section 450-3 and the secondary conductors in accordance with the ampere value shown in Table 430-72(b) for the particular size of the control wires fed by the secondary. This use is limited to transformers with 2-wire secondaries (Fig. 3.41). Because Section 430-72(a) notes that the rules of Section 430-72 apply to control circuits tapped from the motor branch circuit, the rule of Exception No. 3 must be taken as applying to a control transformer installed within the starter enclosure—although the general application may be used for any transformer because it conforms to Section 240-3, Exception No. 5, and to Section 450-3.

Exception No. 4 eliminates any need for control-circuit protection where opening of the circuit would be objectionable, as for a fire-pump motor or other essential or safety-related operation.

MOTOR BRANCH CIRCUITS AND CONTROL CIRCUITS 221

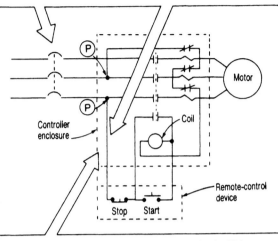

FIGURE 3.39 This is covered by Exception No. 2 to Section 430-72(b).

Section 430-72(c) covers the use of control transformers and requires protection on the primary side. And, again, it must be taken to apply specifically to such transformers used in motor-control equipment enclosures. The basic rule calls for each control transformer to be protected in accordance with Section 450-3 (usually by a primary-side protective device rated not over 125 or 167 percent of primary current, as shown in Fig. 3.41. But, exceptions are given.

Exception No. 1 eliminates any need for protection of any control transformer rated less than 50 VA, provided it is part of the starter and within its enclosure.

Exception No. 2 permits a control transformer with a rated primary current of *less* than 2 A to be protected at up to 500 percent of rated primary current by a protective device in each ungrounded conductor of the supply circuit to the transformer primary, as shown in Fig. 3.42.

In the majority of magnetic motor controllers and contactors, the voltage of the operating coil is the voltage provided between two of the conductors supplying the load, or one conductor and the neutral. Conventional starters are factory wired with coils of the same voltage rating as the phase voltage to the motor. However, there are many cases in which it is desirable or necessary to use control circuits and devices of lower voltage rating than the motor. Such could be the case with high-

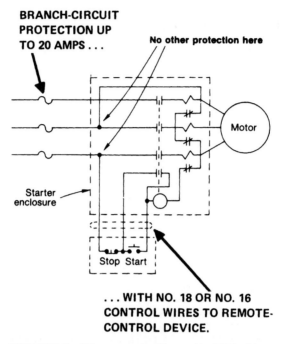

FIGURE 3.40 This was permitted by previous **NEC** editions but is now a violation of Exception No. 2 of Section 430-72(b).

voltage (over 600 V) controllers, for instance, in which it is necessary to provide a source of low voltage for practical operation of magnetic coils. And even in many cases of motor controllers and contactors for use under 600 V, safety requirements dictate the use of control circuits of lower voltage than the load circuit.

Although contactor coils and pilot devices are available and effectively used for motor controllers with up to 600-V control circuits, such practice has been prohibited in applications where atmospheric and other working conditions make it dangerous for operating personnel to use control circuits of such voltage. And certain OSHA regulations require 120- or 240-V coil circuits for the 460-V motors. In such cases, control transformers are used to step the voltage down to permit the use of lower-voltage coil circuits.

The use of secondary protection for control wires is suited to applications in which the transformer is within the starter enclosure. Primary protection, without secondary protection, is suited to use with either internal or external control transformers with 2-wire primaries and 2-wire secondaries and constitutes more effective protection because *both* the transformer and the primary and secondary control wires are protected.

Exception No. 2 of Section 725-12 applies to the protection of remote-control conductors fed by the 2-wire secondary of a separate control transformer supplying the coils of one or more motor starters or magnetic contactors. The rule is the same as that of Exception No. 3 of Section 430-72(b), as described above. Figure 3.43 shows an example of the application of that rule to a 3000-VA control transformer supplying starter-coil circuits. A properly sized circuit breaker or fuses may be used

MOTOR BRANCH CIRCUITS AND CONTROL CIRCUITS

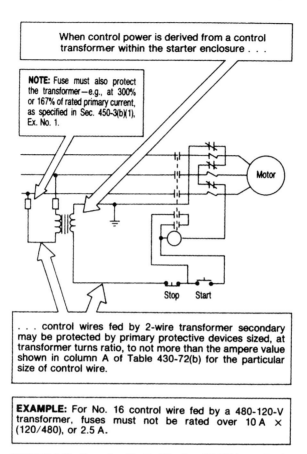

FIGURE 3.41 Exception No. 3 of Section 430-72(b) permits the secondary wires of the coil circuit to be protected by primary-side overcurrent protection.

at the supply end of the circuit that feeds the transformer primary and may provide overcurrent protection for the primary conductors, for the transformer itself, and for conductors of the control circuit that is run from the transformer secondary to supply power to motor starters or other control equipment, as follows:

1. The primary-side protection must not be rated greater than the protection required by Section 450-3(b)(1) for transformers rated up to 600 V. For a transformer rated 9 A or more, the rating of the primary CB or fuses must not be greater than 125 percent of the rated transformer primary current. However, if 125 percent of the rated primary current does not yield a standard rating of fuse or CB, the next higher rated standard protective device may be used. Where the transformer-rated primary current is less than 9 A (as it would be for all the usual control transformers rated up to 5000 VA and stepping 480 V down to 120 V), the maximum permitted rating of the primary protection is 167 percent of the rated primary current. For a transformer with a primary rated less than 2 A, the pri-

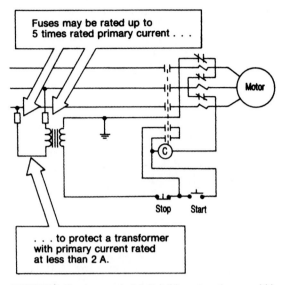

FIGURE 3.42 A control circuit fed by a transformer within the starter enclosure may have overcurrent protection in the primary rated up to 500 percent of the rated primary current of a small transformer.

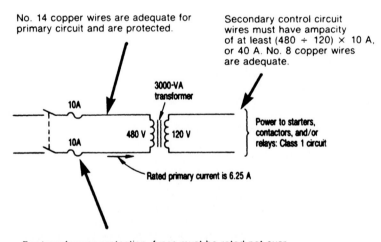

FIGURE 3.43 Protection must be provided for a separate control transformer *and* the circuit wires.

mary protection must never exceed 300 percent of the rated primary current, except as noted above for control-circuit transformers within motor starters. For most control transformers—with primary ratings well below 10 A—fuse protection will be required on the primary because the smallest standard CB rating is 15 A, and that will generally exceed the maximum values of primary protection permitted by Section 450-3(b)(1).

2. The primary protection must not exceed the ampere rating of the primary-circuit conductors. When protection is sized for the transformer as described above, No. 14 copper primary conductors will be protected well within their 15-A rating.

3. Secondary conductors for the control circuit can then be selected to have an ampacity at least equal to the rating of the primary protection times the primary-to-secondary transformer voltage ratio. Of course, larger conductors may be used if needed to keep the voltage drop within limits.

Exception No. 4 of Section 430-72(b) says that the overcurrent protection for a control circuit (other than branch-circuit protection) must be eliminated where opening of the control circuit would create a hazard. An example would be the control circuits for fire-pump motors.

Control Transformers

In the majority of magnetic motor controllers and contactors, the voltage of the operating coil is the voltage provided between two of the conductors supplying the load or between one conductor and the neutral. Conventional starters are factory-wired with coils of the same voltage rating as the phase voltage to the motor. However, in many cases it is desirable or necessary to use control circuits and devices of lower voltage rating than the motor. Such would be the case with high-voltage (over 600 V) controllers, for instance, when it is necessary to provide a source of low voltage for practical operation of magnetic coils. And even with motor controllers and contactors using less than 600 V, safety requirements dictate the use of control circuits of lower voltage than the load circuit.

Although contactor coils and pilot devices are available and are effectively used for motor controllers with up to 550-V control circuits, this practice has been prohibited where atmospheric and other working conditions make such high control-circuit voltages dangerous for operating personnel. And certain OSHA regulations require 120- or 240-V coil circuits for 460-V motors. In such cases, control transformers are used to step the voltage down to permit the use of lower-voltage coil circuits.

Control transformers are relatively small, compact, dry-type potential transformers. They are available in many ratings to meet any common motor control-circuit application. For a very wide range of motor controllers, control transformers are available as accessory equipment to the basic starter types. They can be supplied by manufacturers as separate units with provisions for mounting external to the controller or can be incorporated in the controller enclosure by the manufacturer, wired in with an operating coil of the proper voltage rating. Such transformers can be obtained with fused or otherwise protected secondaries to meet **NEC** requirements on control-circuit overcurrent protection. And extra transformer capacity can be included to permit the operation of a local lighting unit.

For low-voltage motor controllers, typical control transformers have single or double primary and secondary windings to give either a basic transformation (from, say, 480 to 120 V) or a selection of transformations (as 480/240-V primary to 240/120-V secondary). These units range in capacity from 25 VA to as high as 8000 VA. Control transformers for high-voltage controllers for 2300- and 4160-V motors are generally built into controller enclosures.

Selection of the proper control transformer for a controller is a simple matter of matching the characteristics of the control circuit to the specs of the transformer. The line voltage of the supply to the motor determines the required primary rating of the transformer. The transformer secondary must be rated to provide the desired control-circuit voltage to match the voltage of the controller operating coil. The continuous secondary-current rating of the transformer must be sufficient for the magnetizing current of the operating coil and must also be able to handle the inrush current. Of course, if other devices are to be energized from the transformer, additional capacity must be provided. Complete coordination between coil circuit and transformer is provided in factory-assembled controllers.

Grounding of Control-Transformer Circuit.   When a control transformer is used to derive control power, the transformer secondary may have to be operated with one conductor grounded. This may be either one of the conductors where the transformer has a 2-wire secondary or the neutral conductor of a 3-wire, single-phase secondary.

NEC Section 250-5(b) applies to the secondaries of control transformers. According to the rules of this section:

1. Any 120-V, 2-wire circuit must normally have one of its conductors grounded.
2. The neutral conductor of any 240/120-V, 3-wire, single-phase circuit must be grounded.
3. The neutral of a 208/120-V, 3-phase, 4-wire circuit must be grounded.

Those requirements have often caused difficulty when applied to control circuits derived from a control transformer. For instance, there are cases where a ground fault on the hot leg of a grounded control circuit can cause a hazard to personnel by actuating the control-circuit fuse or CB and shutting down an industrial process in a sudden, unexpected, nonorderly way. A metal-casting facility is an example of an installation where sudden shutdown due to a ground fault in the hot leg of a grounded control circuit could be objectionable. Use of a 120-V control circuit without an intentional grounding connection for one of the two circuit conductors eliminates any chance of a control-circuit protective device being activated when a single accidental (fault) ground occurs on the circuit. Because designers often wish to operate such 120-V control circuits ungrounded, Exception No. 3 of Section 250-5(b) permits ungrounded control circuits under certain specified conditions. A 120-V control circuit may be operated ungrounded when all the following exist:

1. The circuit is derived from a transformer that has a primary rating below 1000 V.
2. In a commercial, institutional, or industrial facility, supervision will ensure that only persons qualified in electrical work will maintain and service the control circuits.
3. There is a need to prevent circuit opening on a ground fault—that is, continuity of power is required for safety or for operating reliability.
4. Some type of ground detector is used on the ungrounded system to alert personnel to the presence of a ground fault, enabling them to clear the ground fault in the normal downtime of the system (Fig. 3.44).

MOTOR BRANCH CIRCUITS AND CONTROL CIRCUITS

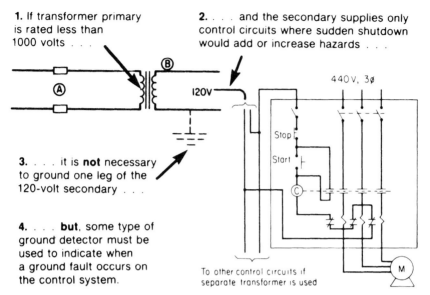

FIGURE 3.44 Under these conditions a 120-V control circuit may be used ungrounded.

Although no mention is made of secondary voltage in this **NEC** rule, the exception permitting ungrounded control circuits is primarily significant for 120-V control circuits. The **NEC** has long permitted 240- and 480-V control circuits to be operated ungrounded. This exception can be applied to a 120-V control circuit derived from a control transformer in an individual motor starter or to a separate control transformer that supplies control power for a number of motor starters or magnetic contactors.

The previously described rules also apply to the hookup shown in Fig. 3.44. Overcurrent protection in each ungrounded leg of the circuit supplying a separate control transformer must be used on the primary side of the transformer (at A) and must be sized at not over 125 percent of the rated primary current [or as permitted by Section 450-3(b)]. The primary conductors must be protected at their ampacity by the primary protection. All conductors on the secondary side (that is, all control-circuit conductors) may be protected by the primary protection if the control conductors are sized so that the rating of the primary protection does not exceed the ampacity of the control conductors as reflected from the secondary to the primary. The primary-protection rating must not exceed the value obtained by multiplying the control-conductor ampacity by the secondary-to-primary voltage ratio. [See Exception No. 1 to Section 430-72(b).]

If the control transformer is supplied as part of an individual starter, within the starter enclosure, that transformer is a component part of the starter and is excluded by Exception No. 2 of Section 450-1 from the requirement for primary protection in Section 450-3(b). However, protection is required on the secondary-side conductors of the transformer by Section 430-72(b) to protect the control conductors against overload in excess of their ampacity. Of course, primary conductor protection must also be provided.

228 CHAPTER THREE

Grounding and Bonding. According to the basic rule of Section 250-26(a), when one of the secondary conductors of a control transformer is grounded, it is necessary to (1) bond the grounded conductor to the metal case of the transformer and (2) run a grounding-electrode conductor from the grounded transformer secondary terminal to nearby grounded building steel or a grounded metal water pipe. But an exception exempts small control transformers from the second of these basic requirements. A Class 1 remote-control transformer that is rated not over 1000 VA simply has to have a grounded secondary conductor bonded to the metal case of the transformer; no grounding-electrode conductor is needed, provided that the metal transformer case itself is properly grounded by grounded metal raceway that supplies its primary or by means of a suitable (Section 250-57) equipment-grounding conductor that ties the case back to the grounding electrode for the primary system.

As shown in Fig. 3.45, the hookup of a control transformer under the two new exceptions consists in simply using a No. 14 copper conductor to bond the grounded leg of the transformer secondary to the transformer frame, leaving the supply conduit to the transformer to provide the path back to the main service ground, but with the connection between neutral and frame providing effective return for clearing faults, as shown.

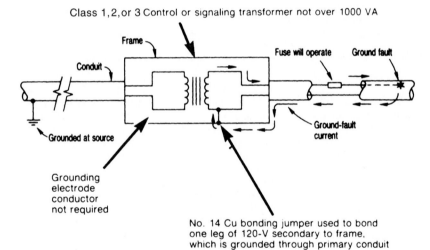

FIGURE 3.45 Small control transformers do not require a grounding-electrode conductor.

Transformer housings must be grounded by connection to grounded cable armor or metal raceway or by use of a grounded conductor run with the circuit conductors (either a bare conductor or a conductor with green-colored covering).

Control Disconnects

NEC Section 430-74(a), on the disconnection of coil-control circuits of magnetic motor starters, says:

MOTOR BRANCH CIRCUITS AND CONTROL CIRCUITS   229

Motor control circuits shall be so arranged that they will be disconnected from all sources of supply when the disconnecting means is in the open position. The disconnecting means shall be permitted to consist of two or more separate devices, one of which disconnects the motor and the controller from the source(s) of power supply for the motor, and the other(s), the control circuit(s) from its power supply. Where separate devices are used, they shall be located immediately adjacent one to each other.

Remote Disconnects. In recognition of the unusual and complex control conditions that exist in many industrial applications—particularly process industries and manufacturing facilities—Exception No. 1 to Section 430-74(a) alters the basic rule that disconnecting means for control circuits must "be located immediately adjacent one to each other." When a piece of motor-control equipment has more than 12 motor-control conductors associated with it, remote locating of the disconnect means is permitted under Exception No. 1. As shown in Fig. 3.46, this permission is applicable only where (1) only qualified persons have access to the live parts and (2) sufficient warning signs are posted on the equipment to locate and identify the various disconnects associated with the control-circuit conductors.

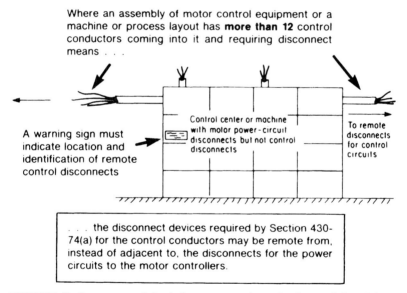

FIGURE 3.46 Remote control-circuit disconnects may be used under guarded conditions.

Exception No. 2 presents another instance in which control-circuit disconnects may be mounted other than immediately adjacent to each other. It notes that where the opening of one or more motor control-circuit disconnects might result in a hazard to personnel or property, remote mounting may be used under the conditions specified in Exception No. 1—that is, that access is limited to qualified persons and that warning signs are located on the outside of the equipment to indicate the location and the identification of each remote control-circuit disconnect.

Similar recognition of the need for remote disconnects in complex industrial layouts is indicated in Section 430-113. The basic rule of this section requires that a disconnecting means be provided for each source of electric energy input to equipment with more than one circuit supplying power to it. And each source is permitted to have a separate disconnecting means. But an exception to this rule states that where a motor receives electric energy from more than one source (such as a synchronous motor receiving both ac and dc energy input), the disconnecting means for the main power supply to the motor is not required to be immediately adjacent to the motor—provided that the controller disconnecting means, which is the disconnect ahead of the motor starter in the main power circuit, is capable of being locked in the open position. If, for instance, the motor-control disconnect can be locked in the open position, it may be remote; but the disconnect for the other energy input circuit would have to be adjacent to the machine itself, as indicated in Fig. 3.47.

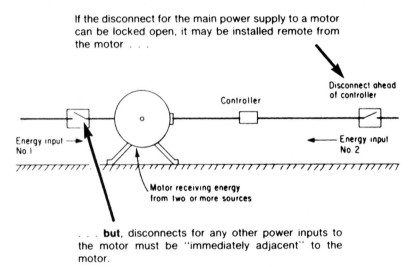

FIGURE 3.47 One power-source disconnect may be installed remote from a motor.

Protection against Accidental Starting

Combinations of ground faults often occur in motor-control circuits, shorting the pilot starting device (pushbutton, limit switch, pressure switch, etc.) and accidentally starting the motor even though the pilot device is in the off position. And because many remote-control circuits are long, possible faults may occur at many points in those circuits. Insulation breakdowns, contact shorts due to the accumulation of foreign matter or moisture, and grounds to conduit are common fault conditions responsible for the accidental starting of motor controllers.

As shown in Fig. 3.48, a magnetic motor controller used on a 3-phase, 3-wire ungrounded system always presents the possibility of accidental starting of the motor. If, for instance, an undetected ground fault exists on one phase of the 3-phase system—even if this ground fault is a long distance from the controller—a second ground fault in the remote-control circuit for the operating coil of the starter can start the motor. Effective circuit design must always be utilized to prevent such accidental occurrences.

Figure 3.49 shows the use of a control transformer to isolate the control circuit and keep it from responding to the combination of ground faults shown in Fig. 3.48. This transformer may be a one-to-one isolating transformer, with the same primary and secondary voltages, or the transformer can step the motor-circuit voltage down to a lower level for the control circuit.

In the hookup shown in Fig. 3.50, a two-pole start button is used in conjunction with two sets of holding contacts in the motor starter. This hookup protects against accidental starting of the motor under the fault conditions shown in Fig. 3.48. The hookup also protects against accidental starting due to two ground faults in the control circuit simply shorting out the start button and energizing the operating coil. This could happen in the circuit of Fig. 3.48 or that of Fig. 3.49.

Another type of motor control-circuit fault can produce a current path through the coil of a closed contactor to hold it closed even if the pilot device operates to open the coil circuit. Again, this can result when a combination of ground faults shorts the stop device. Failure to open can do serious damage to motors in some applications and can be a hazard to personnel. The operating characteristics of contactor coils contribute to the possible failure of a controller to respond to the opening of the stop contacts. It takes about 85 percent of rated coil voltage to operate the armature associated with the coil; but it takes only about 50 percent of the rated voltage to enable the coil to hold the contactor closed once it is closed. Thus, even partial grounds and shorts on control contact assemblies can produce paths for sufficient current flow to cause shorting of the stop position of pilot devices. And faults can short out running overload relays, negating the overcurrent protection of the motor, its associated control equipment, and conductors.

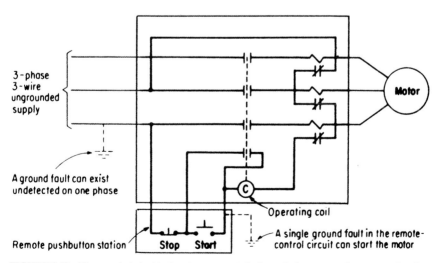

FIGURE 3.48 Ungrounded electrical systems present the hazard of unexpected motor starting due to faults.

232 CHAPTER THREE

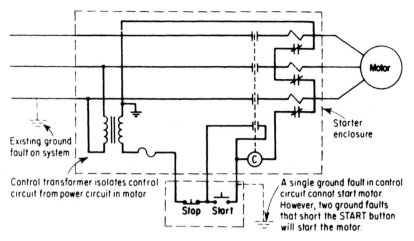

FIGURE 3.49 Control transformer can isolate coil-circuit wires from accidental starting connection.

Figure 3.51 is a modification of the circuit of Fig. 3.50, using a two-pole start button and a two-pole stop button; it protects against both accidental starting and accidental failure to stop when the stop button is pressed. Both effects of ground faults are eliminated.

Figure 3.52 shows another example of a control-circuit installation that should be carefully designed—one that is required by the second sentence of Section 430-73 for any control circuit which has one leg grounded. Whenever the coil is fed from a circuit made up of a hot conductor and a grounded conductor (as when the coil is fed from a panelboard or separate control transformer, instead of from the supply conductors to the motor), care must be taken to place the pushbutton station or other switching control device in the hot leg to the coil and not in the grounded leg to the coil. By switching in the hot leg, starting of the motor through an accidental ground fault can be effectively eliminated.

Section 430-74(a), on disconnection of coil control circuits of magnetic motor starters, was quoted under "Control Disconnects" above. Figures 3.53 and 3.54 show two methods of providing for the disconnection of control power. Note that, in Fig. 3.53, the single disconnect switch kills power to the motor and to the entire control circuit—because the step-down control transformer is installed on the load side of the disconnect. In Fig. 3.54, however, it is necessary to provide some means for ensuring that the control circuit is deenergized when the power circuit to the motor is open. Control circuits operating contactor coils and other devices within controllers present a shock hazard if they are allowed to remain energized when the disconnect is in the off position. Therefore, the control circuit either must be designed in such a way that it is disconnected from the source of supply by the controller disconnecting means or must be equipped with a separate disconnect immediately adjacent to the controller disconnect (Section 430-74) for opening both circuits. The four-pole disconnect shown in the sketch may be a four-pole switch or a three-pole switch with an auxiliary set of contacts to open the hot leg of the control circuit.

All the foregoing is based on rules given in **NEC** Article 430. For the most part, these rules are straightforward and clear. In any case where a question arises about

MOTOR BRANCH CIRCUITS AND CONTROL CIRCUITS 233

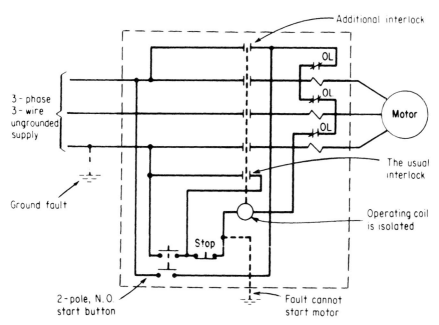

FIGURE 3.50 Two-pole start button protects against accidental starting due to ground faults.

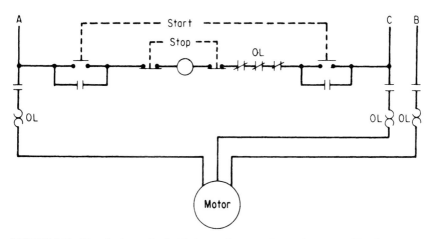

FIGURE 3.51 Use of a two-pole start button and a two-pole stop button provides the greatest safety.

the interpretation of **NEC** rules, the authority responsible for enforcement in the particular area should be consulted. Where interpretations are given in this presentation, they are offered to provide a concrete concept which can be checked out with the local inspector. Such consultation provides an easy way to find out exactly how an inspector will rule on specific fine points.

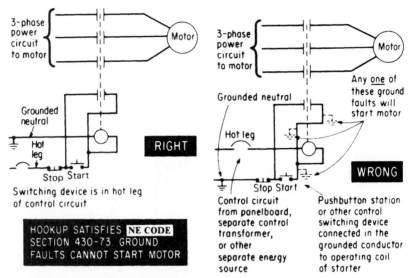

FIGURE 3.52 Switching must be done in the hot leg of a grounded control circuit.

Voltage Regulation

To ensure the proper and efficient operation of motors, the matter of voltage regulation must be considered carefully. Many factors (e.g., number of motors, sizes and types of motors, duty cycles, load densities, type of distribution system, loading of various feeders, and power factor) are related to the design problem of assuring the necessary level and stability of motor voltages. Voltage regulation follows through every step in design and must be accounted for in sizing conductors.

The voltage drop from the sources of voltage to any motor in the system must not exceed 5 percent. Normally, the proper power feeder design will limit the voltage drop to 3 percent, leaving a maximum permissible voltage drop of 2 percent in any motor branch circuit under full-load conditions. However, a 1 percent maximum circuit drop is recommended.

The power factor should be taken into consideration in all calculations; the known power-factor value or an 80 percent assumed value should be used. By raising the power factor, for the same actual power delivered, the current is decreased in the generator, transformers, and lines up to the point where the capacitor is connected. Power-factor capacitors can be connected into motor circuits to neutralize the effect of lagging-power-factor loads, thereby reducing the current drawn for a given kilowatt load. In a distribution system, small capacitor units may be connected at the individual loads, or the total required capacitance (in kilovars) may be grouped at one point and connected to the main. Although the kilovar total is the same in both cases, the use of small capacitors at the individual loads reduces the current all the way from the loads back to the source; this method therefore has a greater power-factor corrective effect than the use of one big unit on the main, which reduces the current only from the point of installation back to the source.

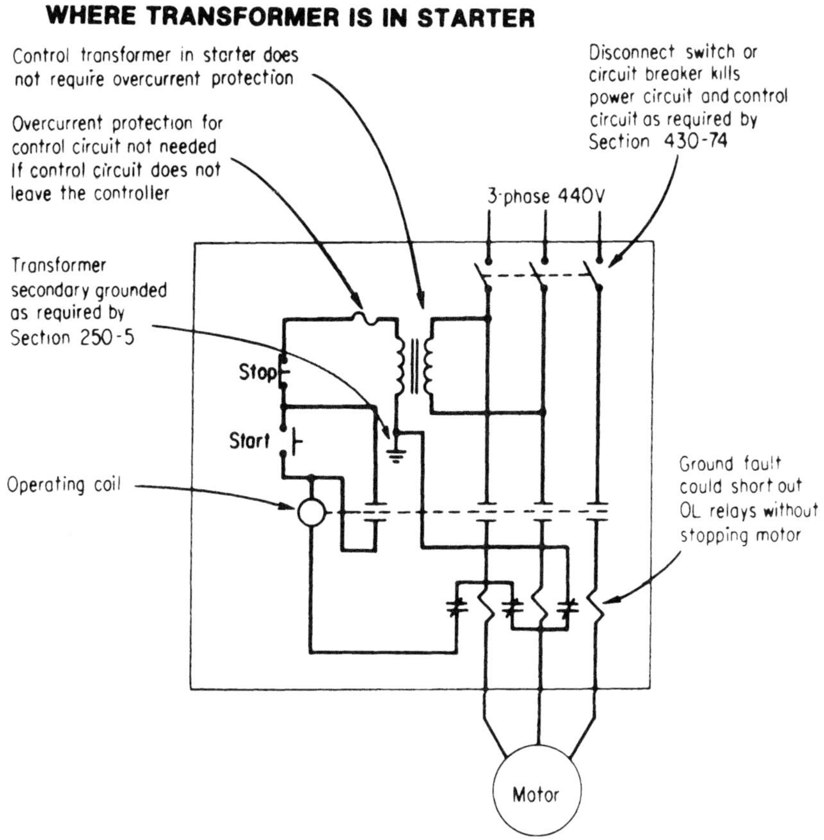

MOTOR BRANCH CIRCUITS AND CONTROL CIRCUITS 235

WHERE TRANSFORMER IS IN STARTER

Control transformer in starter does not require overcurrent protection

Overcurrent protection for control circuit not needed If control circuit does not leave the controller

Transformer secondary grounded as required by Section 250-5

Stop

Start

Operating coil

3-phase 440V

Disconnect switch or circuit breaker kills power circuit and control circuit as required by Section 430-74

Ground fault could short out OL relays without stopping motor

Motor

FIGURE 3.53 One disconnect kills all power to the starter.

Figure 3.55 shows the basic relationships involved in power-factor correction. A typical calculation would be made as follows.

Example. A 3-phase, 460-V, 50-hp motor has a power factor of 70 percent. What rating of capacitor is needed to improve the power factor to 90 percent?
Solution. From **NEC** Table 430-150, the full-load current of the motor is 65 A. That gives a kilowatt load of

$$L = \frac{460 \times 65 \times 1.73}{1000} PF_1$$

$$= \frac{51{,}740}{1000} \times 0.7 = 36.2 \text{ kW}$$

WHERE TRANSFORMER IS OUTSIDE STARTER

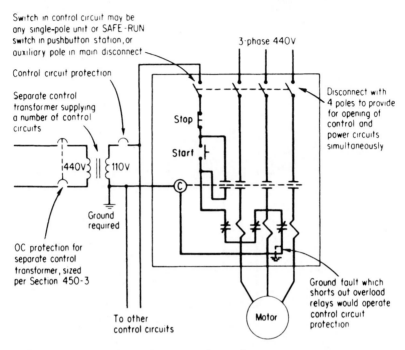

FIGURE 3.54 External control power must have a disconnect.

If $\cos \theta_1 = 0.70$, then $\theta_1 = 45.58°$ and $\tan \theta_1 = 1.0203$; and if $\cos \theta_2 = 0.90$, then $\theta_2 = 25.84°$ and $\tan \theta_2 = 0.4844$. Then

$$KVAR_R = L(\tan \theta_1 - \tan \theta_2)$$

$$= 36.2(1.0203 - 0.4844)$$

$$= 36.2 \times 0.5359 = 19.4 \text{ kvar}$$

Depending upon the relationship between the voltage drop and the power factor, measures for improving the power factor may be designed into motor branch circuits. At individual motor locations, power-factor-correcting capacitors offer improved voltage regulation. As shown in Fig. 3.56, power-factor capacitors installed at terminals of motors provide maximum relief from reactive currents, reducing the required current-carrying capacities of conductors from their point of application all the way back to the supply system. Such application also eliminates extra switching devices, since each capacitor can be switched with the motor it serves. Figure 3.57 shows details included on a typical set of electrical plans to convey the designer's concept for power-factor correction of a large motor. For small, numerous motors that are operated intermittently, however, it is often economically more desirable to install the required capacitor kilovars at the motor load center, as at the right in Fig. 3.55.

In most applications, power-factor capacitors are installed to raise the system power factor for increased circuit or system current-carrying capacity, reduced

TO IMPROVE THE POWER FACTOR OF A CIRCUIT FROM PF_1 TO PF_2 THE REQUIRED RATING OF A CAPACITOR MUST BE

$$KVAR_R = KW(\tan \theta_1 - \tan \theta_2)$$

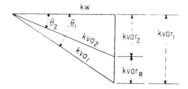

where, in the sketch
$KVAR_R$ = required kilovars rating of capacitor to change from PF_1 to PF_2
KW = the kilowatt value of the circuit load
θ_1 = original phase angle
θ_2 = improved phase angle
PF_1 = power factor before correction
PF_2 = power factor after correction
$KVAR_1$ = reactive kilovolt-amperes at PF_1
$KVAR_2$ = reactive kilovolt-amperes at PF_2

NOTE: The phase angles θ_1 and θ_2 can be determined from a table of trigonometric functions using the following relationships:
θ_1 = The angle which has its cosine equal to the decimal value of the original power factor (e.g., 0.70 for 70% PF; 0.65 for 65%; etc.)
θ_2 = The angle which has its cosine equal to the decimal value of the improved power factor

FIGURE 3.55 Motor-circuit power factor is raised with this procedure.

power losses, and lower reactive-power charges (most utility companies include a power-factor penalty clause in their industrial billing). Additional benefits derived from power-factor-capacitor installation are reduced voltage drop and increased voltage stability. Capacitor manufacturers provide various tables and graphs to facilitate the selection of the proper capacitor for a given motor load.

The **NEC** limits power-factor correction to unity (100 percent, or 1.0) when there is no load on the motor. This results in a power factor of 95 percent or better when the motor is fully loaded. The **NEC** used to recognize the use of capacitors either sized at the value that will produce 100 percent power factor when the motor is running at no load or sized at a value equal to 50 percent of the kilovoltampere rating of the motor input, for motors up to 50 hp and 600 V, as shown in Fig. 3.58. Most motor-associated capacitors are used with low-voltage, 5- to 50-hp, 1800- and 1200-rpm, across-the-line-start motors. In this range, the no-load rule for determining the maximum capacitor kilovars restricts capacitors to the equivalent of less than 50 percent of the horsepower. Noticeable economy can be effected by applying larger capacitors, up to 50 percent of the motor input power, to such motors. This has been done for years, with excellent results and no field trouble.

The no-load power factor of a motor is a design constant for the motor; it may be obtained from the manufacturer of the motor, or it may be measured or calculated. With reference to Fig. 3.59, the known no-load power factor (*PF*) of the motor may

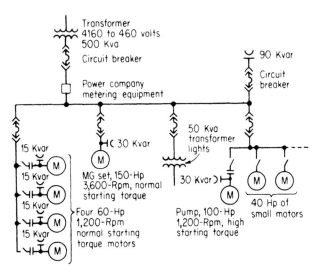

FIGURE 3.56 Installation of the power-factor capacitors at the individual motors offers maximum benefit from the correction.

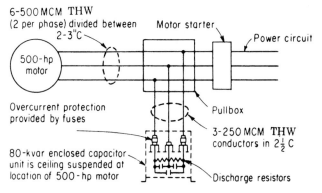

FIGURE 3.57 Design must provide a place for connecting a capacitor unit between motor and starter.

be used to calculate the required power in kilowatts as $kW = PF \times kVA_1$, where kVA_1 is calculated from the circuit voltage and current, as measured with a clamp-on ammeter. Then the kilovar rating of the capacitor required to raise the no-load power factor to 100 percent equals the square root of $(kVA_1)^2 - (kW)^2$.

Capacitors installed as a group or bank at some central point, such as a switchboard, loadcenter, busway, or outdoor substation, usually serve only to reduce the utility-company penalty charges. However, in many instances, installation costs also will be lowered.

Capacitor installations may consist of an individual unit connected as close as possible to the inductive load (e.g., at the terminals of a motor) or of a bank of many

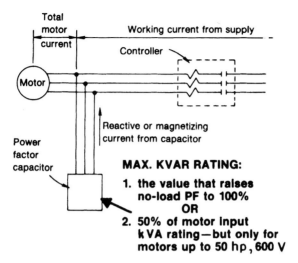

FIGURE 3.58 The NEC sets a maximum value on capacitor rating for safety purposes.

units connected in multiple across a main feeder. Units are available in specific kilovar and voltage ratings. Standard low-voltage capacitor units are rated from about 0.5 to 25 kvar at voltages from 216 to 600 V. For high-voltage applications, standard ratings are 15, 25, 50, and 100 kvar. Available in single-, 2-, or 3-phase configurations, power capacitors may be supplied either unfused or equipped with current-limiting or high-capacity fuses (single-phase units are furnished with one fuse; 3-phase capacitors usually have two fuses). On low-voltage units, fuses may be mounted on the capacitor bushings inside the terminal compartment.

Figure 3.59 is a table for calculating the capacitor kilovar rating required to raise the power factor from an original value to a desired higher value. Its use is illustrated in the following example.

Example. A circuit supplies a load that has a 100-kW value (obtained either by calculation or from a wattmeter) and operates at a power factor of 68 percent. What capacitor kilovar rating is needed to raise the power factor to 96 percent?

Solution. Find the required multiplier in the table of Fig. 3.59 by locating 68 percent in the vertical column at the left and then moving across to the column of multipliers headed 96 percent. The multiplier is 0.788. Then the required rating is

$$100 \text{ kW} \times 0.788 = 78.8 \text{ kvar}$$

Circuit Layout

Based on the foregoing rules and considerations, motor branch circuits can be designed for any type of occupancy. Generally, motor-circuit requirements for commercial and institutional buildings are concerned with known loads which are fairly permanent in size and location. These loads include refrigeration and air-conditioning compressors, pumps, elevators, escalators, blowers, and fans. Such loads can be

CHAPTER THREE

Desired power factor in percentage

	80%	81	82	83	84	85	86	87	88	89	90	91	92	93	94	95	96	97	98	99	100
50%	.982	1.008	1.034	1.060	1.086	1.112	1.139	1.165	1.192	1.220	1.248	1.276	1.303	1.337	1.369	1.402	1.441	1.481	1.529	1.590	1.732
51	.936	.962	.988	1.014	1.040	1.066	1.093	1.119	1.146	1.174	1.202	1.230	1.257	1.291	1.320	1.357	1.395	1.435	1.483	1.544	1.686
52	.894	.920	.946	.972	.998	1.024	1.051	1.077	1.104	1.132	1.160	1.188	1.215	1.249	1.281	1.315	1.353	1.393	1.441	1.502	1.644
53	.850	.876	.902	.928	.954	.980	1.007	1.033	1.060	1.088	1.116	1.144	1.171	1.205	1.237	1.271	1.309	1.349	1.397	1.458	1.600
54	.809	.835	.861	.887	.913	.939	.966	.992	1.019	1.047	1.075	1.103	1.130	1.164	1.196	1.230	1.268	1.308	1.356	1.417	1.559
55	.769	.795	.821	.847	.873	.899	.926	.952	.979	1.007	1.035	1.063	1.090	1.124	1.156	1.190	1.228	1.268	1.316	1.377	1.519
56	.730	.756	.782	.808	.834	.860	.887	.913	.940	.968	.996	1.024	1.051	1.085	1.117	1.151	1.189	1.229	1.277	1.338	1.480
57	.692	.718	.744	.770	.796	.822	.849	.875	.902	.930	.958	.986	1.013	1.047	1.079	1.113	1.151	1.191	1.239	1.300	1.442
58	.655	.681	.707	.733	.759	.785	.812	.838	.865	.893	.921	.949	.976	1.010	1.042	1.076	1.114	1.154	1.202	1.263	1.405
59	.618	.644	.670	.696	.722	.748	.775	.801	.828	.856	.884	.912	.939	.973	1.005	1.039	1.077	1.117	1.165	1.226	1.368
60	.584	.610	.636	.662	.688	.714	.741	.767	.794	.822	.849	.878	.905	.939	.971	1.005	1.043	1.083	1.131	1.192	1.334
61	.549	.575	.601	.627	.653	.679	.706	.732	.759	.787	.815	.843	.870	.904	.936	.970	1.008	1.048	1.096	1.157	1.299
62	.515	.541	.567	.593	.619	.645	.672	.698	.725	.753	.781	.809	.836	.870	.902	.936	.974	1.014	1.062	1.123	1.265
63	.483	.509	.535	.561	.587	.613	.640	.666	.693	.721	.749	.777	.804	.838	.870	.904	.942	.982	1.030	1.091	1.233
64	.450	.476	.502	.528	.554	.580	.607	.633	.660	.688	.716	.744	.771	.805	.837	.871	.909	.949	.997	1.058	1.200
65	.419	.445	.471	.497	.523	.549	.576	.602	.629	.657	.685	.713	.740	.774	.806	.840	.878	.918	.966	1.027	1.169
66	.388	.414	.440	.466	.492	.518	.545	.571	.598	.626	.654	.682	.709	.743	.775	.809	.847	.887	.935	.996	1.138
67	.358	.384	.410	.436	.462	.488	.515	.541	.568	.596	.624	.652	.679	.713	.745	.779	.817	.857	.905	.966	1.108
68	.329	.355	.381	.407	.433	.459	.486	.512	.539	.567	.595	.623	.650	.684	.716	.750	.788	.828	.876	.937	1.079
69	.299	.325	.351	.377	.403	.429	.456	.482	.509	.537	.565	.593	.620	.654	.686	.720	.758	.798	.846	.907	1.049
70	.270	.296	.322	.348	.374	.400	.427	.453	.480	.508	.536	.564	.591	.625	.657	.691	.729	.769	.811	.878	1.020
71	.242	.268	.294	.320	.346	.372	.399	.425	.452	.480	.508	.536	.563	.597	.629	.663	.701	.741	.783	.850	.992
72	.213	.239	.265	.291	.317	.343	.370	.396	.423	.451	.479	.507	.534	.568	.600	.634	.672	.712	.754	.821	.963
73	.186	.212	.238	.264	.290	.316	.343	.369	.396	.424	.452	.480	.507	.541	.573	.607	.645	.685	.727	.794	.936
74	.159	.185	.211	.237	.263	.289	.316	.342	.369	.397	.425	.453	.480	.514	.546	.580	.618	.658	.700	.767	.909
75	.132	.158	.184	.210	.236	.262	.289	.315	.342	.370	.398	.426	.453	.487	.519	.553	.591	.631	.673	.740	.882
76	.105	.131	.157	.183	.209	.235	.262	.288	.315	.343	.371	.399	.426	.460	.492	.526	.564	.604	.652	.713	.855
77	.079	.105	.131	.157	.183	.209	.236	.262	.289	.317	.345	.373	.400	.434	.466	.500	.538	.578	.620	.687	.829
78	.053	.079	.105	.131	.157	.183	.210	.236	.263	.291	.319	.347	.374	.408	.440	.474	.512	.552	.594	.661	.803
79	.026	.052	.078	.104	.130	.156	.183	.209	.236	.264	.292	.320	.347	.381	.413	.447	.485	.525	.567	.634	.776
80	.000	.026	.052	.078	.104	.130	.157	.183	.210	.238	.266	.294	.321	.355	.387	.421	.459	.499	.541	.608	.750
81	—	.000	.026	.052	.078	.104	.131	.157	.184	.212	.240	.268	.295	.329	.361	.395	.433	.473	.515	.582	.724
82	—	—	.000	.026	.052	.078	.105	.131	.158	.186	.214	.242	.269	.303	.335	.369	.407	.447	.489	.556	.698
83	—	—	—	.000	.026	.052	.079	.105	.132	.160	.188	.216	.243	.277	.309	.343	.381	.421	.463	.530	.672
84	—	—	—	—	.000	.026	.053	.079	.106	.134	.162	.190	.217	.251	.283	.317	.355	.395	.437	.504	.645
85	—	—	—	—	—	.000	.027	.053	.080	.108	.136	.164	.191	.225	.257	.291	.329	.369	.417	.478	.620

FIGURE 3.59 Multipliers for calculating capacitor kilovars required to raise the power factors of motor circuits.

served by many types of wiring systems without special provisions for flexibility of power circuiting. Industrial plants, on the other hand, require considerable power-circuiting flexibility to accommodate the moving of motors, shifting of production lines, and expansion of motor loads. To provide this flexibility, motor circuits may be fed from motor-control centers, from plug-in busway, from auxiliary gutters, from junction boxes connected in conduit and raceways which allow changes and expansion in the wiring, or, in the case where many small motors are to be supplied, from an underfloor raceway system. Spare capacity for increased motor loads is a design consideration for motor feeders and subfeeders. Spare capacity should be included in the motor-circuit panelboard where required. By using sectional-type panelboards with interchangeable units and oversized raceway for branch-circuit wiring, future increases in the size of the motors on the circuit and in associated branch-circuit equipment are greatly facilitated.

As shown in Fig. 3.60, wiring to serve two or more motors may be laid out according to several plans:

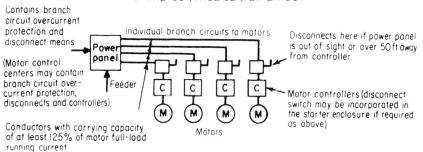

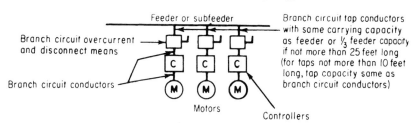

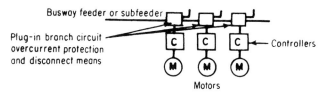

FIGURE 3.60 Motor-circuit layout must suit the particular needs of a given installation.

1. Each motor may be served by a separate branch circuit from a panel or distribution center. Branch-circuit protective devices for motor circuits are commonly grouped with required controller and disconnect equipment. The grouping may be contained in a single enclosure, such as a motor-control center (a so-called MCC), or it may consist of a number of individual enclosures at one location fed from a trough (an auxiliary gutter).

2. Individual circuits to motors may be tapped from a feeder without individual overcurrent protection for the taps, provided they are carried directly to the disconnecting means or combination starter for each motor. Such taps must have the same carrying capacity as the feeder if they are over 25 ft long. If they are under 25 ft long, they must have a carrying capacity at least equal to one-third the carrying capacity of the feeder. In such cases, the motor branch-circuit protection is included with the disconnect means (switch and fuses, or CB) or in a combina-

tion starter, and the branch circuit originates at that point. Taps not over 10 ft long need have only the same capacity as the branch-circuit wires. Taps not over 10 ft long are applied where two or more combination motor starters are nippled out of a gutter where the motor-circuit wires are tapped to the feeder conductors supplying the gutter.

3. Individual branch circuits to motors may originate at disconnects fed directly by a feeder or subfeeder (e.g., tap units on a busway), with disconnect and branch-circuit protection provided by the tap switch (with fuses) or CB.

The diagrams of motor branch circuits in Figs. 3.61 and 3.62 cover the most common layouts of motor circuits, supplied either from panelboards or switchboards or from taps to a feeder or subfeeder as shown. Figure 3.61, a typical layout of motor branch circuits in a fan room on top of a hospital, shows the circuit symbols and design information which must guide the installer in carrying out the plan. Other data on the motors and controls are incorporated in the job specifications. In Fig. 3.61, the distribution diagram for the boiler plant of a large medical research build-

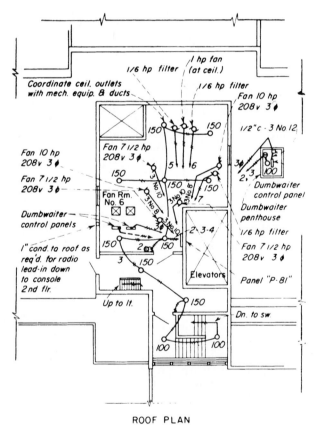

ROOF PLAN

FIGURE 3.61 This kind of layout detail fully conveys to the installer the designer's intent regarding the circuiting of motors.

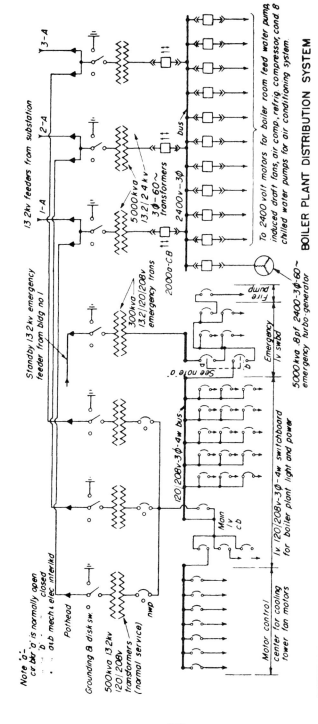

FIGURE 3.62 One-line diagrams on electrical plans should fully detail the power-supply sequence for motor circuits at all voltages.

POWER PANEL SCHEDULE

Panel Designation	Location	Branch Cct. Equip.	Mains Type	Mains Cap	Branch Circuits No.	Poles	Frame	Trip Calibr.	Service Phase	Service Voltg.	Conductor No.	Conductor Size	Conductor Type	Motor Equipment	H P
PHP	Pent House	Cct. Bkr.	Lugs Only	NEMA Std.	1	3	225	125	3	120/208	3	2	R	Inter. Zone Fan	25
						3	100	50	3	"	3	8	R	Exter. Zone Fan	10
					2	3	50	15	3	"	3	12	R	Booster Pump	2
					3	3	50	15	3	"	3	12	R	Toilet Exh. Fan	3/4
					4	3	50	15	3	"	3	12	R	Dehumid. Spray Pump	1/2
					5	3	50	15	3	"	3	12	R	Power Roof Ventilator	1
					6	3	50	15	1	"	3	12	R	" "	1
					7	2	50	20	1	"	3		R	Ltg. Panel LPPH1A	
					8	3	50	15	3	120/208	3	12	R	Roof Vent	13/8
					9	3	50	Spares	1						
					10,11,12										
PP3	3rd Fl Mech. Rm.	Cct. Bkr.	Lugs Only	NEMA Std.	1				3	120/208	3	12	R	Dehum. Spray Pump	1/2
					2	3	225	15	3	"	3	4	R	A C Unit "C"	20
					3	3	50	100	3	"	2	12	R	Filter Motor	1/6
					4	3	50	15	1	"	2	12	R	Bd. Rm. Exh. Fan	1/4
					5	3	50	15	3	120/208	3	12	R	A C Unit "J"	1/2
					6	3	50	Spare							
					7	3	50	Spare							
					8	3	50	Spare							
BPP	Bsm't Power	Cct. Bkr.	Lugs Only	NEMA Std.	1				3	120/208	3	12	R	Evap. Cond. Fan	1
					2	3	50	15	3	"	3	12	R	Ch. D.W. Pump	1 1/2
					3	3	50	15	3	"	3	12	R	Ch. D.W. Pump	2
					4	3	50	15	3	"	3	12	R	D.W. Compr.	3
					5	3	50	30	3	"	3	12	R	D.W. Compr.	3
					6	3	50	30	1	"	2	12	R	Evap. Cond. Spray Pump	1/3
					7	1	50	20	1	"	2	12	R	D.W. Pump	1/4
					8	1	50	20	1	"	2	12	R	" "	1/4
					9	3	50	20	3	120/208	2	12	R	Boiler Cont.	1/4
					10,11,12	3	50	Spares							

FIGURE 3.63 Power-panel schedule should summarize the ratings and layout of all motor-circuit equipment.

ing, the motor branch circuits are derived from a 3-phase, 4-wire, 120/208-V motor-control center for fan motors; from a 120/208-V switchboard for general-purpose motor applications; and from a 2400-V, 3-phase circuit-breaker switchboard supplying 2400-V circuits to motors for pumps, compressors, air-conditioning systems, etc. The successful design of motor branch circuits depends upon a thorough understanding of the ways in which the required types of protection may be combined.

Where a particular installation involves a large number of motors, the electrical plans should include one or more clear, detailed motor-circuit schedules. As shown in Fig. 3.63, power panels or motor control centers supplying motor branch circuits should be described as to both their feeds and the circuits they serve. Data on branch-circuit protection include frame size and trip calibration for CBs.

CHAPTER 4
FEEDERS FOR LIGHTING AND POWER

In any electrical system, the distribution system comprises the equipment and methods used to carry power from the service equipment to the overcurrent devices protecting the branch circuits. This system is a layout of equipment designed to provide the right amount of current at the right voltage to each utilization outlet. The distribution system carries power to lighting panelboards, power panelboards, motor-control centers, and the branch-circuit protective devices for individual motor or power loads.

Depending upon the type of building, the size and nature of the total load, various economic factors, and local conditions, a distribution system may operate at a single voltage level or may involve one or more transformations of voltage. A distribution system may also incorporate changes in the frequency of ac power or rectification from ac to dc power.

Figure 4.1 shows the basic layout of a distribution system and the terms used to refer to various parts of the system. Typical distribution systems can vary widely in size and complexity and will commonly involve the range of system components shown in Fig. 4.2. Because the terminology of distribution systems is frequently misused or misunderstood, it is best to begin this section with several definitions:

- *Mains* are the conductors extending from the utility service terminals at the building wall (or generator or converter bus) to the service switch or to the main distribution center.
- A *feeder* is a set of conductors originating at a main distribution center and feeding one or more subdistribution centers, one or more branch-circuit distribution centers, one or more branch circuits (as in the case of plug-in busway or motor-circuit taps to a feeder), or a combination of these. It may be a primary- or secondary-voltage circuit, but its function is always to deliver a block of power from one point to another point at which the power capacity is apportioned among a number of other circuits.
- A *lighting feeder* is a feeder to a load which is made up primarily of lighting circuits.
- A *power feeder* is a feeder to a load of branch circuits for motors, heating, or other power loads.

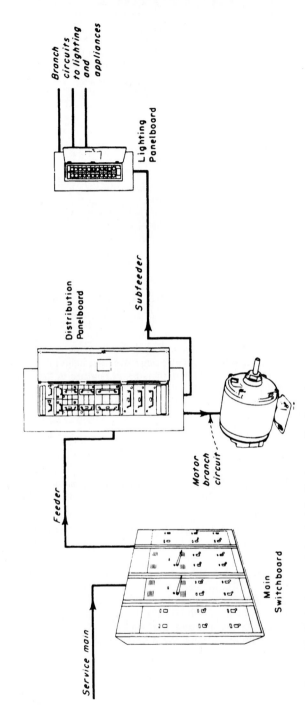

FIGURE 4.1 Basic elements of electrical distribution at utilization voltage.

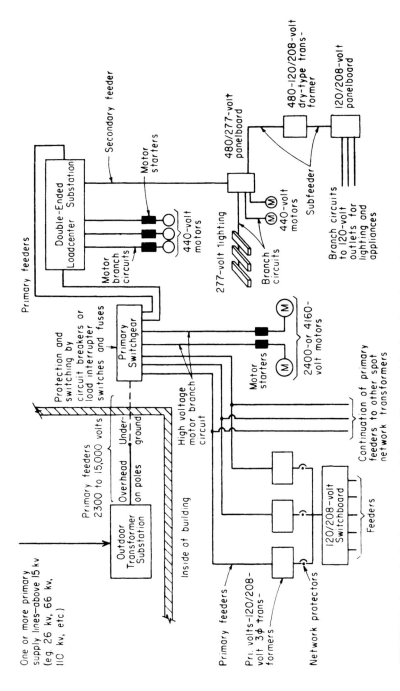

FIGURE 4.2 Large distribution systems may include a wide range of circuit configurations and voltages.

- A *subfeeder* is a set of conductors originating at a distribution center other than the main distribution center and supplying one or more other distribution panelboards, branch-circuit panelboards, or branch circuits.
- A *switchboard* is a large single panel, frame, or assembly of panels with switches, overcurrent and other protective devices, and usually instruments mounted on the front and/or the back. Switchboards are generally accessible from the back as well as from the front and are not intended to be installed in cabinets.
- A *panelboard* is a single panel or a group of panel units assembled in the form of a single panel. It contains busses tapped by fuse holders, with or without switches, or by circuit breakers, providing protection for—and, if switches or circuit breakers are used, also providing control of—circuits for light, heat, or power. These circuits may be branch circuits or subfeeders. A panelboard is designed to be placed in a cabinet or cutout box placed in or against a wall or partition and accessible only from the front.

It should be noted that manufacturers commonly refer to light-duty panelboards as *loadcenters*. But **NEC** and UL regulations do not distinguish between light- and heavy-duty panelboards. All those regulations apply identically to panelboards and so-called loadcenters.

Distribution Voltage Levels

Distribution systems are basically classified according to the voltage levels used to carry the power either directly to the branch circuits or to loadcenter transformers or substations at which feeders to branch circuits originate. Figure 4.3 shows three basic layouts utilizing the loadcenter concept—carrying a feeder to the center of a layout of utilization loads and then subdividing the feeder capacity among branch circuits for motor and/or lighting loads.

Common distribution systems operate at the following voltage levels:

1. *120/240-V, 3-wire, single-phase combination light and power distribution (also called the 115/230-V system).* This system is restricted to applications where the total load is small and is primarily lighting and receptacle circuits. It operates with the neutral wire grounded.

The most common application is in residential occupancies—individual homes or multifamily dwellings. This system is also used in stores, small schools, and other small commercial occupancies. In most small commercial buildings, however, the use of 120/208-V, 3-phase distribution offers greater economy owing to the higher operating efficiency of 3-phase circuits.

In those cases where 120/240-V feeders are used as the basic distribution method, the service to the premises is made at that voltage. Of course, 120/240-V distribution is frequently an effective and economical system for lighting subfeeder distribution in electrical systems which use a higher-voltage basic distribution system with loadcenter step-down to utilization voltages for local and incidental lighting and receptacle circuits.

2. *120/208-V wye, 3-phase, 4-wire light and power distribution.* This is the most common type of system for commercial buildings, institutional occupancies, and industrial shops with limited electrical loads. Such a system operates with the neutral conductor grounded; it offers substantial economy over the 120/240-V system with regard to the amount of conductor material required to carry a given amount of

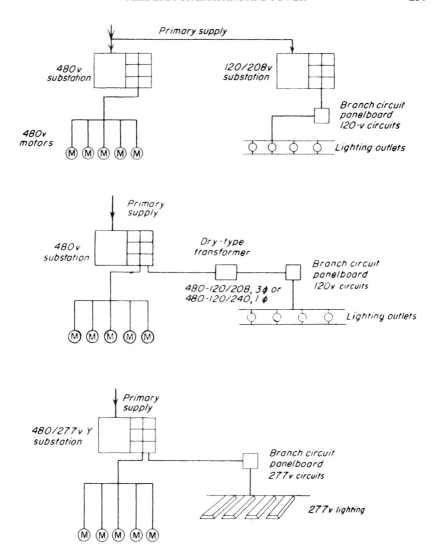

FIGURE 4.3 Feeders to loadcenter substations, transformer distribution centers, and/or panels provide an effective, economical design technique.

power to a load. This combination system provides 120 V phase to neutral for lighting and single-phase loads, and 208 V phase to phase for single- or 3-phase motor or other power loads.

The system is used as the basic distribution system in occupancies in which the service to the building is of this voltage. It is also the most common subdistribution system for lighting and receptacle circuits in occupancies using higher-voltage distribution to loadcenters. Figure 4.4 shows two examples of 120/208-V systems.

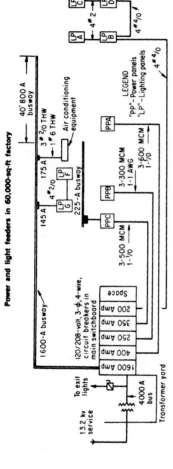
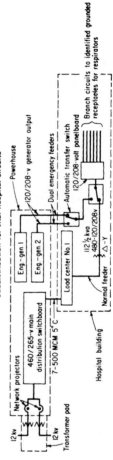

FIGURE 4.4 Utilization-equipment voltage requirements dictate extensive use of 120/208-V distribution.

3. *240-V, 3-phase, 3-wire distribution.* This is a common system for power loads in commercial and industrial buildings. It may operate ungrounded or with one of the phase legs grounded—in which case it is called a "corner-grounded delta system." In such cases, service to the premises is 240-V, 3-phase power. Feeders carry the power to panelboards or gutters supplying branch circuits for motor loads. Lighting loads are then handled by a separate single-phase service to the building or through a step-down transformer.

This system offers economical application where the power load is large compared to the lighting load. In some 240-V, 3-phase systems, a tap on one of the phases is brought out as a neutral to provide 120 V for lighting and receptacle circuits, forming a 4-wire system.

4. *480-V, 3-phase, 3-wire distribution.* This system is commonly used in commercial and industrial buildings with substantial motor loads. It is generally operated ungrounded. Such a system provides a high level of power continuity because a ground fault on one phase will not cause automatic operation of a protective device to open the circuit, as in grounded systems. But it can be operated with one phase leg grounded—as a corner-grounded delta system. Service to the building may be made at this voltage, with the 480-V feeders carried to motor loads and to step-down transformers for lighting and receptacle circuits.

In many cases, 480-V feeders will be derived from loadcenter substations within the building and carried to motor loads or power panels (Fig. 4.5).

5. *480Y/277-V (460Y/265-V), 3-phase, 4-wire combination power and light distribution.* This is a very popular basic system for use in commercial and industrial buildings. In office and other commercial buildings, the 480-V, 3-phase, 4-wire feeders are carried to each floor, where 480-V, 3-phase, 3-wire power is tapped to a power

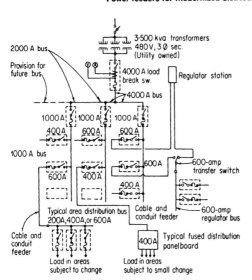

FIGURE 4.5 480-V, 3-phase, 3-wire ungrounded distribution offers a high degree of power continuity.

254   CHAPTER FOUR

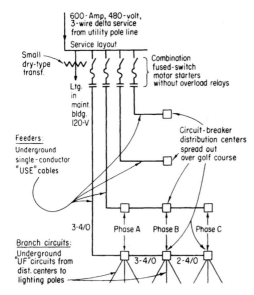

FIGURE 4.5 *(Continued)*

panel or to motors; general area fluorescent lighting using 277-V ballasts is connected between each phase leg and the neutral. Required 120/280-V, 3-phase, 4-wire circuits are derived from step-down transformers for local lighting and appliance and receptacle circuits.

This system offers the advantage of economy over a 120/208-V system when fewer than about half the load devices require 120- or 208-V power. Where the 480Y system can be used, it costs less than the 120/208-V system, owing to savings on the smaller sizes of conductors, reduced raceway needs, reduced installation labor costs, and the lower cost of system elements, which have lower current capacities.

If the required amount of 120- or 208-V power is over half the total load in a building, the cost of the step-down transformers required to supply these circuits will generally offset the savings in the 480-V circuiting. The costs of switchgear and panelboards also affect the cost comparison between 120/208- and 480/277-V systems.

The 480Y system is generally more advantageous in multifloor buildings than in buildings of only a few floors and is more advantageous in large or long single-level areas than in small areas. The savings are directly related to reductions in the sizes of long feeders, allowing more kilovoltamperes to be delivered per pound of installed system. (See Figs. 4.6 and 4.7).

6. *2400-V, 3-phase delta distribution.* This is an industrial-type system used to feed heavy motor loads directly, and motor and lighting loads through loadcenter substations and lighting transformers.

7. *4160Y/2400-V, 3-phase, 4-wire distribution with a grounded neutral.* This system is more common than the 2400-V delta-connected system. Originally developed primarily for industrial applications, this system of high-voltage distribution has gained wide use for large-area, spread-out commercial and institutional buildings, such as shopping centers, schools, and motels.

POWER AND LIGHT FEEDERS
IN NEW INDUSTRIAL PLANT

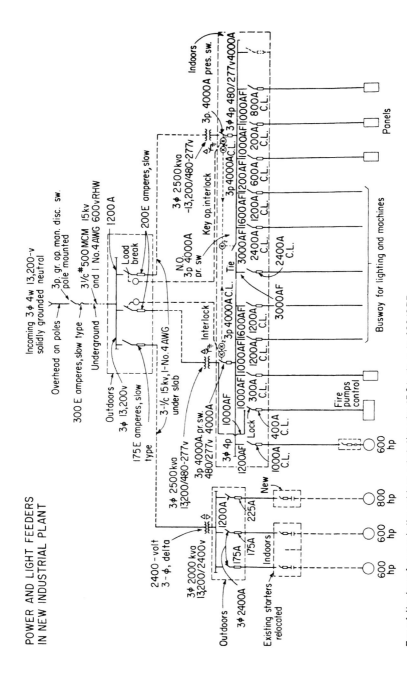

Two of the transformers in the outdoor substation step 13.2 kv down to 480/277.
The third outdoor transformer steps 13.2 down to 2400 volts, 3-phase, 3-wire delta for circuits to three large machine motors.

FIGURE 4.6 480/277-V distribution plus 2400-V delta serves lighting and heavy power loads.

255

256 CHAPTER FOUR

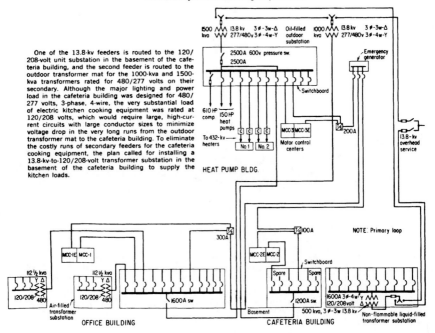

FIGURE 4.7 480/277-V plus 13.8-kV distribution provide effective design for long feeders.

The 4160Y/2400-V system also appears to have become the most popular high-voltage system for inside distribution. Economy in the use of cables and switchgear seems to be greatest for this particular primary system. In many installations, it offers the advantages of loadcenter primary distribution for power and lighting loads with direct connection of heavy motor loads operating at 4160 V.

This system is widely used to supply loadcenter substations in which the voltage is stepped to 480 V to feed motors and lighting transformers for 120/240- and/or 120/208-V circuits. It may also be used for distribution to substations stepping the voltage directly to 120/208 V. (See Figs. 4.8 and 4.9).

8. *4800-V distribution.* This is a delta-connected industrial system for layouts of substations supplying motors and lighting transformers.

9. *7200-V, 3-phase distribution.* This is another industrial system used with substations for stepping voltage to lower levels for power and lighting.

10. *13.2Y/7.2-kV (or 13.8-kV), 3-phase, 4-wire distribution.* This is a modern distribution system, widely used for large industrial plants. Power at this voltage is delivered to substations which step it to 480 V for motor loads and which supply 480-to-120/240-V or 480-to-120/208-V transformers for lighting. Or, 480Y/277-V substations may be used to supply motor loads and 277-V incandescent, fluorescent, or mercury-vapor lighting for office and industrial areas. Lighting transformers are then used to supply 120 V for lighting and receptacles. Figure 4.10 shows how this voltage level is used for distribution in a high-rise apartment building.

Feeders to basement and roof vaults in high-rise apartment building

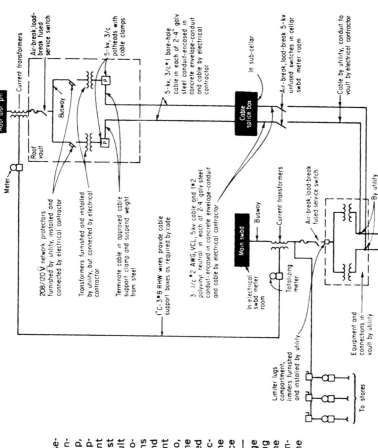

The basement vault supplies basement loads—oil burners, air-conditioning equipment, a 75-hp fire pump, garage loads, pool loads, laundry equipment, etc.—and risers to apartment panels, up to and including the first 21 floors of the building. The roof vault supplies typical roof loads, elevator motors, exhaust fans, cooling tower fans and mechanical auxiliary motors—and feeders running down to apartment panels from the 33rd floor down to, and including, the 22nd floor. Thus, the entire building load is clearly divided between the two vaults. The only secondary-voltage connection between the basement service and the roof service consists of two small conduit circuits—one to connect the secondary-voltage electric meter on the roof to a totalizing meter at the basement service, and one control conduit between air-conditioning equipment in the basement and the cooling tower on the roof.

FIGURE 4.8 4160-V distribution minimizes voltage drop and affords great equipment economy.

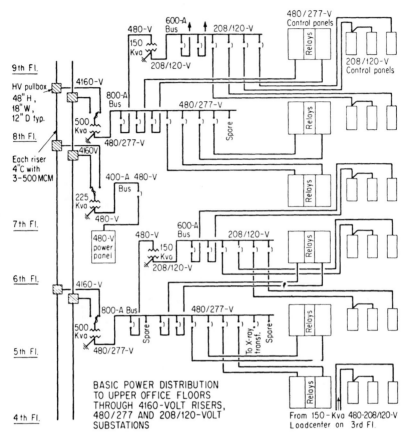

FIGURE 4.9 4160-V risers to loadcenter transformers provide low losses and low voltage drop.

The voltage values given for these distribution systems are subject, of course, to the usual variation or spreads due to distance of transmission and distribution, local conditions of utility supply, and settings of transformer taps. Figure 4.11 shows how a high-voltage distribution system can be derived from a 120/208-V service. This is the service arrangement for a multibuilding suburban hospital facility. The utility company would supply power only at 120/208 V for this modernization project, but the engineer provided for immediate step-up of the 120/208 V to 4160Y/2400 V to supply an underground distribution system. The cost of the step-up transformers was more than offset by the savings due to high-voltage distribution.

There are times when the selection of distribution secondary voltages other than those described offers a better application, or is required. (See Figs. 4.12 and 4.13.) Other high-voltage systems may operate at 6.6, 8.3, 11, and 12 kV.

In all systems, operating at various voltage levels, there may be a need for equipment to modify the electrical characteristics of circuits. The characteristics of the distribution voltage often have to be modified or refined for such purposes as those indicated in Fig. 4.14. In such cases, the designer must analyze the load served from

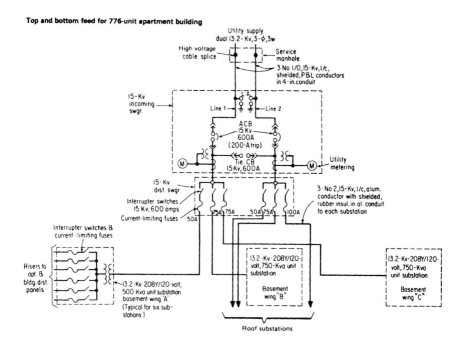

FIGURE 4.10 13.2-kV distribution supplies upper and lower halves of high-rise building.

the various types of equipment and determine the magnitude and characteristics of each load as it is reflected back into the distribution system. Manufacturers' instructions should be carefully followed, and the engineering assistance of the manufacturer should be sought in extensive or complex installations.

Of the high-voltage (over 600 V) distribution systems, 4160- and 13,200-V systems are the most common; they represent good design selection and economy of application for most installations. However, selection of the primary voltage depends upon the size and layout of the load and the supply voltages which the utility makes available. Consultation with the utility must precede any design for high-voltage distribution.

Modern design practice has favored the use of 13-kV systems over other high-voltage distribution systems for large industrial plants, and the use of 4160/2400-V systems for spread-out commercial and institutional buildings. The total system load, layout, and load density will dictate the best high voltage for any specific system.

For any given system voltage configuration, the industry has more or less standardized on three different voltage values which may be used to describe the system: (1) the transformer voltage, (2) the nominal system voltage, and (3) the utilization voltage. For a given system these three levels account for the voltage drop from the source to the load. As shown in Fig. 4.15, for instance, the 480/277-V system has 480 V at the source or transformer; 460 V is the nominal system voltage; and 440 V is delivered to the 440-V-rated motor loads.

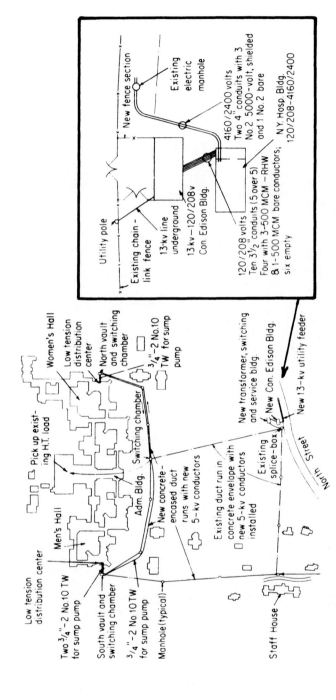

FIGURE 4.11 Low-voltage service can often be stepped up to provide a high-voltage distribution system.

Two-stage step-down transformer hookup serves the ungrounded circuits formerly required in a hospital operating room. The **NE CODE** generally requires that circuits in areas used for anesthetizing—operating rooms, delivery rooms, anesthesia rooms and corridors and utility rooms used for administering flammable anesthetics—must be supplied from an ungrounded distribution system which is isolated from any distribution system supplying areas other than anesthetizing locations [**NE CODE** Section 517-104 of the '81 Code.] Such isolation is generally provided by means of two-winding transformers with no electrical connection between primary and secondary (that is, primary to secondary energy transfer must be made only by magnetic induction). It would seem, just from this, that the circuits could be provided by simply installing single-phase transformers for 480 to 120/240 volts or 3-phase transformers for 480-volt delta to 120/208-volt wye. But there is a further complication.

Section 517-104 of the **NE CODE** used to state that: "Circuits supplying primaries of isolating transformers [as required by Section 517-104] shall operate at not more than 300 volts between conductors..." That ruled out the use of any transformer with a 480-volt primary for stepping down to the ungrounded distribution system for anesthetizing areas. As a result, a two-stage transformation was necessary to obtain the ungrounded system from the 480-volt delta system. This transformation was made in one of the roof penthouses where a 480-volt 3-wire feeder from the main switchboard comes into a hookup of three 5-kva, single-phase transformers connected for 480-volt delta down to 240-volt delta. The secondary conductors are brought into a trough from which single-phase, 2-wire subfeeds are tapped to single-phase transformers (2 and 3 kva sizes) connected for 240 volts down to 120

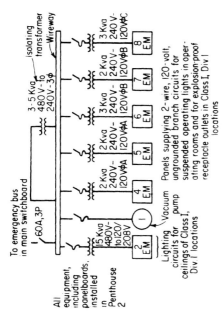

volts. These small transformers provide required isolation of their secondary, 2-wire ungrounded subfeeds to panels for the ungrounded branch circuits. The small single-phase transformers are divided among the three phases of the 240-volt delta secondary of the 3-phase transformer bank.

Now, however, such double transformation is *not* required. The rules covering isolated power systems, covered in Sec. 517-160 (a)(2), now permit primary supply circuits to be rated at up to 600v between conductors.

FIGURE 4.12 Special distribution voltage levels are sometimes dictated by job conditions, such as a hospital's need for 240-V, 3-phase ungrounded feeds to isolation transformers.

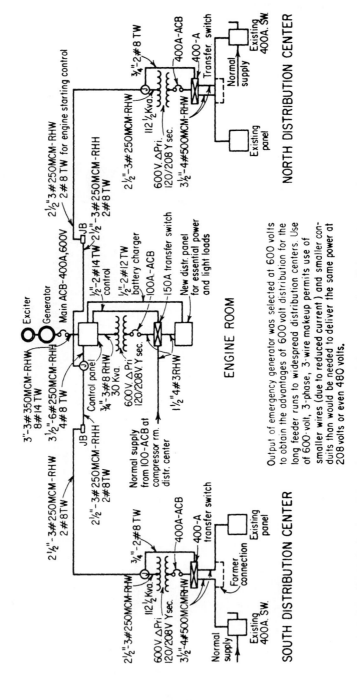

FIGURE 4.13 600-V distribution offers maximum economy and minimum current by utilizing conductors and system equipment at their 600-V operating limit.

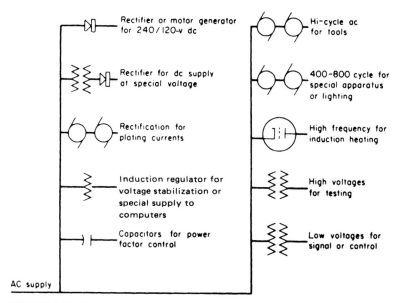

FIGURE 4.14 Characteristics of the building distribution system must often be altered to supply special load equipment.

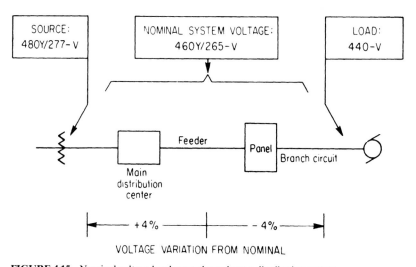

FIGURE 4.15 Nominal voltage levels vary throughout a distribution system.

Distribution Design

The design of a distribution system is a matter of selecting circuit layouts and equipment to accomplish electrical actions and operations necessary for the conditions of voltage, current, and frequency required by the loads. This means that such factors as service voltage, distribution voltage or voltages, conductors, transformers, converters, switches, protective devices, regulators, and power-factor corrective means must be related to economy, load conditions, continuity of service, operating efficiency, and future power requirements. The factors of capacity, accessibility, flexibility, and safety, which were discussed previously, must be carefully considered in designing for distribution.

Basic rules for distribution design are as follows:

1. Determine the magnitude and characteristics of all individual loads and load groupings.
2. Locate the one or more electric supply points as near as possible to building loadcenters.
3. Select and arrange feeders and other distribution equipment to provide the power continuity required by the commercial or industrial functions of the building.
4. Constantly relate the building's electrical requirements to the system characteristics of flexibility, accessibility, and regulation.
5. Provide a calculated amount of spare capacity in all system components from the supply to the load devices, carefully correlating feeder and subfeeder spare capacities to realistic demand expectations.
6. Use modern loadcenter layouts where possible.
7. In design calculations, observe minimum and maximum **NEC** values for conductors, conduit, protective devices, switches, and control equipment. Note that **NEC** standards must frequently be surpassed to obtain required convenience, flexibility, and effectiveness, and greater safety.

In an electrical system, the *feeders* are those conductors (usually insulated conductors in conduit, EMT, or busway) that carry electric power from the service equipment (or generator switchboard, where power is generated on the premises) to the overcurrent protective devices for branch circuits supplying the various loads. *Main feeders* originate at the service equipment and may supply switchboards or panelboards where either subfeeders or branch circuits originate. *Subfeeders* originate at a switchboard, panelboard, or distribution center other than the service equipment or generator switchboard and supply one or more other distribution panelboards, branch-circuit panelboards, or branch circuits.

Feeders and subfeeders are sized to provide sufficient power to the circuits they supply. For the given circuit voltage, they must be capable of carrying the amount of current required by the load, plus any current which may be required in the future. The selection of the size of a feeder depends upon the size and nature of the known load, computed from branch-circuit data, anticipated future load requirements, and voltage drop. This represents one of the most important engineering tasks in electrical design work. The economy and efficiency of system operation and maintenance depend heavily on the selection of the proper size of feeders.

Experience with today's electrical modernization work has revealed feeder capacity as the big bottleneck in rewiring old buildings—and even some that are not so old. In those buildings, the cost of raising the capacity of the electrical system to meet modern load requirements would be considerably less if the original design of the feeders had been based on sound study of known and anticipated future loads. All this experience confirms the importance of careful sizing of the feeders, through calculations related to the particular conditions of the job and not just by a mechanical procedure of adding up load watts and dividing by volts to get the required current-carrying capacity.

MINIMUM FEEDER CAPACITY

According to the **NEC**, each feeder must be sized to carry a computed load current that is not less than the sum of the branch-circuit load currents it supplies, with certain qualifications:

1. For general illumination, a feeder must have sufficient capacity to carry the total load of the lighting branch circuits, determined as part of the lighting design; however, regardless of the actual connected branch-circuit loads, the feeder conductors must have an ampacity that is not less than the minimum branch-circuit load determined on a watts-per-square-foot basis from the table given in **NEC** Section 220-3(b), as shown in Fig. 4.16 (See Chapter 9 of this handbook for tables of unit loads in watts per square foot for hundreds of general lighting applications in commercial and industrial buildings.) Of course, if the actual connected load supplied by a feeder is greater than the total minimum required branch-circuit load computed on a watts-per-square-foot basis, then the feeder conductors must have ampacity at least equal to the actual connected load.

When a feeder supplies a continuous load or a combination of continuous and noncontinuous loads, the rating of the feeder overcurrent device must not be less than the noncontinuous load plus 125 percent of the continuous load. However, if the overcurrent protection for the feeder is an assembly approved for operation at 100 percent of its rating, the feeder protective device must simply be at least equal to the sum of the continuous load and the noncontinuous load.

NEC Section 220-4 makes clear that a feeder supplying a branch-circuit panelboard must have a minimum ampacity sufficient to serve the calculated total load of lighting, appliances, motors, and other loads supplied. And the amount of feeder and panel ampacity required for the general lighting load must not be less than the current value determined from the circuit voltage and the total power (watts) resulting from multiplying the minimum unit load from Table 220-3(b) (in watts per square foot) by the area of the occupancy supplied by the feeder—even if the actual connected load is less than the calculated load determined on a watts-per-square-foot basis. But, as noted above, if the connected load is greater than that calculated on a watts-per-square-foot basis, the greater value of load must be used in determining the number of branch circuits, the panelboard capacity, and the feeder capacity. (Refer back to the discussion, under "Branch-circuit panelboards," covering the minimum sizing of feeder conductors where the total "computed" load is greater than the actual connected load.)

It should be carefully noted that the first sentence of Section 220-4(d) states, "Where the load is computed on a watts-per-square-foot (0.093 sq m) basis, the wiring system up to and including the branch-circuit panelboard(s) shall be provided to serve not less than the calculated load." Use of the phrase "wiring system up to and including" requires that a feeder must have capacity for the total minimum branch-circuit load, determined from the area in square feet times the minimum unit load [voltamperes per square foot from Table 220-3(b)]. And that phrase clearly

Table 220-3(b). General Lighting Loads by Occupancies

Type of Occupancy	Unit Load per Square Foot (Volt-Amperes)
Armories and Auditoriums	1
Banks	3½**
Barber Shops and Beauty Parlors	3
Churches	1
Clubs	2
Court Rooms	2
Dwelling Units*	3
Garages — Commercial (storage)	½
Hospitals	2
Hotels and Motels, including apartment houses without provision for cooking by tenants*	2
Industrial Commercial (Loft) Buildings	2
Lodge Rooms	1½
Office Buildings	3½**
Restaurants	2
Schools	3
Stores	3
Warehouses (storage)	¼
In any of the above occupancies except one-family dwellings and individual dwelling units of two-family and multifamily dwellings:	
Assembly Halls and Auditoriums	1
Halls, Corridors, Closets, Stairways	½
Storage Spaces	¼

For SI units: one square foot = 0.093 square meter.

*All general-use receptacle outlets of 20-ampere or less rating in one-family, two-family, and multifamily dwellings and in guest rooms of hotels and motels [except those connected to the receptacle circuits specified in Sections 220-4(b) and (c)] shall be considered as outlets for general illumination, and no additional load calculations shall be required for such outlets.

**In addition, a unit load of 1 volt-ampere per square foot shall be included for general-purpose receptacle outlets where the actual number of general-purpose receptacle outlets is unknown.

FIGURE 4.16 Feeder conductors must always have capacity for the **NEC** minimum branch-circuit load.

requires that amount of capacity to be allowed in every part of the distribution system supplying the load. The required capacity would, for instance, be required in a subfeeder to the panel, in the main feeder from which the subfeeder is tapped, and in the service conductors supplying the whole system.

Reference to the "wiring system" in Section 220-4(d) presents a requirement that goes beyond the heading "Branch Circuits Required" of Section 220-4 and, in fact, constitutes a requirement on feeder capacity that supplements the rule of the second sentence of Section 220-10(a). This requires a feeder to be sized to have enough capacity for "the computed load" as determined by Part **A** of this article (which means as computed in accordance with Section 220-3).

2. In sizing feeder conductors on the basis of a voltamperes-per-square-foot calculation of the minimum required ampacity of a feeder, the demand factors given in **NEC** Table 220-11 may be applied to the total calculated branch-circuit load to get a reduced value of minimum required feeder capacity for lighting (but those demand factors must not be used in calculating branch-circuit capacity; Fig. 4.17).

Table 220-11. Lighting Load Feeder Demand Factors

Type of Occupancy	Portion of Lighting Load to which Demand Factor Applies (volt-amperes)	Demand Factor Percent
Dwelling Units	First 3000 or less at	100
	From 3001 to 120,000 at	35
	Remainder over 120,000 at	25
Hospitals*	First 50,000 or less at	40
	Remainder over 50,000 at	20
Hotels and Motels — Including Apartment Houses without Provision for Cooking by Tenants*	First 20,000 or less at	50
	From 20,001 to 100,000 at	40
	Remainder over 100,000 at	30
Warehouses (Storage)	First 12,500 or less at	100
	Remainder over 12,500 at	50
All Others	Total volt-amperes	100

*The demand factors of this table shall not apply to the computed load of feeders to areas in hospitals, hotels, and motels where the entire lighting is likely to be used at one time, as in operating rooms, ballrooms, or dining rooms.

FIGURE 4.17 Demand factors may be applied to the total branch-circuit load in sizing feeders.

3. If show-window lighting is supplied by the feeder, capacity must be included in the feeder to handle 200 W per linear foot of show-window length.

4. In single-family dwellings, in individual apartments of multifamily dwellings with provisions for cooking by tenants, or in a hotel or motel suite with cooking facilities or a serving pantry, at least 1500 W for each 2-wire, 20-A small-appliance circuit (to handle the small-appliance load in kitchen, pantry, and dining areas) must be added to the general lighting load on each feeder. The total small-appliance load may be added to the general lighting load and may be reduced by the demand factors given in **NEC** Table 220-11.

5. A feeder load of at least 1500 W must be added for each 2-wire, 20-A laundry circuit installed as required by Section 220-4(c). And that load may also be added to the general lighting load and reduced according to the demand factors of Table 220-11.

6. In dwelling units, receptacle outlets of 20 A or less, other than receptacles on the two or more required 20-A small-appliance circuits in kitchens and dining rooms, are considered to be outlets for general lighting. In other than dwelling occupancies, the branch-circuit load for receptacle outlets that are taken as a load of 180

VA per outlet may be added to the general lighting load and may also be reduced by the demand factors in Table 220-11.

Prior to the 1978 **NEC**, Section 220-13 required a feeder to have ampacity for a load equal to 180 VA times the number of general-purpose receptacle outlets that the feeder supplies, and a feeder supplying such a load had to have a capacity for 100 percent of that total load in occupancies that come under the heading "All Others" at the bottom of Table 220-11 (such as office buildings, stores, and schools). But the **NEC** no longer requires that approach. Recognizing that there is great diversity in the use of receptacles in office buildings, stores, schools, and the occupancies that come under "All Others" in Table 220-11, Section 220-13 contains a table to permit reduction of the feeder capacity for receptacle loads on feeders in such occupancies (Fig. 4.18). The minimum required feeder capacity for a typical case where, say, an office-building feeder supplies two or more panelboards feeding a total of 500 receptacles is shown in Fig. 4.19.

ARTICLE 220 — BRANCH-CIRCUIT, FEEDER, AND SERVICE CALCS **70–55**

**Table 220-13.
Demand Factors for Nondwelling Receptacle Loads**

Portion of Receptacle Load to which Demand Factor Applies (volt-amperes)	Demand Factor Percent
First 10 kVA or less	100
Remainder over 10 kVA at	50

FIGURE 4.18 Calculated receptacle load may be reduced by demand factors.

Although the calculations of Fig. 4.19 cannot always be taken as realistically related to usage of receptacles, it is relief from the 100 percent demand factor, which presumed that all receptacles were supplying 180-VA loads simultaneously in those occupancies that came under "All Others."

If the 500 receptacles in Fig. 4.19 were in a hospital instead of an office building, the capacity of a feeder supplying that total receptacle load could be determined with the demand factors of Table 220-11. As shown, the feeder capacity for that load in a hospital would have to be only 28,000 VA, instead of 50,000 VA.

7. Feeder capacity must be allowed for "household" electric cooking appliances rated over 1¾ kW, in accordance with **NEC** Table 220-19. Feeder demand loads for a number of cooking appliances on a feeder may be obtained from Table 220-19.

8. "Commercial" electric cooking loads must comply with Section 220-20 and its table of demand factors.

9. For fixed appliances (fastened in place) other than ranges, clothes dryers, air-conditioning equipment, and space-heating equipment, feeder capacity in dwelling occupancies must be provided for the sum of their loads, and the total load of four or more such appliances may be reduced with a demand factor of 75 percent (**NEC** Section 220-17).

10. **NEC** Article 430 (Sections 430-24 to 430-26) gives the required capacity of a feeder that supplies motor loads as well as appliance and lighting loads as described above. The feeder capacity for continuous-duty motor loads is taken at 125 percent

Take the total number of general-purpose receptacle outlets fed by a given feeder. . .

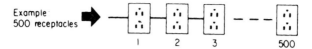

. . . multiply the total by 180 voltamperes [required load of Section 220-3(c) (6) for each receptacle]. . .

500 × 180 VA = 90,000 VA

Then apply the demand factors from Table 220-13:

First 10 kw or less @ 100% demand   = **10,000** VA
Remainder over 10 kw @ 50% demand
= (90,000 − 10,000) × 50%
= 80,000 × 0.5   = **40,000** VA

Minimum demand-load total = **50,000** VA

Therefore, any feeder (or service-entrance conductors) that supplies those receptacles as part of its total load must include a current-carrying capacity of at least 50 kVA for the total receptacle load. Required minimum ampacity for that load is then determined from the voltage and phase makeup (single or 3 phase) of the feeder.

But if the 500 receptacles are in a hospital:

500 receptacles × 180 VA = 90,000 VA

Then apply the demand factors from Table 220-11:

First 50 kW or less @ 40% demand   = **20,000** VA
Remainder over 50 kW @ 20% demand
= (90,000 − 50,000) × 20% = 40,000 × 20%   = **8,000** VA

Minimum demand-load total = **28,000** VA

FIGURE 4.19 How demand factors are applied to office or other nondwelling areas.

of the full-load current rating of the largest motor supplied plus the sum of the full-load currents of the other motors supplied. For a motor load in which some or all of the motors do not operate continuously (such as motors on short-time, intermittent, periodic, or varying-duty cycles), the feeder capacity must be sized according to the procedure set forth in Section 430-24 or 430-25.

11. The capacity of a feeder supplying fixed electric space-heating equipment is determined on the basis of a load equal to the total connected load on all branch circuits served from the feeder. Under conditions of intermittent operation or where all units cannot operate at the same time, permission may be granted for use of less than a 100 percent demand factor in sizing the feeder. Sections 220-30 to 220-32 permit alternative calculations of electric heating loads in dwelling units.

12. Feeder capacity must be included for air-conditioning equipment loads. Figure 4.20 shows unit loads that can be used in determining the total load, and that is usually taken as a continuous load at 100 percent demand factor. When dissimilar loads (such as space heating and air cooling) are supplied by the same feeder, the smaller of the two loads may be omitted from the total capacity required for the feeder if it is "unlikely" that the two loads will operate at the same time. Note that the rule of **NEC** Section 220-21, which permits elimination of the smaller of two loads in calculating minimum required feeder capacity, does not require that a double-throw switch or similar control arrangement be provided to prevent simultaneous connection of, say, a heating load and an air-conditioning load to the feeder where the feeder is sized for only the larger of the two loads. The rule simply refers to "dissimilar loads" and requires only that it be "unlikely" that both loads are connected at the same time. Of course an "either-or" control arrangement can prevent overload and undesired opening of the feeder protective device, as shown in Fig. 4.21.

Type of Building	Watts per Square Foot of Floor Area of Conditioned Space		
	Low	Average	High
Banks	4	5	7
Department Stores			
Basement	4	5	10
Main Floor	5	7	12
Upper Floors	3	4	6
Hotel Guest Rooms (per room)	1	1.5	2
Office Buildings	3	4	6
Office Suites	3	3.5	5
Stores	2 to 6	4 to 7	5 to 10
Restaurants	10	15	20
Theatres (per seat)	65	75	80

FIGURE 4.20 Air-conditioning unit loads can facilitate load calculations and feeder sizing.

13. The feeder load for each household electric clothes dryer in a dwelling unit must be 5000 W or the nameplate rating, whichever is larger. The demand factors in Table 220-18 may be used where a feeder supplies a number of such dryers.

It should be noted that all the foregoing are general **NEC** requirements that cover minimum load conditions. This is particularly true of the voltamperes-per-square-foot unit loads given in the table. Feeder sizes determined solely on the basis of **NEC** requirements include no provision for the particular operating conditions of an occupancy, such as voltage stability and power losses, and no allowance for future growth in the load served by the feeder.

Although certain feeder applications can be satisfactorily sized with the **NEC** method, modern feeder design must carefully incorporate such factors as voltage drop, power factor, detailed analysis of watts-per-square-foot loads, realistic analysis and application of demand factors, and spare-capacity requirements.

FEEDERS FOR LIGHTING AND POWER   271

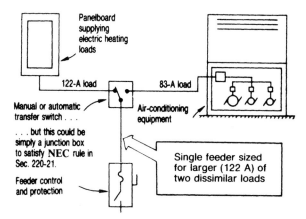

FIGURE 4.21 Transfer switching hookup (such as "heating/air conditioning") ensures proper circuit operation where a single feeder supplies dissimilar loads that should not be energized simultaneously.

Adding Branch-Circuit Loads

Based on the foregoing **NEC** rules, the required capacity of any feeder can be determined by adding up the branch-circuit loads supplied by the feeder. In earlier discussions of branch circuits, procedures were given for determining branch-circuit loads—for general lighting, fixed appliances, special receptacles, and general-use convenience receptacles—and for determining the number and characteristics of circuits required to handle the loads.

Whether general lighting loads are determined on a watts-per-square-foot basis or from a lighting design layout, the load for general illumination is established, and capacity for this load must be designed into the feeder, modified by the applicable demand factor from Table 220-11. Other circuit loads (such as for local and/or special lighting, motors, electric cooking appliances, and fixed appliances) are also generally known and can be accounted for in a feeder. Loads for convenience receptacle circuits and spare circuits in the panelboard, however, are not known and must be estimated in sizing a feeder.

From the known and estimated branch-circuit loads, the minimum capacity for a feeder to a number of branch circuits can be determined as follows:

1. For each multioutlet branch circuit supplying general lighting, allow voltamperes equal to the load on the circuit and not less than the minimum branch-circuit load set by the minimum watts per square foot of branch-circuit capacity—as described previously for **NEC** Sections 220-3 and 220-4(d).

2. For each multioutlet plug-receptacle branch circuit, allow voltamperes equal to the load on the circuit as used in branch-circuit design. The minimum allowance for each 120-V receptacle circuit must equal the number of receptacles (count a duplex receptacle as one) times 1.5 A times 120 V, 180 VA, unless a demand factor is considered acceptable for the expected use of the receptacles and is applied in accordance with Section 220-13 as described above.

Where multioutlet assemblies (plug-in strip or raceway) are used, each 5 ft or fraction thereof of each separate and continuous length is taken as a load equal to 1.5 A. And when such assemblies are located where a number of appliances are likely to be used simultaneously, each 1 ft or fraction thereof is taken as a load of 1.5 A.

3. For each multioutlet branch circuit supplying heavy-duty lampholders, allow voltamperes equal to the load on the circuit as used in branch-circuit design. The minimum allowance for each heavy-duty lampholder circuit must equal the number of outlets times 600 W (which is equal to 600 VA at unity power factor).

4. For each multioutlet branch circuit supplying local or special lighting, allow voltamperes equal to the load on the circuit. **NEC** Section 220-3(c) establishes the minimum loads that must be allowed in computing the minimum required branch-circuit capacity for general-use receptacles and "outlets not used for general illumination." Item (3) of that section requires that the actual voltampere rating of a recessed lighting fixture be taken as the amount of load that must be included in branch-circuit capacity. This permits local and/or decorative recessed lighting fixtures to be included at their actual load values rather than as "other outlets," which would require a load allowance of "180 VA per outlet"—even if such a fixture were lamped at, say, 25 W. (On the other hand, where a recessed fixture contained a 300-W lamp, an allowance of only 180 VA would be inadequate.)

5. For each individual branch circuit supplying a fixed appliance, allow voltamperes equal to the rating of the appliance.

6. For the sum of the loads of individual branch circuits supplying electric cooking appliances, allow kilowatts equal to the values given in Table 220-19 for the particular number and sizes of cooking appliances.

7. For each individual branch circuit supplying a motor, allow a voltampere load equal to 1.25 times the motor full-load current (from Tables 430-147 to 430-150 according to the horsepower of the motor) times the circuit voltage. For 3-phase motors, this product must be multiplied by 1.732 to obtain the correct load. In sizing conductors for required load currents, the voltampere load should always be used for devices operating at less than unity power factor; this will ensure the required current capacity, which is greater than would be required to supply only the wattage (effective power) represented by such loads.

8. For each branch circuit supplying more than one motor, allow voltamperes by proceeding as follows: Take 1.25 times the full-load current rating of the highest-rated motor; add to this the full-load current ratings of all the other motors; then multiply the sum by the circuit voltage. Again, this product must be multiplied by 1.732 to obtain the correct voltampere product for 3-phase motors.

9. For each branch circuit supplying show-window lighting, allow 200 W for each linear foot (measured horizontally) of show window served by the circuit.

By adding up these computed values, a total number of voltamperes is obtained which is the initial minimum feeder load. Of course, the majority of feeders used in commercial, institutional, and industrial buildings will serve only a few of the above-described loads. And feeders in different buildings will handle different combinations of loads.

For each spare branch circuit supplied by a feeder, voltamperes equal to the values given above for multioutlet general lighting branch circuits should be added to the feeder load.

The table in **NEC** Section 220-11 allows limited use of demand factors for general lighting circuits and multioutlet plug-receptacle circuits to reduce the feeder design load; however, most occupancies require a demand factor of 100 percent of the total load as obtained by adding all the circuit loads.

Using Demand Factors

Two terms constantly used in the literature of modern electrical design are "demand factor" and "diversity factor." Because there is a very fine difference between the meanings for the words, the terms are often confused.

- The *demand factor* is the ratio of the maximum demand of a system (or part of a system) to the total connected load on the system (or part of the system). This factor is always less than unity.
- The *diversity factor* is the ratio of the sum of the individual maximum demands of the various subdivisions of a system (or part of a system) to the maximum demand of the whole system (or the part of the system under consideration). This factor generally varies between 1.00 and 2.00.

Demand and diversity factors are used in system design. For instance, the sum of the connected loads supplied by a feeder is multiplied by the demand factor to determine the load which the feeder must be sized to serve. This load is termed the *maximum demand* of the feeder. The sum of the maximum demand loads for a number of subfeeders divided by the diversity factor for the subfeeders gives the maximum demand load to be supplied by the feeder from which the subfeeders are derived.

Tables of demand and diversity factors have been developed from experience with various types of load concentrations and various layouts of feeders and subfeeders supplying such loads. **NEC** Table 220-11 presents common demand factors for feeders to general lighting loads in various types of buildings, as shown in Fig. 4.16.

It is common and preferred practice in modern electrical design to take unity as the diversity factor in main feeders to loadcenter substations, to provide a measure of spare capacity. Main secondary feeders are also commonly sized on the full value of the sum of the demand loads of the subfeeders supplied.

From power-distribution practice, however, basic diversity factors have been developed. These provide a general indication of the way in which main feeders can be reduced in capacity below the sum of the demands of the subfeeders they supply. On a radial feeder system, the diversity of the demands of a number of transformers reduces the maximum load which the feeder must supply to some value less than the sum of the transformer loads. Typical diversity factors for main feeders are as follows:

- Lighting feeders: 1.10 to 1.50
- Power and light feeders: 1.50 to 2.00 (or higher)

The selection of a diversity factor in any particular case must be based on a study of load characteristics and operating cycles. Very accurate data on demand diversity are recorded by operating personnel in many types of commercial and industrial buildings.

Another factor used to analyze systems is the *load factor*, which is the ratio of the average load over a particular period of time to the peak load occurring in that period of time.

Figure 4.22 illustrates the use of demand and diversity factors in sizing system components. It should be noted that the kilovoltampere rating of each of the subfeeders must have the minimum value as described in this section. And any feeder or subfeeder (even the main feeder in the diagram) can be sized directly by totaling all connected loads fed by that particular circuit and applying demand factors where conditions make it necessary or desirable.

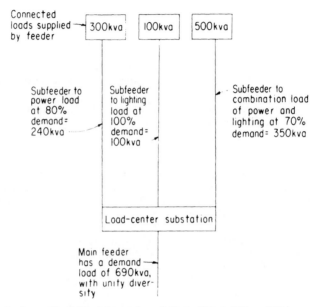

1. Sum of individual demands = 240 + 100 + 350 = 690 kva.
2. Sizing the substation at unity diversity, the required

$$\text{kva} = \frac{690}{1.00} = 690 \text{ kva.}$$

3. To meet this load, use a 750-kva substation.
4. If analysis dictates the use of a diversity factor of 1.4, the

$$\text{required kva} = \frac{690}{1.40} = 492 \text{ kva.}$$

5. To meet this load, use a 500-kva substation.
6. Primary feeder to unit substation must have capacity to match the substation load.

FIGURE 4.22 Application of demand and diversity factors must not conflict with provision of spare capacity.

Feeder Spare Capacity for Growth

The total feeder load obtained by summing all the branch-circuit loads fed by the feeder represents only the initial provision for branch-circuit loads. Future utilization of the spare capacity designed into the branch circuits (they are loaded only to 50 percent) is not accounted for in that total. And there is no provision for the future possibility of exceeding the maximum capacity of the panelboard itself; therefore, additional panelboard capacity—either a second panel or a new larger panel—is required.

Present experience with electrical modernization indicates that a little forethought and study will often reveal the advisability of providing spare feeder capacity—perhaps not in all or even most cases, but certainly in many cases. And only careful consideration of the nature of electrical utilization in a particular occupancy will accurately indicate possible future load requirements. But in all electrical designs, at least some spare feeder capacity must be provided.

Allowance for load growth in feeders should begin with the spare capacity designed into the branch circuits. In each circuit loaded to 50 percent of capacity, it can be assumed that there is an allowance for load growth equal to 30 percent of the circuit capacity. This allowance is based on the **NEC** limitation of 80 percent load on circuits that are in operation for long periods of time (that is, in continuous loading, for 3 hours or more). On circuits that supply motor-operated appliances in addition to other appliances and/or lighting, if an initial noncontinuous load is limited to no more than 50 percent of the circuit capacity (its ampere rating), there is at least spare capacity equal to 30 percent. The 30 percent growth allowance in all branch circuits should be converted to voltamperes and added to the load on the feeder as calculated above.

This grand total—the sum of the actual or calculated branch-circuit loads plus the allowable load growth—then represents the required feeder capacity to handle the full-circuit loads on the panelboard. The next step is to provide capacity in the feeder for anticipated load growth beyond the capacity of the circuits already provided (plus some amount for possible unforeseen future requirements). Based on experience, modern design practice dictates sizing to allow an increase of 25 to 50 percent in the load on a feeder, to provide for the addition of branch circuits where analysis reveals any load-growth possibilities.

In general, spare-capacity allowances for feeders depend upon the type of building, the work performed, the plans or expectations of management with respect to expansion of facilities or growth of business, the type of distribution system used, the locations of centers of loads, the permanence of various load conditions, and economic conditions. Thorough study of all pertinent factors should indicate the advisability of providing spare capacity for each feeder in a distribution system. Where such study indicates extra feeder capacity, a substantial growth allowance—50 percent—is generally essential to provide economy in future electrical expansion. Skimpy upsizing of feeder conductors or raceways has proved to be a major shortcoming of past electrical design work. This is particularly true in tall office buildings, apartment houses, and other commercial buildings in which the elimination of riser bottlenecks represents a large part of the total modernization cost.

Spare feeder capacity may be provided in one or more of several ways. If it is anticipated that the feeder load will be increased in the near future, the extra capacity should be included in the conductor sizing and installed as part of the initial electrical system. In many small office or commercial buildings and apartment houses, extra capacity for load growth should automatically be included in the feeder conductors when they are initially installed. Here are other ways of providing for growth in feeder loads:

1. Selecting raceways larger than required by the initial size and number of feeder conductors. This allows the conductors to be replaced with larger sizes if required at a later date. This method should be carefully considered for risers or underground or concealed feeder raceway runs and similar permanent installations, where it would be very costly to add conduit runs at a later date. In some cases, a compromise can be made between providing spare capacity in the feeder conductor size and increasing the size of the feeder raceway.

2. Including spare raceways in which conductors may be installed at a later date. Arrangements of multiple raceways must be carefully laid out and related to the existing overall system. The types of feeder distribution centers or main switchboards and the layout of local branch-circuit panelboards must be able to accommodate future expansion of distribution capacity based on the use of spare raceways, with modification and regrouping of feeder loads.

Figure 4.23 shows the several elements involved in sizing feeder conductors for the initial load plus two types of spare capacity. The feeder, consisting of three phase legs and the neutral, is run in conduit, so the neutral must be counted as a current-carrying conductor because of the third-harmonic current present in a circuit supplying fluorescent lighting [Note 10(c) to **NEC** Table 310-16]. Then Note 8 to the same table requires wire ampacity derating because there are four current-carrying conductors in the conduit. Although No. 4/0 THW copper wire has an allowable ampacity of 230 A (Table 310-16) and might appear satisfactory for the ultimate feeder load of 224 A, Note 8 derates ampacity of No. 4/0 to only 0.8 230 A or 184 A—which is inadequate. The minimum size of conductor that will have a derated ampacity at least equal to the required value of 224 A is 300-kcmil THW copper, with an ampacity of 228 A (0.8 × 285 A) under the given conditions. And note how the continuous nature (3 hours or more of operation) of the feeder load involves Sections 220-10(b) and 384-16(a) on panelboard main protection.

Feeder Sizing versus Panelboard Rating

The sizing of feeder conductors for continuous loading (such as lighting in schools, office buildings, stores, and other occupancies where lighting is on all day long) raises problems in panelboard sizing. As discussed under the heading "Branch-circuit panelboards" in Chapter 2 of this handbook, **NEC** Section 384-13 requires that any panelboard have busbar ampacity at least equal to the **NEC** minimum feeder conductor capacity, required by Article 220. In effect, the clear, straightforward rule of Section 384-13 calls for a panelboard to have a busbar ampacity that is not less than the minimum ampacity required for the feeder conductors that supply the sum of the branch-circuit loads fed from the panel. When the minimum required feeder ampacity is calculated for the total of branch-circuit loads in accordance with Section 220-10, the panelboard supplied by that feeder must have busbars with an ampacity rating at least equal to the minimum feeder ampacity. But here is where a controversy starts for feeders with continuous (3 hours or more) loading:

1. Section 220-10(b) sets two different possible minimum values for feeder ampacity, depending upon the type of overcurrent device protecting the feeder.
2. If the circuit breaker or fusible switch protecting the feeder conductors is not "listed" (such as by UL) for continuous operation at 100 percent of its current rating, the CB or fuses in the switch must be rated at least at 125 percent of the continuous load current; the feeder conductors must also have ampacity not less than 125 percent of the continuous load current or be of such a size as to be properly protected by the 125 percent rated protective device in accordance with Section 240-3(b). If the feeder protective device is listed for 100 percent continuous loading, then the minimum required feeder conductor ampacity is simply the feeder load current.

FEEDERS FOR LIGHTING AND POWER

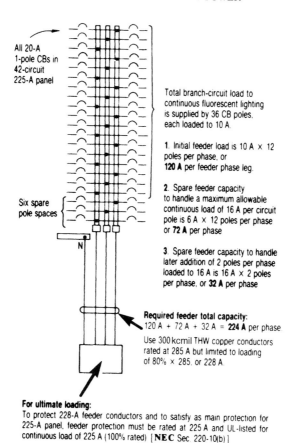

FIGURE 4.23 For real spare capacity, the feeder and all circuit components must be sized to carry the ultimate load.

3. As a result of the interaction of Sections 384-13 and 220-10(b), the minimum required panel busbar rating depends on the type of protective device used in the feeder that supplies the panel.

4. As shown previously in Fig. 3.34, a continuous minimum feeder load current of 176 A must have feeder conductors (and therefore a panelboard busbar rating) of not less than 125 percent of 176 A, or 220 A, if the CB or fusible switch to the panel is not listed for 100 percent continuous loading. Only a very small percentage of feeder protective devices are so listed, and then only in CB and switch ratings of 600 A and above.

5. As shown in Fig. 4.24, the impact of these rules on panelboard busbar rating is to produce illogical conditions, because panelboard sizing is not related to the reason behind the 125 percent multiplier in Section 220-10(b). A circuit breaker or fusible-switch assembly that is not listed for 100 percent continuous loading would generate excessive heat within its own parts if it were operated at its full current rating continuously, and such heat would damage conductor insulation.

For that reason, the 125 percent multiplier of Section 220-10(b) establishes the same limitation as the general UL rule that (unless marked otherwise) a circuit breaker must not be loaded continuously to more than 80 percent of its rating (which is the reciprocal of having a rating of at least 125 percent of the continuous load current). Circuit breakers and fusible switches that are listed for 100 percent continuous loading are constructed and/or ventilated to prevent the generation of excessive heat.

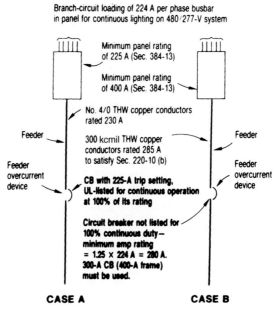

FIGURE 4.24 Sizing of feeder conductors and overcurrent protection affects the sizing of panelboards.

6. Although the letter of the **NEC** rules requires the panel busbar ratings shown in Fig. 4.24, there is no safety compromise in sizing the busbars at no more than 100 percent of continuous load, because application of the panelboard is independent of the conditions that dictate the 125 percent multiplier for feeder protection and feeder conductors. In Fig. 3.34, for instance, sizing of the panel busbars at 125 percent of 176 A calls for a panel of not less than 1.25 × 176 A, or 220-A, busbar capacity—dictating the use of a 225-A panel. But safety and the rule of **NEC** Section 220-4(d) would be fully served by using a panel with busbars rated for 100 percent of the minimum "computed" load of 176 A (although a 225-A panel is the next larger standard size above 176 A). The use of a panel rated at 125 A (on the basis that the actual load is only 116 A) would be contrary to the requirement of Section 220-4(d), calling for the "wiring system" (including the panelboard) to be sized as a minimum on a watts-per-square-foot basis.

IMPORTANT: In applications such as described above, it is extremely important to distinguish between a "lighting and appliance" panelboard (as defined in

Section 384-14) and other panelboards. (See "Panel main protection" in Chapter 2 of this handbook.) Because a lighting and appliance panelboard requires main overcurrent protection (on the supply side of the panel) rated not over the panel busbar rating, when the rating of the feeder overcurrent device is increased to 125 percent of the continuous feeder load, a panel for which the feeder overcurrent device serves as main protection must have a busbar rating not lower than the current rating of the feeder protection—as shown in case B of Fig. 4.24. The feeder CB may not be rated less than 300 A, and that value of protection would not satisfy as "mains protection" for a 225-A panel at the load end of the feeder. The next standard size of panelboard above 225 A is 400 A, and that would be properly protected by the 300-A feeder CB (or fuses). Of course, a 225-A panel with its own main 225-A CB or fused switch could be used in case B, instead of the 400-A panel. This is a typical example of the need to interrelate and satisfy all **NEC** rules bearing on each design situation.

In case B of Fig. 4.24, if the panel fed by the feeder were not a lighting and appliance panel but a panel made up of multipole CBs feeding, say, continuous electric-heating loads connected phase to phase, the panel would not require main protection; and a panel rated at 225 A could be used for a total load of 224 A per phase busbar (although that would be contrary to the exact wording of the rule of Section 384-13).

Figure 4.25 shows how the rules of Sections 220-10(b) and 384-13 combine to dictate the minimum busbar rating of a branch-circuit or feeder panel that is not subject to the mains protection requirement for a lighting and appliance panel. The load on the panel is exactly the same in cases A and B. But, because the feeder protection in case A is not rated for 100 percent continuous loading, the sizes of the feeder protection, feeder conductors, and panel must be increased because of the continuous load. In case B, the 125 percent multiplier is not required. Because the panel in case A does not require a main protective device at its rating, there would be no problem in having an 800-A panel there instead of the 1200-A (or 1000-A) panel—even though the 900-A feeder protection exceeds the 800-A panel busbar rating. The size of the panel should not be required to be increased to 125 percent of the continuous load simply because of the unrelated heat characteristics of a remotely located feeder protective device. However, the wording of Section 384-13 does dictate such upsizing of the panel busbar rating.

Selecting Conductor Sizes

When the total design power (watts) or apparent power (voltamperes) has been determined and includes all the foregoing provisions, it and the voltage and other electrical characteristics of the feeder (or the total load in amperes) are used to obtain the required current-carrying capacity of the conductors. In addition to the circuit voltage, the feeder conductor size for a particular total load will depend upon the power factor and the voltage drop in the conductors.

In calculations of the feeder design load, care should be taken to distinguish between circuit load values in *watts* and those in *voltamperes*. When load values are in watts, the power factor must be taken into consideration.

The sizes of conductors are related to the total current, and they must provide the capacity to carry the current at the given value of power factor. Of course, at low power factor, the conductors must have more current-carrying capacity to supply a particular load wattage than at high power factor.

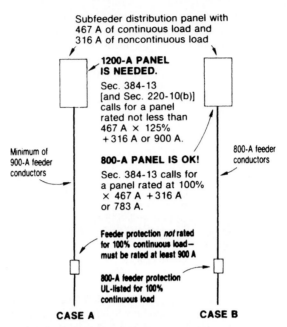

FIGURE 4.25 NEC rules produce puzzling conflicts in sizing panels.

The required size of a feeder conductor is found by calculating the feeder load current at the feeder voltage:

$$\text{Feeder current rating} = \frac{\text{feeder load (kW or kVA)}}{\text{circuit voltage}}$$

This general formula is expressed exactly in the following two forms:

$$I = \frac{\text{load watts}}{k \times E \times \text{pf}} = \frac{\text{load voltamperes}}{k \times E}$$

where I = current in any line wire which the feeder must be rated to carry (check tables of conductor current ratings), amps
 k = 1.0 for 2-wire dc or single-phase ac; 1.73 for 3-wire, 3-phase ac; 2.0 for 3-wire dc or single-phase ac; 3.0 for 4-wire, 3-phase ac
 E = voltage between ungrounded wire and neutral or, if no neutral exists, between any two ungrounded wires, volts
 pf = power factor of load (decimal value)

NOTE: Whenever a voltage value is used in a formula to determine the current rating, the lowest voltage at which the current might operate should be used; this provides the highest current value which might flow in the circuit and will ensure the selection of conductors with fully adequate carrying capacity.

Example. What is the minimum required feeder load current to be used in sizing a feeder for a 20,000-W lighting load that will operate continuously, fed by a 208/120-V, 3-phase, 4-wire feeder circuit and operating at 100 percent power factor?
  Solution. Using the lower value of voltage, 115 V, the formula gives

$$I = \frac{20,000}{3 \times 115 \times 1} = \frac{20,000}{345} = 58 \text{ A}$$

Then, as required by **NEC** Section 220-10(b), that value must be multiplied by 1.25 because the load is continuous (unless a 100 percent rated protective device is used for the circuit):

$$1.25 \times 58 \text{ A} = 73 \text{ A}$$

To meet this capacity requirement, any of the following feeder conductors (from **NEC** Table 310-16) could be used:

Copper. No. 3 TW, THW, RHW, XHHW, RHH, or THHN. (Because of the UL rule, the conductor must be sized for a 60°C ampere rating.)

Aluminum and copper-clad aluminum. No. 2 TW, THW, RHW, XHHW, RHH, or THHN.

Example. What feeder current rating is needed in sizing the minimum 480-V, 3-phase, 3-wire circuit to supply a 100-kVA load?
  Solution. From the formula,

$$I = \frac{100,000}{1.73 \times 440} = \frac{100,000}{761} = 132 \text{ A}$$

To meet this capacity requirement, any of the following feeder conductors (from **NEC** Table 310-16) could be used:

Copper. No. 2/0 TW or No. 1/0 THW, RHW, RHH, XHHW, or THHN

Aluminum and copper-clad aluminum. No. 4/0 TW or No. 2/0 THW, RHH, XHHW, or THHN

Figure 4.26 shows the formulas that can be used to calculate power loss due to I^2R heating in feeder conductors. With today's electric-utility rates, it is often economical to shift to a larger size of conductor when the resulting reduction in energy costs over the first year or two covers the extra cost. (See Chapter 9 of this handbook for a table of formulas used for electrical calculations.)

Sizing the Feeder Neutral

NEC Section 220-22 covers requirements for sizing the neutral conductor in a feeder. It states that the neutral feeder load shall be the "maximum unbalance" of the feeder load and defines the maximum unbalance as "the maximum connected load between the neutral and any one ungrounded conductor." In a 3-wire, 240/120-V, single-phase feeder, the neutral must have a current-carrying capacity at least equal to the current drawn by the total 120-V load connected between the more heavily loaded hot leg and the neutral.

Assuming resistivity of conductor metal to be as follows:
K = 12 for circuits loaded to more than 50% (copper)
K = 11 for circuits loaded to less than 50% (copper)
K = 18 for aluminum conductors
For a 2-wire circuit (direct-current or single-phase):

$$P = \frac{2 \times K \times L \times I^2}{CM}$$

For a 3-wire, 3-phase circuit (assuming balanced load):

$$P = \frac{3 \times K \times L \times I^2}{CM}$$

in which, P = power lost in the circuit (watts)
K = resistivity of conductor metal (circular mil-ohms/foot)
I = current in each wire of the circuit (amps)
L = one-way length of the circuit (ft)
CM = cross-section area of each of the wires in circular mils
When the resistance of conductors is determined from a table, the power loss is calculated from the relation

$$P = I^2 R$$

FIGURE 4.26 These calculations show power loss in feeder conductors, which may dictate the use of the next larger size of conductor.

Because 240-V loads, connected between the two hot legs, do not place any load on the neutral, the neutral conductor of such a feeder must be sized to make up a 2-wire, 120-V circuit with the more heavily loaded hot leg. Actually, the 120-V circuit loads on such a feeder would be considered as balanced on both sides of the neutral. The neutral, then, would be the same size as each of the hot legs if only 120-V loads also were supplied by the feeder. If 240-V loads also were supplied, the hot legs would be sized for the total load, but the neutral would be sized for only the total 120-V load connected between one hot leg and the neutral. Figure 4.27 shows how phase-to-neutral loading affects the size of the neutral but phase-to-phase loads impose no neutral load.

Section 220-22 contains three other provisions relative to the sizing of neutral conductors:

1. When a feeder supplies household electric ranges, the neutral conductor may be smaller than the hot conductors but must have a carrying capacity at least equal to 70 percent of the current capacity required in the ungrounded conductors to handle the load (i.e., 70 percent of the load on the ungrounded conductors). Table 220-19 gives the demand loads to be used in sizing feeders that supply electric ranges and similar cooking appliances. Figure 4.28 shows this application.

2. For feeders of three or more conductors (3-wire dc; 3-wire, single-phase; and 4-wire, 3-phase) a further demand factor of 70 percent may be applied to that portion of the unbalanced load in excess of 200 A. That is, in a feeder supplying only 120-V loads evenly divided between each ungrounded conductor and the neutral, the neutral must be sized at the same load as each ungrounded conductor up to 200 A capacity; however, its size may be reduced from the size of the ungrounded conductors for loads above 200 A by adding to the 200 A only 70 percent of the amount of load current above 200 A (Fig. 4.29). It should be noted that this 70

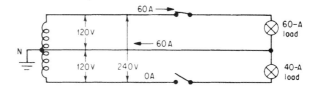

Under unbalanced conditions, with one hot leg fully loaded to 60 amps and the other leg open, the neutral would carry 60 amps and must have the same rating as the loaded hot leg. Thus No. 4 TW hot legs would require No. 4 TW neutral (copper).

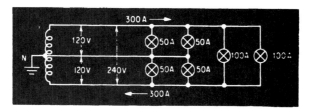

The neutral here must carry only the unbalance of the two 50-amp, hot-to-neutral loads and has nothing to do with the two straight 240-volt, 100-amp loads. Neutral must be sized for a maximum of 100 amps—the maximum unbalance from hot to neutral.

FIGURE 4.27 Neutral sizing is determined only by neutral-connected loads.

```
H─────────────────────
115/230-V  N─────────────────────   } To 8 electric ranges
feeder                                 rated 10 kw each
H─────────────────────
```

DEMAND LOAD for 8 10-kw ranges = 23 kw
LOAD ON EACH UNGROUNDED LEG

$$= \frac{23{,}000 \text{ W}}{230 \text{ V}} = 100 \text{ amps} \quad \text{(e.g. No. 3 THW)}$$

$$\frac{\text{Required minimum}}{\text{Neutral capacity}} = 70\% \times 100 \text{ amps} = 70 \text{ amps}$$

FIGURE 4.28 Because of diversity in the use of heater elements, a range neutral may be reduced.

percent demand factor is applicable to the "unbalanced load" in excess of 200 A and not simply to the total load, which in many cases may include 240-V loads on 240/120-V, 3-wire, single-phase feeders or 3-phase loads or phase-to-phase-connected loads on 3-phase feeders.

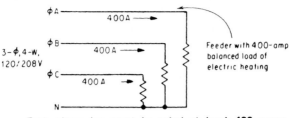

Each phase leg must be rated at least 400 amps.
Neutral must be rated at least

200 amps + (70% × 200 amps) = 340 amps

FIGURE 4.29 Feeder neutral size may be reduced for resistance loads over 200 A.

3. This reduction of the neutral load to 200 A plus 70 percent of the current over 200 A does not apply to electric-discharge lighting. Figure 4.30 indicates the difference between feeder-neutral sizing for incandescent lighting and that for electric-discharge lighting (fluorescent, metal-halide, etc.). In a feeder supplying ballasts for electric-discharge lamps, there must not be a reduction of the neutral capacity for the part of the load that consists of electric-discharge light sources. For feeders supplying only electric-discharge lighting, the neutral conductor load must be the same as the phase conductors, no matter how large the total load may be. And for feeders supplying motor loads as well as lighting loads, any loads that do not connect to the neutral are not involved in sizing of the neutral, as shown for the 3-phase motor loads in Fig. 4.31.

Full sizing of the neutrals of such feeders is required because, in a balanced circuit supplying ballasts, neutral current approximating the phase current is produced by third (and other odd-order) harmonics developed by the ballasts (Fig. 4.32). for large electric-discharge lighting loads, this factor affects the sizing of neutrals all the way back to the service. It also affects the rating of conductors in conduit because such a feeder circuit consists of four current-carrying wires, which requires application of an 80 percent wire-ampacity derating factor. [See Notes 8 and 10(c) to **NEC** Tables 310-16 through 310-19.]

It should be noted that the wording in **NEC** Section 220-22 prohibits reduction of the size of the neutral when electric-discharge lighting is used, even if the feeder supplying the electric-discharge lighting load over 200 A happens to be a 120/240-V, 3-wire, single-phase feeder. In such a feeder, however, the third-harmonic currents in the hot legs are 180° out of phase with each other and, therefore, would not be additive in the neutral as they are in a 3-phase, 4-wire circuit. In the latter circuit, the third-harmonic components of the phase currents are in phase with each other and add together in the neutral instead of canceling out. Figure 4.33 shows a 120/240-V circuit.

In the case of a feeder supplying, say, 200 A of fluorescent lighting and 200 A of incandescent, there can be no reduction of the neutral below the required 400-A capacity of the phase legs, because the 200 A of fluorescent lighting load cannot be used in any way to take advantage of the 70 percent demand factor on that part of the load in excess of 200 A.

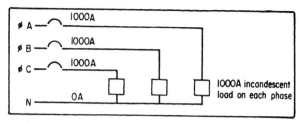

Incandescent lighting only

Serving an incandescent load, each phase conductor must be rated for 1000 amps. But neutral only has to be rated for 200 amps plus (70% x 800 amps) or 200 + 560 = 760 amps.

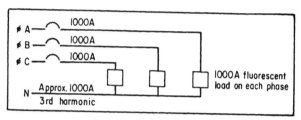

Electric discharge lighting only

Because load is electric discharge lighting, there can be no reduction in the size of the neutral. Neutral must be rated for 1000 amps, the same as the phase conductors, because the third harmonic currents of the phase legs add together in the neutral. This applies also when the load is mercury-vapor or other metallic-vapor lighting.

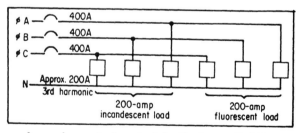

Incandescent plus electric discharge lighting

Each phase leg carries a total of 400 amps to supply the incandescent load plus the fluorescent load. But because there can be no reduction of neutral capacity for the fluorescent and because the incandescent load is not over 200 amps, the neutral must be sized for the maximum possible unbalance, which is 400 amps.

FIGURE 4.30 Type of lighting load affects neutral-conductor sizing.

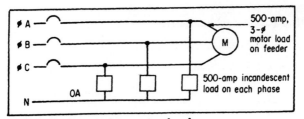

Incandescent plus motor load

Although 1000 amps flow on each phase leg, only 500 amps is related to the neutral. Neutral, then, is sized for 200 amps plus (70% x 300 amps) or 200 + 210 = 410 amps. The amount of current taken for 3-phase motors cannot be "unbalanced load" and no capacity has to be provided for this in the neutral.

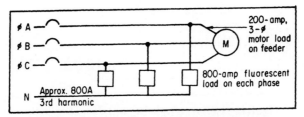

Electric discharge lighting plus motor load

Here, again, the only possible load that could flow on the neutral is the 800 amps flowing over each phase to the fluorescent lighting. But because it is fluorescent lighting there can be no reduction of neutral capacity below the 800-amp value on each phase. The 70% factor for that current above 200 amps DOES NOT APPLY in such cases.

FIGURE 4.31 Phase-to-phase loads do not affect neutral sizing.

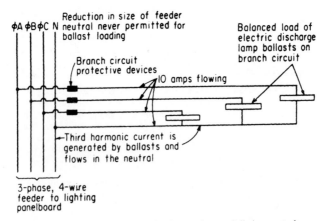

FIGURE 4.32 Feeder to ballast loads must have a full-size neutral.

No reduction of neutral capacity even with zero neutral current?

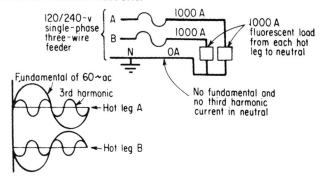

As shown, both the fundamental and harmonic currents are 180° out of phase and both cancel in the neutral. This action is the same as it would be for an incandescent load. Under balanced conditions, the neutral current is zero. But the literal wording of Section 220-22 says there can be no reduction in neutral capacity when fluorescent lighting is supplied. As a result, there should be no use of the 70% factor for current over 200 amps as there would be for incandescent loading. Neutral here must be rated for 1000 amps.

FIGURE 4.33   NEC rule does not permit reduction of neutral sizing in this case.

Common Neutral versus Separate Neutrals

Over recent years, electrical systems supplying data-processing equipment and electric-discharge lighting have been plagued with severe overheating of branch-circuit and feeder neutral conductors due to harmonic currents that are *additive* in the neutral of 3-phase, 4-wire circuits supplying such loads. This problem has been widely experienced, often with dangerous and destructive effects. The big question today is, "What can be done from a design and/or installation standpoint to reduce or eliminate the heating problem?"

As we all know, in a 3-phase, 4-wire branch circuit or feeder derived from a wye-connected transformer secondary, there will be no current flow on the circuit neutral conductor if the circuit supplies the same amount of resistive load connected from each phase leg to neutral—as shown in Fig. 4.34. With, say, 15A of current being supplied from each phase leg of the circuit, there will be zero current in the neutral because the 120° phase displacement of the current vectors produces a zero resultant in the neutral. That is a basic and typical condition in any such circuit where the loads are incandescent lighting, electric resistance heating, or other resistive loads. *But,* the very same circuit used to supply "non-linear" loads (like electric-discharge lighting and data-processing equipment or other electronic equipment with switching-mode power supplies) will have current flowing on the neutral conductor *even* with balanced (the same value of) load current on each of the phase legs.

288   CHAPTER FOUR

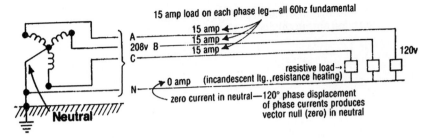

FIGURE 4.34

When a sine-wave voltage source is applied to a nonlinear load connected from each phase leg to neutral on a 3-phase, 4-wire branch circuit, the *load* itself will draw a current wave made up of the 60-Hz fundamental frequency of the voltage source *plus* third and other odd-order harmonics (multiples of the 60-Hz fundamental frequency)—all of which are "generated" or introduced by the "reactive" nature of the load. The current flow through each phase-to-neutral nonlinear load will have a wave shape that is a composite of the 60-Hz fundamental sine wave plus sine waves of the harmonic frequencies. The effect is to produce a resultant wave shape as shown in Fig. 4.35.

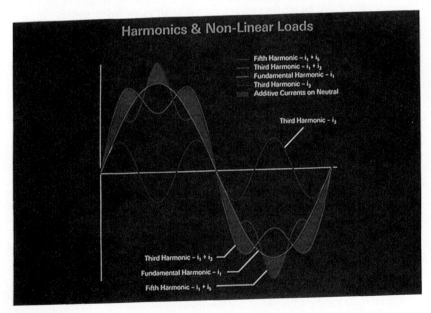

FIGURE 4.35

When an equal nonlinear load is connected from each phase-to-neutral on a 3-phase, 4-wire branch circuit, the 60-Hz fundamental current supplied by each phase leg will be canceled out in the circuit neutral because the 120° phase displacement of the fundamental currents will produce a vector sum of zero in the neutral. *But,* third and other odd-order harmonics (fifth, seventh, ninth, etc.) of the load current supplied by each phase conductor of the circuit are *not* canceled out. Because of their phase relationship to the fundamental, the harmonic currents become additive in the circuit neutral. The amount of third harmonic current in the neutral will be equal in magnitude to *3* times the amount of third harmonic in each phase leg (or equal to the sum of the harmonic currents on the three phase legs). And other odd-order (fifth, seventh, ninth, etc.) harmonic currents also become additive in the neutral.

As a simplified version of the concept, Fig. 4.36 shows how the fundamental and third harmonic currents are related. At any time on those wave forms, the sum of the 60-Hz fundamental currents in the three phases adds up to zero—meaning no neutral current at the fundamental 60-Hz frequency. *But,* because the third harmonics fall in-phase with each other, the neutral carries third harmonic current equal to the sum of the third harmonic currents on the three phase legs.

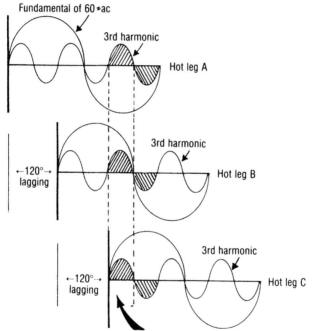

At any given instant—or in any given time frame, the 60hz currents on the three phase legs have a vector resultant of zero and cancel in the neutral. *BUT,* the third (and other odd-order harmonics) on the phase legs are in phase and become additive in the neutral.

FIGURE 4.36

In any particular 3-phase, 4-wire branch circuit supplying balanced nonlinear loads on each phase leg, the actual amount of *additive* harmonic current in the neutral will depend upon the amount of harmonic current generated by the particular nonlinear load. And nonlinear load devices—whether they are electric-discharge lighting ballasts or some type of electronic data-processing equipment—vary very widely in their harmonic generating characteristics. The amount of harmonic current generated depends on the actual construction and components of the specific ballast or data-processing equipment. Some load devices produce small amounts of harmonics, others produce moderate amounts, and still others produce very large amounts of harmonics. Because there is not generally available data or information cataloging the harmonic characteristics of the almost endless variety of equipment in this category, the circuit design task of keeping I^2R heating within safe limits is, at best, extremely difficult.

From tests, it has been determined that the usual harmonic content of load current for electric discharge lighting (with a conventional non-solid-state ballast) will produce neutral harmonic current rms value that "approximates" the phase total rms current (fundamental and harmonics) drawn by one of the phase legs in a balanced 3-phase, 4-wire circuit. As shown in Fig. 4.37, if a circuit has a phase load of 15 A to a balanced loading of, say, fluorescent lighting on the three phases, the total harmonic current in the neutral can safely be taken to be about 15 A also. And that approximate relationship between phase loading and neutral loading is the basis for the rules of Section 220-22 in the **NEC** calling for the neutral load to be equal to the phase load—under balanced conditions—if "the load consists of electric discharge lighting, electronic computer/data processing equipment, or similar equipment." But conditions can be considerably more troublesome for data-processing loads than for electric discharge lighting.

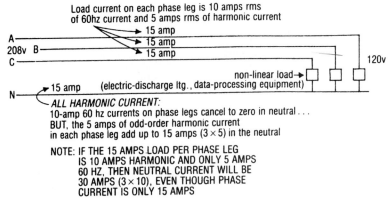

FIGURE 4.37

Widespread, repetitive experience has shown that many 3-phase, 4-wire circuits supplying balanced phase loading of computer equipment will have neutral harmonic currents with rms values considerably greater than the phase currents—up to 2 or more times the phase current. (Again, the actual level of harmonic neutral current will depend on the load devices themselves.) And the commonly experienced problem in such situations has been that severe overheating of neutral conductors results from the conductors being subjected to load currents in excess of their ampacities.

The overheating of neutral conductors in such circuits arises, of course, from very high I^2R heat losses in the conductor when the conductor is carrying current in excess of its **NEC** ampacity, producing temperatures that exceed the insulation temperature rating (e.g., 75°C of THW, etc.). And the whole heating problem is exaggerated by "skin effect," the effect of increased resistance in a given conductor due to the phenomenon that alternating current tends to flow on the outer part (the "skin") of a conductor's round cross section. The skin effect raises resistance by actually reducing the effective cross section of a conductor (like making a No. 12 have the effective cross section of a No. 14 or even a No. 16). And the skin effect increase in conductor resistance is directly proportional to the frequency of current alternations—thereby producing higher heating losses in a wire carrying higher frequency current (the harmonics).

Although it has been extremely difficult to provide any useful quantitative classification of heat losses for the varying magnitudes and frequencies—which can and do vary "all over the ballpark"—some general techniques have been devised to address the general nature of the problem, techniques which can be readily and easily applied.

First, as a general attempt to limit and control heating of harmonic current loading on neutrals, a larger than usual neutral conductor will be selected to reduce the resistance of the conductor as a means of reducing heating. Some designers will specify use of a neutral conductor at least one size larger than the required phase conductors for a 3-phase, 4-wire branch circuit supplying computer/data-processing equipment. Others may go up two sizes. This is a very difficult problem because circuit design is generally done well in advance of any firm knowledge of the relative offending nature (harmonic generation) of equipment, which will be selected and applied at a later date. Such rules of thumb are better than nothing and will provide relief against excessive heating; but they do not guarantee proper, safe, and/or reliable operation.

Another heat-control technique that has been widely utilized is providing a separate neutral conductor for each phase leg of the circuit. So instead of being a 3-phase, 4-wire circuit (three phase legs and one neutral), some circuits are designed as 3-phase, 6-wire branch circuits—based on the concept that in a 3-phase, 6-wire circuit, each neutral will carry only one-third the amount of harmonic current that would flow on a single common neutral of a 3-phase, 4-wire circuit. If a single, common neutral conductor of 0.1 Ω resistance for its entire length carried 15 A of harmonic current (5 A per phase leg), the heating effect (I^2R) would be 15 × 15 × 0.1 = 22.5 W. But if the 15 A of harmonic current are divided among three neutral conductors of 0.1 Ω resistance each, then each conductor would have a heat loss (I^2R) of 5 × 5 × 0.1 = 2.5 W. And the three neutrals together would have a heat loss of 3 × 2.5 W, or 7.5 W instead of 22.5).

But, when a separate neutral is used for each phase leg of the branch circuit, there is an entirely different condition with respect to the 60-Hz fundamental current—because the 60-Hz neutral current is not canceled, as in the 3-phase, 4-wire circuit. Then a question arises as to the relative heating effects of the 60-Hz current *plus* the harmonic currents as they both flow in the separate neutrals—compared to the heating effect in a common neutral, which carries *only* the harmonic current (because the 60-Hz currents cancel to zero in a common neutral).

Fig. 4.38 shows the comparison of heating effects for the two different circuit makeups. In those basic diagrams, there are factors that will alter the indicated heating effects. The alterations will be slight or very substantial depending upon the amount of harmonic current that makes up the total of 15 A on each phase leg. In both the *common neutral* circuit and the *separate neutrals* circuits, the heating effect of the phase legs will be identical. The *total* heating effect for each depends upon the conditions of the neutral(s).

In the common neutral circuit of Fig. 4.38, the resistance of the No. 12 copper neutral will be higher than the 0.1 Ω shown, because the 15 A of all-harmonic cur-

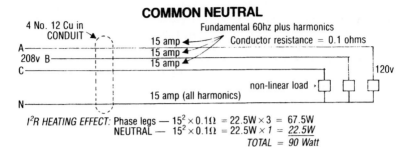

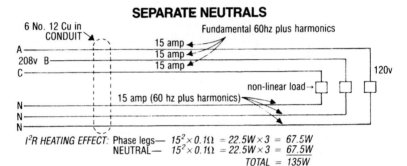

FIGURE 4.38

rent on the neutral will cause skin-effect increase in the neutral conductor resistance. The actual resistance of that conductor will depend upon the amplitudes and frequencies of the harmonics.

In the separate neutrals arrangement of Fig. 4.38, the resistance of each neutral will be the same as the resistance of each phase leg (0.1 Ω), because *all* the conductors are No. 12 copper and *all* have exactly the same loading of 60 Hz and harmonic currents.

A comparison of the conditions of heating for the two different circuit makeups suggests that the common neutral circuit with a larger neutral (say, No. 10 or 8) provides the most effective and economical reduction of heating in circuits with small to moderate amounts of harmonic content. As harmonic current loading becomes heavy, it is possible that very large increases in resistance due to skin effect and "proximity effect" would dictate use of separate neutrals—especially where the ratio of harmonic to fundamental is very high.

Based on the general information given here, each installation must be carefully evaluated for its own particular characteristics to make a judgment on common versus separate neutrals.

Minimizing Voltage Drop

Voltage drop must be carefully considered in sizing feeder conductors, and calculations should be made for peak load conditions. The voltage drop must be calculated on the basis that full system voltage (for example, 120, 240, or 480 V) is available at the service entrance or transformer secondary supplying the feeder. The voltage drop is the amount of reduction from the nominal supply voltage. Feeder conductors should be sized such that the voltage drop up to the branch-circuit panelboards or the point at which the branch circuits originate is not more than 1 percent for lighting loads or combined lighting, heating, and power loads, and not more than 2 percent for power or heating loads. Local codes may impose lower limits on voltage drop. Voltage-drop limitations are shown in Fig. 4.39, as follows:

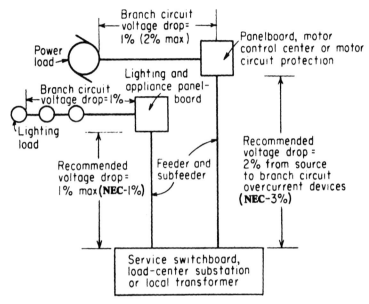

FIGURE 4.39 Feeder voltage drop must be carefully calculated to be within the limits of good design.

1. For combinations of lighting and power loads on feeders and branch circuits, use the voltage-drop percentages for lighting loads (at the left in Fig. 4.39).
2. The word "feeder" here refers to the overall run of conductors carrying power from the source (the service entrance, loadcenter substation, or local transformer at which the full value of system voltage is available) to the point of final branch-circuit distribution, including feeders, subfeeders, sub-subfeeders, etc.
3. The voltage-drop percentages are based on nominal circuit voltage being available at the load terminals of the source of each voltage level in the distribution system.

The voltage drop in any set of feeders can be calculated with the formulas given in Fig. 4.40. From this calculation, it can be determined if the conductor size initially selected to handle the load will be adequate to maintain the voltage drop within given limits. If it is not, the size of the conductors must be increased (or other steps taken where conductor reactance is not negligible) until the voltage drop is within the prescribed limits.

Two-Wire, Single-Phase Circuits (Inductance Negligible):

$$V = \frac{2k \times L \times I}{d^2} = 2R \times L \times I \qquad d^2 = \frac{2k \times I \times L}{V}$$

V = drop in circuit voltage (volts)
R = resistance per ft of conductor (ohms/ft)
I = current in conductor (amps)

Three-Wire, Single-Phase Circuits (Inductance Negligible):

$$V = \frac{2k \times L \times I}{d^2}$$

V = drop between outside conductors (volts)
I = current in more heavily loaded outside conductor (amperes)

Three-Wire, Three-Phase Circuits (Inductance Negligible):

$$V = \frac{2k \times I \times L}{d^2} \times 0.866$$

V = voltage drop of 3-phase circuit

Four-Wire, Three-Phase Balanced Circuits (Inductance Negligible):
For lighting loads: Voltage drop between one outside conductor and neutral equals one-half of drop calculated by formula for 2-wire circuits.
For motor loads: Voltage drop between any two outside conductors equals 0.866 times drop determined by formula for 2-wire circuits.

In above formulas
L = one-way length of circuit (ft)
d^2 = cross-section area of conductor (circular mils)
k = resistivity of conductor metal (cir mil-ohms/ft) (k = 12 for circuits loaded to more than 50% of allowable carrying capacity; k = 11 for circuits loaded less than 50%; k = 18 for aluminum conductor)

FIGURE 4.40 These basic formulas indicate resistive voltage drop in feeders.

In application, the loadings and lengths of feeders can be adjusted to accommodate voltage-drop requirements, with additional advantages as shown in Fig. 4.41 for a modern office building. A total of nine risers supply lighting and receptacle circuits in all of the building's office space—floors 2 through 22. Each of these floors has three electric closets. At each feeder-riser location, one feeder supplies the bottom eight floors, the second feeder supplies seven floors in the middle of the building's height, and the third feeder supplies the top six floors. With the design demand load the same for each of the electric closets, the feeder serving the bottom eight closets has a heavier load than the one serving the middle seven closets, which in turn is more heavily loaded than the feeder to the top six closets. Because each feeder is designed to deliver its kilovoltampere capacity with only a 2 percent drop in voltage from the switchboard to any panel, the unbalanced loading of the feeders made possible the use of a single makeup—four 500-kcmil THW conductors in 3½-in conduit—for all the feeders. This one size of feeder can supply a total load of eight closets with a 2 percent voltage drop for the distance from the switchboard to the lowest load, but it can only supply a total of six closets with a 2 percent drop when the feeder has to run to the top of the building.

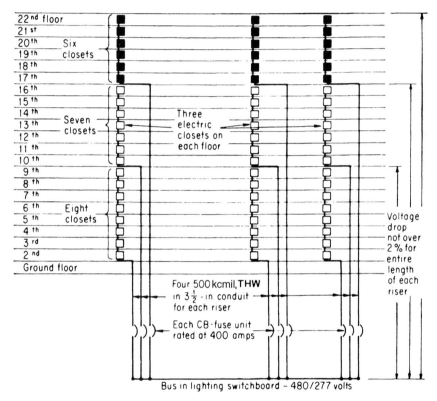

FIGURE 4.41 Feeder loads may be varied to balance voltage drop against circuit length and provide uniform circuit makeup.

The use of a single basic feeder makeup for all feeders provides two distinct advantages. First, four 500-kcmil in 3½-in conduit is a highly economical and efficient makeup, based on apparent power delivered, voltage drop, and cable power loss and measured against the installed cost of the feeder. Second, the use of a single feeder makeup minimizes construction costs. The use of only 500-kcmil in 3½-in conduit permitted a high degree of standardization and mechanization in installing the conduit, pulling the conductors, coupling to panel enclosures, making taps, installing lugs, etc.

There are many cases in which the above-mentioned limits of voltage drop (1 percent for lighting feeders, etc.) should be relaxed in the interest of reducing the prohibitive costs of conductors and conduits that may be required by such low drops. In many installations, a 5 percent drop in feeders is not critical or unsafe. A recommended plan for realistic voltage drop in apartment houses in shown in Fig. 4.42, where the slightly greater drop is not considered detrimental to the lighting and appliance loads in the apartments.

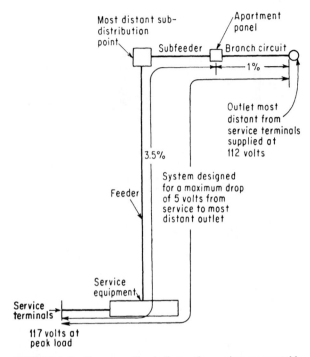

FIGURE 4.42 Economy often indicates the maximum acceptable voltage drop for noncritical loads.

Figure 4.43 shows how loadcenter distribution improves voltage level and regulation by deriving full-voltage feeders from transformers located at the centers of load groupings, thereby shortening the secondary feeders and reducing the voltage drop.

Voltage Regulators. In general, when overcabling is needed to comply with **NEC** rules on voltage drop, the alternative of voltage regulators plus normal cabling should be considered, for it may turn out to be the most economical method.

Voltage regulators are used wherever it is necessary to eliminate objectionable variations in the voltage level of a supply circuit. Over- and undervoltage have undesirable effects on the operation of a wide range of electrical and electronic load devices. In particular, electronic data-processing equipment and communication, instrumentation, and control devices require a fairly constant operating voltage.

Voltage regulators are particularly effective in high-ampere (thousands of amperes) distribution circuits for modern commercial and industrial properties,

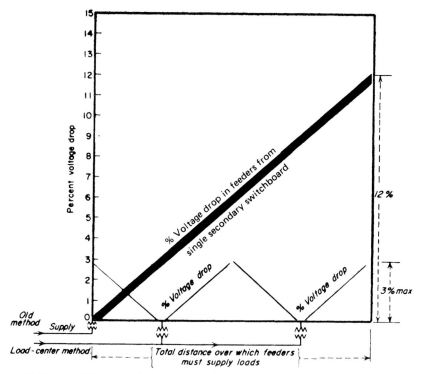

In the case of the load-center layout, shorter feeders cover the same distance over the plant as the long secondary feeders from the single switchboard, but the voltage drop is greatly reduced in the load-center system. Improved voltage regulation adds up to overall increase in operating efficiency of the system.

FIGURE 4.43 Voltage drop can be greatly reduced by deriving feeders from loadcenter substations or transformer distribution centers.

where long secondary-voltage runs (typically, 120/208 V) would require many conductors per phase leg to keep voltage drops within reasonable limits. They are extremely economical when used on long high-ampere circuits with constantly varying load demands (such as in apartment houses, schools, and hospitals), where they permit the use of only the amount of cable necessary for the load current, instead of that needed to maintain a low voltage drop.

Figure 4.44 shows the use of a voltage-regulator installation on the campus of a university. The utility company supplied secondary voltage from the transformer bank at 208Y/120 V, nominal, at +5 percent and −2 percent from the 208-V base. The local code limited the voltage drop between service-entrance equipment and the final distribution point to 2½ percent (or 5.2 V) on a 208-V base.

Calculations to determine the number and size of conductors that would be needed to serve the three buildings assigned approximately half the allowable voltage drop to the supply feeders and the remaining half to the building distribution feeders. Since the utility transformer bank was close to the classroom building and the administration building, the supply feeders were relatively short, and the allowable voltage drop could be achieved by normal cabling. The library, however,

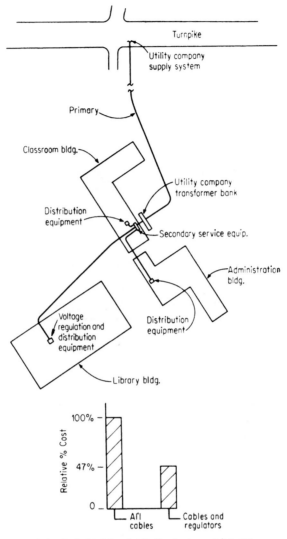

FIGURE 4.44 Voltage regulators also provide a solution to extreme voltage-drop conditions.

required a 450-ft supply feeder from the transformer bank to the distribution center located in the basement. In addition, the library had an estimated load of 1900 A.

Oversizing of the feeders to meet the voltage-drop requirement (half the allowable drop taking place in the supply feeder over a run of 450 ft at 1900 A) would have required twenty 500-kcmil conductors per phase. The cost of materials and labor, the amount of space required, and the difficulty of installation made this method impractical.

Voltage regulators were located at the load end of the library feeder, to compensate for all the voltage drop in the supply feeder. With this design, only six 500-kcmil conductors per phase were needed, since they could be selected to carry only the load current. Because the regulators would maintain nominal voltage at the distribution center in the library basement, the full 2½ percent voltage drop allowed by code could be assigned to the building's distribution feeders. The amount of cabling space was thus kept to a minimum.

The regulators maintain nominal voltage at the distribution center not only when the load is heavy and the utility supply voltage is low but also when the load is light and the utility supply voltage is high. This results in the utilization equipment in the library building always operating at peak efficiency with maximum life.

The short-circuit current available from the utility for a 3-phase bolted fault was 116,000 rms symmetrical amperes. The impedance introduced by the all-cable method would have reduced the short-circuit current at the distribution equipment to approximately 40,000 A. However, with the regulator-cable installation, the additional impedance of the smaller cables reduces the short-circuit current to approximately 19,000 A—an added bonus.

Figures 4.45 and 4.46 show sets of curves correlating size of conductor, ampere load, and voltage drop for 3-phase circuits in steel conduit. The curves are for a 1-V drop and 90 percent power factor, based on ampacities for types RHW and THW (75°C) 600-V insulation. Many such graphs and tabulated data on voltage drop are available in handbooks and from manufacturers.

Figures 4.47 and 4.48 are handy tables for determining maximum feeder-circuit lengths for 1 percent voltage drop on single-phase and 3-phase feeders.

The table in Fig. 4.47 gives the one-way lengths for 115/230-V circuits under balanced loading. With a balanced 3-wire load, the drop will be 1.15 V for each 1.15 V circuit—that is, for each side of the 115/230-V circuit. For a 230-V, 2-wire circuit, the given lengths produce a voltage drop of 2.3 V. An example is shown below the table.

The table in Fig. 4.48 gives the one-way lengths for 3-phase, 230-V delta ac load circuits with 85 percent power factor. For balanced load conditions, the conductors and lengths shown produce voltage drops of 2.3 V (1 percent of 230 V). For other voltage-drop percentages (2 percent, 3 percent, etc.) the lengths shown can be doubled, tripled, etc. Thus, if, as shown, a 90-A-loaded 3-phase circuit can be run 113 ft using No. 1/0 conductors for a 2.3-V drop, then the same load could be carried twice as far (226 ft) if double the voltage drop (4.6 V) were acceptable. The table may be used for calculating other feeder lengths for 1 percent voltage drop:

- For 208-V, 4-wire wye feeders, multiply the given lengths by 0.9.
- For 230-V, single-phase feeders, multiply the given lengths by 0.85.
- For 460-V, 3- or 4-wire feeders, multiply the given lengths by 2.
- For aluminum wire, multiply the given lengths by 0.7 or use the length of copper wire which is two sizes smaller than the aluminum size under consideration.

Multiple Conductors

In making up high-current-capacity feeders, there are distinct advantages to the use of multiple conductors in each leg. This may be accomplished in accordance with **NEC** Section 310-4, which reads as follows:

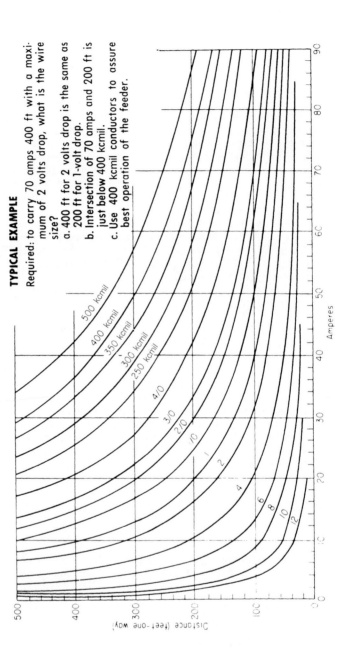

FIGURE 4.45 Maximum lengths of 3-phase feeder circuits for a 1-V drop at loads up to 90 A.

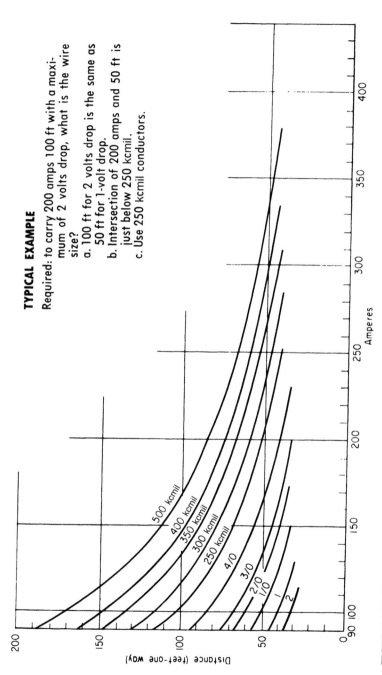

FIGURE 4.46 Maximum lengths of 3-phase feeder circuits for a 1-V drop at loads from 90 to 380 A.

SINGLE-PHASE AC LOADS

115/230 volts, 60 cycles, 100% PF

Amp load	Wire size—circular mils								Wire size—B & S or AWG					
	500	400	350	300	250	4/0	3/0	2/0	1/0	1	2	3	4	6	8
40	1106	898	788	669	558	475	378	299	239	188	150	119	94	59	38
50	885	719	630	535	447	380	303	240	191	150	120	91	75	47	30
60	737	599	525	446	372	317	252	200	159	125	100	79	62	39	
70	632	513	450	382	319	271	216	171	136	107	86	68	53	34	
80	553	449	394	334	279	238	189	150	119	94	75	59	47		
90	491	399	350	297	248	211	168	133	106	83	67	53	42		
100	442	359	315	267	223	190	151	120	95	75	60	47			
110	402	327	286	243	203	173	138	109	87	68	55				
120	369	299	263	223	186	158	126	100	79	63					
130	340	276	242	206	172	146	116	92	73	58					
140	316	257	225	191	159	136	108	86	68						
150	295	240	210	178	149	127	101	80	64						
160	276	225	197	167	140	119	95	75	60						
170	260	211	185	157	131	112	89	70							
180	246	200	175	148	124	106	84	66							
190	233	189	166	140	117	100	80								
200	221	180	157	134	112	95	76								
210	211	171	150	127	106	90									
220	201	163	143	122	101	86									
230	192	156	137	116	97	83									
240	184	150	131	111	93										
250	177	144	126	107	89										
260	170	138	121	103	80										
270	164	133	117	99											
280	158	128	112	96											
290	152	124	109	92											
300	147	120	105												
310	143	116	102												
320	138	112													
330	134	109													
340	130	106													

Calculations based on copper resistance of 12.5 ohms per CM-ft at 50C (122F).

Reactance and impedance losses calculated for each wire.

Conductors closely grouped in metallic conduit.

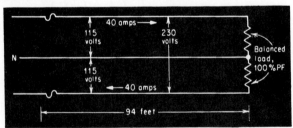

REFER TO TABLE: For a load of 40 amps, it shows that No. 4 conductors must be used if the circuit length is between 60 and 94 ft. Under the condition shown in the diagram, there will be a 1.15-volt drop in each 115-volt circuit.

FIGURE 4.47 Circuit lengths (in feet) for a 1 percent voltage drop in single-phase feeders.

3-PHASE DELTA AC LOADS

230 volts, 60 cycles, 85% PF

Amp load	Wire size—circular mils									Wire size—B & S or AWG					
	500	400	350	300	250	4/0	3/0	2/0	1/0	1	2	3	4	6	8
40	710	625	584	530	475	429	364	303	253	208	173	139	113	75	49
50	568	500	467	424	380	343	291	242	203	167	139	111	90	60	39
60	473	417	389	353	317	286	243	202	169	139	115	93	75	50	
70	406	357	333	303	271	245	208	173	145	119	99	79	64	43	
80	355	312	292	265	238	214	182	151	127	104	87	69	56		
90	316	278	259	235	211	191	162	134	113	93	77	62	45		
100	284	250	233	212	190	172	146	121	101	83	69	55			
110	258	227	212	193	173	156	132	110	92	76	63				
120	237	208	195	177	158	143	121	101	84	69	58				
130	218	192	180	163	146	132	112	93	78	64					
140	203	179	167	151	136	123	104	86	72						
150	189	168	156	141	127	114	97	81	67						
160	177	156	146	132	119	107	91	76							
170	167	147	137	125	112	101	86	71							
180	158	139	130	118	106	95	81	67							
190	149	132	123	112	100	90	77								
200	142	125	117	106	95	86	73								
210	135	119	111	101	90	82									
220	129	114	106	96	86	78									
230	123	109	101	92	83	75									
240	118	104	97	88	79										
250	114	100	93	85	76										
260	109	96	90	81	73										
270	105	93	86	78											
280	101	89	83	76											
290	98	86	80	73											
300	95	83	78												
310	92	81	75												
320	89	78													
330	86	76													
340	83	73													
350	81														
360	79														
370	77														
380	75														

Calculations based on copper resistance of 12.5 ohms per CM-ft at 50C (122F).

Reactance and impedance losses calculated for each wire.

Conductors closely grouped in metallic conduit.

FIGURE 4.48 Circuit lengths (in feet) for a 1 percent voltage drop in 3-phase, 3-wire feeders.

310-4. Conductors in Parallel. Aluminum, copper-clad aluminum, or copper conductors of size No. 1/0 and larger, comprising each phase, neutral or grounded circuit conductor, shall be permitted to be connected in parallel (electrically joined at both ends to form a single conductor).

The paralleled conductors in each phase, neutral or grounded circuit conductor shall:
(1) Be the same length;
(2) Have the same conductor material;
(3) Be the same size in circular mil area;

(4) Have the same insulation type;
(5) Be terminated in the same manner.
Where run in separate raceways or cables, the raceways or cables shall have the same physical characteristics.

(FPN): Differences in inductive reactance and unequal division of current can be minimized by choice of materials, methods of construction, and orientation of conductors. It is not the intent to require that conductors of one phase, neutral or grounded circuit conductor be the same as those of another phase, neutral or grounded circuit conductor to achieve balance.

Where equipment grounding conductors are used with conductors in parallel, they shall comply with the requirements of this section except that they shall be sized in accordance with Section 250-95.

This specific **NEC** provision is intended to allow a practical means of installing large-capacity feeders and services. For instance, six conductors might be used per phase and neutral to obtain a 2000-A capacity per phase, which simply could not be accomplished without parallel conductors. To ensure equal division of the total current among all conductors involved, all the multiple conductors must be of the same length, the same material, the same circular-mil area, and the same insulation type; they must be terminated in the same manner; and, where multiple conductors are run in separate raceways or cables, the raceways or cables must have the same physical characteristics. The reason for the last condition is explained in the FPN. It basically says that the impedance of a circuit in a nonferrous raceway is different from the impedance of the same circuit in a ferrous raceway; the use of raceways or enclosures with different characteristics would unbalance currents, which could result in overloading one of the paralleled conductors.

From the **NEC** tables that give the current-carrying capacities of various sizes of conductors, it can be seen that small conductors carry more current per circular mil of cross section than large conductors. This results from the rating of conductor capacity according to temperature rise. The larger the cable, the smaller the radiating surface per circular mil of cross section. Losses due to skin effect (the apparent higher resistance of conductors to alternating current than to direct current) are also higher in the larger conductor sizes. And larger conductors cost more per ampere than smaller conductors.

All these factors point to the advisability of using a number of smaller conductors in multiple to obtain a particular carrying capacity, rather than using a single conductor of that capacity. In many cases, multiple conductors for a feeder provide distinct operating advantages and are more economical than the equivalent-capacity single-conductor makeup. But it should be noted that the reduced total cross section of a conductor resulting from the use of multiple conductors per leg produces higher resistance and a greater voltage drop than in the same length of an equal-ampacity single conductor per leg. Voltage drop may therefore be a limitation.

Figure 4.49 shows a typical application of copper conductors in multiple, with the advantages of such use. Where more than three conductors are located in a single conduit, the ampacity (i.e., load-current rating) of each conductor must be derated from the ampacity value shown in **NEC** Table 310-16. The four circuit makeups show:

1. Without derating for conduit occupancy, circuit 2 would be equivalent to circuit 1.
2. A circuit of six 400-kcmil can be made equivalent in current-carrying capacity to a circuit of three 2000-kcmil by dividing the 400-kcmil between two conduits (three conductors per 3-in conduit). If three different phases are run in each of two 3-in conduits for this circuit, the multiple circuit would not require derating,

FEEDERS FOR LIGHTING AND POWER 305

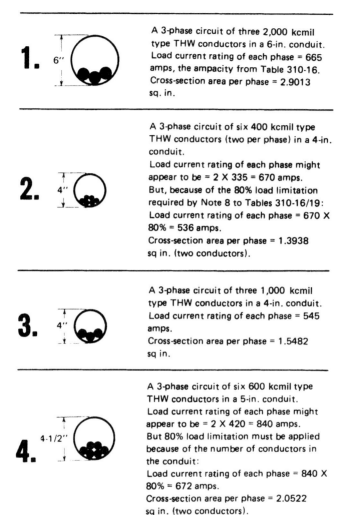

FIGURE 4.49 Multiple-conductor feeder makeups must be related to conductor ampacity derating (Note 8, Table 310-16).

and its 670-A rating would exceed the 665-A rating of circuit 1.
3. Circuit 2 is almost equivalent to circuit 3 in current rating.
4. Circuit 4 is equivalent to circuit 1 in current rating but uses less conductor copper and a smaller conduit. And these advantages are obtained even with occupancy derating.

Where large currents are involved, it is particularly important that the separate phase conductors be located close together to avoid excessive voltage drop and ensure equal division of current. It is also essential that all phases and the neutral (and grounding wires, if any) be run in each conduit, even where the conduit is of nonmetallic material.

Figure 4.50 shows a handy way of evaluating single-conductor feeder makeup versus multiple-conductor feeder makeup, based on achieving the maximum current rating (even with ampacity derating according to Note 8 to Table 310-16), lowest voltage drop and power loss, and maximum economy—all in one conduit.

Except where the conductor size is governed by voltage-drop considerations, it is seldom economical to use conductors of sizes larger than 1000 kcmil, because above this size the increase in ampacity is very small in proportion to the increase in the size of the conductor. Thus, for a 50 percent increase in the conductor size (e.g., from 1000 to 1500 kcmil), the ampacity of a Type THW conductor increases only 80 A, or less than 15 percent; for an increase in size from 1000 to 2000 kcmil, a 100 percent size increase, the ampacity increases only 120 A, or about 20 percent. In any case where single conductors larger than 500 kcmil would be required, it is worthwhile to compute the total installation cost using single conductors and the cost using two or more conductors in parallel.

It is important to note that when multiple conductors are used per circuit phase leg, more space may be required at equipment terminals to bend and install the conductors.

Exception No. 1 to Section 310-4 clearly indicates long-time **NEC** acceptance of the practice of paralleling conductors smaller than No. 1/0 for use in traveling cables of elevators, dumbwaiters, and similar equipment.

Exceptions No. 2 and 3 of Section 310-4 permits parallel circuit makeup using conductors smaller than No. 1/0, but all the given conditions must be observed. These exceptions permit the use of smaller conductors in parallel for certain applications where it is necessary to reduce the conductor capacitance effect or to reduce the voltage drop over long circuit runs.

A Common Neutral for a Feeder. A frequently discussed **NEC** requirement is that of Section 215-4, covering the use of a common neutral with more than one set of feeders. This section says that a common neutral feeder may be used for two or three sets of 3-wire feeders, or two sets of 4-wire feeders. It further requires that all conductors of feeder circuits employing a common neutral feeder must be within the same enclosure when the enclosure or raceway containing them is metal.

A common neutral is a single neutral conductor used as the neutral for more than one set of feeder conductors. It must have a current-carrying capacity equal to the sum of the neutral-conductor capacities that would be required if an individual neutral conductor were used with each feeder set. Figure 4.51 shows a typical example of a common neutral—here, used for three-feeder circuits.

A common neutral may be used only with feeders; it may never be used with branch circuits. A single neutral of a multiwire branch circuit is not a "common neutral." Rather, it is the neutral of only a single circuit, even though the circuit may consist of three or four wires. A feeder common neutral is used only with more than one feeder circuit.

OVERCURRENT PROTECTION

The most important task in modern feeder design is the selection of overcurrent protection. Certainly, the vital relation between effective overcurrent protection and

POWER FEEDER SPECIFICATION DATA

P	Cable Size	Cables Per Phase	Conduit Size	Cost in Dollars Per 100 Feet	Ampere Rating NE Code	Dollars Per 100 Feet Per Unit P	Economic Choice
2.0	4	1	1¼	$136.00	70	$68.00	
3.0	2	1	1¼	151.00	95	50.30	
3.7	1	1	1½	189.00	110	51.10	
4.6	0	1	2	237.00	125	51.50	
5.5	2/0	1	2	260.00	145	47.20	
6.6	3/0	1	2	286.00	165	43.40	
7.4	1	2	2½	342.00	176	46.30	B
7.7	4/0	1	2½	356.00	195	46.30	A
8.2	250	1	2½	421.00	255	51.40	B
9.2	0	2	2½	378.00	200	41.10	A
9.5	350	1	3	532.00	310	56.00	C
11.0	2/0	2	3	468.00	232	42.50	A
11.0	500	1	3	622.00	380	56.60	C
11.2	1	3	3	472.00	231	42.20	B
13.2	3/0	2	3	521.00	264	39.50	A
13.8	0	3	3	526.00	262	38.10	A
15.4	4/0	2	3	586.00	312	38.00	A
16.4	250	2	3½	755.00	408	46.00	B
16.5	2/0	3	3½	634.00	304	38.40	A
19.8	3/0	3	3½	713.00	346	36.00	A
23.1	4/0	3	4	848.00	410	36.70	A
24.5	250	3	4½	1085.00	535	44.30	A

Wiring in steel conduit for up to 600-volt three-phase 30 C ambient NE Code ratings.

Size 250 kcmil and above — Heat and moisture resistant type cable insulation with 75C rating.

Below 250 kcmil — Thermoplastic cable insulation with 60 C rating.

A — Preferred   B — Second Choice   C — Uneconomical

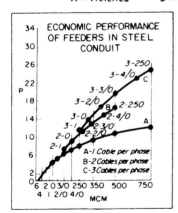

A suitable unit for comparing the performance of cables is P, which is defined as (1)

$$P = LxI/dV \times 1000$$

where, $L =$ feeder length in feet

$I =$ load current in amperes

$dV =$ feeder voltage drop

(1000 is included for brevity)

Based on the average value of load power factor, 0.8 to 0.95, it can be assumed that voltage drop in a feeder is equal to the impedance drop, so that dV in Equation (1) becomes $LIZ\sqrt{3}/1000$, Equation (1) then simplifies to $P = 1/(Z\sqrt{3})$

where, $Z =$ ohms to neutral impedance per 1000 feet of feeder.

The higher the value of "P," the lower the voltage drop and power losses and the higher the extra capacity.

FIGURE 4.50 "P factor" gives figure of merit for comparing single and multiple feeder-conductor makeup.

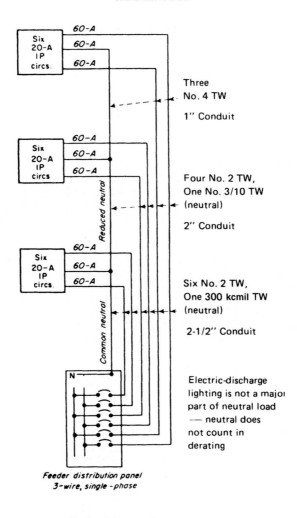

FIGURE 4.51 Single neutral conductor may be used with two or more sets of feeder phase legs.

the safety of personnel and property must be fully recognized. With electrical application continuing to grow at a rapid pace, experience has shown that carelessness or inadequacy in the selection of overcurrent protection can lead to shock and fire hazards. Of only slightly less importance is the need for full and thorough overcurrent protection for long, reliable life of equipment and systems.

In this total-electric age, all modern facilities are dependent upon electrical distribution systems for efficient, economical, and productive performance of activities within the facility. Office buildings, schools, stores, factories, hotels, hospitals—all modern buildings—have an almost total dependence on artificial internal environments created by air conditioning, heat, light, elevators, pumps, etc. Their electrical systems must be based on overcurrent protection that fully and surely protects against overloads up to the maximum possible short-circuit currents, that is carefully coordinated to minimize the extent of electrical outages, and that eliminates nuisance or unwanted opening by operating only when and as required for proper protection of life and property.

In the design of an electrical system, required current-carrying capacities are determined for the various feeders and subfeeders. Then these required capacities are translated into standard circuit conductors which have sufficient current-carrying capacities, based on the size of the conductors, the type of insulation on the conductors, the ambient temperature at the place of installation, the number of conductors in each conduit, the type and continuity of the load, and judicious inclusion of spare capacity to provide for future load growth. Or, if busway, armored cable, or other cable assemblies are to be used, similar considerations go into the selection of conductors with the required current-carrying capacities. In any case, then, the next step is to provide overcurrent protection for each and every circuit.

The overcurrent device for conductors or equipment must automatically open the circuit it protects if the current flowing in that circuit reaches a value which will cause an excessive or dangerous temperature in the conductors or conductor installation.

Overcurrent protection for conductors must also be rated for safe operation at the level of fault current obtainable at the point of application. Every fuse and circuit breaker for short-circuit protection must be applied in such a way that the fault current produced by a bolted short circuit on its load terminals will not damage or destroy the device. Specifically this requires that a short-circuit overcurrent device have a proven interrupting capacity at least equal to the current which the electrical system can deliver into a short on its load terminals.

But safe application of a protective device does not stop with adequate interrupting capacity to allow its use at the point of installation in the system. The speed of operation of the device must then be analyzed in relation to the thermal and magnetic energy which the device will permit to flow in the faulted circuit. A very important consideration is the provision of conductor sizing to meet the potential heating load of short-circuit currents in cables. With expanded use of circuit-breaker overcurrent protection, coordination of protection from loads back to the source has introduced time delays in the operation of overcurrent devices. Cables in CB-protected systems must be able to withstand any impressed short-circuit currents for the duration of the overcurrent delay. For example, a motor circuit to a 100-hp motor might be required to carry as much as 15,000 A for a number of seconds. To limit damage to the cable due to heating, a much larger size of conductor than necessary for the load current alone may be required.

A device may be able to break a given short-circuit current without damaging itself in the process; however, in the time it takes to open the faulted circuit, enough energy may get through to damage or destroy other equipment in series with the fault. This other equipment might be a cable or busway or a switch or motor controller—any circuit component which simply cannot withstand the few cycles of high-amperage short-circuit current that flows in the period of time between initiation of the fault and interruption of the current flow.

Basic Rules. Conductors in an electric circuit—whether it is a branch circuit or a feeder—must be protected against excessive current flow (Fig. 4.52). Such protection must be provided in accordance with the current-carrying capacities of the conductors, except where certain higher settings or ratings of protective devices are permitted or required—as in the case of motor branch circuits or motor feeders.

PROTECTIVE DEVICE IN SERIES

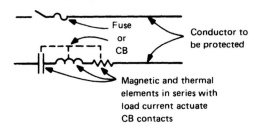

CT-RELAY RESPONDS TO LOAD CURRENT

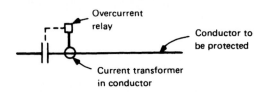

FIGURE 4.52 Protective device must respond automatically to excess feeder current.

NEC Section 240-12, "Electrical System Coordination," is aimed at "industrial locations" where hazard to personnel would result from sudden disorderly shutdown of electrical equipment under fault conditions. For such systems, it is permissible to oversize protective devices so that conductors are not protected at their ampacities, and the circuit will not open on overload. But to satisfy the rule, the circuit protective devices must provide proper short-circuit protection (with selective coordination of the time-current characteristics of protective devices in series from the service to any load) so that, automatically, any fault will actuate only the short-circuit protective device closest to the fault on the line side of the fault, thereby minimizing the extent of an electrical outage due to a fault. In addition, each circuit must be provided with "overload indication based on monitoring systems or devices." In the event of an overload, although the circuit will not open, a visible and/or audible signal must be given to warn of the overload and indicate the need for proper attention to the matter. The rule says that "The monitoring system may cause the condition to go to alarm allowing corrective action or an orderly shutdown thereby minimizing personnel hazard and equipment damage." (See Fig. 4.53.)

An overcurrent device must not be used in a permanently grounded conductor, except where the device simultaneously opens all conductors of the circuit *or* where three fuses are used in a 3-phase, 3-wire circuit fed from a corner-grounded delta system and a fuse is required in the grounded leg by **NEC** Section 430-36 because the fuses are providing running overload protection (as well as short-circuit protection) for a motor.

Overcurrent protection for a conductor must be located at the point where the conductor receives its supply. This means that a conductor run of a particular current-carrying capacity must be protected at the point at which it is fed by a conductor of higher current-carrying capacity. In a feeder run, then, a change (reduction) in

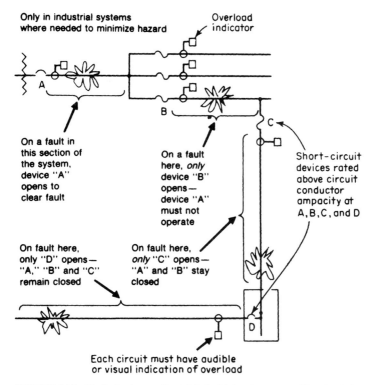

FIGURE 4.53 If selectively coordinated, industrial systems may utilize short-circuit protection *only*.

conductor size must be accompanied by protection for the smaller conductor (Fig. 4.54). Exceptions to this rule are made in the case of taps, which are discussed later in this section under "Feeder Taps." Another exception is where the protection of the larger conductor meets the requirements for protection of the smaller conductor, that is, the rating or setting of the fuse or CB protecting the larger wires is not in excess of the ampacity of the smaller conductors or is suitable to protect the smaller tap wires in accordance with some specific **NEC** rule.

For feeder circuits other than motor feeders, the overcurrent device that protects against short circuits and grounds can be sized as follows:

1. If the allowable current-carrying capacity of a conductor does not correspond to the rating of a standard-size fuse, the next larger rating of fuse may be used only where the rating of the fuse does not exceed 800 A.

2. Non-adjustable-trip breakers must be rated in accordance with the current-carrying capacity of the conductors they protect—except that a higher-rated circuit breaker may be used if the current-carrying capacity of the conductor does not correspond to a standard rating. In such a case, the next higher standard rating and setting may be used only where the rating is 800 A or less. An adjustable-trip breaker must be set to protect the feeder conductors against overload current.

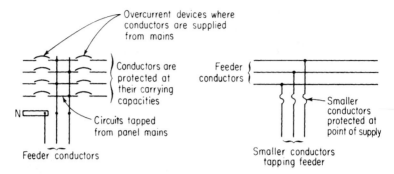

FIGURE 4.54 Effective design protects feeder conductors where they receive their supply.

The general rule, in both these cases, is that the device must be rated to protect conductors in accordance with their safe, allowable current-carrying capacities. The current rating of a protective device is commonly called the *continuous current rating*. The word "continuous" means an uninterrupted current flow for a period of 3 hours or more. It represents the maximum value of current that the device will carry without tripping open or blowing, under certain specified conditions. Of course, there will be cases where standard ampere ratings and settings of overcurrent devices will not correspond with conductor capacities. In such cases, the next larger standard size of overcurrent device may be used where the rating of the protective device is 800 A or less. A basic guide to effective selection of the ampere ratings of overcurrent devices for any feeder or service application is given in the following:

If a circuit conductor of, say, 500-kcmil THW copper (not more than three in a conduit at not over 86°F ambient) satisfies design requirements and **NEC** rules for a particular load current not in excess of the conductor's table ampacity of 380 A, then the conductor may be protected by a fuse or CB rated at 400 A. **NEC** Section 240-6, which gives the "standard ampere ratings" of protective devices, shows devices rated at 350 and 400 A, but none at 380 A. In such a case, the **NEC** accepts 400 A as "the next higher standard device rating" above the conductor ampacity of 380 A.

But such a 400-A device would permit a load increase above the 380 A that is the safe maximum limit for the conductor. It would be more effective practice to use protection rated at 350 A and prevent such overload.

For the application of fuses and CBs:

1. If the ampacity of a conductor does not correspond to the rating of a standard-size fuse, the next higher rating of fuse may be used only where that rating is 800 A or less. Over 800 A, the next lower fuse rating must be used.

2. A non-adjustable-trip breaker (one without overload trip adjustment above its rating—although it may have adjustable short-circuit trip) must be rated in accordance with the current-carrying capacity of the conductors it protects, except that the next higher standard rating of CB may be used if the ampacity of the conductor does not correspond to a standard rating. However, in such a case, the next higher standard rating may be used only where the rating is 800 A or less.

An example of such an application is shown in Fig. 4.55, where a nonadjustable CB with a rating of 1200 A is used to protect the conductors of a feeder circuit, which are rated at 1140 A. As shown, the use of a 1200-A CB to protect a circuit rated at 1140 A (3 × 380 A = 1140 A) clearly violates **NEC** Section 240-3(c), because the CB rating is the next higher rating above the ampacity of the conductors—on a circuit rated over 800 A. With a feeder circuit as shown (three 500-kcmil THW conductors, each rated at 380 A), the CB must not be rated over 1140 A. An FPN following Section 240-6 makes clear that for the purpose of practical application, the use of "nonstandard" ratings is permitted. So, even though, a 1000-A-rated device is actually the next lower *standard* rating, the use of an 1100-A-rated device is intended to be permitted. A standard 1100-A CB would satisfy the **NEC** rule. Of course, if 500-kcmil THHN or XHHW conductors were used instead of THW conductors, then each would be rated at 430 A; three per phase would give the circuit an ampacity of 1290 A (3 × 430 A); the 1200-A CB would satisfy the basic rule in the first sentence of Section 240-3; and Exception No. 1 would not be involved.

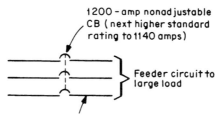

FIGURE 4.55 This use of "the next higher standard rated protective device" is *not* acceptable.

In the design of electrical systems for industrial manufacturing and process operations, special or unusual conditions often require suspension of the normal application of **NEC** rules on overcurrent protection. To meet such needs, **NEC** Article 685 allows the customized design and layout of overcurrent protection for "integrated electrical systems." In effect, the rules of this article exempt the interconnected and/or interrelated electrical components of manufacturing assembly or production lines and related industrial process operations from the conventional rules on overcurrent protection. Section 685-2 lists those **NEC** rules that allow special treatment for installations where orderly shutdown is required but would not be provided by the standard rules.

In the **NEC**, the scope of Article 685 is described as follows:

This article covers integrated electrical systems, other than unit equipment, in which orderly shutdown is necessary to assure safe operation. An integrated electrical system as used in this article is a unitized segment of an industrial wiring system where all of the following conditions are met: (1) an orderly shutdown is required to minimize personnel hazard and equipment damage; (2) the conditions of maintenance and supervision assure that qualified persons will service the system; and (3) effective safeguards, acceptable to the authority having jurisdiction, are established and maintained.

Sizing Feeder Protection

As noted earlier under "Feeder Sizing versus Panelboard Rating," **NEC** Section 220-10(b) must be fully satisfied when feeder overcurrent protective devices are sized. To repeat the basic concept, for emphasis: When a feeder supplies continuous loads or any combination of continuous and noncontinuous loads, the rating of overcurrent devices protecting the feeder and the ampacity of the feeder conductors shall not be less than the sum of the noncontinuous load plus 125 percent of the continuous load, unless the CB or fused switch protecting the feeder is UL-listed for continuous operation at 100 percent of its load rating. This is shown in Fig. 4.56.

To correspond to that rule, the UL provides clear information about available circuit breakers and fusible switches for use with continuous loading (maximum current flowing continuously for 3 hours or more).

THE RULE

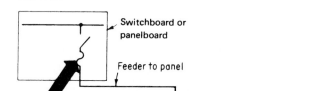

FEEDER OVERCURRENT DEVICE must be rated not less than 125% of the continuous load, and the feeder conductors must be sized on the same basis to assure them effective protection in accordance with their ampacity

EXAMPLE

For this feeder, with conductors rated at 380 amps, the maximum load permitted for a conventional fused switch is 80% of 380 amps, or 304 amps.

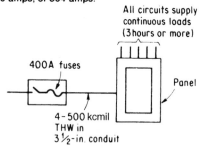

Rating of fuses and ampacity of feeder conductors are both at least equal to 125% times the continuous load.

FIGURE 4.56 Basic approach to the sizing of protection for a continuous-load feeder.

The nominal or theoretical continuous current rating of a CB generally is taken to be the same as its trip setting—the value of current at which the breaker will open, either instantaneously or after some intentional time delay. But the real continuous current rating of a CB—the value of current that it can safely and properly carry for periods of 3 hours or more—frequently is reduced to 80 percent of the nameplate value by rules of codes and standards.

The UL *Construction Materials Directory* contains a clear, simple rule in the instructions under "Circuit Breakers, Molded-Case." It says: "Unless otherwise marked, circuit breakers should not be loaded to exceed 80% of their current rating, where in normal operation the load will continue for three or more hours."

A load that continues for three hours or more is a continuous load. If a breaker is marked for continuous operation, it may be loaded to 100 percent of its rating and operated continuously.

Circuit breakers are available for continuous operation at 100 percent of their current rating, but they must be used in the mounting and enclosure arrangements established by UL for 100 percent rating. Molded-case CBs of the 100 percent continuous type are made in frame ratings from 600 A up, with trip settings from 225 A up. Information on the use of breakers with 100 percent ratings is given on their nameplates (Fig. 4.57). The nameplates clearly indicate that ventilation may or may not be required. Because most switchboards have fairly large interior volumes, the "minimum enclosure" dimensions shown on these nameplates (45 by 38 by 20 in) usually are readily achieved. But, special UL tests must be performed if these dimensions are not achieved. Where busbar extensions and lugs are connected to the circuit breakers within the switchboard, the caution about copper conductors does not apply, and aluminum conductors may be used.

If a switchboard does not provide the ventilation pattern and the enclosure size specified on the nameplate, the circuit breaker must be applied at 80 percent of its rating. Switchboard manufacturers have UL tests conducted with a circuit breaker installed in a specific enclosure; the enclosure can receive a listing for operation at 100 percent of rating even though the ventilation pattern or overall enclosure size does not meet the specifications. In cases where the breaker nameplate specifications are not provided by the switchboard, the customer must request a letter from the manufacturer certifying that a 100 percent listing has been received. Otherwise, the breaker must be applied at 80 percent.

It is essential to check the instructions given in the UL listing to determine if and under what conditions a circuit breaker (or a fuse in a switch) is rated for continuous operation at 100 percent of its current rating.

Care must also be exercised in selecting fuse protection for feeders. In general, the rating of a fuse is taken as 100 percent of the rated nameplate current when the fuse is enclosed by a switch or panel housing. But, because of the heat generated by many fuses, the maximum continuous load permitted on a fused switch is restricted by a number of NEMA, UL, and **NEC** rules to 80 percent of its rating. The limitation of circuit load current to no more than 80 percent of the current rating of fuses in equipment protects the switch or other piece of equipment from the heat produced in the fuse element; it also protects attached circuit wires from excessive heating close to the terminals. The fuse itself can actually carry 100 percent of its current rating continuously without damage to itself, but its heat is conducted into the adjacent wiring and switch components.

NEMA standards require that a fused, enclosed switch be marked, as part of the electrical rating, "Continuous load current not to exceed 80 percent of the rating of fuses employed in other than motor circuits." That derating compensates for the extra heat produced by continuous operation. Motor circuits are excluded from that

EXAMPLE 1

The ventilation applies only to 100% rated applications.

Because of the relatively large volume in the switchboard structure, this is normally not a problem. Special UL tests must be performed if these specifications are not met.

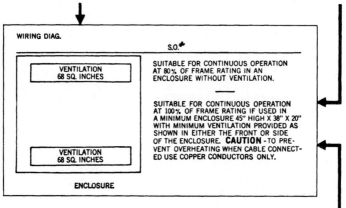

EXAMPLE 2

Generally, bus bar extensions and lugs are connected to the circuit breaker within the switchboard. In that case, aluminum cable may be used.

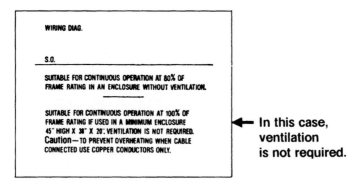

In this case, ventilation is not required.

FIGURE 4.57 Circuit breakers that may be used for continuous loading to 100 percent of their rating are marked in detail.

rule, but a motor circuit is required by the **NEC** to have conductors rated at least 125 percent of the motor full-load current; this, in effect, limits the load current to 80 percent of the conductor ampacity and limits the load on the fuses rated to protect those conductors. But, the UL *Construction Materials Directory* does recognize fused bolted-pressure switches and high-pressure butt-contact switches for use at 100 percent of their ratings on circuits with available fault currents up to 100,000, 150,000, or 200,000 rms symmetrical amperes—as marked. (See "Fused Power Circuit Devices" in that UL publication.) The ratings of fuses for feeder protection, therefore, are determined by the type of switch used and its suitability for continuous operation at 100 percent loading (Fig. 4.58).

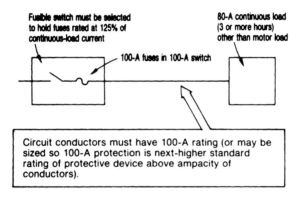

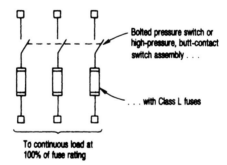

FIGURE 4.58 Rating of feeder fuse protection depends on the switch assembly.

Figure 4.59 shows how the sizing of feeder overcurrent protection and conductor size differs with the type of CB used. In practice, the ratings of the two breakers would be increased to the next highest standard trip ratings—2500 instead of 2255 A, and 2000 instead of 1804 A.

The selection of a circuit breaker or fuse of the proper current rating must be carefully based on a determination of the real "continuous" rating—whether it is the nameplate value or some lesser value because of derating to supply loads that operate steadily for periods of 3 hours or more. In addition, that same selection task must be related to the need to ensure that the CB or fuse is correctly rated to automati-

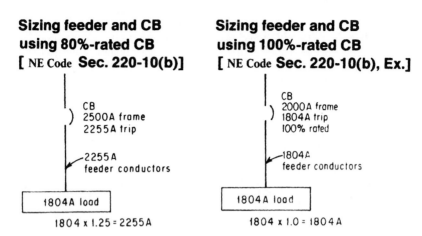

FIGURE 4.59 Continuous-rated CB permits reduced sizing of feeder conductors and protection.

cally open the circuit if the current reaches a value that will cause an excessive or dangerous temperature in the conductor or conductor insulation or in other equipment being protected.

Fuses. Fuses comprise a simple and effective means of providing overcurrent protection. Containing only a fusible element and having no moving parts, a fuse requires no maintenance, although the periodic inspection of fuseholders is recommended to ensure good electrical contact. The fusible element of a fuse opens at a precisely determined time in proportion to the amount of current that flows through the fuse. This time-current characteristic varies with the size and type of fuse.

The cartridge fuses used for today's feeder-protection equipment are made in a wide range of types, sizes, and ratings. Various fuse types are designated by NEMA and UL standards. In broad terms, these fuses are classified as (1) one-time, (2) renewable, (3) dual-element, (4) current-limiting, and (5) high-interrupting-capacity.

By present NEMA and UL standards, the standard (so-called **NEC**-type) cartridge fuses of the one-time and renewable types are designated as Class H fuses. Such fuses have interrupting ratings of 10,000 A. Fuses in this category are used in applications where available short-circuit currents are relatively low.

The same standards classify cartridge fuses with interrupting-current ratings above 10,000 rms symmetrical amperes in Classes G, J, K, L, R, and T. These are high-interrupting-capacity or current-limiting fuses. The term "high-interrupting-capacity" indicates an interrupting rating between 10,000 and about 200,000 rms symmetrical amperes, depending upon the particular fuse. A "current-limiting" fuse is one that safely interrupts all available currents within its interrupting rating and limits the peak let-through current I_p and the total ampere-squared-seconds I^2t to specified values.

The term "current-limiting" indicates that the fuse, when tested on a circuit capable of delivering a specific short-circuit current (rms symmetrical amperes) at rated voltage, will start to melt within 90 electrical degrees and will clear the circuit within 180 electrical degrees (one-half cycle). Because the time required for a fuse to melt depends on the available current, a fuse that is current-limiting when subjected to a specific short-circuit current (rms symmetrical amperes) may not be current-limiting on a circuit with lower maximum available current.

FEEDERS FOR LIGHTING AND POWER   319

Class J and L Fuses.   Class J (0 to 600 A, 600 V ac) and Class L (601 to 6000 A, 600 V ac) fuses are current-limiting, high-interrupting-capacity fuses. The interrupting ratings are 100,000 and 200,000 rms symmetrical amperes, and the designated rating is marked on the label of each Class J or L fuse. Class J and L fuses are also marked "Current limiting."

Class J fuse dimensions are different from those for standard Class H cartridge fuses of the same voltage rating and ampere classification. They require special fuseholders that do not accept non-current-limiting fuses. This arrangement complies with the last sentence of **NEC** Section 240-60(b), which reads, "Fuseholders for current-limiting fuses shall not permit insertion of fuses that are not current limiting."

Class K Fuses.   These are subdivided into Classes K-1, K-5, and K-9. Class K fuses have the same dimensions as Class H (standard **NEC**) fuses and are interchangeable with them. Class K-1, K-5, and K-9 fuses have different degrees of current limitation but are not permitted to be labeled "Current-limiting" because their physical characteristics allow them to be interchanged with non-current-limiting types. Classes R and T have been developed to provide current limitation and prevent interchangeability with non-current-limiting types.

Class R Fuses.   These fuses are made in two designations: RK1 and RK5. UL data are as follows: Fuses marked "Class RK1" or "Class RK5" are high-interrupting-capacity types and are marked "Current-limiting." Although these fuses will fit into standard fuseholders that take Class H and Class K fuses, special rejection-type fuseholders designed for Class RK1 and RK5 fuses will not accept Class H and Class K fuses. In that way, circuits and equipment protected in accordance with the characteristics of RK1 and RK5 fuses cannot have that protection reduced by the insertion of other fuses of a lower protective level.

Other Fuse Data.   Other UL applications data that affect the selection of fuses are as follows:

- Fuses designated as Class G (0 to 60 A, 300 V ac) are high-interrupting-capacity types and are marked "Current limiting." They are not interchangeable with other fuses mentioned above and below.
- Fuses designated as Class T (0 to 600 A, 250 and 600 V ac) are high-interrupting-capacity types and are marked "Current limiting." They are not interchangeable with other fuses mentioned above.
- Class K-1, K-5, and K-9 fuses are marked, in addition to their regular voltage and current ratings, with an interrupting rating of 200,000, 100,000 or 50,000 A (rms symmetrical).
- Class RK1, RK5, J, L, and T fuses are marked, in addition to their regular voltage and current ratings, with interrupting rating up to 200,000 A (rms symmetrical).
- Class G fuses are marked, in addition to their regular voltage and current ratings, with an interrupting rating of 100,000 A (rms symmetrical).
- Class L fuses are designed for use in equipment to which line and load connections are made by means of solid busbars. For this reason, temperature rises on Class L fuse blades may exceed those observed in connection with other cartridge-fuse designs. Terminal connections for wires in such equipment must be designed to avoid excessive temperatures on the wire insulation.

Some manufacturers provide fuses that are advertised and marked to indicate that they have "time-delay" characteristics. In the case of Class G, H, K, and RK fuses, their time-delay characteristics (minimum blowing time) have been investigated. Class G fuses, which can carry 200 percent of their rated current for 12 s or more, and Class H, K, and RK fuses, which can carry 500 percent of their rated current for 10 s or more, may be marked "D" or "Time delay," or with some equivalent designation. Class L fuses are permitted to be marked "Time delay" but have not been evaluated for such performance. Class J and T fuses are not permitted to be marked "Time delay."

Figure 4.60 shows the variation in basic operating characteristics of types of fuses commonly used for feeder protection in modern systems.

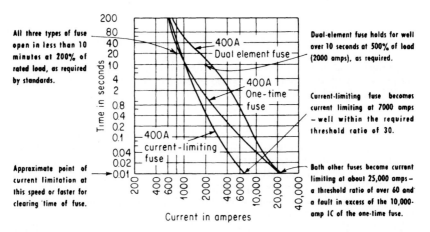

FIGURE 4.60 Each type of fuse is suited to particular feeder protective functions.

Circuit Breakers. Circuit breakers must be chosen to suit the application, whether they are molded-case breakers or low-voltage air CBs. Molded-case breakers may be of the thermal-magnetic or fully magnetic type; air CBs may be installed in conjunction with current-limiting fuses. Circuit breakers may incorporate electrical operation, adjustable time delays, and protective relay hookups; a variety of other features, such as trips for ground-fault protection, may be required and should be carefully selected from manufacturers' data.

The two basic types of breakers, according to general construction, are the molded-case CB and the air (ac power) CB. There is some overlap in the continuous-current ratings of these devices, but the air CB is a larger, heavier-duty device than the molded-case breaker, has generally higher short-circuit interrupting ratings, and is available with current ratings up to 6000 A, whereas the molded-case device is commonly used only up to 3000 A. The use of current-limiting fuses in molded-case or power CBs extends the application of both types with respect to short-circuit interrupting ratings.

The availability of molded-case CBs in sizes up to 3000 A has stimulated their use where air CBs would have been required in past years. Up to 3000 A, the selection of a molded-case CB instead of an air CB offers substantial economy in space and equipment costs. Required interrupting duty (symmetrical short-circuit current rating) is an important selection criterion that may often favor air CBs over molded-case CBs where either may be used. But the use of current-limiting fuses in conjunction with molded-case CBs can provide fully adequate interrupting capacity.

The selection of a molded-case CB often involves a choice of CB trip arrangement. For feeders up to 400 A, the most commonly used molded-case CB that fully meets the needs of circuit protection is the unit that has a noninterchangeable trip assembly made up of a fixed value of thermal overload pickup current with time delay, and a fixed magnetic trip element that provides instantaneous (no intentional delay) opening of the CB on values of current of 10 or more times its continuous current rating (its thermal-trip setting).

Where adjustable magnetic (instantaneous) trip setting is required, a molded-case CB can be selected with a noninterchangeable, built-in trip assembly that provides adjustment of the magnetic trip setting from, say, 3 to 10 times the continuous rating (the thermal-trip setting). And CB units with interchangeable trip assemblies allow the continuous current rating of installed CBs to be changed from one value to another, up to the CB-frame ampere rating. And each trip-assembly unit has an adjustable magnetic trip element. Such CBs offer flexibility of current rating to meet changes in circuit load conditions or conductor makeup.

Although economy indicates the use of molded-case CBs over air CBs for those circuit sizes where either might be used, a basic concern that affects selection is the presence of an instantaneous trip element in the molded-case CB. Because of this element, a series layout of such CBs can induce cascade operation of two or more breakers if all the CBs are subjected to a fault current above their instantaneous trip settings (Fig. 4.61). This potential for excessive system outage commonly dictates the use of an air CB as a main ahead of molded-case branch CBs; the main air CB has long- and short-time-delay trip characteristics but no instantaneous trip element. Thus, on any faults in the molded-case-CB feeders, the air CB will remain closed and allow only a molded-case CB to clear the fault. Selection of CBs for this highly desirable selective coordination depends on careful study of time-current characteristics, interrupting ratings, and available circuit fault current.

Although electromechanical tripping mechanisms have long been the most widely used in molded-case and power circuit breakers, solid-state (electronic) tripping circuitry has grown rapidly in both capability and popularity. Solid-state-trip

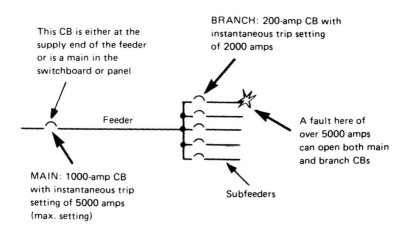

NOTE: If the main is an air CB or encased CB without an instantaneous trip, the main can ride through the fault and remain closed while the instantaneous trip of the branch CB opens the faulted subfeeder.

FIGURE 4.61 Lack of selective coordination knocks out all subfeeders when a fault develops on one subfeeder.

molded-case circuit breakers are made with ratings of 100 A and higher. Offering adjustable pickup and time settings for continuous current rating, long time delay, short time delay, and the value of instantaneous operation, these solid-state breakers may also incorporate ground-fault protection. These "smart" CBs are fast becoming the standard type in ratings of 1200 A and above.

Another growing capability of both molded-case and power (air) CBs is that of current limitation. Short-circuit-duty ratings up to 200,000 A can provide the same fast clearing and current limitation as current-limiting fuses.

Hybrid-type molded-case CBs, called *encased* or *insulated-case* CBs, combine the characteristics of molded-case and air CBs. Such units, which very closely approach the capabilities of air CBs, provide for fully selective application with conventional load-side molded-case units. Such CBs include a solid-state trip assembly that practically eliminates the effect of instantaneous operation described above. With a short time delay of 18 cycles, such units without instantaneous trip (actually an instantaneous-trip override) afford selective coordination that has previously been available only from the larger, more expensive open-type air CBs.

Low-voltage, open-type air circuit breakers are heavy-duty switching and protective devices with high interrupting capacities. Such CBs have rugged contact assemblies, provisions for arc suppression, and either manually or electrically powered operation. They are available either as individually mounted units—in wall or floor mounting enclosures—or in switchboards or substations, as follows:

1. In ratings from 15 A up to 6000 A continuous current, with voltage ratings of 240, 480, and 600 V.
2. With interrupting ratings that vary with size and voltage rating from 15,000 to 150,000 rms asymmetrical amperes (and beyond 200,000 A in combination with suitable current-limiting fuses).
3. With fully magnetic tripping characteristics based on as many as three individual elements: long time delay, short time delay, and instantaneous trip. These separate elements may be used singly or in any combination to accomplish time-current curve shaping for any selective coordination scheme including omission of the instantaneous trip setting to assure full selectivity with load-side devices having instantaneous trip. Adjustments are provided for the long- and short-time-delay pickup settings. The instantaneous trip may be set at the factory or may be adjustable.

Low-voltage power circuit breakers provide for circuit-breaker application above the range of molded-case CBs (3000 A). They are suited to applications where high switching frequency demands a rugged unit to handle heavy current: for such applications, the air CB is the best choice even within the current ratings of molded-case CBs.

Short-Circuit Protection

The selection of circuit overcurrent devices must include careful consideration of the ability of the devices to operate properly and safely on short-circuit faults—phase-to-ground and phase-to-phase faults in grounded wiring systems and phase-to-phase faults in ungrounded wiring systems. In all cases, the following points are important:

1. Under normal operation, a circuit draws current in proportion to the voltage applied and the impedance of the load. When a short circuit occurs, the source voltage no longer encounters the opposition to current flow that the normal load had pre-

sented. Instead, the voltage is applied across a load of much lower impedance, made up of the impedance of the conductors from the source of the voltage to the short-circuit fault, the impedance of the transformer from which the circuit is derived, and any other impedances due to equipment interposed in the circuit between the transformer and the fault.

In the top part of Fig. 4.62, the amount of current which will flow through the short circuit is determined from the system voltage and the total impedance connected in the path of current flow from the source to the fault. This impedance includes resistance and reactance in the conductors, in one or more transformers (going from the fault back to the source), and in any equipment connected in the path of current flow. If, as shown, a source of infinite capacity is assumed on the primary side of the transformer—as if it were a source of voltage with no internal impedance—then the impedance that determines the amount of short-circuit current on the secondary of the transformer consists of the impedance of the transformer itself and the impedance of the secondary conductors up to the fault.

In a transformer circuit, the "impedance" of the unit is involved in the ability of the transformer to supply short-circuit currents into faults on the load side of the transformer. The impedance of a transformer is the opposition which the transformer presents to the flow of short-circuit current through it.

As shown at the bottom of Fig. 4.62, under normal conditions of operation (a), a transformer winding can be considered to be a source of voltage E_r. This source is made up of a generator producing open-circuit voltage E_∞ and having internal resistance R'' and inductive reactance X_L''. Under normal conditions, load current I_L flows, determined by the transformer impedance, the impedance of the circuit conductors, and the impedance of the load. When a short occurs (b), the transformer open-circuit voltage E_∞ is connected across a total load made up only of the transformer impedance and whatever part of the circuit-conductor impedance is in the short circuit. If the supply to the transformer primary can deliver the necessary primary current, the secondary short-circuit current is equal to E/Z. Thus the transformer impedance is the limiting factor for short-circuit current.

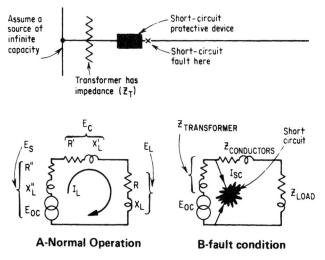

FIGURE 4.62 Elements of a short-circuit condition.

All transformers have impedance, which generally is expressed as a percentage of the normal rated primary voltage. That percentage of the primary voltage must be applied to the transformer to cause full-load rated current to flow in the short-circuited secondary. For instance, if a 480- to 120-V transformer has an impedance of 5 percent, then 5 percent of 480 V, or 24 V, applied to the primary will cause rated load current to flow in the short-circuited secondary. If 5 percent of the primary voltage will cause such current, then 100 percent of the primary voltage will cause 100/5 (or 20) times the full-load rated secondary current to flow through a solid short circuit on the secondary terminals.

From the foregoing, it can be seen that the lower the impedance of a transformer of given kilovoltampere rating, the higher will be the short-circuit current which it can deliver. Take two transformers, each rated at 500 kVA. Assume the rated secondary load current is the same for both transformers. If one transformer is rated at 10 percent impedance, it can supply 100/10 (or 10) times the rated secondary current into a short circuit on its secondary terminals. If the other transformer is rated at 2 percent impedance, it can supply 100/2 (or 50) times the rated secondary current into a short circuit on its secondary terminals. Thus, the second transformer can supply 5 times as much short-circuit current as the first, even though they have the same load-handling ability.

Figure 4.63 presents a simple calculation that yields the symmetrical, single-phase, short-circuit current. The short-circuit protective device must be capable of safely interrupting this value of current and the asymmetrical current value that is obtained by applying some multiplier (such as 1.25) to the symmetrical value. In a more rigorous analysis, this value would be reduced by all impedances in the circuit.

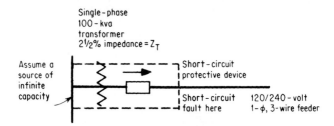

Assuming negligible line and other impedances between the transformer and the fault:

1. Transformer full-load secondary current $= \dfrac{100{,}000 \text{ va}}{240 \text{ v}} = 417 \text{ amps}$

2. Maximum short-circuit current based on transformer impedance $=$

$$\dfrac{100\%}{\%Z_T} \times \text{secondary current}$$

$$\text{Max. } I_{sc} = \dfrac{100}{2.5} \times 417 = 16{,}680 \text{ amps symmetrical}$$

FIGURE 4.63 Basic idea behind short-circuit calculations.

2. Every overcurrent protective device, in a circuit at any voltage level, must be capable of interrupting the maximum possible short-circuit current that might be delivered by the system into a solid short on the load terminals of the device—without destroying itself in the process. This is shown in Fig. 4.64, and the short-circuit protective device in Fig. 4.63 must be rated for at least the 16,680-A calculated current.

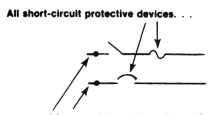

All short-circuit protective devices. . .

. . . must have an interrupting rating at least equal to the maximum fault current that the circuit could deliver into a short circuit on the *line side* of the device.

NOTE: That means that the fault current "available" at the line terminals of all fuses and circuit breakers *must be known* in order to assure that the device has a rating sufficient for the level of fault current.

FIGURE 4.64 All feeder fuses and circuit breakers must have adequate interrupting capacity.

3. Coordinated selective protection for modern circuits provides fast, effective isolation of any faulted section of a system but does not interrupt service to any other section. A careful study of the time-current characteristics of protective devices along with correct application can ensure clearing of a fault by the device nearest to the fault on its supply side, as shown in Fig. 4.65.

Selective coordination of overcurrent protection, in which every fault is cleared by the closest protective device on the line side of the fault, is widely considered to be a particular capability of modern CB equipment—either CBs alone or CBs in conjunction with load-side fusible equipment. Selective coordination ensures fault-initiated opening at only one point in the system, such as on a branch circuit or a feeder; it eliminates cascade operation, where, say, both a branch device and a feeder device open on a branch-circuit fault.

Curve-shaping of the time-current trip characteristics of both molded-case and air CBs by means of long-time-delay, short-time-delay, and/or instantaneous-trip elements is the technique by which sophisticated selective coordination can be achieved, even through a number of voltage transformations. Only the faulted circuit is opened, and the extent of the outage is minimized. But, again, such coordination must be carefully and precisely related to all other design, installation, operational, and economic factors.

4. Let-through current is another factor in the effective application of protective devices for short circuits. A given protective device may be able to interrupt the maximum short-circuit current at its point of installation safely, but it may take so much time to open the faulted circuit that damage will occur to equipment or devices connected in series with the fault. Unless the device operates quickly enough, the tremendous rupturing stresses created by short circuits can cause severe damage. The

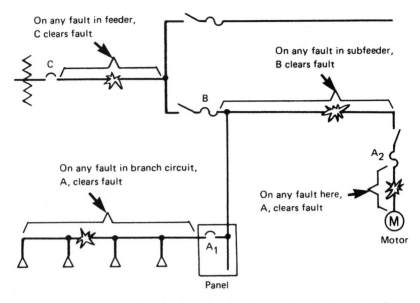

NOTE: Any fault is cleared by nearest device on supply side of fault, and all other devices in series with the fault do not operate.

FIGURE 4.65 Selection coordination of overcurrent devices minimizes the extent of an outage due to a fault.

system components must be related to the let-through current, the current which flows from the time the fault develops until the circuit is opened. That is required by **NEC** Section 110-10, as shown in Fig. 4.66. Current-limiting fuses, for instance, open a short circuit in much less than half a cycle, thereby squelching let-through current.

Because a short-circuit fault is an accidental condition that occurs at random with respect to the alternating voltage wave of a circuit, the fault can occur at any point in the wave form of a voltage cycle. The two extremes of the wave form are where the voltage wave passes through the "zero" axis and where the voltage is a maximum. Figure 4.67 shows these two extreme conditions at which short-circuit current can start and continue to flow if the circuit is not opened by fuses or breakers. Maximum asymmetry in the flow of short-circuit current is produced when the short circuit occurs at that point in the voltage wave for which the short-circuit current is exactly at its positive or negative peak value, as determined by the ratio of reactance to resistance in the short circuit; this ratio establishes the phase-angle difference between the voltage wave and the short-circuit current wave. In the case shown in Fig. 4.67a, with the short circuit occurring as shown on the voltage wave, the phase angle is such that the short-circuit current is exactly at its negative peak. Maximum asymmetry is thus produced. The instantaneous change in phase relation between current and voltage (from the normal condition to fault condition) produces an offset in the ac current wave which acts like a dc transient.

From the immediate condition of asymmetry, the dc component of the short-circuit current wave decays to zero at a rate determined by the ratio of reactance to resistance in the short circuit. For a purely resistive short circuit (zero ratio of reac-

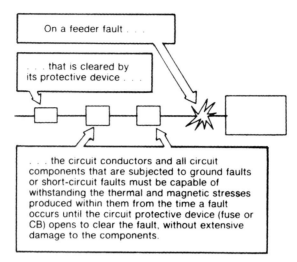

FIGURE 4.66 Operating speed of short-circuit protection must be related to equipment "withstandability" ratings.

tance to resistance), which is a completely theoretical condition that can never be achieved in an ac circuit, the decay to zero of the dc component will be instantaneous. If the short circuit is purely reactive (infinite ratio of reactance to resistance), which is also a practical impossibility, the dc component never decays to zero, the current remains offset, and the asymmetrical rms current is equal to 1.73 times the symmetrical rms current. Of course, actual circuits have ratios of reactance to resistance up to about 20, although most cases have ratios of less than 7. This means the dc offset current will decay to zero in several cycles.

In Fig. 4.67b, a short-circuit current wave of complete symmetry flows when the short occurs at that point in the voltage wave for which the instantaneously initiated short-circuit current is at its zero value. Again the exact point in the voltage wave which will correspond in time to a zero value for the short-circuit current wave is determined by the phase angle (or power factor, or ratio of reactance to resistance) of the short circuit.

Although short circuits can start at the two extremes shown in Fig. 4.67, chances are greatest that a fault will occur at some intermediate degree of asymmetry. But because design must account for the worst possibility, short-circuit calculations should be performed to determine the asymmetrical current which represents the most destructive current that the protective device might be called upon to break.

From the character of the asymmetry, it can be seen that the ratio of reactance to resistance is important in applying protective devices because it indicates how fast the direct current will decay and how much current the device will be called upon to handle in an attempt to open the circuit in the first half cycle or in the second cycle or the third, etc.—since the current decreases with each succeeding cycle. A fast-acting fuse such as a current-limiting fuse, which opens the short circuit in the first half cycle, will break almost the maximum value of the asymmetrical wave and must be rated for that current value. On the other hand, a circuit breaker which will not act until the fourth cycle of the short-circuit current wave will have to interrupt only the level of current to which the asymmetrical wave has decayed.

A—MAXIMUM ASYMMETRY

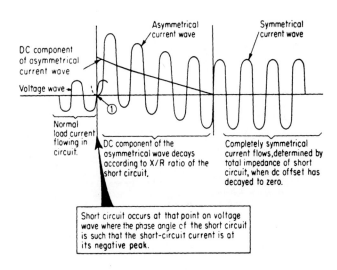

B—COMPLETE SYMMETRY

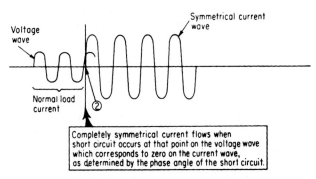

FIGURE 4.67 How short-circuit currents could flow if they were not opened by overcurrent protection.

On a fully asymmetrical short, the maximum instantaneous current—the peak asymmetrical current—occurs at the end of the first half cycle of the current wave. Owing to decay of the dc component, subsequent peaks are lower in value. Various values for this half-cycle asymmetrical current have been determined from study of actual circuits and are shown in Fig. 4.68.

The foregoing discussion explains the use of multipliers to determine the asymmetrical current from the calculated symmetrical value, such as that arrived at in Fig. 4.68. For fuses, which operate relatively fast, a multiplier of 1.4 is commonly applied to the rms symmetrical short-circuit current to get the rms asymmetrical value. For circuit breakers which operate almost as fast as some fuses, the same multiplier would be used in determining the possible asymmetrical current when the breaker

FEEDERS FOR LIGHTING AND POWER

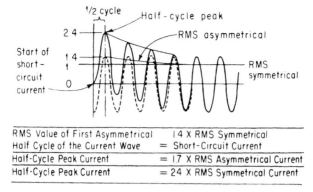

RMS Value of First Asymmetrical Half Cycle of the Current Wave	1.4 X RMS Symmetrical = Short-Circuit Current
Half-Cycle Peak Current	= 1.7 X RMS Asymmetrical Current
Half-Cycle Peak Current	= 2.4 X RMS Symmetrical Current

FIGURE 4.68 Fast-acting fuses and CBs must be rated for the current value in the first half cycle.

opens the circuit. But for breakers which operate more slowly, say two or three cycles after the fault starts, a multiplier of only 1.25 or 1.1 will give the value to which the asymmetrical current has decayed when the circuit is opened.

The value of 1.4, mentioned above as the multiplier for determining the first-half-cycle value of the rms asymmetrical current, has been established for a ratio of reactance to resistance which is not exceeded in the majority of cases. Thus, this multiplier yields the value to which the asymmetrical current has decayed in the first half cycle. But in circuits with higher-than-normal ratio of reactance to resistance, the decay will not be as great in the first half cycle. A multiplier of, say, 1.5 or 1.6 would be required to compute the possible value of the first-half-cycle rms asymmetrical current in such circuits. In all applications, when multipliers are used, the X/R ratio to which the multiplier applies should be known, as well as the X/R ratio of the circuit being protected.

Figure 4.69 shows the application of current-limiting fuses to quickly clear ground and short-circuit faults and to limit the energy let-through on circuits which could deliver fault currents over 100,000 or 200,000 A. On a given short circuit, a current-limiting fuse may be called upon to interrupt either of the two extreme conditions shown here, or some intermediate condition. In the fully asymmetrical case, the fuse would limit both rms current and peak available current if the melting time of the fuse were anything less than one-half cycle. But within its current-limiting range, a true current-limiting fuse must clear the circuit in a time interval that is not greater than the duration of the first symmetrical current loop, and it must limit the peak let-through current to a value less than the peak available current. Thus, it is the action of a fuse in clearing a symmetrical short circuit that determines its current-limiting ability, as shown in the lower part of the sketch. From the upper part of the sketch, it can be seen that a fuse which has a melting time almost up to the peak of the asymmetrical first half cycle will limit current let-through on an asymmetrical fault but will not limit let-through if the short happens to occur as a completely symmetrical wave.

The rms value of the triangular wave of let-through current through the fuse is

$$I_{LT} = \frac{I_p}{1.7}$$

Where I_{LT} = rms let-through current, A
I_p = peak let-through current, A

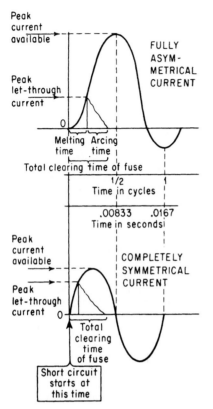

FIGURE 4.69 Current-limiting operation of fuse protection.

The application of standard thermal-magnetic or fully magnetic circuit breakers should be based on a correlation of their time-delay tripping curves and instantaneous trip setting with the requirements for overload protection, the current characteristics of the circuit being protected, and the need for coordination in the operation of overcurrent devices connected in series (such as a feeder device in series with a subfeeder device in series with a branch-circuit device). Breaker operation is readily analyzed in relation to a particular circuit by reference to the manufacturer's curves on the breaker.

Figure 4.70 shows a typical set of curves for a molded-case CB. The curves show trip time versus current for a 225-A breaker with interchangeable trip units. Each trip unit consists of a fixed thermal trip element and an independent adjustable magnetic trip element. The various trip units give the breaker its particular continuous current rating and are available in standard settings: 70, 100, 125, 150, 175, 200, and 225 A. Then the adjustable magnetic element associated with each thermal trip can be adjusted to from 3 to 10 times the current rating of the thermal trip. The curves are shown as bands to cover the tolerances to which the elements are manufactured and within which they will operate. A band can be taken to be a large number of individual curves laid one against another. The curves show time plotted against multiples of any current rating (i.e., multiples of 70 A, or 150 A, etc.). The two operating bands shown below 100 s represent the two extreme conditions of setting of the adjustable magnetic trip—3 times the CB rating at left, and 10 times the CB rating at right. Intermediate settings of the magnetic trip would produce bands between the two shown here.

Figure 4.71 shows the application of current-limiting fuses in combination with a power circuit breaker. Current-limiting fuses can be combined with both molded-case and power circuit breakers. A typical application is represented by the curves shown. The fuses extend the maximum interrupting capacity of the CB to values above 200,000 rms symmetrical amperes.

The three-pole CB is an electrically operated low-voltage power circuit breaker with a continuous rating of 3000 A and an interrupting capacity of 75,000 rms asymmetrical amperes at 480 V. The CB has three magnetic tripping characteristics in a 2000-A trip device. The first, the long-time-delay trip, is adjusted to pick up at 2000 A. The second, the short-time-delay trip, is adjusted to pick up at 6000 A. The third trip is the instantaneous trip, set for 12 times the coil rating, or 24,000 A. The tolerance on pickup settings is ± 15 percent.

FEEDERS FOR LIGHTING AND POWER 331

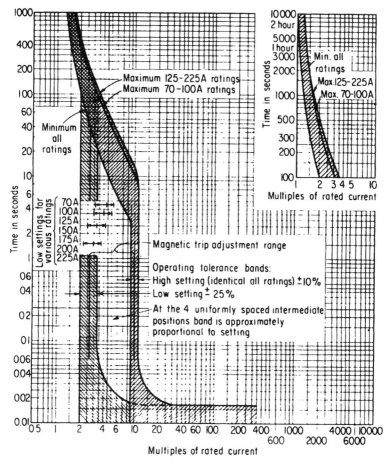

FIGURE 4.70 CB time-current curves reveal overload and short-circuit characteristics.

The time-delay trips are factory adjusted for required delays. The fuses used are rated at 3000 A continuous carrying capacity with interrupting capacity of 200,000 A. The coordination between the fuse operating band and the breaker tolerance band is shown in the graph. The combination provides CB operation on faults up to about 60,000 A. Both devices operate above that.

Fuse or CB?

In modern electrical systems, both fuses and circuit breakers are used to provide required overcurrent protection for circuits and equipment. The choice between fuses and a circuit breaker for any protective application involves careful evaluation of the cost, capabilities, and characteristics of the particular sizes and types to be used—all related to the characteristics of the load and the thermal and magnetic damage limits of equipment to be protected. Fuses and circuit breakers each have advantages and

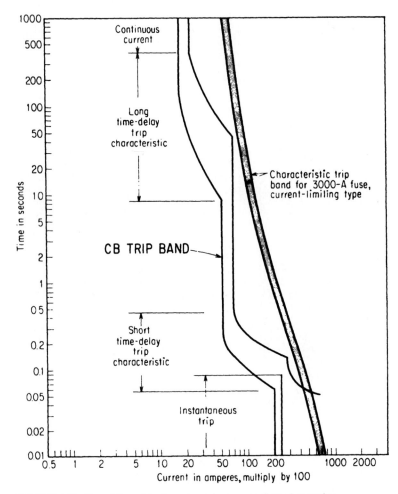

FIGURE 4.71 Fuse-CB combinations depend upon coordinated operation.

limitations, and different electrical users tend to prefer one device over the other, for various reasons. Although the many factors that are involved in a choice between fuses and CBs reveal almost as much art as science in the selection process, certain **NEC** rules and UL regulations dictate the use of only fuses in certain applications.

The basic concerns in choosing between fuses and breakers are obvious, but the choice should be carefully evaluated. A CB, for instance, is able to reclose immediately after a fault trip, because it usually is not damaged in the fault interruption and does not require maintenance or repair. Although there are exceptions to that ability, such immediate reclosure is often viewed as an important advantage over fuses, which must be replaced after blowing and which thereby entail labor and replacement costs as well as some loss of time in restoring the circuit. However, when a fault in a circuit opens a protective device, time and effort are usually required to repair the fault condition, so immediate reclosing of the protective device is often not even possible. And the initial economy of fusible equipment relative to CBs, particularly in applications demanding high short-circuit capacity, commonly far outweighs the costs of labor and fuse replacements over the life of the overall system—even where the number of fault operations might be greater than average.

Certainly a major factor today in choosing between fusible protection and CBs is the matter of interrupting capacity (or short-circuit duty). With the continual beefing up of utility supply systems and the increasing use of transformers with higher kilovoltampere ratings to handle the growth in average electrical loads, overcurrent protection rated above 50,000 rms symmetrical amperes has become commonplace. For such applications, the ready availability of fuses rated at 100,000 and 200,000 A promotes the operating (as well as the economic) advantage of using fuses or fuse devices—either instead of CBs or in conjunction with CBs.

However, there are applications where the adjustable trip characteristics of circuit breakers—long time delay, short time delay, and instantaneous—provide the only means for precisely selective coordination in elaborate electrical systems. In such systems, the use of molded-case CBs with extremely versatile solid-state trip mechanisms (including ground-fault protection) can provide highly engineered protection. And the availability of current-limiting circuit breakers with 200,000 A short-circuit-duty ratings makes possible high-capacity total overcurrent protection using only circuit breakers, even for the largest and most sophisticated electrical systems.

Current-limiting circuit breakers of both the molded-case and power types can completely interrupt a fault current in less than one-quarter cycle. The action is similar to that shown for a current-limiting fuse in Fig. 4.69. Such breakers can sustain many such fault interruptions in circuits with up to 200,000 A available and may be restored to the closed position as soon as the fault is removed. A typical 100-A molded-case current-limiting CB, rated for 100,000 rms symmetrical amperes, can operate in a circuit with a prospective peak fault current (the current that could flow if the fuse were not in the circuit) of 230,000 A, and the CB will limit the current to not over 25,000 A. Such breakers are listed by UL for panelboard main protection where the branch CBs have 10,000-A interrupting capacity but the entire panelboard assembly is rated for 100,000-A short-circuit duty.

Locating Protection

All feeder overcurrent protective devices must satisfy **NEC** Section 240-24 with respect to installation location. The basic rule requires that fuses and circuit breakers must be readily accessible. In accordance with the definition of "readily accessible," they must be "capable of being reached quickly for operation, renewal, or inspections, without requiring those to whom ready access is requisite to climb over or remove obstacles or to resort to portable ladders, chairs, etc." (Fig. 4.72).

Although the **NEC** gives no maximum heights below which overcurrent protective devices are considered readily accessible, some guidance can be obtained from Section 380-8, which provides detailed requirements for the location of switches and CBs. This section states that switches and CBs must be installed so that the center of the grip of the operating handle, when in its highest position, is not more than 6½ ft above the floor or working platform.

An exception to the basic rule covers any case where an overcurrent device is used in a busway plug-in unit to tap a branch circuit from the busway. Section 364-12 requires that such devices consist of an externally operable CB or an externally operable fusible switch. These devices must be capable of being operated from the floor by means of ropes, chains, or sticks.

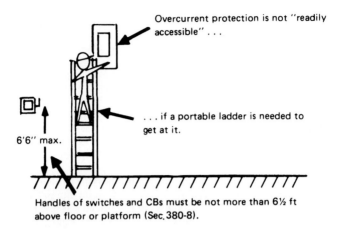

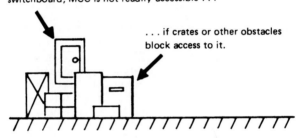

FIGURE 4.72 Feeder overcurrent devices should be installed so as to be "readily accessible."

Section 240-24 clarifies the use of plug-in overcurrent protective devices on busway for the protection of circuits tapped from busway. After the general rule that overcurrent protective devices must be readily accessible (capable of being reached without stepping on a chair or table or resorting to a portable ladder), an exception notes that it is not only permissible to use busway protective devices up on the busway but it is required by Section 364-12. Such devices on high-mounted busway are not "readily accessible" (not within reach of a person standing on the floor). The wording of Section 364-11 makes clear that this requirement for overcurrent protection in the device on the busway applies to subfeeders tapped from the busway as well as branch circuits tapped from the busway. Figures 4.73 and 4.74 cover the rules of **NEC** Sections 240-24, 364-12, and 380-8—all of which are related.

Overcurrent protection—either a fused switch or a circuit breaker—is usually required in each subfeeder tapping power from a busway feeder. This is necessary to protect the lower current-carrying capacity of the subfeeder, and the device should be placed at the point at which the subfeeder connects into the feeder. However, the **NEC** provides that overcurrent protection may be omitted "at points where busways are reduced in size, provided that the smaller busway does not extend more than 50

FEEDERS FOR LIGHTING AND POWER 335

Plug-in connection for tapping-off branch circuit shall contain overcurrent protection, which does not have to be within reach of person standing on floor.

Branch circuit tap from busway

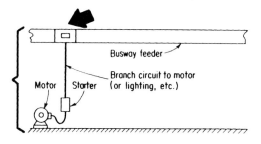

Plug-in connection for tapping-off a feeder or subfeeder shall contain overcurrent protection, which does not have to be within reach of person standing on floor.

Feeder or subfeeder tap from busway

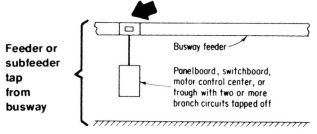

FIGURE 4.73 These overcurrent devices are not required to be readily accessible.

ft and has a current rating at least equal to one-third the rating or setting of the overcurrent device next back on the line" (Fig. 4.75).

Figure 4.76 summarizes requirements regarding the location of switches and circuit breakers.

Feeder Taps

Although basic **NEC** requirements dictate the use of an overcurrent device at the point at which a conductor received its supply, subparts (b) through (n) effectively present exceptions to this rule in the case of taps to feeders. That is, to meet the practical demands of field application, certain lengths of unprotected conductors may be used to tap energy from protected feeder conductors.

These "exceptions" to the rule for protecting conductors at their points of supply are made in the case of 10-, 25-, and 100-ft taps from a feeder, as described in Section 240-21, parts (b), (d), and (m). Application of the tap rules should be made carefully to effectively minimize any sacrifice in safety. The taps are permitted without overcurrent protective devices at the point of supply.

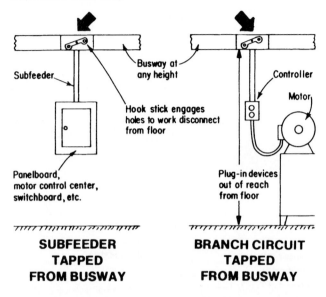

FIGURE 4.74 Circuit tapped from busway must have protection on the busway.

Unprotected taps not over 10 ft long (Fig. 4.77) may be made from feeders or transformer secondaries provided:

1. The smaller conductors have a current rating that is not less than the combined computed loads of the circuits supplied by the tap conductors and must have ampacity of *not less than the rating of the "device" supplied by the tap conductors,* or *not less than the rating of the overcurrent device (fuses or CB) that might be installed at the termination of the tap conductors.*

Important Limitation: For any 10-ft unprotected feeder tap installed in the field, the rule limits its connection to a feeder that has protection rated *not* more than 1000 percent of (10 times) the ampacity of the tap conductor. Under the rule, unprotected No. 14 tap conductors are not permitted to tap a feeder any larger than 1000 percent of the 20-A ampacity of No. 14 copper conductors—which would limit such a tap for use with a maximum feeder protective device of not over 10×20 A, or 200 A.

FEEDERS FOR LIGHTING AND POWER

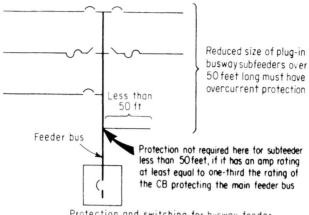

FIGURE 4.75 Busway subfeeder not over 50 ft long does not *require* protection at industrial facilities only.

2. The tap does not extend beyond the switchboard, panelboard, or control device which it supplies.

3. The tap conductors are enclosed in conduit, EMT, metal gutter, or other approved raceway when not a part of the switchboard or panelboard.

Section 240-21(b) specifically recognizes that a 10-ft tap may be made from a transformer secondary in the same way it has always been permitted from a feeder. In either case, the tap conductors must not be over 10 ft long and must have ampacity not less than the ampere rating of the switchboard, panelboard, or control device—or the tap conductors may be terminated in an overcurrent protective device rated not more than the ampacity of the tap conductors. In the case of an unprotected tap from a transformer secondary, the ampacity of the 10-ft tap conductors would have to be related through the transformer voltage ratio to the size of the transformer primary protective device—which in such a case would be "the overcurrent device on the line side of the tap conductors."

Taps not over 25 ft long (Fig. 4.78) may be made from feeders, as noted in Section 240-21 (c) provided:

1. The smaller conductors have a current rating at least one-third that of the feeder overcurrent device rating or of the conductors from which they are tapped.

2. The tap conductors are suitably protected from mechanical damage. In previous NEC editions, the 25-ft feeder tap without overcurrent protection at its supply end simply had to be "suitably protected from physical damage"—which could accept use of cable for such a tap. Now, the rule requires such tap conductors to be "enclosed in a raceway"—just as has always been required for 10-ft tap conductors.

3. The tap is terminated in a single CB or set of fuses which will limit the load on the tap to the ampacity of the tap conductors.

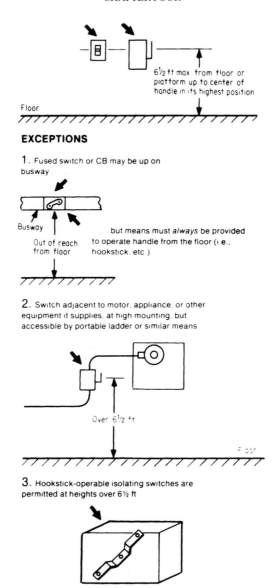

FIGURE 4.76 Switches and CBs must have ready access for effective operation.

Examples of Taps. Figure 4.79 shows use of a 10-ft feeder tap to supply a single motor branch circuit. The conduit feeder may be a horizontal or a vertical run, such as a riser. If the tap conductors are of such size that they have a current rating at least one-third that of the feeder conductors (or protection rating) from which they are tapped, they could be run a distance of 25 ft without protection at the point of tap-off from the feeder because they would comply with the rules of Section 240-21(c)

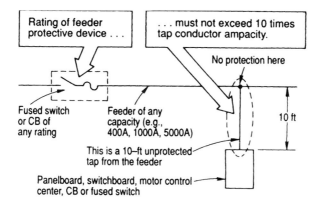

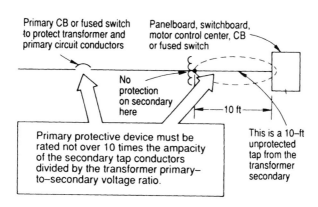

EXAMPLE: If the above transformer is stepping 480V single phase down to 240/120V single phase, and the 10-ft tap conductors on the transformer secondary have an ampacity of 100A, the primary feeder CB must be rated not more than 10 times 100A divided by two (480/240), or 500A.

FIGURE 4.77 Ten-ft taps may be made from a feeder or a transformer secondary.

which permit a 25-ft tap if the conductors terminate in a single protective device rated not more than the conductor ampacity. Because Section 364-12 requires that any busway used as a feeder must have overcurrent protection on the busway for any subfeeder or branch circuit tapped from the busway, the use of a cable-tap box on a busway without overcurrent protection (as shown in the conduit installation of Fig. 4.79) would be a violation. Refer to Sections 240-24 and 364-12.

A common application of the 10-ft tap exception is the supply of panelboards from conduit feeders or busways, as shown in Fig. 4.80. The case shows an interesting requirement that arises from Section 384-16, which requires that lighting and appli-

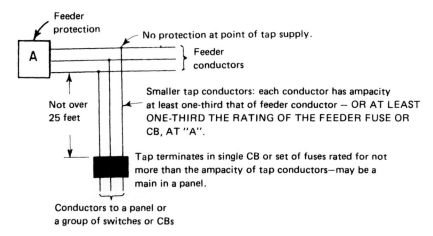

FIGURE 4.78 Sizing feeder taps not over 25 ft long. (Sec. 240-21.)

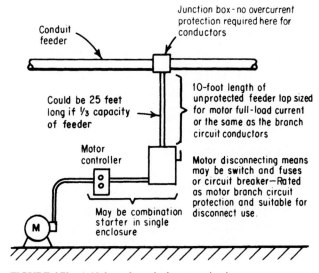

FIGURE 4.79 A 10-ft tap for a single motor circuit.

ance panelboards be protected on their supply side by overcurrent protection rated not more than the rating of the panelboard busbars. If the feeder is a busway, the protection must be placed (a requirement of Section 364-12) at the point of tap on the busway. In that case a 100-A CB or fused switch on the busway would provide the required protection of the panel, and the panel would not require a main in it. But, if the feeder circuit is in conduit, the 100-A panel protection would have to be in the panel or just ahead of it. It could not be at the junction box on the conduit because that would make it not readily accessible and therefore a violation of Section 240-24. With a conduit feeder, a fused-switch or CB main in the panelboard could be rated up to the 100-A main rating.

FEEDERS FOR LIGHTING AND POWER

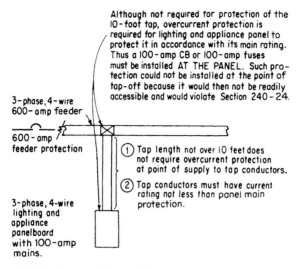

FIGURE 4.80 A 10-ft tap to lighting panel with unprotected conductors.

For transformer applications, typical 10- and 25-ft tap considerations are shown in Fig. 4.81.

Figure 4.82 shows application of Section 240-21(d) in conjunction with the rule of Section 450-3(b)(2), covering transformer protection. As shown in Example 1, the 100-A main protection in the panel is sufficient protection for the transformer and the primary and secondary conductors when these conditions are met:

1. Tap conductors have ampacity at least one-third that of the 125-A feeder conductors.
2. Secondary conductors are rated at least one-third the ampacity of the 125-A feeder conductors, based on the primary-to-secondary transformer ratio.
3. Total tap is not over 25 ft, primary plus secondary.
4. All conductors are in conduit.
5. Secondary conductors terminate in the 100-A main protection that limits the secondary load to the ampacity of the secondary conductor and simultaneously provides the protection required by the lighting panel and is not rated over 125 percent of transformer secondary current.
6. Primary feeder protection is not over 250 percent of transformer rated primary current, as recognized by Section 450-3(b)(2), and the 100-A main breaker in the panel satisfies as the required "overcurrent device on the secondary side rated or set at not more than 125 percent of the rated secondary current of the transformer."

In Example 2, each set of tap conductors from the primary feeder to each transformer may be the same size as primary feeder conductors *or* may be smaller than primary conductors if sized in accordance with Section 240-21(d), which permits a 25-ft tap from a primary feeder to be made up of both primary and secondary tap conductors. The 25-ft tap may have any part of its length on the primary or secondary but must not be longer than 25 ft and must terminate in a single CB or set of fuses.

10-FT TAP

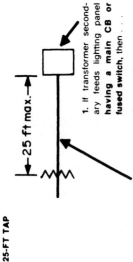

25-FT TAP

1. A 10-ft tap may be made from transformer secondary to a panel, switchboard, MCC, etc.

2. If this is a lighting panel that requires main protection, a fused switch or CB must be installed as a main protective device in the panel or just ahead of it, at the end of the 10-ft tap.

3. If the panel is *not* a lighting panel (such as a panel with 240-volt or 480-volt heating circuits or other makeup that does not make it a lighting panel as specified in Section 384-14), then main protection is not required at all, and the 10-ft tap conductors terminate in main lugs of the panel switchboard, or other equipment.

1. If transformer secondary feeds lighting panel **having a main CB or fused switch**, then ....

2. ... secondary tap conductors from transformer may be 25 ft long, as permitted by Section 240-21(d), but *only* where the tap terminates in a single CB or set of fuses.

3. Or, a 25-ft tap may be made from a transformer to a CB or fused switch in an individual enclosure or serving as a main in a switchboard or MCC.

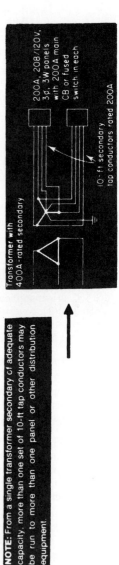

NOTE: From a single transformer secondary of adequate capacity, more than one set of 10-ft tap conductors may be run to more than one panel or other distribution equipment.

FIGURE 4.81 Taps from transformer secondaries.

EXAMPLE 1:

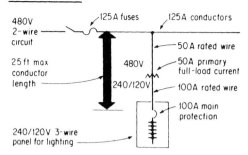

EXAMPLE 2:

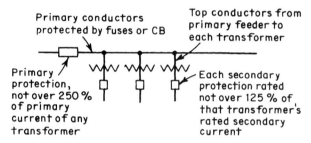

FIGURE 4.82 Feeder tap of primary-plus-secondary not over 25 ft long.

Figure 4.83 shows another example of Section 240-21(d) and Section 450-3(b)(2). Because the primary wires tapped to each transformer from the main 100-A feeder are also rated 100 A and are therefore protected by the 100-A feeder protection, all the primary circuit to each transformer is excluded from the allowable 25 ft of tap to the secondary main protective device. The 100-A protection in each panel is not over 125 percent of the rated transformer secondary current. It, therefore, provides the transformer protection required by Section 450-3(b)(2). The same device also protects each panel at its main busbar rating of 100 A.

Figure 4.84 compares the two different 25-ft tap techniques covered by Parts (c) and (d).

As shown in Fig. 4.85, Section 240-21(i) gives permission for unprotected taps to be made from generator terminals to the first overcurrent device it supplies—such as in the fusible switch or circuit breakers used for control and protection of the circuit that the generator supplies. As the rule is worded, no maximum length is specified for the generator tap conductors. But because the tap conductors terminate in a single circuit breaker or set of fuses rated or set for the tap-conductor ampacity, tap conductors up to 25 ft long would comply with the basic concept given in Section 240-21(c) for 25-ft taps. And Section 445-5, which is referenced, requires the tap conductors to have an ampacity of at least 115 percent of the generator nameplate current rating.

Part (e) is another departure from the rule that conductors must be provided with overcurrent protection at their supply ends, where they receive current from other larger conductors or from a transformer. Section 240-21(e) permits a longer length

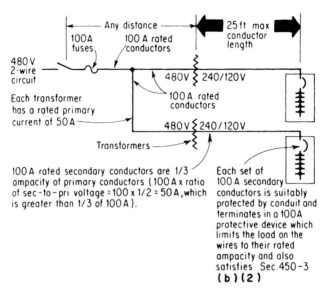

FIGURE 4.83 Sizing a 25-ft tap and transformer protection.

than the 10-ft unprotected tap of Part (b) and the 25-ft tap of Parts (c) and (d). Under specified conditions that are similar to the requirements of the 25-ft-tap exception, an unprotected tap up to 100 ft in length may be used in "high-bay manufacturing buildings" that are over 35 ft high *at the walls*—but only "where conditions of maintenance and supervision assure that only qualified persons will service the system." Obviously, that last phrase can lead to some very subjective and individualistic determinations by the authorities enforcing the **NEC**. And the phrase "35 ft high at the walls" means that this Exception cannot be applied where the height is over 35 ft at the peak of a triangular or curved roof section but less than 35 ft at the walls.

The 100-ft-tap exception must meet specific conditions:

1. From the point at which the tap is made to a larger feeder, the tap run must not have more than 25 ft of its length run horizontally, and the sum of horizontal run and vertical run must not exceed 100 ft. Figure 4.86 shows some of the almost limitless configurations of tap layout that would fall within the dimension limitations.
2. The tap conductors must have an ampacity equal to at least one-third of the rating of the overcurrent device protecting the larger feeder conductors from which the tap is made.
3. The tap conductors must terminate in a circuit breaker or fused switch, where the rating of overcurrent protection is not greater than the tap-conductor ampacity.
4. The tap conductors must be protected from physical damage and must be installed in a metal or nonmetallic raceway.
5. There must be no splices in the total length of each of the conductors of the tap.
6. The tap conductors must not be smaller than No. 6 copper or No. 4 aluminum.
7. The tap conductors must not pass through walls, floors, or ceilings.
8. The point at which the tap conductors connect to the feeder conductors must be at least 30 ft above the floor of the building.

FEEDERS FOR LIGHTING AND POWER

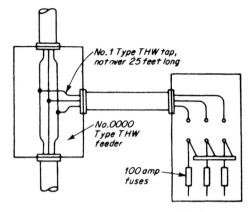

25-ft tap — Sec. 240-21 (c)

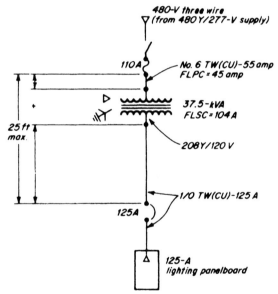

Taps protected from physical damage
Secondary-to-primary voltage ratio = 208:480 = 1:2.3

25-ft tap — Sec. 240-21 (d)

FIGURE 4.84 Examples show difference between the two types of 25-ft taps covered by Parts **C** and **D**.

As shown in Fig. 4.86, the tap conductors from a feeder protected at 1200 A are rated at not less than one-third the protection rating, or 400 A. Although 500-kcmil THW copper is rated at 380 A, that value does not satisfy the minimum requirement for 400 A. But if 500-kcmil THHN or XHHW copper, with an ampacity of 430 A, were used for the tap conductors, the rule would be satisfied. However, in such a

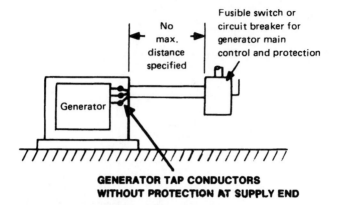

FIGURE 4.85 Unprotected tap may be made from a generator's output terminals to the first overcurrent device.

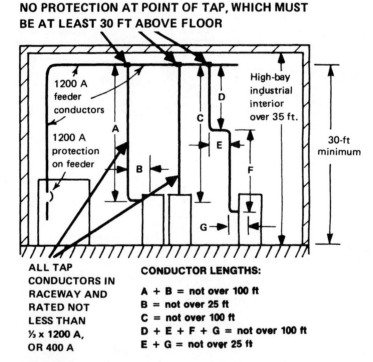

FIGURE 4.86 Unprotected taps up to 100 ft long may be used in "high-bay manufacturing buildings."

case, those conductors would have to be used as if their ampacity were 380 A for the purpose of load calculation because of the general UL rule of 75°C conductor terminations for connecting to equipment rated over 100 A—such as the panelboard, switch, motor-control center, or other equipment fed by the taps. And the conductors for the main feeder being tapped could be rated less than the 1200 A shown in the sketch if the 1200-A protection on the feeder was selected in accordance with Sections 430-62 or 430-63 for supplying a motor load or motor and lighting load. In such cases, the overcurrent protection may be rated considerably higher than the feeder conductor ampacity. But the tap conductors must have ampacity at least equal to one-third the *feeder protection rating*.

Section 240-21(j) applies exclusively to industrial electrical systems. Conductors up to 25 ft long may be tapped from a transformer secondary without overcurrent protection at their supply end and without need for a single circuit breaker or set of fuses at their load end. Normally, a transformer secondary tap over 10 ft long and up to 25 ft long must comply with the rules of Section 240-21(c) and (d)—which call for such a transformer secondary tap to be made with conductors that require no overcurrent protection at their supply end but are required to terminate at their load end in a single CB or single set of fuses with a setting or rating not over the conductor ampacity. However, Section 240-21(j) permits a 10- to 25-ft tap from a transformer secondary without termination in a single main overcurrent device—but it limits the application to "Industrial Installations." The tap conductor ampacity must be at least equal to the transformer's secondary current rating and must be at least equal to the sum of the ratings of overcurrent devices supplied by the tap conductors. The conductors could come into main-lugs-only of a power panel if the conductor ampacity is at least equal to the sum of the ratings of the protective devices supplied by the busbars in the panel. Or the tap could be made to an auxiliary gutter from which a number of individually enclosed circuit breakers or fused switches are fed from the tap—provided that the ampacity of the tap conductors is at least equal to the sum of the ratings of protective devices supplied.

An example of the application of Section 240-21(j) is shown at the top of Fig. 4.87. If that panel contains eight 100-A circuit breakers (or eight switches fused at 100 A), then the 25-ft tap conductors must have an ampacity of at least 8×100 A, or 800 A. In addition, the tap conductor ampacity must be not less than the secondary current rating of the transformer. The layout of a similar application at an auxiliary gutter is shown at the bottom of Fig. 4.87.

Ground-Fault Protection (GFP)

Fuses and circuit breakers, applied as described above, are sized to protect conductors in accordance with their current-carrying capacities. If the current exceeds the rating of the fuse or the trip setting of the CB, the device opens the circuit—either instantaneously or after some time delay, depending upon the actual level of the fault current. This excessive current might be caused by an operating overload, by a phase-to-ground fault, or by a short circuit between two circuit conductors. Thus, a 1000-A fuse will blow if current in excess of that value flows over the circuit. It will blow earlier on heavy overcurrent and later on lower overcurrent. But it will blow, and the circuit and equipment will be protected against damage due to the overcurrent.

Another type of fault condition is very common in grounded systems but will not be cleared by conventional overcurrent devices. That is the phase-to-ground fault (invariably by arcing contact between an energized circuit conductor and a grounded metal enclosure or conduit) having a current value below the rating of the overcurrent device.

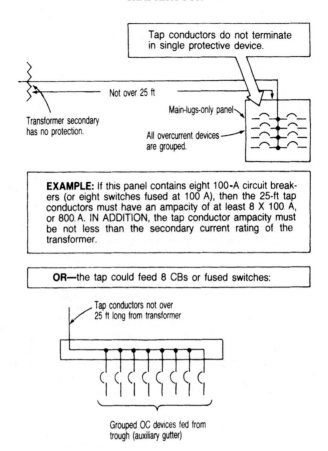

FIGURE 4.87 These tap applications are permitted for transformer secondaries only in "industrial" electrical systems.

Most modern grounded electrical systems, for which huge sums of money are expended on meticulously designed systems of conventional overcurrent protection, are completely unprotected against the most common type of electrical failure—low-current ground faults (from one phase leg to a grounded equipment enclosure). The fuses and/or CBs might be selected to have the highest interrupting capacities required by potential short-circuit currents, and such factors as time delay and current limitation may be perfectly tailored to the needs of the particular system. But these vast sums of money are expended to protect very expensive electrical systems (and valuable industrial processes or commercial operations) against the type of electrical fault which almost never happens—the bolted 3-phase short circuit on the load terminals of a protective device. At the same time, in spite of these expensive overcurrent provisions, the grounded system is totally unprotected against the fantastically destructive effects of the very common phase-to-ground fault (Fig. 4.88).

On a high-capacity feeder, a line-to-ground fault (i.e., a fault from a phase conductor to a conduit, to a junction box, or to some other metallic equipment enclosure) can and frequently does draw current whose value is below the rating or

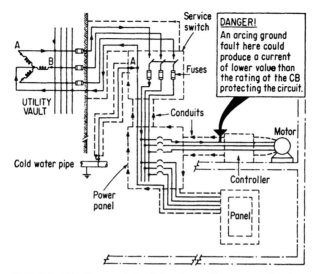

FIGURE 4.88 Fuses and circuit breakers will not respond to low-level arcing ground faults.

setting of the circuit protective device. For instance, a 500-A ground fault on a 2000-A protective device that has only a 1200-A load will not be cleared by the device. If such a fault is a "bolted" line-to-ground fault (a highly unlikely fault), a certain amount of heat will be generated by the I^2R effect of the current; but this will usually not be dangerous, and such fault current will merely register as additional operating load, with wasted energy (wattage) in the system. But, bolted phase-to-ground faults are very rare. The usual phase-to-ground fault exists as an intermittent or arcing fault, and an arcing fault of the same current rating as the essentially harmless bolted fault can be fantastically destructive because of the intense heat of the arc.

Of course, any ground-fault current (bolted or arcing) above the rating or setting of the circuit protective device will normally be cleared by the device. In such cases, the bolted fault current will be eliminated. But, even when the protective device eventually operates (in the case of a heavy ground-fault current that adds to the normal circuit load current to produce a total current in excess of the rating of the circuit protective device), the time delay of the device may be minutes or even hours; this is more than enough time for the arcing fault current to burn out conduit and enclosures, acting just like a torch, and even to propagate flames to create a fire hazard.

In spite of increasingly effective application of conventional overcurrent protective devices, the problem of ground faults continues to persist and even grows with expanding electric usage. In the interest of safety, definitive engineering design must include protection against such faults. Phase overcurrent protective devices are normally limited in their effectiveness because (1) they must have a time delay and a setting somewhat higher than full load to ride through normal inrushes, and (2) they are unable to distinguish between normal currents and low-magnitude fault currents which may be smaller than full-load currents.

Dangerous temperatures and magnetic forces are proportional to current during overloads and short circuits; therefore, overcurrent protective devices usually are adequate protection against such faults. However, the temperatures of arcing faults are, generally, independent of current magnitude; and arcs of great and extensive destructive capability can be sustained by currents that do not exceed the settings of overcurrent devices. Other means of protection are therefore necessary: A ground-detection device that responds only to ground-fault current can be coupled to an automatic switching device to open all three phases when a line-to-ground fault exists on the circuit.

With a ground-fault protection system of the type shown in Fig. 4.89 at the service entrance, a ground fault anywhere in the system is immediately sensed by the ground-relay system; however, action to open the circuit usually is delayed to allow a conventional overcurrent device near the point where the fault will open if it can. As a practical measure, this time delay is designed to be only a few cycles or seconds, depending on the voltage of the circuit, the time-current characteristics of the overcurrent devices in the system, and the location of the ground-fault relay in the distribution system. If the conventional overcurrent protective devices fail to clear the circuit during the predetermined time delay, and if the fault continues, the ground-fault protective relays will open the circuit. This provides added overcurrent protection not available by any other means.

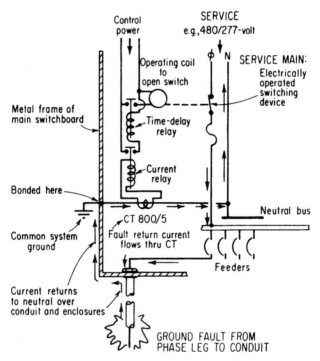

FIGURE 4.89 Ground-fault sensing may be performed on a bonding jumper from ground bus to neutral bus.

As shown in Fig. 4.89, the ground-fault protective relays are located in the main switchboard and will thus detect any ground faults that may develop at any point in the wiring system, from the furthest point back to—and including—the switchboard.

FEEDERS FOR LIGHTING AND POWER 351

If desired, another set of ground-fault protective relays can be located at any point downstream from the service equipment.

The ground-fault protection hookup in Fig. 4.89 makes use of ground-strap sensing; this scheme is applicable only to the service-entrance switchboard to disconnect the main service switch or CB on a ground fault somewhere in the system. Or, it may be applied on the secondary side of a transformer where the neutral and equipment grounding bus are bonded together. It includes a ground-sensing current transformer (CT) on the bonding strap that connects the frame or grounding bus of the switchboard to the neutral—all within the switchboard enclosure. The current transformer is a standard ratio-type CT and not a zero-sequence type CT. On any ground fault in the system, fault current will return to the neutral by flowing back over the conduit and other enclosures and then through the bonding strap to the neutral. Such current flow causes a current output from the CT, energizing a ground relay that, in turn, energizes the operating coil or shunt-trip coil of the main switch or CB, causing it to open.

This method of detecting ground-fault current in the neutral-frame bonding strap can only be used at service equipment, where such a connection is made. Moreover, selective coordination with downstream overcurrent devices is not easily accomplished. A ground fault on any branch circuit, feeder, or subfeeder—anywhere in the system—will cause current to flow on the bonding strap. To minimize nuisance tripping and the chance of disconnecting the entire service unnecessarily, the trip setting of the CT relay circuit must be high, and the time delay in its operation must be fairly long to give downstream protection a chance to clear the fault with minimal circuit outage. For instance, it could be necessary to set the CT response at 400 A, with a 5-s delay in the ground relay. On large feeders (say, 400 A or larger), conventional overcurrent devices would not clear a 400-A ground fault in the 5-s delay time, and the service main would open. Although this might not always be completely intolerable, it is generally not desirable to shut down the entire service because of a branch-circuit or subfeeder fault. Selective ground-fault protection is the obvious answer to the problem. But selective coordination between ground-fault protection and conventional protective devices (fuses and CBs) at service, main-building, and feeder disconnects is a very clear and specific task as a result of **NEC** Section 230-95(a), which calls for a maximum time delay of 1 for any ground-fault current value of 3000 A or more.

Field experience has repeatedly shown that ground-fault interrupting equipment should be set as low as practical so that virtually no ground-fault current will remain unrecognized; introducing a time-delay action will permit the normal overcurrent protective device to operate. For a 4000-A circuit, the relay setting should usually be about 800 A. Ground-fault current of that magnitude would not be cleared by similarly sized overcurrent devices, but such current would be cleared automatically by the ground-fault protective system. On the other hand, downstream overcurrent devices with lower settings of, say, 200 A and less would open by themselves at such low ground-fault values.

Ground-fault-current values are highly unpredictable because of the many variable factors that are involved. As a consequence, the current settings of ground-fault protection equipment and the time delays to be allowed are matters of judgment.

The basic hookup offering widest application for ground-fault protection of individual circuits anywhere in a system is the zero-sequence sensing shown in Fig. 4.90; this hookup makes use of a current transformer and relay to actuate electrically operated switching devices. The CTs are "doughnut"-type or rectangular zero-sequence current transformers that readily fit over large groups of cables, switchboard busbars, or circuit-breakers studs. Such CTs are able to detect milliamperes of current leaking out of systems with loads of many thousands of amperes.

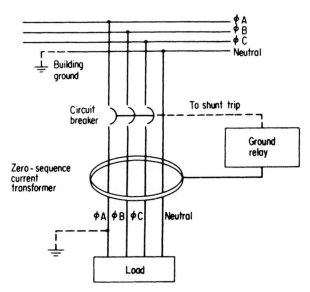

FIGURE 4.90 Zero-sequence sensing provides for ground-fault protection on any circuit.

This system allows selective protection for individual circuits. Each branch circuit, subfeeder, feeder, or main can be equipped with its own separate detection and operation hookup to deenergize only the circuit with the ground fault—with adjustable time delay in each relay to coordinate its operation with other line-side relays to ensure minimum outage on a ground fault. If the fault is on a branch circuit, the branch-circuit device will operate, and not the subfeeder or other device supplying the branch circuit further upstream.

For zero-sequence sensing, all conductors of a circuit being protected (including the neutral, if one is used) are run through the sensing current transformer. As long as the current associated with the circuit flows out and returns over those circuit conductors, the net flux in the CT is zero, and the CT has no output. Under such conditions, the CT indicates that no current is taking an unauthorized or fault path. That is, there is no current flowing to ground through grounded conduit or other metal enclosures. No current is generated in the CT winding, regardless of circuit balance, current magnitude, or any third or other odd-order harmonics (as from fluorescent ballasts or other magnetostrictive load devices). But, when a fault to ground does develop on one phase, the return of fault current over a path that is not enclosed by the CT (such as fault current on the conduit) causes an unbalance in the CT flux and generates an output signal from the CT to the relay, which then actuates the tripping mechanism of the associated electrically operated switching device—such as the shunt-trip of a CB.

The zero-sequence-CT method permits ground-fault protection to be as extensive or limited as dictated by the design and economics of any system. And, because separate GFP hookups can be used on individual circuits, with adjustable current trip settings and time-delay values, completely selective coordinated arrangements are possible to confine circuit outage to the faulted circuit.

Complete systems of zero-sequence CTs and related relays with adjustable time delay are available for application throughout a system. Figure 4.91 shows the idea behind a "total protection" scheme that uses a zero-sequence CT and associated relay on each circuit to be provided with protection against low-level ground faults. Each CT-relay hookup will detect a ground fault on its load side and operate the shunt-trip of a circuit breaker, the shunt-trip of a bolted pressure switch, or the coil of an electrically operated switch (such as by breaking the holding-coil circuit of a contactor) to quickly clear the fault. The time delays of the relays, which are all responding to the same ground-fault current, can be used to coordinate operation so that any ground fault is cleared by the GFP device nearest to the fault on its line side. The individual GFP setups are unaffected by operating overloads and short-circuit currents, which cancel in the CT. The CT-relay control responds only to ground faults.

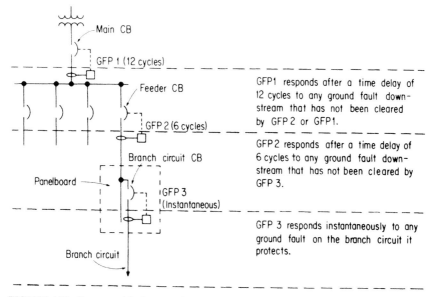

FIGURE 4.91 For ground-fault protection system, time-delay variations can provide selectivity of operation.

The system operates very quickly, to minimize the damage due to ground faults; the circuit opening speed is limited only by the need to coordinate the operating speeds of the relays to ensure that each ground fault will be cleared by the CT relay nearest to the fault on the line side of the fault. But the protection is completed with fuses or CBs that will open any circuit in which there is an operating overload or short-circuit fault above the continuous rating of these fuses or CBs.

Figure 4.92 shows series GFP layouts using a zone-selective system based on the use of a feedback lockout signal that enables each GFP device to be set for instantaneous operation for any fault on its load side and within its zone. The instantaneous trip at each step of the series-connected GFP hookups ensures the minimum amount of damage due to ground faults.

A less complicated way to include GFP on larger circuits is to use a molded-case CB with built-in ground-fault protection. Available in ratings from 400 to 3000 A, such CBs contain all the elements for GFP within the molded case; they do not

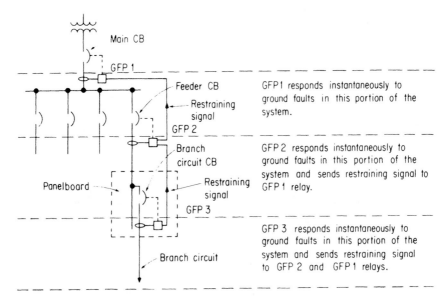

FIGURE 4.92 Selectivity with maximum protection is provided by a zone-selective instantaneous system.

require an external shunt-trip, zero-sequence CT, or relay. Air and insulted-case CBs also are equipped with GFP.

Another sensing hookup that has been used for ground-fault detection in service switchboards is called the *residual method*. Four conventional CTs are used—one on each phase leg (with polarities carefully observed) and one on the neutral. The CT secondaries are hooked up to provide a signal output on ground fault and direct opening of the disconnect switch or CB.

In **NEC** Section 230-95, the rule requiring ground-fault protection for any service disconnect rated 1000 A or more (on 480/277-V services) specifies a maximum time delay of 1 s for ground-fault currents of 3000 A or more (Fig. 4.93). The maximum setting (the current at and above which the GFP will be actuated) of any service ground-fault protection hookup is 1200 A, but the time-current trip characteristic of the relay must ensure opening of the disconnect in not more than 1 s for any ground-fault current of 3000 A or more. That **NEC** rule is intended to establish a specific level of protection in GFP equipment by setting a maximum limit on I^2t for fault energy.

The reasoning behind the maximum time delay of 1 s on faults of 3000 A or more is as follows: The amount of damage done by an arcing fault is directly proportional to the time it is allowed to burn. Commercially available GFP systems can easily meet the 1-s limit. Some users, however, request time delays up to 60 s to allow downstream overcurrent devices plenty of time to trip thermally before the GFP on the main disconnect trips. But an arcing fault lasting 60 s can virtually destroy a service-equipment installation. Coordination with downstream overcurrent devices can and should be achieved by adding GFP on feeder circuits where needed, and the **NEC** should require a reasonable time limit for GFP. The value of 3000 A is 250 percent of 1200 A, and 250 percent of the setting is the calibrating point specified in ANSI 37.17. Specifying a maximum time delay starting at this current value will

FEEDERS FOR LIGHTING AND POWER   355

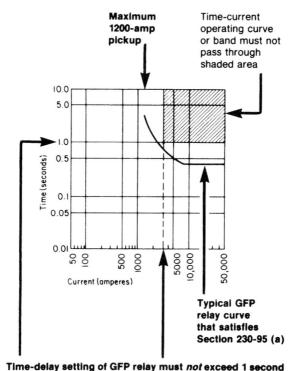

FIGURE 4.93 Time delay is limited for GFP on a service disconnect.

allow either flat or inverse time-delay characteristics for ground-fault relays with approximately the same level of protection.

Equipment ground-fault protection is also mandated for every feeder disconnect switch or circuit used on a 480Y/277-V, 3-phase, 4-wire feeder where the disconnect is rated 1000 A or more, as shown in Fig. 4.94. This is a very significant **NEC** requirement for ground-fault protection of the same type that has long been required by Section 230-95 for every *service* disconnect rated 1000 A or more on a 480Y/277-V service.

An exception notes that feeder ground-fault protection is not required on a feeder disconnect if equipment ground-fault protection is provided on the supply (line) side of the feeder disconnect.

The substantiation submitted as the basis for the addition of this new rule stated as follows:

> Substantiation: The need for ground-fault equipment protection for 1000 amp or larger 277/480 grounded system is recognized and required when the service equipment is 277/480 volts. This proposal will require the same needed protection when the service equipment is not 277/480 volts. Past proposals attempted to require these feeders be treated as services in order to achieve this protection, but treating a feeder like a service created many other concerns. This proposal only addresses the feeder equipment ground-fault protection needs when it is not provided in the service equipment.

356 CHAPTER FOUR

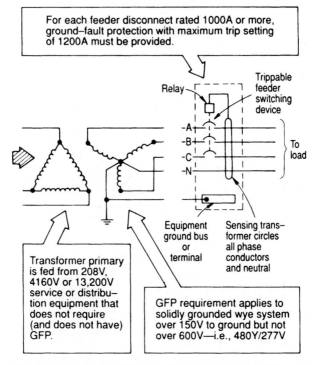

FIGURE 4.94 A 480Y/277-V feeder disconnect rated 1000 A or more must have ground-fault protection if there is not GFP on its supply side.

As noted, this requirement calls for this type of feeder ground-fault protection when ground-fault protection is not provided on the supply side of the feeder disconnect, such as where a building has, say a 13.2-kV medium-voltage service or has, say, a 208Y/120-V service with load-side transformers stepping up the voltage to 480Y/277 V—because a service at any voltage other than 480Y/277 V does *not* require GFP equipment.

Equipment ground-fault protection—of the type required for 480Y/227-V service disconnects—is now also required for each disconnect rated 1000 A or more that serves as a main disconnect for a building or structure. Like Section 215-10, Section 240-13 expands the application of protection against destructive arcing burndowns of electrical equipment. The intent is to equip a main building disconnect with GFP whether the disconnect is technically a service disconnect or a building disconnect on the load side of service equipment located elsewhere. This was specifically devised to cover those cases where a building or structure is supplied by a 480Y/277-V feeder from another building or from outdoor service equipment. Because the main disconnect (or disconnects) for such a building serves essentially the same function as a service disconnect, this requirement makes such disconnects subject to all of the rules of Section 230-95, covering GFP for services (Fig. 4.95).

FEEDERS FOR LIGHTING AND POWER 357

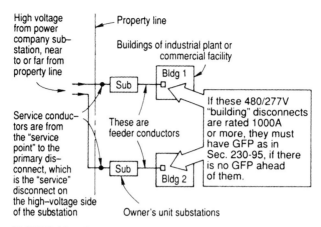

FIGURE 4.95 Ground-fault protection is required for the feeder disconnect for each building—either at the building or at the substation secondary.

The first exception here excludes the need for such GFP from disconnects for critical processes where automatic shutdown would introduce additional or different hazards. And as with service GFP, the requirement does not apply to fire-pump disconnects.

Exception No. 3 also suspends the need for GFP on a building or structure disconnect *if* such protection is provided on the upstream (line) side of the building disconnect. The rule also stipulates that there must not be any desensitizing of the ground-fault protection because of neutral and/or grounding electrode connections that return current to the neutral on the load side of the service ground-fault sensing hookup.

GFP for Health-Care Facilities. At least one additional level of ground-fault protection is required for hospitals and health-care facilities where ground-fault protection is used on service equipment or feeder disconnects (**NEC** Section 230-95 or 215-10). Where ground-fault protection is installed on the normal service disconnecting means, each main feeder must be provided with similar protective means. This requirement is intended to minimize the chance of extensive outage. By applying appropriate selectivity at each level, the ground fault can be limited to a single feeder, and service to the balance of the health-care facility may be maintained.

As shown in Fig. 4.96, if there is a GFP hookup on the service, then a GFP hookup must be put on each feeder derived from the service. The tripping time of the main GFP must be such that each feeder GFP will operate to open a ground fault on its feeder without opening the service GFP. And a time interval of not less than 0.1 s (i.e., the time equivalent of six cycles) must be provided between the feeder GFP trip and the service GFP trip. As shown, if the feeder GFP relays are set for instantaneous operation, the relay on the service GFP must have at least a 0.1-s time delay. A zone-selective GFP system with a feedback lockout signal to an instantaneous relay on the service could satisfy the rule on selectivity.

NOTE: For additional information on ground-fault protection at service entrances, as required by **NEC** Section 230-95, refer to Chapter 7 of this handbook.

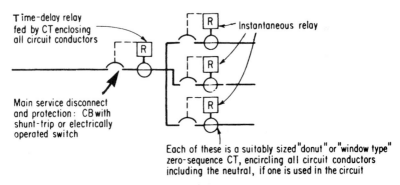

FIGURE 4.96 Hospital systems must have GFP on the next level of feeders fed by the service disconnect or feeder, where the service or feeder disconnects are required to be provided with GFP.

SYSTEM AND EQUIPMENT GROUNDING

One of the most important, but least understood, considerations in the design of electrical systems is grounding. The name "grounding" comes from the fact that the technique involves making a low-resistance connection to the earth or ground. For any given piece of equipment or circuit, this connection may be a direct wire connection to a grounding electrode which is buried in the earth; or it may be a connection to some other conductive metallic element (such as a conduit or switchboard enclosure) that is connected to a grounding electrode. The purpose of grounding is to protect personnel, equipment, and circuits by eliminating the possibility of dangerous or excessive voltages.

There are two distinct considerations in the grounding of electrical systems: (1) the grounding of one of the conductors of the wiring system, and (2) the grounding of all metal enclosures that contain electrical wires or equipment to eliminate the possibility that an insulation failure in such an enclosure might place a potential on the enclosure and constitute a shock or fire hazard.

Wiring-System Ground. This consists of the grounding of one of the wires of the electrical system to limit the voltage that might otherwise occur in the circuit through exposure to lightning or other voltages higher than that for which the circuit is designed (Fig. 4.97). Another purpose in grounding one of the wires of the system is to limit the maximum voltage to ground under normal operating conditions. Also, a system that operates with one of its conductors intentionally grounded will provide for automatic opening of the circuit if an accidental or fault ground occurs on one of its ungrounded conductors.

The selection of the wiring-system conductor to be grounded depends upon the type of system. In 3-wire, single-phase systems, the midpoint of the transformer winding—the point from which the system neutral is derived—is grounded. In grounded 3-phase wiring systems, the neutral point of the wye-connected transformer (or transformers) or generator is usually the point connected to ground. In delta-connected transformer hookups, grounding can be effected by grounding one of the three phase legs, by grounding a center-tap point on one of the transformer windings (as in the 3-phase, 4-wire "red-leg" delta system), or by using a special grounding transformer that establishes a neutral point of a wye connection, which is grounded.

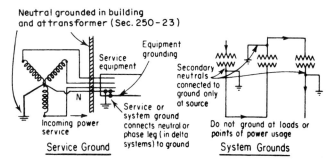

FIGURE 4.97 "Grounded electrical system" operates with one of its circuit conductors connected to a grounding electrode.

According to the **NEC**, all interior ac wiring systems must be solidly grounded (not through an intentional resistance or impedance) if they can be grounded in such a way that the maximum voltage to ground does not exceed 150 V. This rule makes it mandatory that, except for control circuits and isolated circuits in hospital operating suites (as covered in the exception to **NEC** Section 250-5), the following systems or circuits operate with one conductor grounded:

- A 120-V, 2-wire system or circuit must have one of its circuit wires grounded.
- A 240/120-V, 3-wire, single-phase system or circuit must have its neutral conductor grounded.
- A 208/120-V, 3-phase, 4-wire, wye-connected system or circuit must be operated with the neutral conductor grounded.

In all these systems or circuits, the neutrals must be grounded because the maximum voltage to ground from any other conductor of the system does not exceed 150 V when the neutral conductor is grounded.

In addition, **NEC** Section 250-5(b) requires that the neutral conductor of a 240/120-V, 3-phase, delta-connected, 4-wire system (with the neutral taken from the midpoint of one phase) be grounded at the transformer. It is also mandatory that a 480Y/277-V, 3-phase, 4-wire interior wiring system have its neutral grounded if the neutral is to be used as a circuit conductor—such as for 277-V lighting. And if 480-V autotransformer-type fluorescent or mercury-vapor ballasts are to be supplied from a 480/277-V system, then the neutral conductor must be grounded at the voltage source, even though the neutral is not used as a circuit conductor. Of course, it should be noted that 480/277-V systems are usually operated with the neutral grounded to obtain the automatic fault clearing of a grounded system. Figure 4.98 shows these rules.

According to these **NEC** rules, any 120-V, 2-wire circuit must normally have one of its conductors grounded. These requirements have often caused difficulty when applied to control circuits derived from the secondary of a control transformer that supplies power to the operating coils of motor starters, contactors, and relays. For instance, there are cases where a ground fault on the hot leg of a grounded control circuit can cause a hazard to personnel by actuating the control-circuit fuse of CB and shutting down an industrial process in a sudden, unexpected, nonorderly way. A metal-casting facility is an example of an installation where sudden shutdown due to a ground fault in the hot leg of a grounded control circuit could be objectionable.

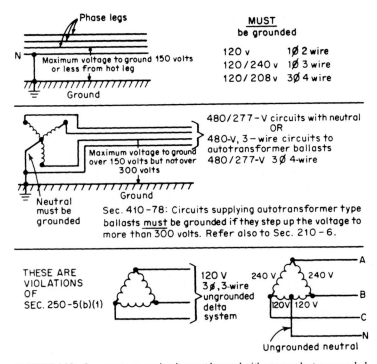

FIGURE 4.98 Some systems or circuits must be used with one conductor grounded.

Because designers often wish to operate such 120-V control circuits ungrounded, Exception No. 3 of Section 250-5(b) permits ungrounded control circuits under certain specified conditions.

A 120-V control circuit may be operated ungrounded when all the following exist:

1. The circuit is derived from a transformer that has a primary rating below 1000 V.
2. Whether in a commercial, institutional, or industrial facility, supervision will ensure that only persons qualified in electrical work will maintain and service the control circuits.
3. There is a need to prevent circuit opening on a ground fault; i.e., continuity of power is required for safety or for operating reliability.
4. Some type of ground detector is used on the ungrounded system to alert personnel to the presence of a ground fault, enabling them to clear the ground fault in the normal downtime of the system (Fig. 4.99).

Although no mention is made of secondary voltage in this rule, Exception No. 3 is significant primarily for 120-V control circuits. The **NEC** has long permitted 240- and 480-V control circuits to be operated ungrounded. This exception can be applied to any 120-V control circuit derived from a control transformer in an individual motor starter or magnetic contactor. Of course, the exception could also be used to permit ungrounded 277-V control circuits under the same conditions.

An application where ungrounded 120-V circuits are not only permitted but are specifically required is shown in Fig. 4.100. Any electrical circuit within an anes-

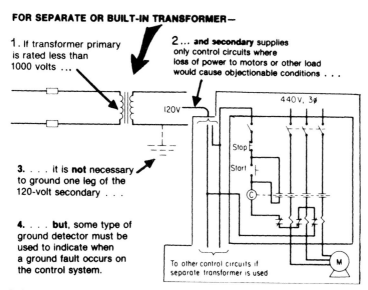

FIGURE 4.99 Ungrounded 120-V circuit may be used for control purposes.

thetizing location must be supplied from an isolated ungrounded distribution system (**NEC** Section 517-160). An anesthetizing location is any area in a hospital in which flammable or nonflammable anesthetics are or may be administered to patients—operating rooms, delivery rooms, anesthesia rooms, and any corridors, utility rooms, or other areas used for the administration of anesthetics.

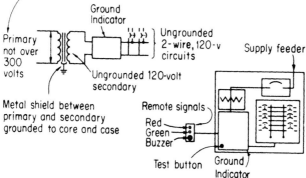

FIGURE 4.100 Any 120-V circuit *must* be operated ungrounded in an anesthetizing location within a hospital.

The **NEC** does recognize the use of ungrounded circuits or systems operating at less than 50 V. But grounding of circuits under 50 V is required for the cases shown in Fig. 4.101.

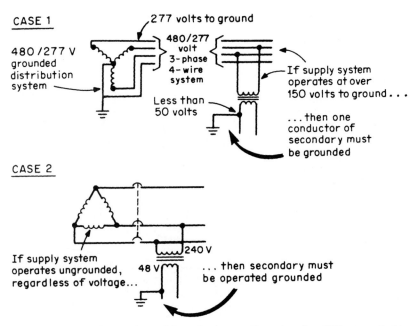

FIGURE 4.101 Under certain conditions, circuits operating at less than 50 V are required to be grounded.

Although the **NEC** does not require grounding for electrical systems in which the voltage to ground would exceed 150 V, it does recommend that ground-fault detectors be used with ungrounded systems which operate at more than 150 V and less than 1000 V. Such detectors indicate when an accidental ground fault develops on one of the phase legs of an ungrounded system. The indicated ground fault can be removed during the normal downtime of the industrial operation—i.e., when the production machinery is not running.

For industrial electrical systems (for manufacturing plants and process industries), many designers prefer ungrounded or resistance-grounded systems to solidly neutral-grounded systems such as the 480Y/277-V solidly grounded system. Ungrounded or resistance-grounded operation of an electrical system is preferred because the occurrence of a ground fault on only one phase of such a system will not cause the automatic operation of a fuse or CB and the interruption of power to important loads, as it does for solidly grounded systems. There is also much strong opinion that solidly grounded systems result in greater production losses, decreased reliability, and higher maintenance costs. And arcing ground faults are widely considered to be much less of a concern in ungrounded or resistance-grounded systems.

The use of resistance grounding is preferred where a ground fault is to be detected but the circuit is not to be automatically shut down. Although this condition can be (and frequently is) obtained by using an ungrounded system with suitable ground detectors, some designers believe there is greater advantage in the use of resistance grounding, which can very effectively be used with GFP equipment to give better protection against arcing-type faults. Resistance grounding is most often selected for industrial plants, where lighting may be no more than 10 percent of the total load, and the greater concern is for continuous power to production or process equipment—particularly where unplanned shutdowns of such equipment can cost hundreds or thousands of dollars per hour in lost production. In these systems, detected faults can be cleared during planned shutdowns of equipment.

The major operating difference between a grounded and an ungrounded system is that a single ground fault will automatically cause opening of the circuit in a grounded system but will not interrupt an ungrounded system. However, the presence of a single ground fault on an ungrounded system exposes the system to the very destructive possibility of a phase-to-phase short if another ground fault should simultaneously develop on a different phase leg of the system (Fig. 4.102).

UNGROUNDED SYSTEMS

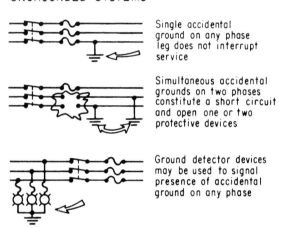

FIGURE 4.102 Ungrounded system will not automatically operate overcurrent protection on a single ground fault.

Grounded-neutral systems are generally recommended for high-voltage (over 600 V) distribution. Although ungrounded systems do not undergo a power outage as the result of a one-phase ground fault, the time and money spent in tracing faults indicated by ground detectors and other disadvantages of ungrounded systems have favored the use of grounded-neutral systems. Grounded systems are more economical in operation and maintenance. In such systems, if a fault occurs, it is isolated immediately and automatically.

Grounded-neutral systems have many other advantages. The elimination of multiple faults caused by undetected restriking grounds greatly increases service and equipment reliability. The lower voltage to ground which results from grounding the neutral offers greater safety for personnel and requires lower equipment voltage ratings. And, on high-voltage (above 600 V) systems, residual relays can be used to detect ground faults before they become phase-to-phase faults, which have substantial destructive ability.

OSHA and System Grounding. One of the most consistently troublesome and controversial areas of modern electrical design is application of the **NEC** rules and OSHA regulations concerned with system grounding. For both new electrical systems and existing systems, many questions are raised about the need to operate with the neutral or one of the phase legs intentionally grounded at the service or transformer secondary. When must one of the circuit conductors be a grounded conductor? (Note that this is a matter of system grounding and is separate from the question of equipment grounding or the effective tying together and grounding of equipment enclosures, housings, and similar non-current-carrying exposed metal parts of electrical equipment.)

Several **NEC** sections relate to this very important subject:

- Section 210-6(a) regulates the use of lampholders, lighting fixtures, and receptacles on circuits with various voltages to ground. Although such load devices are basically limited to use on circuits rated not over 150 V to ground (240/120-V, 3-wire, single-phase and 208/120-V, 3-phase, 4-wire systems), incandescent and electric-discharge lighting may be used on circuits rated up to 300 V to ground in industrial, commercial, and institutional occupancies under the conditions given. The "voltage to ground" is the maximum voltage between an ungrounded circuit conductor and an intentionally grounded conductor (such as a grounded neutral or phase leg). For an ungrounded system, the **NEC** specifies "voltage to ground" as the maximum voltage between any two of the circuit conductors (which would occur if one of the circuit conductors became accidentally grounded).
- Section 210-6(b) permits electric-discharge lighting units on ungrounded circuits of not more than 600 V "between conductors"—but only for outdoor use or in tunnels.
- Section 410-78 requires that only a grounded system (one circuit conductor intentionally grounded) may be used to supply lighting equipment that contains an autotransformer-type ballast that raises the voltage to more than 300 V.
- Section 250-5 specifies when electrical systems must be operated with one of the circuit conductors grounded.

OSHA has made the rules of Section 250-5 retroactive, so all existing installations that do not conform with that section must be altered to comply with its rules.

The effect of the foregoing rules on various types and voltages of electrical systems are as follows:

1. Any new or existing circuits rated 120 V must have one conductor grounded. And this applies to the secondary of even the smallest lighting transformer that might be used for a light at a machine or other local application. Exceptions permitting ungrounded 120-V circuits are made in Exception No. 3 of Section 250-5(b) for control circuits and in Section 517-104 for circuits in anesthetizing locations in hospitals and health-care facilities. Any new or existing 240/120-V, 3-wire, single-phase system or 208/120-V, 3-phase, 4-wire system must have its neutral grounded. For instance, an ungrounded 120-V, 3-wire delta system violates Section 250-5(b)(1) and must be converted to a corner-grounded delta or grounded wye (through a zigzag grounding autotransformer; see **NEC** Section 450-4).

2. Any new or existing 480/277-V wye system must have the neutral point of the wye grounded if a neutral conductor is derived from that point and used as a circuit conductor to supply any kind of load. But if a 480/277-V transformer secondary is used to supply only 480-V loads and no neutral conductor is used, the

neutral point of the wye does not have to be grounded to satisfy the **NEC**. Utility-company rules may, however, require grounding of the wye, and such rules must be checked.

3. A 480-V ungrounded delta system may not be used to supply indoor lighting of any kind for a new system. In such a system, the voltage to ground is taken as 480 V, which would be a violation of Exceptions No. 1 and 2 of Section 210-6(a). A 480-V ungrounded delta system may be used for outdoor and tunnel lighting as permitted by Section 210-6(b), but in such cases only winding transformer ballasts (not autotransformer-type ballasts) must be used to satisfy Section 410-78.

4. Existing 480-V ungrounded delta systems that supply 480-V indoor lighting fixtures (with either 2-winding or autotransformer-type ballasts) are not in violation of retroactive Section 250-5(b) and therefore do not have to be converted to grounded operation. Such hookups are, however, contrary to the rules of Section 210-6(a) and may violate Section 410-78 if autotransformer ballasts are used. But OSHA does not require retroactive application of the rules contained in **NEC** Sections 210-6(a) and 410-78. Of course, it is always good practice to update existing systems and bring them into compliance, even if this is not required.

5. An existing 480-V, 3-phase, 3-wire ungrounded system derived from a 480/277-V wye transformer secondary and supplying 480-V indoor lighting connected phase to phase is not in violation of retroactive Section 250-5(b) and does not have to have its neutral point grounded. Again, such a hookup could not be used without the neutral grounded for new indoor lighting, because it violates Section 210-6(a). But it could be used for new outdoor and/or tunnel lighting as in Section 210-6(b). And, again, Section 410-78 must be observed for new installations.

6. New or existing 240-V, 3-phase, 3-wire ungrounded systems may supply incandescent or electric-discharge lighting in accordance with Section 210-6(a). In such a system, the voltage to ground is 240 V, which does not exceed the 300-V maximum in Exceptions No. 1 and No. 2 of Section 210-6(a). And a 240-V ungrounded delta system is not in violation of the retroactive rules of Section 250-5(b). But the use of an ungrounded system to supply autotransformer ballasts that raise the voltage to more than 300 V violates Section 410-78.

Obviously, in all design work, it is imperative to observe all applicable regulations and carefully relate them to each other.

Equipment Ground. This is the permanent and continuous bonding together (i.e., connecting together) of all non-current-carrying metal parts of equipment enclosures—conduit, boxes, cabinets, housings, frames of motors, and lighting fixtures—and the connection of this interconnected system of enclosures to the system grounding electrode, either at the service equipment of a premises or on the secondary side of a transformer within a system (Fig. 4.103). All metal enclosures must be interconnected to provide a low-impedance path for fault-current flow along the enclosures, to ensure the operation of overcurrent devices that will open a circuit in the event of a fault. By opening a faulted circuit, these devices prevent dangerous voltages from being present on equipment enclosures that could be touched by personnel, with consequent electric shock to such personnel.

Simply stated, the grounding of all metal enclosures containing electric wires and equipment prevents the occurrence of any potential above ground on the enclosures. Such bonding together and grounding of all metal enclosures is required for both grounded electrical systems (those systems in which one of the circuit conductors is intentionally grounded) and ungrounded electrical systems (systems with none of the circuit wires intentionally grounded).

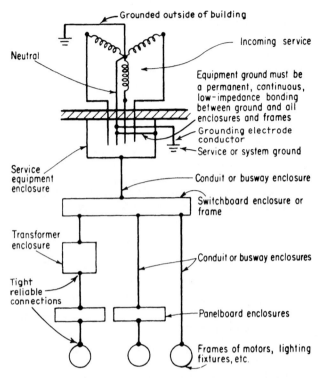

FIGURE 4.103 For either a grounded or ungrounded electrical system, all metallic enclosures of conductors and other operating components must be interconnected and grounded at the service-system source.

But effective equipment grounding is extremely important in grounded electrical systems, to provide the automatic fault clearing that is one of the important advantages of grounded systems. A low-impedance path for fault current is necessary to permit enough current to flow to operate the fuses or circuit breaker protecting the circuit, and low impedance is specifically required by **NEC** Section 250-51.

In a grounded electrical system with a high-impedance equipment ground-return path, if one of the phase conductors of the system (i.e., one of the ungrounded conductors of the wiring system) should accidentally come in contact with one of the metal enclosures in which the wires are run, it is possible that not enough fault current would flow to operate the overcurrent devices. In such a case, the faulted circuit would not automatically open, and a dangerous voltage would be present on the conduit and other metal enclosures. This voltage would present shock and fire hazards through possible arcing or sparking from the energized conduit to some grounded pipe or other grounded metal.

In a grounded system with a high-impedance equipment ground-return system, a ground fault will not open the faulted circuit; a phase-to-phase fault must develop to operate the overcurrent device. Such a condition is quite hazardous.

For effective protection against common ground faults in grounded electrical systems, therefore, low impedance for the equipment bonding system is even more important than low impedance for the grounding-electrode connection to earth. And in long runs of magnetic-material conductor enclosures, the ground-circuit impedance should be taken as 10 times the dc resistance to allow for the many variables.

Ground Connections

When an electrical system is to be operated with one conductor grounded—either because it is required by the **NEC** (as in a 240/120-V, single-phase system) or because it is desired by the system designer (as in a 240-V, 3-phase, corner-grounded system)—a connection to a suitable grounding electrode must be made at the service entrance. That is, the neutral conductor or other conductor to be grounded must be connected at the service equipment to a conductor that runs to a grounding electrode or "grounding electrode system," as described in **NEC** Section 250-81. The conductor that runs to the grounding electrode is defined as the *grounding-electrode conductor* in the **NEC**.

The **NEC** rule says that the connection of the grounding-electrode conductor to the system conductor that is to be grounded must be made "at any accessible point from the load end of the service drop or service lateral" to the service disconnecting means. This means that the grounding-electrode conductor (which runs to building steel and/or water pipe or to a driven ground rod) must be connected to the system neutral or other system wire to be grounded either in the enclosure for the service disconnect or in some enclosure on the supply side of the service disconnect. Such connection may be made, for instance, in the main service switch or CB in a service panel- or switchboard. Or, the grounding-electrode conductor may be connected to the system grounded conductor in a gutter, CT cabinet, or meter housing on the supply side of the service disconnect. The utility company should be checked on grounding connections in meter sockets or other metering equipment (Fig. 4.104).

In addition to the grounding connection for the grounded system conductor at the point of service entrance to the premises, it is further required that another grounding connection be made to the same grounded conductor at the transformer that supplies the system. This means, for example, that a grounded service to a building must have the neutral point of the supplying transformer connected to a grounded electrode at the utility transformer outside the building, as well as having the neutral grounded to a water pipe and/or other suitable electrode at the building, as shown in Fig. 4.105. And, in the case of a building served from an outdoor transformer pad or mat installation, the conductor that is grounded in the building must also be grounded at the transformer pad or mat.

The above grounding details refer to connections at the service-entrance equipment of systems fed by a utility service, as distinguished from grounding connections for a "separately derived" system, such as one fed from a local transformer or from an on-site generator. (The latter are covered under "Transformer Grounding" in Chapter 6 of this handbook.)

These design rules are concerned with the grounding of a utility-fed ac electric supply circuit that has one of its conductors intentionally grounded—such as a 240/120-V single-phase system or 208/120-V, 3-phase, 4-wire system or 480/277-V, 3-phase, 4-wire system that is required to have its neutral conductor grounded.

The basic concepts are as follows:

1. For such grounding connection, a grounding-electrode conductor must be connected to the grounded conductor (the neutral conductor) anywhere from the "load end of the service drop or lateral" to the neutral block or bus within the

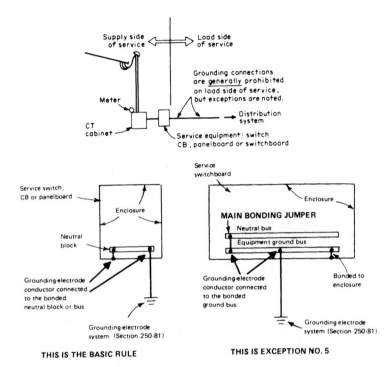

FIGURE 4.104 Grounding connection for a system grounded neutral must be made on the supply side of the service equipment.

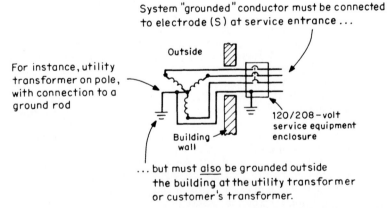

FIGURE 4.105 Grounding connection for a grounded service conductor must be made at two points.

enclosure for the service disconnect—which includes a meter socket, a CT cabinet, or an auxiliary gutter or other enclosure ahead of the service disconnect, panelboard, or switchboard.

2. The grounding-electrode conductor that is connected to the grounded neutral (or grounded phase leg) must be run to a "grounding electrode system," as described later in Chapter 7 on "Services."
3. Exception No. 5 of **NEC** Section 250-23(a) permits the grounding-electrode conductor to be connected to the equipment grounding bus in the service-disconnect enclosure—instead of to the neutral block or bus—where such connection is considered necessary to prevent the desensitization of a service GFP hookup that senses fault current through a CT-type sensor on the ground strap between the neutral bus and the ground bus (see the bottom of Fig. 4.104). However, in any particular installation, the choice between connecting to the neutral bus and connecting to the ground bus will depend on the number and types of grounding electrodes, the presence or absence of grounded building structural steel, bonding between electrical raceways and other metal piping on the load side of the service equipment, and the number and locations of bonding connections. The grounding-electrode conductor may be connected to either the neutral bus (or terminal lug) or to the equipment grounding block or bus in any system that has a conductor or a busbar bonding the neutral bus or terminal to the equipment grounding block or bus. But, where the neutral is bonded to the enclosure simply by a bonding screw, the grounding-electrode conductor must not be connected to the equipment grounding block or bus. It must be connected only to the neutral-conductor terminal block or bus. This rule is based on the idea that a lightning strike on the neutral conductor outside would cause the lightning current to flow through the bonding screw in going to ground through a grounding-electrode conductor that is connected to the equipment ground terminal. The conductivity of a screw-type bond is not adequate to effectively pass the very high lightning currents to earth.

One of the most important and widely discussed considerations in the design of electrical distribution systems is the matter of making a grounding connection to the system grounded neutral or grounded phase wire. The **NEC** says, "Grounding connections shall not be made on the load side of the service disconnecting means." Once a neutral or other circuit conductor is connected to a grounding electrode at the service equipment, the general rule is that the neutral or other grounded leg must be insulated from all equipment enclosures and other grounded parts on the load side of the service. That is, bonding of subpanels (or any other connection between the neutral or other grounded conductor and equipment enclosures) is prohibited.

There are some exceptions to that rule, but they are few and are very specific:

1. In a system, even though it is on the load side of the service, when voltage is stepped down by a transformer, a grounding connection must be made to the secondary neutral, as described in Chapter 6 under "Transformer Grounding."
2. When a circuit is run from one building to another, it may be necessary (or simply permissible) to connect the system "grounded" conductor to a grounding electrode at the other building, as covered below under "Grounding at separate buildings."
3. The frames of ranges, wall ovens, countertop cook units, and clothes dryers may be "grounded" by connection to the grounded neutral of their supply circuit (Section 250-60).

It is clear violation to bond the neutral block in a panelboard to the panel enclosure in other than a service panel. In a panelboard used as service equipment, the neutral block (terminal block) is bonded to the panel cabinet by the bonding screw provided. And such bonding is required to tie the grounded conductor to the interconnected system of metal enclosures for the system (i.e., service-equipment enclosures, conduits, busway, boxes, panel cabinets, etc.). It is this connection which provides for the flow of fault current and operation of the overcurrent device (fuse or breaker) when a ground fault occurs. But, there must not be any connection between the grounded system conductor and the grounded metal enclosure system at any point on the load side of the service equipment, because such connection would constitute a connection of the grounded system conductor to a grounding electrode (through the enclosure and raceway system to the water pipe or driven ground rod). Such connections, like the bonding of subpanels, can be dangerous, as shown in Fig. 4.106.

This **NEC** rule prohibiting connection of the grounded system wire to a grounding electrode on the load side of the service disconnect must not be confused with the rule of Section 250-60 which permits the grounded system conductor to be used for grounding the frames of electric ranges, wall ovens, counter-mounted cooking units, and electric clothes dryers. The connection referred to in Section 250-60 is that of an ungrounded metal enclosure to the grounded conductor for the purpose of grounding the enclosure.

Grounding at Separate Buildings

As described above, bonding of a panel neutral block (or the neutral bus or terminal in a switchboard, switch, or circuit breaker) to the enclosure is required in service equipment. Exception No. 2 of that section permits bonding of the neutral conductor on the load side of the service equipment in those cases where a panelboard (or switchboard, switch, etc.) is used to supply circuits in a building and the panel is fed from another building. This is covered in Section 250-24 which says that where two or more buildings are supplied from a common service to a main building, a grounding electrode at each other building shall be connected to the ac system grounded conductor on the supply side of the building disconnecting means of a grounded system as shown in Fig. 4.107 or connected to the metal enclosure of the building disconnecting means of an ungrounded system. Those are the basic rules covered in Parts (a) and (b) of this section. But Exception No. 1 to Part (a) and Exception No. 1 to Part (b) note that a grounding electrode at a separate building supplied by a feeder or branch circuit is not required where only one branch circuit is supplied and there is no non-current-carrying equipment in the building that requires grounding. An example would be a small residential garage with a single lighting outlet or switch with no metal boxes, faceplates, or lighting fixtures within 8 ft vertically or 5 ft horizontally from a grounded condition.

Exception No. 2 to Part (a) states that the grounded circuit conductor of a feeder to a separate building does not have to be bonded and grounded to a grounding electrode if an equipment grounding conductor is run with the circuit conductors for grounding any non-current-carrying equipment, water piping, or building metal frames in the separate building. (Fig. 4.108). And, as shown at the bottom of that illustration, the need for a grounding electrode at the outbuilding is eliminated because the words "equipment grounding conductor" as used in Exception No. 2 of Section 250-24(a) are understood to include metal "conduit" as indicated by Section 250-91(b), which recognizes an "equipment grounding conductor . . . enclosing the

FEEDERS FOR LIGHTING AND POWER 371

1. THIS CONDITION WILL EXIST... AND...

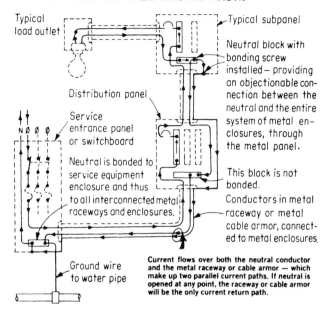

2. THIS HAZARD COULD DEVELOP

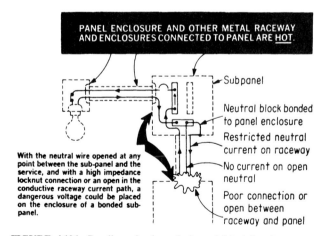

FIGURE 4.106 Bonding of subpanels is prohibited for the reason shown here.

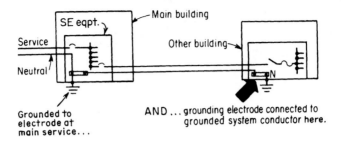

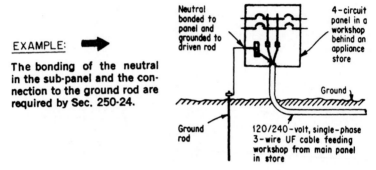

FIGURE 4.107 Grounded conductor (e.g., a neutral) must be grounded at each building.

circuit conductors." If the separate building has an approved grounding electrode and/or interior metallic piping system, the equipment grounding conductor shall be bonded to the electrode and/or piping system. However, if the separate building does not have a grounding electrode—that is, does not have 10 ft or more of underground metal water pipe, does not have grounded structural steel, and does not have any of the other electrodes recognized by Section 250-81—then at least one grounding electrode must be installed. That would most likely be a *made* electrode—such as a driven ground rod—and it must be bonded to the equipment ground terminal or equipment grounding bus in the enclosure of the panel, switchboard, circuit breaker, or switch in which the feeder terminates (Fig. 4.108).

When a grounding electrode connection is made to a grounded system conductor (usually a neutral) at a building that is fed from another building, the necessity for bonding the neutral block in such a subpanel is based on Section 250-24 and 250-54. The latter section says, "Where an AC system is connected to a grounding electrode in or at a building as specified in Sections 250-23 and 250-24, the same electrode shall be used to ground conductor enclosures and equipment in or on that building." Although the **NEC** permits and even requires bonding at both ends, if the feeder circuit is in conduit, neutral current flows on the conduit because it is electrically in parallel with the neutral conductor, being bonded to it at both ends.

At one time, the rule of Exception No. 2 of Part (a) of this section required bonding and grounding of the neutral conductor at any outbuilding in which livestock was housed. That bonding and grounding was required even if an equipment grounding conductor was run from the main building to the livestock building. That rigid rule has been removed from the **NEC**, and it is no longer mandatory that the neutral

FEEDERS FOR LIGHTING AND POWER

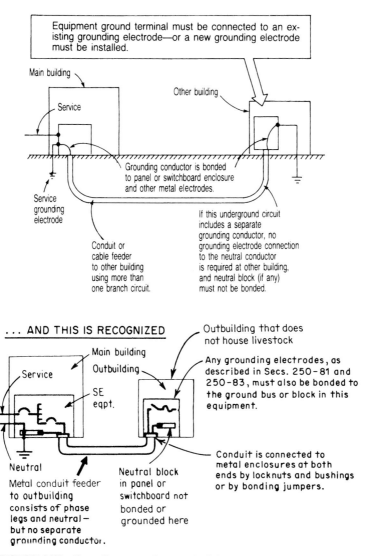

FIGURE 4.108 Grounding connection at outbuildings may be eliminated.

must always be bonded and connected to a grounding electrode at an outbuilding housing livestock.

As required in the 1981 and previous **NEC** editions, the neutral had to be connected to a grounding electrode and bonded to the disconnect enclosure (and, as a result, bonded to the entire equipment grounding system of interconnected metal raceways and housings in the livestock building being fed). When a neutral is bonded in that way, without connection to a grounding electrode at the livestock building, any flow of normal unbalanced load current on the neutral produces a voltage drop on the neutral from the main disconnect in the livestock building back to the ground reference of the grounding electrode at the service of the main building. That voltage drop then appears as a potential to ground from equipment housings in the livestock building and can have an adverse effect on livestock. The reasoning behind elimination of the mandatory rule on neutral bonding in a livestock building was given as follows:

> When livestock are housed in Building No. 2, and when a separate equipment grounding conductor is run from Building No. 1 to Building No. 2 for the purpose of grounding all metal equipment and parts, it shall be permissible to isolate the grounding conductor from the neutral in Building No. 2 if neutral-to-earth voltages cause distress to the confined livestock.
>
> *Substantiation:* Neutral-to-earth voltages (stray voltages) are caused by many factors. One of the primary causes is voltage drop on secondary circuits due to circuit imbalance and long circuits. Second, voltage drops are imposed on neutral busbars in a building service entrance, which are in turn transmitted to metal grounding conductors, conduit, or panel grounding to a metallic water system. Livestock, particularly dairy animals and swine, are very sensitive to AC voltages that can occur when part of their body makes contact with the described metal equipment and part of their body is in contact with true earth. AC voltages over 1 V are known to cause dairy animals to go out of milk production and to inhibit the growth rate of swine.
>
> (Note attached paper on stray voltage problems, page 11, item 7.) You will note that an alternate solution to resolving stray voltage problems is to isolate the neutral from the grounding conductors in the barn panel and run a separate fourth wire either back to the transformer or metering location (main farm service entrance).
>
> By adding the proposed exception to **NEC** Sec. 250-24, it would make it legal or permissible to run the fourth wire and isolate the neutral and grounding wires at the panel serving building No. 2, which is assumed to be a grounded system having a grounding electrode connected to the neutral.

The last sentence of Exception No. 2 to Part (b) does require that the equipment grounding conductor that is run with the feeder to the disconnect enclosure in a livestock building be an "insulated or covered copper" conductor and may not simply consist of metal conduit enclosing the feeder conductors. For such a feeder, the equipment grounding conductor must be insulated or covered copper only for an underground circuit or for any part of the circuit that is run underground. The purpose of that rule is to assure a more reliable equipment grounding path for such buildings to prevent potentials on equipment that would threaten livestock because of their great sensitivity to even very low voltages—as described earlier. An equipment grounding conductor run underground—directly buried or in a raceway—must be insulated or covered copper to prevent corrosion. But an overhead feeder to a barn building may be aluminum or copper multiplex cable with a bare messenger wire used as the equipment grounding conductor. Figure 4.109 shows the considerations in the Exception: (3) applies to any outbuilding; (1) and (2) apply only to a separate building housing livestock.

Figure 4.110 shows another condition in which a grounding electrode connection must be made at the other building, as specified in the basic rule of Section 250-24(b). For an ungrounded system, when, as shown in the sketch, an equipment grounding conductor is not run to the outbuilding, then a grounding electrode conductor must be run from the ground bus or terminal in the outbuilding disconnect to a suitable grounding electrode which must be provided.

According to Section 250-24(b), Exception No. 2, a circuit of an ungrounded system does not require a grounding electrode connection to the equipment ground terminal at an outbuilding disconnect only where the outbuilding uses not more than one branch circuit, where an equipment grounding conductor is run with the circuit

FEEDERS FOR LIGHTING AND POWER

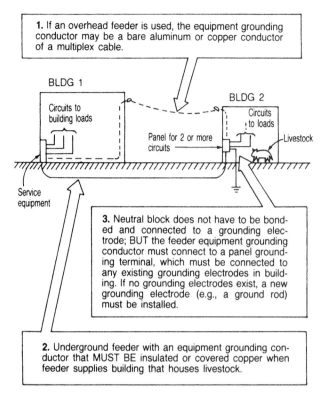

FIGURE 4.109 Building for livestock must be carefully grounded.

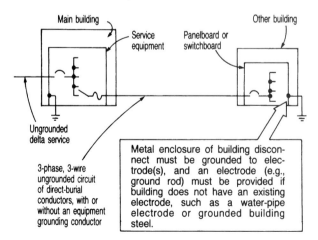

FIGURE 4.110 Grounding connection for an ungrounded supply to an outbuilding.

conductors to the outbuilding, and where there are no existing electrodes at the outbuilding. If the ungrounded circuit to the outbuilding is a feeder supplying more than one branch circuit, then a connection must be made from the equipment grounding terminal in the outbuilding disconnect to either an existing ground electrode at the building (water pipe, building steel, etc.) or to a new ground electrode installed for the purpose. And such a connection to a ground electrode must be made even if an equipment grounding conductor is run with the feeder conductors to the outbuilding. If the building houses livestock, that portion of the equipment grounding conductor that is run underground to the outbuilding must be insulated or covered copper.

In Fig. 4.110, if the 3-phase, 3-wire, ungrounded feeder circuit to the outbuilding had been run with a separate equipment grounding conductor which effectively connected the metal enclosure of the disconnect in the outbuilding to the grounding electrode conductor in the service equipment of the main building, a connection to a grounding electrode would not be required provided that conditions a, b, c, and d of Exception No. 2 are satisfied. But if all those conditions are not satisfied and a grounding electrode connection would be required at the outbuilding, then the equipment grounding conductor run to the outbuilding would have to be bonded to any grounding electrodes that were "existing" at that building—such as an underground metal water service pipe and/or a grounded metal frame of the building. All grounding electrodes that exist at the outbuilding must be bonded to the ground bus or terminal in the disconnect at the outbuilding, whether or not an equipment grounding conductor is run with the circuit conductors from the main building.

Part (c) covers design of the grounding arrangement for a feeder from one building to another building when the main disconnect for the feeder is at a remote location from the building being supplied—such as in the other building where the feeder originates. The rule prohibits grounding and bonding of a feeder to a building from another building *if* the disconnect for the building being fed is located in the building where the feeder originates. Part (c) correlates the grounding concepts of Section 250-24 with the disconnect requirements of Section 225-8(b), Exceptions No. 1 and No. 2. This is most easily understood by considering the following sequence of **NEC** rules involved:

1. First, in Section 250-23(a), bonding of a panel neutral block (or the neutral bus or terminal in a switchboard, switch, or circuit breaker) to the enclosure is required in service equipment. Exception No. 2 of that section permits bonding of the neutral conductor on the load side of the service equipment in those cases where a panelboard (or switchboard, switch, etc.) is used to supply circuits in a building and the panel or switchboard in that building is supplied by a feeder from another building. This is covered in Section 250-24(a), which says that where two or more buildings are supplied from a common service to a main building, a grounding electrode at each other building shall be connected to the ac system grounded conductor on the supply side of the building disconnect means of a grounded system and the neutral must be bonded to the metal enclosure of the disconnect (the enclosure of the panelboard or switchboard or switch or circuit breaker) in the building being supplied. The basic rule calls for grounding and bonding within the building disconnect for the building being fed. That is, the feeder to the second building is treated like a service to the building.

2. But, Exception No. 2 of Section 250-24(a) gives the option of treating the feeder to a building from another building exactly like a feeder that did not leave the main building—that is, running an equipment grounding conductor and not grounding and bonding the feeder neutral in the building disconnect. Exception No. 2 to Part (a) of Section 250-24 states that the grounded circuit conductor of a

feeder to a separate building does not have to be bonded and grounded to a grounding electrode if an equipment grounding conductor is run with the circuit conductors for grounding any non-current-carrying equipment, water piping, or building metal frames in the separate building. If the separate building has an approved grounding electrode and/or interior metallic piping system, the equipment grounding conductor shall be bonded to the electrode and/or piping system. However, if the separate building does not have a grounding electrode—that is, does not have 10 ft or more of underground metal water pipe, does not have grounded structural steel, and does not have any other electrodes recognized by Section 250-81—then at least one grounding electrode must be installed. That would most likely be a *mode* electrode—such as a driven ground rod—and it must be bonded to the equipment ground terminal or equipment grounding bus in the enclosure of the panelboard, switchboard, circuit breaker, or switch in which the feeder terminates.

3. Although the use of an equipment grounding conductor and elimination of the need for a bonded connection to a grounding electrode, as described in (2) above, is given as an optional alternative to (1) above, Part (c) of Section 250-24 makes elimination of a bonded connection to an electrode a *mandatory* method when the disconnect for the feeder is in the other building where the feeder originates, as covered by Section 225-8 (Fig. 4.111).

4. The basic rule of Section 225-8(b) says that for a group of buildings under single management, disconnect means must be provided for each building. This rule requires that the conductors supplying each building in the group be provided with a means for disconnecting all ungrounded conductors from the supply. And the rule specifically permits the disconnect for each building or structure to be the same kind as permitted for a service disconnect—that is, up to six switches or CBs, as covered in Section 230-71.

For large-capacity multibuilding industrial premises with a single owner, Exception No. 1 of Section 225-8(b) permits use of the feeder switch in the main building as the only disconnect for a feeder to an outlying building, provided the disconnect in the main building is accessible to the occupants of the outlying building. Now, when that option of Exception No. 1 of Section 225-8(b) is selected, eliminating a main disconnect for the feeder within the outbuilding being supplied, Part (c) of Section 250-24 requires the following conditions to be met:

1. The grounded (neutral) conductor of the feeder *must not* be connected to a grounding electrode at the outbuilding and the grounded conductors must *not* be bonded to the panelboard or switchboard enclosure in the outbuilding.

2. An equipment grounding conductor *must* be run with the feeder circuit conductors to the outbuilding to ground metal equipment enclosures, metal conduit, and other non-current-carrying metal parts of electrical equipment in the outbuilding. Any interior metal piping and building and structural metal frames in the outbuilding must be bonded to the equipment grounding conductor. In addition, any grounding electrodes at the outbuilding must be bonded to the feeder equipment grounding conductor. And if the outbuilding contains no grounding electrodes as covered in Section 250-81, a made electrode (a ground rod, etc.) must be installed and bonded to the equipment grounding conductor of the feeder.

3. The bond between the feeder equipment grounding conductor and any grounding electrodes at the outbuilding *must* be made in a junction box "located immediately inside or outside the second building or structure."

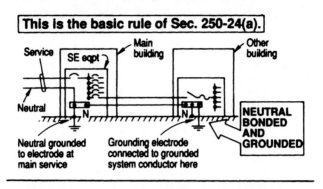

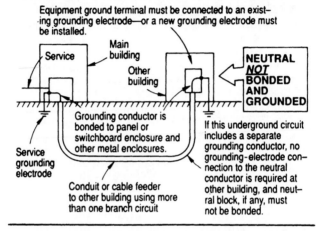

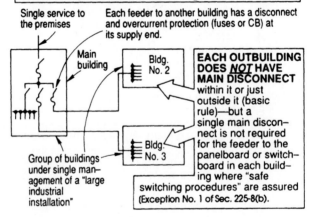

FIGURE 4.111 Part **C** correlates grounding for multiple buildings under a single management with rule of Section 225.8(b) on location of feeder disconnect for an outbuilding.

As follow-up to Part (c), the rule of Part (d) says the conductor used to bond the feeder equipment grounding conductor to the one or more existing or new grounding electrodes at an outbuilding must be sized from Table 250-95, based on the rating of the overcurrent device protecting the ungrounded conductors of the feeder. This is a rule to clarify the sizing of the bonding conductor to building electrodes as required by Exception No. 2 of Section 250-24(a) and Part 2 of Section 250-24(c). In effect, the wording of this rule in Part (d) calls for the bonding conductor to be the same size as the equipment grounding conductor of the feeder, based on Table 250-95. And the same size of conductor would be used to connect to the interior metal piping systems and structural metal frame of the outbuilding.

Exception No. 1 notes that the grounding conductor would never have to be larger than the circuit ungrounded conductors. And Exception No. 2 says the grounding conductor would not have to be larger than No. 6 copper or No. 4 aluminum where it connects to a driven ground rod or other made electrode.

Equipment Grounding

The **NEC** requires that metal equipment enclosures, boxes, and cabinets be grounded; the grounding connection may be made by metal cable armor, by the metal raceway that supplies such enclosures (rigid metal conduit, intermediate metal conduit, EMT, flex, or liquid-tight flex as permitted by Section 350-5 or 351-7), or by an equipment grounding conductor, such as where the equipment is fed by rigid nonmetallic conduit. When a separate equipment grounding conductor (i.e., other than the metal raceway or metal cable armor) is used for ac circuits, it must be contained within the same raceway, cable, or cord or otherwise run with the circuit conductors (Fig. 4.112).

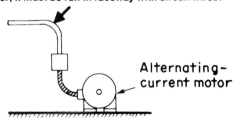

FIGURE 4.112 Equipment grounding conductors must be run with the circuit wires *within* rigid conduit, IMC, or EMT and within flexible metal conduit or liquid-tight flex when the length is over 6 ft.

External grounding of equipment enclosures, frames, or housings is a violation of the **NEC** for ac equipment. It is not acceptable, for instance, to feed an ac motor with a nonmetallic conduit or cable, without a grounding conductor in the conduit or cable, and then to provide grounding of the metal frame through a grounding conductor connected to the metal frame and run to building steel or to a grounding-grid conductor. An equipment grounding conductor must be run with the circuit conductors, within the raceway, for all conduit runs over 6 ft long.

Keeping an equipment grounding conductor physically close to ac-circuit supply conductors is absolutely essential to ensure minimum impedance in grounding current paths, to provide the most effective clearing of ground faults. When an equipment grounding conductor is kept physically close to any circuit conductor that would be supplying the fault current (that is, the grounding conductor is kept in the "same raceway, cable, or cord or otherwise run with the circuit conductors"), the impedance of the fault circuit has minimum inductive reactance and minimum ac resistance because of mutual cancellation of the magnetic fields around the conductors and reduced skin effect. Under such a condition of "sufficiently low impedance," the voltage to ground is limited to the greatest possible extent, the fault current is higher because of the minimized impedance, and the circuit overcurrent device will operate at an earlier point on its time-current characteristic to ensure maximum fault-clearing speed; the overall effect is to facilitate the operation of the protective devices in the circuit.

Direct-current circuits do not present this need to keep the grounding conductor close to the circuit conductors. Because there are no alternating magnetic fields around dc conductors, there is no inductive reactance or skin effect in dc circuits. The only impedance to current flow in a dc circuit is resistance—which is the same for a dc ground-fault path whether or not the equipment grounding conductor is placed physically close to the circuit conductors which would supply the fault current in the event of a ground fault. External equipment grounding by connection to grounded building steel or to an external ground grid is, therefore, acceptable for dc equipment, provided the external grounding path is effectively tied back to the grounded conductor of the dc system.

The arrangement shown in Fig. 4.113 violates the basic concepts of **NEC**-accepted grounding because the lighting fixture, which must be grounded, is not grounded by an equipment grounding conductor contained within the cord.

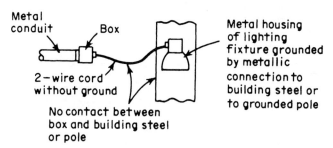

FIGURE 4.113 Alternating-current equipment enclosures that are required to be grounded must be fed by a circuit that includes an equipment grounding conductor.

Except for dc circuits and for isolated, ungrounded power sources, an equipment grounding conductor of any type must not be run separately from the circuit conductors. The engineering reason for keeping the ground return path and the phase legs in close proximity (that is, in the same raceway) is to minimize the impedance of the fault circuit by placing conductors so their magnetic fields mutually cancel each other, keeping inductive reactance down and allowing sufficient current to flow to facilitate the operation of the circuit protective devices.

The hookup in Fig. 4.113 also violates the rule of Section 250-58(a), which prohibits the use of building steel as the equipment grounding conductor for ac equipment.

NOTE: Care must be taken to distinguish between an "equipment grounding conductor" as covered by **NEC** Section 250-57 and an "equipment bonding jumper" as covered by Section 250-79(e). A "bonding jumper" may be external to equipment, but it must not be over 6 ft long.

Grounding Conductors

Equipment grounding conductors that are run within nonmetallic raceways (or even in metal raceways where the designer wants the benefit of a dedicated grounding conductor in addition to the ground path through the metal raceway) may be bare conductors, or they may be insulated conductors provided that the insulation is green or green with one or more yellow stripes. But there are exceptions to that rule for wires larger than No. 6.

A conductor of a color other than green may be used as an equipment grounding conductor if the conductor is stripped for its exposed length within an enclosure, so it appears bare, or if green coloring, green tape, or a green label is used on the conductor at its termination. As shown in Fig. 4.114, the phase legs may be any color other than white, gray, or green. The neutral must be white or gray (or any color other than green if it is larger than No. 6 and if white tape, marking, or paint is applied to the neutral near its terminations). The grounding conductor, as noted, may be green or may be any color conductor if all insulation is stripped off for the exposed length. Alternatives to stripping the black insulated conductor used for equipment grounds include (1) coloring the exposed insulation green or (2) marking the exposed insulation with green tape or green adhesive labels.

EXAMPLE

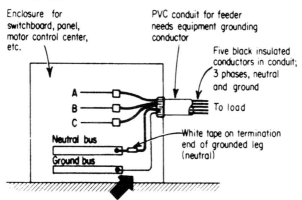

Black insulated conductor used as equipment grounding conductor has all insulation stripped from entire length exposed in enclosure

FIGURE 4.114 Equipment grounding conductor larger than No. 6 does not have to be colored green under certain conditions.

On-the-job identification is permitted for an insulated conductor used as an equipment grounding conductor in a multiconductor cable. Such a conductor, regardless of size, may be identified in the manner described above for conductors larger than No. 6 used in raceway. The conductor may be stripped bare or colored green to indicate that it is a grounding conductor. But such usage is recognized only for industrial-type systems under conditions of competent maintenance.

Building-Frame Ground

The **NEC** clearly prohibits the use of structural building steel as an equipment grounding conductor for equipment mounted on or fastened to the building steel if the supply circuit to the equipment operates on alternating current. But structural building steel that is effectively grounded and bonded to the grounded circuit conductor of a dc supply system may be used as the equipment grounding conductor for the metal enclosures of dc-operated equipment that is conductively attached to the building steel.

As shown in Fig. 4.115, the use of building steel as a grounding conductor provides a long fault return path of very high impedance because the path is separated from the feeder-circuit hot legs. Ground-fault current returning over building steel to the point where the building steel is bonded to the ac system neutral (or other grounded) conductor is separated from the circuit conductor that is providing the fault current. The impedance is, therefore, elevated, and the conditions are less than optimal, so that the grounding cannot be counted on to facilitate operation of the fuse or CB protecting the faulted circuit. The current may not be high enough to provide fast and certain clearing of the fault.

Conduit terminations for metal raceways are important in ensuring a continuous low-impedance ground path through the metal raceway-and-enclosure system that forms the equipment grounding conductor for an electrical distribution system. Single locknut-and-bushing terminations are permitted for 120/240-V and 120/208-V systems. A 480/277-V grounded system, 480-V ungrounded system, or higher-voltage system must have double locknut-and-busing terminals on rigid metal conduit and IMC (Fig. 4.116). Where good electrical continuity is desired on installations of rigid metal conduit or IMC, two locknuts are always specified so that the metal of the box can be solidly clamped between the locknuts, one on the outside and one on the inside. The reason for not relying on the bushing (in place of the inside locknut) is that both conduit and box may be secured in place, and if the conduit is placed so that it extends into the box a distance greater than the thickness of the bushing, the bushing will not make contact with the inside surface of the box. But that possible weakness in the single-locknut termination does not exclude it from use on systems up to 250 V to ground.

When an individual equipment grounding conductor is used in a raceway (either in a nonmetallic raceway, as required, or in a metal raceway—where such a conductor is used for grounding reliability even though metal raceway is a suitable grounding conductor), the grounding conductor must have the minimum size shown in **NEC** Table 250-95. The minimum acceptable size for an equipment grounding conductor is based on the rating of the overcurrent device (fuse or CB) protecting the circuit that is run in the same raceway and for which the equipment grounding conductor is intended to provide a path of ground-fault current return, as shown in Fig. 4.117. Each size of grounding conductor in the table can carry enough current to blow a fuse or trip a CB of the corresponding rating (in the left-hand column of the table). In Fig. 4.117, if the fuses are rated at 60 A, Table 250-95 shows that the equipment grounding conductor must be at least No. 10 copper or No. 8 aluminum or copper-clad aluminum.

FEEDERS FOR LIGHTING AND POWER   383

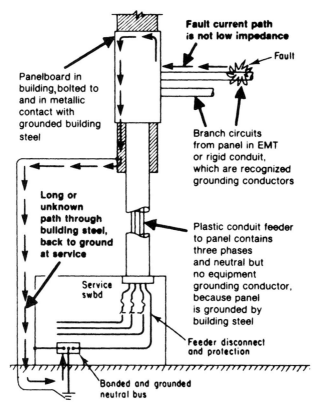

FIGURE 4.115 Grounded metal frame of a building is not an acceptable grounding conductor for ac equipment.

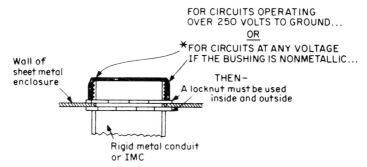

FIGURE 4.116 Conduits carrying circuits operating above 250 V to ground must have double-locknut terminations.

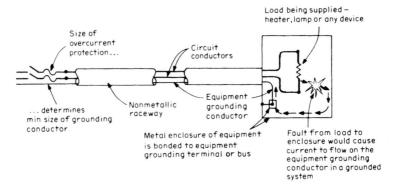

FIGURE 4.117 Equipment grounding conductor must be of sufficient size to carry the current required to operate the circuit protective device under fault condition.

Whenever an equipment grounding conductor is used for a circuit that consists of only one conductor for each hot leg (or phase leg), the grounding conductor is sized simply and directly from Table 250-95, as described. When a circuit is made up of parallel conductors per phase, as in an 800-A circuit with two conductors per phase, an equipment grounding conductor is sized in the same way; in that case, at least No. 1/0 copper or No. 3/0 aluminum would have to be used, as indicated alongside 800 A in Table 250-95. But, if such a circuit is made up using two conduits—that is, three phase legs and a neutral in each conduit—an individual grounding conductor must be run in each of the conduits, and *each* of the two grounding conductors would have to be at least No. 1/0 copper or No. 3/0 aluminum, as shown in Fig. 4.118.

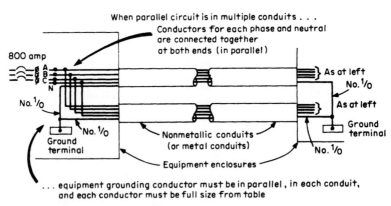

FIGURE 4.118 Equipment grounding conductor must be installed in each conduit for multiconduit, parallel-conductor feeder makeup.

Another example is shown in Fig. 4.119, where a 1200-A protective device on a parallel circuit calls for No. 3/0 copper or 250-kcmil aluminum grounding conductor. [Note in that example that each 500-kcmil XHHW circuit conductor has an ampacity of 430 A (Table 310-16), and three per phase gives a circuit ampacity of 3 × 430 A = 1290 A. Use of a 1200-A protective device satisfies the basic rule of Section 240-3, protecting each phase leg within its ampacity. Because the load on the circuit is continuous (over 3 hours), the circuit is loaded to not over 960 A, satisfying Section 220-10(b), which requires a continuous load to be limited to no more than 80 percent of the circuit conductor ampacity and no more than 80 percent of the protection rating. Each circuit conductor is actually made up of three 500-kcmil XHHWs, with a total per-phase ampacity of 3 × 430 A = 1290 A. But the load is limited to 0.8 × 1200 A = 960 A per phase. Each 500-kcmil is then carrying 960 A ÷ 3 = 320 A. Because that value is less than 380 A, which is the ampacity of 500-kcmil THW copper, the use of XHHW conductors does comply with the UL requirement that, for equipment rated over 100 A, conductors connected to the equipment be rated at not over 75°C (such as THW) or, if 90°C conductors are used (such as XHHW), they must be used at no more than the ampacity of 75°C conductors of the same size. However, some authorities object to that usage on the grounds that 1200-A protection would *not* be acceptable according to Section 240-3(c) if 75°C (THW) conductors were used, with an ampacity of only 3 × 380 A = 1140 A per phase, because the rating is over the 800-A level, and it is not permitted to go to the next larger size of protective device. Therefore, they note, the XHHW conductors are not actually being used as 75°C conductors; the load current could later be increased above the 75°C ampacity; and the application does violate the letter and intent of the UL rule, thereby violating **NEC** Section 110-3(b).]

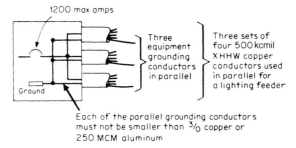

FIGURE 4.119 Grounding conductor in each conduit of a multi-conduit feeder must be sized to correspond to the full value of the circuit protection.

The same **NEC** rule points out that the equipment grounding conductor need not be larger than the circuit conductors. The main application of this exception is to motor circuits, where short-circuit protective devices are usually rated considerably higher than the motor branch-circuit conductor ampacity, as permitted in Section 430-52 (up to 400 percent), to allow starting of a motor without opening the circuit on inrush current. In such cases, the literal use of Table 250-95 could result in the use of grounding conductors that are larger than the circuit conductors.

Exception No. 3 points out that metal raceways and cable armor are recognized as equipment grounding conductors, and Table 250-95 does not apply to them.

Figure 4.120 shows details of a controversy that often arises. When two (or more) circuits are located in the same conduit, it is logical to conclude that a single equipment grounding conductor within the conduit may serve as the required grounding conductor for both circuits if it satisfies Table 250-95 for the circuit with the highest-rated overcurrent protection. The common contention is that if a single metal conduit is adequate as the equipment grounding conductor for all the contained circuits, a single grounding conductor can serve the same purpose when installed in a nonmetallic conduit that connects two metal enclosures (such as a panel and a homerun junction box) where both circuits are within both enclosures. As shown, a No. 12 copper conductor satisfies Table 250-95 as an equipment grounding conductor for the circuit protected at 20 A. The same No. 12 also may serve for the circuit protected at 15 A, for which a grounding conductor must not be smaller than No. 14 copper. Such use is specifically permitted by the **NEC**. Although this will have primary application for PVC conduit, where an equipment grounding conductor is required, it may also apply to circuits in EMT, IMC, or rigid metal conduit when an equipment grounding conductor is run with the circuit conductors to supplement the metal raceway as an equipment grounding return path.

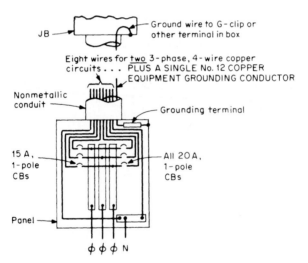

FIGURE 4.120 Single equipment grounding conductor may serve for two or more circuits run in a single raceway.

Minimum Size of Equipment Grounding Conductor

For many years now, there has been considerable discussion among electrical designers and installers regarding proper sizing for an equipment grounding conductor as covered by the **NEC** in Section 250-95 and Table 250-95. That discussion has centered around the meaning of the word "minimum" found in the table heading. But what is the "minimum" if high fault currents are available?

Over the years, most designers and installers have considered the minimum size of equipment grounding conductor given in Table 250-95 to be adequate regardless of the short-circuit current available at that point in the distribution system. That is, whether there were 5000 or 50,000 A of available fault current, generally the same size of equipment grounding conductor would be selected. While this may not have been an issue years ago—very few installations had available short-circuit currents than were greater than 5000 or 10,000 A—today, with much larger distribution systems and lower-impedance transformers (some below 1 percent), it is no longer unusual to see available fault currents in excess of 100,000, and in some cases, over 200,000 A. It is those installations where high levels of short-circuit current are available that pose the real problem and require additional consideration.

The concern for providing protection for the equipment grounding conductor was recognized a number of years ago and was explained by Eustace Soares in his now famous work entitled, *Grounding Electrical Distribution Systems for Safety*. In that book, Mr. Soares indicates that the "validity" of a conductor as a fault return path is maintained only where loaded not in excess of a specific current value for a specified amount of time based on the conductor's cross-sectional circular mils. That is, for a copper conductor, its integrity or validity is maintained where the amount of current for each 30 circular mils of cross-sectional area is not greater than 1 A and does not persist for more than 5 s. That validity is related to the amount of energy that would cause the copper conductor to become loosened at its point of attachment after the copper cools to ambient. That validity rating is based on raising the temperature of the copper conductor from 75 to 250°C.

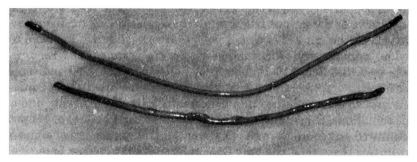

FIGURE 4.121 Here are two No. 12 AWG, THHN, copper conductors after being exposed to 40,000 A. The one No. 12 (at top) was protected in accordance with ICEA limits by a current-limiting overcurrent protective device. The other was not. As can be seen, in the first case, the conductor was not damaged because the current-limiting protective device opened the circuit before insulation temperature limits were exceeded. In the other conductor (at bottom), the operating characteristics of the overcurrent protective device resulted in a value of current flow that severely damaged the THHN (90°C-rated) insulation.

Another recognized method looks at the amount of energy that will produce damage to the conductor's insulating material; it is covered in the International Cable Engineers Association publication P-32–382. That method calculates the energy required to raise the conductor's temperature from 75 to 150°C, which will cause damage to the conductor's insulation. And a third method promoted by Onderdonk calculates the amount of energy necessary to cause the temperature of the conductor material to rise from 75 to 1083°C, which will cause the copper to melt. That is, the Onderdonk method calculates the conductor melting point. Values of short-circuit current and withstand ratings calculated in accordance with each of those recognized methods are shown in Fig. 4.122.

But, does the **NEC** *require* that equipment grounding conductors be evaluated with regard to the available fault currents? In answer to that question, consider the following.

COMPARISON OF EQUIPMENT GROUNDING CONDUCTOR SHORT CIRCUIT WITHSTAND RATINGS.

CONDUCTOR SIZE	5 SEC RATING (AMPS)			I^2t RATING $\times 10^6$ (AMPERE Squared Seconds)		
	ICEA P32-382 INSULATION DAMAGE 150 C	SOARES 1 AMP/30 cm VALIDITY 250 C	ONDERDONK MELTING POINT 1083 C	ICEA P32-382 INSULATION DAMAGE 150 C	SOARES 1 AMP/30 cm VALIDITY 250 C	ONDERDONK MELTING POINT 1083 C
14	97	137	253	.047	.094	.320
12	155	218	401	.120	.238	.804
10	246	346	638	.303	.599	2.03
8	391	550	1015	.764	1.51	5.15
6	621	875	1613	1.93	3.83	13.0
4	988	1391	2565	4.88	9.67	32.9
3	1246	1754	3234	7.76	15.4	52.3
2	1571	2212	4078	12.3	24.5	83.1
1	1981	2790	5144	19.6	38.9	132.
1/0	2500	3520	6490	31.2	61.9	210.
2/0	3150	4437	8180	49.6	98.4	331.
3/0	3972	5593	10313	78.9	156.	532.
4/0	5009	7053	13005	125.	248.	845.
250	5918	8333	15365	175.	347.	1180.
300	7101	10000	18438	252.	500.	1700.
350	8285	11667	21511	343.	680.	2314.
400	9468	13333	24584	448.	889.	3022.
500	11835	16667	30730	700.	1389.	4721.
600	14202	20000	36876	1008.	2000.	6799.
700	16569	23333	43022	1372.	2722.	9254.
750	17753	25000	46095	1576.	3125.	10623.
800	18936	26667	49168	1793.	3556.	12087.
900	21303	30000	55314	2269.	4500.	15298.
1000	23670	33333	61460	2801.	5555.	18867.

FIGURE 4.122 Values of current (at left) and the let-through energy (I^2t) withstand ratings for copper conductors as determined in accordance with three industry-recognized methods. Designers and installers should be aware that equipment grounding conductors selected in accordance with Section 250-95 and Table 250-95 of the NEC may *not* be adequate where large amounts of short-circuit current are available. In such applications, safety

NEC Section 240-1, which indicates the "Scope" of Art. 240, is followed by an FPN that says, "overcurrent protection is provided to open the circuit if the current reaches a value that will cause an excessive or dangerous temperature in conductors or conductor insulation." Although FPNs do not present a rule, they are included to provide "explanatory" information regarding certain requirements (see **NEC** Section 110-1). The FPN following Section 240-1 indicates the overall concept behind overcurrent protection. Although an equipment grounding conductor is not normally carrying current, under fault conditions, it becomes part of "the circuit" and really should not be damaged any more than the phase or neutral conductors of a given circuit. But that is not the only reference that applies here.

The **NEC** in Section 250-51 requires that each and every equipment grounding conductor be capable of carrying any fault currents likely to be imposed upon it. And compliance with that requirement would necessitate some evaluation of the selected size of grounding conductor with respect to the available fault current. That is, in addition to assuring compliance with Section 250-95 on the minimum sizing, the conductor selected in accordance with that **NEC** section must have sufficient cross-sectional area "to conduct safely" the fault current it will be required to carry.

Although the phrase "to conduct safely" is not very clear, use of one of the three preceding methods—either the Soares validity of connection method, the ICEA insulation damage method, or the Onderdonk melt-point method—should serve to satisfy the wording of Section 250-51. Of course, if the authority having jurisdiction prefers one method over the other, then the method preferred by the local inspector should be used.

Which is the more desirable and realistic approach? That depends on the application. For example, in most applications it would seem that Soares' validity of connection method is adequate. But, for isolated ground applications, the ICEA insulation damage method might be more appropriate to assure the desired isolation of the ground-return path even after a short-circuit. The melt-point method would appear to be the least desirable but could be construed as satisfying the wording of Section 250-51(2). Regardless of which method is used, certainly some evaluation of the grounding conductor's fault current-carrying capability must be performed where there are large values of available short-circuit current to assure compliance with the **NEC** and to assure continuity of the ground-return path, which provides for automatic clearing of a faulted circuit.

How does one analyze the equipment grounding conductor for fault-carrying capability?

First, you will need to refer to the short-circuit analysis—which should have been performed to determine the minimum interrupting rating of the selected protective devices—and find the available fault current at that point in the system where the circuit originates. Next, using the protective device manufacturer's operating characteristics data, determine the amount of short-circuit current and the amount of let-through energy (I^2t) that the grounding conductor will be exposed to during a bolted fault. Then, refer to the data given in Fig. 4.122 to verify that the equipment grounding conductor, selected in accordance with Section 250-95 and Table 250-95, will be capable of carrying that value of current and withstanding the let-through energy (I^2t). If not, then a larger-sized conductor—that is, one capable of sustaining that value of current and withstanding the I^2t value—must be selected.

Although the problem of excessive fault current for the size of grounding conductor is more of a concern for feeders, branch circuits that originate in close proximity to the service equipment—such as at motor-control centers located near the service equipment—should also be evaluated. Only through such an evaluation is it possible to assure compliance with the **NEC** and to assure that the equipment

FIGURE 4.123 Careful consideration should be given to sizing of the equipment grounding conductor where large amounts of short-circuit current are available. The equipment grounding conductors selected for the lighting branch circuits supplied above (arrow), which are fed from a panel in close proximity to the service equipment, should be evaluated to assure the selected grounding conductors will be capable of withstanding the value of fault current they are likely to carry.

grounding conductor will, in fact, be capable of facilitating operation of the circuit overcurrent protective device and provide automatic clearing of faulted conductors.

Bonding Jumpers

Any bonding jumper on the load side of the service should conform to the requirements for equipment grounding conductors. As a result, the bonding of conduits for a parallel circuit makeup would have to comply with the **NEC** rule that requires equipment grounding conductors to be run in parallel "where conductors are run in parallel in multiple raceways." That rule would then be taken to require that bonding jumpers also must be run in parallel for multiple-conduit circuits. In the case of the 800-A circuit run in two conduits, as described above, instead of a single No. 1/0 copper jumper from one bushing lug to the other bushing lug and then to the ground

FEEDERS FOR LIGHTING AND POWER 391

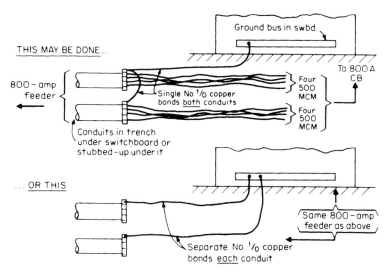

FIGURE 4.124 Equipment bonding jumpers for multiple-conduit feeders provide better protection when run in parallel, as at bottom here.

bus, it would be better to use a separate No. 1/0 copper jumper from each bushing to the ground bus so that the jumpers are run in parallel, as required for equipment grounding conductors. Figure 4.124 shows the two possible arrangements. Bonding jumpers on the load side of service equipment are sized and routed in the same way as equipment grounding conductors, because such bonding jumpers and equipment grounding conductors serve identical functions. And Section 250-95 requires the equipment grounding conductor for each of the conduits of a parallel circuit to be the full size determined from the circuit rating. Here, No. 1/0 copper for each conduit would be based on the 800-A rating of the feeder protective device.

[NOTE: Although it is not recommended practice, **NEC** Sec. 250-79(d) does permit the above described use of a single No. 1/0 jumper to bond both conduits to the equipment ground bus. In fact the **NEC** rule permits any number of metal conduits to be bonded to a switchboard or other equipment enclosure by a "single common continuous" jumper if it is sized from Table 250-95 for "the largest overcurrent device supplying" any of the circuits contained in all of the conduits.]

CHAPTER 5
MOTOR FEEDERS

When a load consists only of motors or of motors plus lighting, heating, and/or other types of loads, the feeder conductors and overcurrent protection must be carefully selected on the basis of all applicable design considerations.

The sizing of feeder conductors must satisfy applicable **NEC** rules and, among other considerations, ensure that voltage drop and copper loss are kept to reasonable values. The initial design procedure is based on the following requirements:

1. The current-carrying capacity of conductors supplying several motors must be at least equal to 125 percent of the full-load current of the highest rated motor plus the sum of the full-load current of the other motors supplied by the feeder (**NEC** Section 430-24).

2. The current-carrying capacity of feeder conductors supplying a single motor plus other loads must be at least equal to 125 percent of the full-load current of the motor plus other load current.

3. The current-carrying capacity of feeder conductors supplying a motor load and a lighting and/or appliance load must be sufficient to handle the lighting and/or appliance load as determined from the procedure for sizing lighting feeders, plus the motor load as determined from items 1 and 2 above.

The **NEC** permits inspectors to authorize the use of demand factors for motor feeders—based on reduced heating of conductors that supply motors operating intermittently, motors on duty cycles, or motors not operating together. Where necessary, the authority enforcing the **NEC** should be consulted to ensure that the conditions and operating characteristics are deemed suitable for reduced-capacity feeders.

Sizing Conductors

The **NEC** allows the sizing of motor feeders (and feeders supplying combination power and lighting loads) on the basis of maximum demand running current, calculated as follows:

$$\text{Running current} = (1.25 \times I_f) + (DF \times I_t)$$

where I_f = full-load current of largest motor, A
DF = demand factor
I_t = sum of full-load currents of all motors except largest, A

But modern design must include provisions for limiting the voltage drop under conditions of high motor-current inrush and therefore dictates the use of the maximum demand starting current in sizing conductors for improved voltage stability on the feeder. Thus current is calculated as follows:

$$\text{Starting current} = I_s + \text{DF} \times I_t$$

where I_s is the average starting current of the largest motor. (Use the percentage of motor full-load current given for fuses in Table 430-152.)

Voltage Drop and Power Factor

Voltage drop and I^2R loss must be carefully considered in sizing motor feeders. The design voltage-drop percentage may vary with the particular operating conditions and layout of motor loads; however, it should never exceed a 3 percent drop from the service entrance to the point of origin of motor branch circuits. And a maximum voltage drop of only 2 percent from the service to the motor branch-circuit protection is widely used for motor feeder design.

Both the reactance and the resistance of the feeder conductors must be included in voltage-drop calculations, as both contribute to the drop. Power factor must also be accounted for in these calculations. And voltage drop in a feeder must be analyzed in terms of the number of motors supplied and the size and operating duty of each motor.

When a number of motors may be starting simultaneously, or when several motors driving sluggish loads may be started at or near the same time, the voltage drop in the feeder could be large, unless it is sized to account for high load currents. Of course, such conditions, when analyzed, will often clearly indicate the need for further subdivision or adjustment of feeder loads, as well as the best types of motors and/or controllers to select for use. The initial value of the starting current—the locked-rotor current—must be used in studying the effect of motor loads on voltage drop.

The I^2R loss in a motor feeder—the power (watts) lost in the conductors due to heat developed by current flow through the conductors—is equal to the square of the total current drawn through the conductors times the total resistance of the conductors. This loss may frequently be substantial, even when the voltage drop in the feeder is within recommended limits. All voltage-drop studies should include the consideration of an increase in conductor size beyond that necessary to limit the voltage drop, for the purpose of limiting I^2R losses and the cost of the power they waste.

The power factor of a motor feeder is improved by placing a capacitor of the proper kilovar rating at the supply end of the feeder; this reduces both the required current flow to a given load and the voltage drop, by eliminating the reactive current from the feeder conductors.

The power factor of a circuit supplying several motors is determined as follows:

1. For each motor, multiply its horsepower by its power factor at 75 percent of rated load.
2. Add these products for all the motors.
3. Divide the sum obtained in step 2 by the total horsepower connected to the circuit, to obtain the approximate (but accurate enough for most calculations) power factor of the circuit.

MOTOR FEEDERS 395

The circuit can then be corrected to raise the power factor to a desired level by selecting a capacitor of the proper kilovar rating for the given load as directed by the literature of capacitor manufacturers.

Figure 5.1 is a handy table commonly used for computing the capacitor kilovars (reactive kilovoltamperes) required to raise the power factor from one value to another for a given kilowatt load.

Figure 5.2 presents data on the application of power-factor capacitors on plug-in busway supplying motor loads. The busways were used in a modern industrial plant to supply plug-in circuits to motor loads, and the data were used as follows: The total horsepower of the connected motor load was determined for each busway run. A demand factor was applied, based on past experience and metering, reducing the demand on the busway run. The horsepower ratings were then corrected for the efficiency of the motors, which was assumed to be 95 percent. This corrected horsepower was converted to kilowatts, and a corresponding kilovoltampere value was established for an 80 percent power factor, for each busway. Reactive kilovoltampere (kilovar) values were also determined. Then kilovoltampere and kilovar values were calculated for the given kilowatt values, based on an assumed power factor of

Original Power Factor in Percentage	Desired Power Factor in Percentage																				
	80%	81	82	83	84	85	86	87	88	89	90	91	92	93	94	95	96	97	98	99	100
50%	.982	1.008	1.034	1.060	1.086	1.112	1.139	1.165	1.192	1.220	1.248	1.276	1.303	1.337	1.369	1.403	1.441	1.461	1.529	1.590	1.732
51	.936	.962	.988	1.014	1.040	1.066	1.093	1.119	1.146	1.174	1.202	1.230	1.257	1.291	1.323	1.357	1.395	1.435	1.483	1.544	1.686
52	.894	.920	.946	.972	.998	1.024	1.051	1.077	1.104	1.132	1.160	1.188	1.215	1.249	1.281	1.315	1.353	1.393	1.441	1.502	1.644
53	.850	.876	.902	.928	.954	.980	1.007	1.033	1.060	1.088	1.116	1.144	1.171	1.205	1.237	1.271	1.309	1.349	1.397	1.458	1.600
54	.809	.835	.861	.887	.913	.939	.966	.992	1.019	1.047	1.075	1.103	1.130	1.164	1.196	1.230	1.268	1.308	1.356	1.417	1.559
55	.769	.795	.821	.847	.873	.899	.926	.952	.979	1.007	1.035	1.063	1.090	1.124	1.156	1.190	1.228	1.268	1.316	1.377	1.519
56	.730	.756	.782	.808	.834	.860	.887	.913	.940	.968	.996	1.024	1.051	1.085	1.117	1.151	1.189	1.229	1.277	1.338	1.480
57	.692	.718	.744	.770	.796	.822	.849	.875	.902	.930	.958	.986	1.013	1.047	1.079	1.113	1.151	1.191	1.239	1.300	1.442
58	.655	.681	.707	.733	.759	.785	.812	.838	.865	.893	.921	.949	.976	1.010	1.042	1.076	1.114	1.154	1.202	1.263	1.405
59	.618	.644	.670	.696	.722	.748	.775	.801	.828	.856	.884	.912	.939	.973	1.005	1.039	1.077	1.117	1.165	1.226	1.368
60	.584	.610	.636	.662	.688	.714	.741	.767	.794	.822	.849	.878	.905	.939	.971	1.005	1.043	1.083	1.131	1.192	1.334
61	.549	.575	.601	.627	.653	.679	.706	.732	.759	.787	.815	.843	.870	.904	.936	.970	1.008	1.048	1.096	1.157	1.299
61	.515	.541	.567	.593	.619	.645	.672	.698	.725	.753	.781	.809	.836	.870	.902	.936	.974	1.014	1.062	1.123	1.265
63	.483	.509	.535	.561	.587	.613	.640	.666	.693	.721	.749	.777	.804	.838	.870	.904	.942	.982	1.030	1.091	1.233
64	.450	.476	.502	.528	.554	.580	.607	.633	.660	.688	.716	.744	.771	.805	.837	.871	.909	.949	.997	1.058	1.200
65	.419	.445	.471	.497	.523	.549	.576	.602	.629	.657	.685	.713	.740	.774	.806	.840	.878	.918	.966	1.027	1.169
66	.388	.414	.440	.466	.492	.518	.545	.571	.598	.626	.654	.682	.709	.743	.775	.809	.847	.887	.935	.996	1.138
67	.358	.384	.410	.436	.462	.488	.515	.541	.568	.596	.624	.652	.679	.713	.745	.779	.817	.857	.905	.966	1.108
68	.329	.355	.381	.407	.433	.459	.486	.512	.539	.567	.595	.623	.650	.684	.716	.750	.788	.828	.876	.937	1.079
69	.299	.325	.351	.377	.403	.429	.456	.482	.509	.537	.565	.593	.620	.654	.686	.720	.758	.798	.840	.907	1.049
70	.270	.296	.322	.348	.374	.400	.427	.453	.480	.508	.536	.564	.591	.625	.657	.691	.729	.769	.811	.878	1.020
71	.242	.268	.294	.320	.346	.372	.399	.425	.452	.480	.508	.536	.563	.597	.629	.663	.701	.741	.783	.850	.992
72	.213	.239	.265	.291	.317	.343	.370	.396	.423	.451	.479	.507	.534	.568	.600	.634	.672	.712	.754	.821	.963
73	.186	.212	.238	.264	.290	.316	.343	.369	.396	.424	.452	.480	.507	.541	.573	.607	.645	.685	.727	.794	.936
74	.159	.185	.211	.237	.263	.289	.316	.342	.369	.397	.425	.453	.480	.514	.546	.580	.618	.658	.700	.767	.909
75	.132	.158	.184	.210	.236	.262	.289	.315	.342	.370	.398	.426	.453	.487	.519	.553	.591	.631	.673	.740	.882
76	.105	.131	.157	.183	.209	.235	.262	.288	.315	.343	.371	.399	.426	.460	.492	.526	.564	.604	.652	.713	.855
77	.079	.105	.131	.157	.183	.209	.236	.262	.289	.317	.345	.373	.400	.434	.466	.500	.538	.578	.620	.687	.829
78	.053	.079	.105	.131	.157	.183	.210	.236	.263	.291	.319	.347	.374	.408	.440	.474	.512	.552	.594	.661	.803
79	.026	.052	.078	.104	.130	.156	.183	.209	.236	.264	.292	.320	.347	.381	.413	.447	.485	.525	.567	.634	.776
80	.000	.026	.052	.078	.104	.130	.157	.183	.210	.238	.266	.294	.321	.355	.387	.421	.459	.499	.541	.608	.750
81	—	.000	.026	.052	.078	.104	.131	.157	.184	.212	.240	.268	.295	.329	.361	.395	.433	.473	.515	.582	.724
82	—	—	.000	.026	.052	.078	.105	.131	.158	.186	.214	.242	.269	.303	.335	.369	.407	.447	.489	.556	.698
83	—	—	—	.000	.026	.052	.079	.105	.132	.160	.188	.216	.243	.277	.309	.343	.381	.421	.463	.530	.672
84	—	—	—	—	.000	.026	.053	.079	.106	.134	.162	.190	.217	.251	.283	.317	.355	.395	.437	.504	.645
85	—	—	—	—	—	.000	.027	.053	.080	.108	.136	.164	.191	.225	.257	.291	.329	.369	.417	.478	.620

Example: Total kw input of load from wattmeter reading 100 kw at a power factor of 60%. The leading reactive kva necessary to raise the power factor to 90% is found by multiplying the 100 kw by the factor found in the table, which is .849. Then 100 kw × 0.849 = 84.9 kva. Use 85 kva.

FIGURE 5.1 Capacitor kilovars required to improve a load to a desired power factor.

95 percent. The difference in kilovars was then noted for each busway, to establish the kilovars required to bring the power factor of each busway up to about 95 percent. The last column in Fig. 5.2 shows the amounts of capacitor equipment that were required, based on the use of 15-kvar unit capacitors rated at 575 V for the 12 sections of 600-A, 480-V, 3-phase, 4-wire busway. The capacitor equipment was mounted on the busway sections.

Protecting Motor Feeders

Overcurrent protection for a feeder to several motors must have a rating or setting not greater than the largest rating or setting of the branch-circuit protective device for any motor of the group plus the sum of the full-load currents of the other motors supplied by the feeder. It is possible for motors of different horsepower ratings to have the same rating as the branch-circuit protective device, depending upon the type of motor and the type of protective device. If two or more motors in the group are of different horsepower ratings but the rating or setting of the branch-circuit protective device is the same for both motors, then one of the protective devices should be considered as the largest for the calculation of feeder overcurrent protection.

Because **NEC** Table 430-152 recognizes many different ratings of branch-circuit protective devices (based on the use of fuses or circuit breakers and depending upon the particular type of motor), it is possible for two motors of equal horsepower rating to have vastly different ratings of branch-circuit protection. For instance, for a 25-hp motor protected by non-time-delay fuses, Table 430-152 gives 300 percent of the full-load motor current as the maximum rating or setting of the branch-circuit device. Thus, 250-A fuses would be used for a motor with a 78-A full-load rating. But for a motor of the same horsepower (and even of the same type), time-delay fuses must be rated at only 175 percent of 78 A, which would mean using 150-A fuses for branch-circuit protection (Fig. 5.3). If the two 25-hp motors were of different types—one being a wound-rotor motor and the other a squirrel-cage induction motor—it would still be necessary to base the selection of the feeder protection on the largest rating or setting of a branch-circuit protective device, regardless of the horsepower rating of the motor.

In large-capacity installations where extra feeder capacity is provided for load growth or future changes, the feeder overcurrent protection may be calculated on the basis of the rated current-carrying capacity of the feeder conductors. In some cases, such as where two or more motors on a feeder may be started simultaneously, feeder conductors may have to be larger than usually required for feeders to several motors.

The **NEC** calculation for selecting the size of a feeder overcurrent protective device is concerned with establishing the maximum setting or rating of the CB or fuse. If a lower value of protection is suitable, it may be used. Conversely, the **NEC** rules for feeder conductor sizing are aimed at establishing the minimum acceptable size, and larger conductors may be used.

Example. In Fig. 5.4, 250-A fuses are used to protect the feeder conductors to a group of four motors; the fuses were selected as follows: The four motors supplied by the 3-phase, 440-V, 60-Hz feeder are not marked with a code letter (**NEC** Table 430-152) and are:

- One 50-hp squirrel-cage induction motor (full-voltage starting)
- One 30-hp wound-rotor induction motor
- Two 10-hp squirrel-cage induction motors (full-voltage starting)

MOTOR FEEDERS

Bus duct	Total hp	40% Demand	95% Eff.	Kw	Kva 80%	Kvar 80%	Kva 95%	Kvar 95%	Kvar diff.	Kvar use
1 A	364	146	153	109	136	82	115	35	47	45
1 B	374	150	157	117	142	85	124	38.4	46.6	45
1 C	461	185	194	145	181	109	153	47	62	60
2 A	735	(30%) 220	230	172	215	130	181	56	74	75
2 B	1300	(30%) 390	410	306	382	230	322	100	130	135
3 A	729	291	307	229	286	172	241	75	97	105
3 B	275	110	116	87	109	65	92	29	36	45

FORMULAS: Power Factor = $\frac{KW}{KVA}$ KW output of motor = HP x .746 $\cos \phi_1 = .8$ $\cos \phi_2 = .95$

$KVA = \frac{1.73 \times E \times I}{1000}$ KW input to motor = $\frac{HP \times .746}{Efficiency}$ $\sin \phi_1 = .6$ $\sin \phi_2 = .312$

KVA input to induction motor = $\frac{HP \times .746}{P.F. \times Eff.}$ $KVA = \frac{KW}{\cos \phi}$ KVAR = KVA x sin ø

Power Factor = cos ø

> Power factor can be increased with capacitors or with synchronous motors. The following shows how a 75 kVA synchronous motor load (which draws current that "leads" the voltage instead of current that "lags" the voltage, as with an induction motor) was used to provide 45 kvar (reactive kilovolt amperes) to offset part of the 135 kvar of the induction motor load. The kW components add up to 290 kW, and the lagging kvar is reduced to 90 kvar.

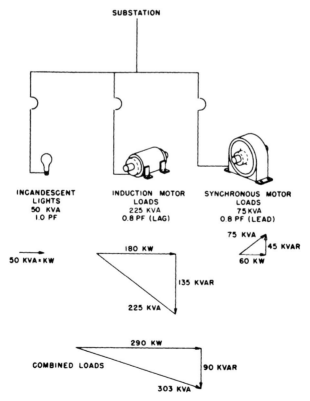

FIGURE 5.2 In an industrial system, these design data were used to determine capacitor power-factor correction for plug-in busway to motor loads.

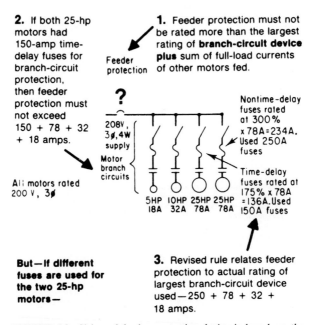

FIGURE 5.3 Sizing of feeder protective device is based on the largest branch-circuit protective device, regardless of the actual motor horsepower ratings.

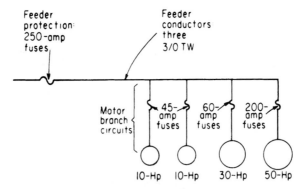

FIGURE 5.4 Rating of feeder protection and size of feeder conductors are based on motor full-load currents.

MOTOR FEEDERS 399

Step 1: Branch-Circuit Loads. From **NEC** Table 430-150, the motors have full-load current ratings as follows:

50-hp motor: 65 A
30-hp motor: 40 A
10-hp motor: 14 A

Step 2: Conductors. The feeder conductors must have a carrying capacity that is computed as follows:

$$1.25 \times 65 \text{ A} = 81 \text{ A}$$

$$81 \text{ A} + 40 \text{ A} + (2 \times 14 \text{ A}) = 149 \text{ A}$$

The feeder conductors must be at least No. 3/0 TW, 1/0 THW, or 1/0 RHH, XHHW, or THHN (copper) wires.

Step 3: Branch-Circuit Protection. The required overcurrent (branch-circuit) protection (from **NEC** Table 430-152 and Section 430-52) using non-time-delay fuses is found as follows:

1. The 50-hp motor must be protected at not more than 200 A (since 300 percent of 65 A is 3×65 A = 195 A, and the next larger standard size of fuse is 200 A).
2. The 30-hp motor must be protected at not more than 60 A (150 percent of 40 A).
3. Each 10-hp motor must be protected at not more than 45 A (300 percent of 14 A = 42 A).

Step 4: Feeder Protection. Based on step 3, the maximum rating or setting for the overcurrent device protecting such a feeder must not be greater than the largest rating or setting of branch-circuit protective device for one of the motors of the group plus the sum of the full-load currents of the other motors. From the above, then, the maximum allowable size of feeder fuses is 200 + 40 + 14 + 14 = 268 A.

This calls for a maximum rating of 250 A for the motor feeder fuses, since 250 A is the nearest standard fuse rating that does not exceed the maximum allowable value of 268 A.

NOTE: *There is no provision* in **NEC** Section 430-62 for the use of the next higher size, rating, or setting of protective device for a motor feeder when the calculated maximum rating does not correspond to a standard size of device.

Example. Figure 5.5 shows a feeder that supplies five motors as follows:

- One 40-hp wound-rotor induction motor
- Two 25-hp squirrel-cage induction motors (full-voltage start)
- One 10-hp squirrel-cage induction motor (full-voltage start)
- One 5-hp squirrel-cage induction motor (full-voltage start)

The design procedure involves these steps:
Step 1. The **NEC** full-load running currents of these motors are as follows:

40-hp motor: 119.6 A (use 120 A)
25-hp motors: 78.2 A (use 78 A)
10-hp motor: 32.2 A (use 32 A)
5-hp motor: 17.5 A (use 18 A)

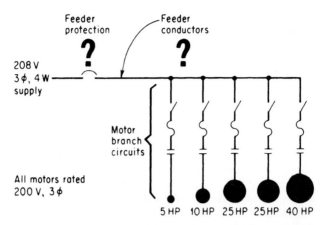

FIGURE 5.5 Rating of a circuit breaker for motor feeder protection is determined by the largest rating of branch-circuit fuses in a case like this.

These current values are derived from **NEC** Table 430-150, by using the multiplier of 1.15 to increase the values given in the table for 230-V motors to the correct current ratings for 200-V motors (as required in the footnote to the table).

Step 2. The minimum permissible ampacity of the feeder conductors supplying this group of continuously operating motors is the sum of the full-load running currents of all the motors plus 25 percent of the full-load current of the motor with the highest rated running current. (However, if some of the motors were operating on short-time, intermittent, periodic, or varying duty instead of all operating continuously, the sizing of the feeders would be modified as explained in **NEC** Section 430-24.)

Step 3. Based on the full-load running currents of these continuous-duty motors, the feeder conductors here must have an ampacity of not less than 356 A, that is, 120 A + 78 A + 78 A + 32 A + 18 A + (0.25 × 120 A).

Step 4. If THW aluminum conductors are used in a conduit for the feeder, the smallest conductor size that would properly satisfy the need here is 700 kcmil (375 A).

Step 5. The maximum permissible rating of a standard inverse-time CB used for the overcurrent protection of this feeder is 225 percent of the full-load current rating of the 40-hp motor plus the sum of the full-load currents of the other four motors of the group, when time-delay fuses are used for branch-circuit protection of the motors and a rating higher than 150 percent is needed to allow for inrush current. (See Exception No. 2*b* of **NEC** Section 430-52.)

Step 6. Assuming that the fuses protecting the branch circuits to the motors are of the dual-element (time-delay) type, and the largest motor is fully capable of starting with its branch-circuit fuses sized according to Table 430-152, the maximum acceptable setting of the feeder CB is (1.50 × 120 A) + 78 A + 78 A + 32 A + 18 A, or 368 A.

Step 7. The maximum setting or rating of a standard, time-delay-type, molded-case CB that may be used for the protection of this feeder is 350 A, since Section 430-62 does not recognize the use of the next larger size of CB.

Power and Lighting Feeders

The conductors and protection for a feeder to both a motor load and a lighting and/or appliance load must be sized on the basis of both loads. The rating or setting of the overcurrent device must be sufficient to carry the lighting and/or appliance load plus the rating or setting of the motor branch-circuit protective device if only one motor is supplied *or* plus the highest rating or setting of branch-circuit protective device for any one motor plus the sum of the full-load currents of the other motors if more than one motor is supplied.

Example. Figure 5.6 shows a feeder arrangement for a combination power and lighting load. As shown, the feeder supplies a 277-V lighting panel and four 3-phase motors. Such considerations as voltage drop, I^2R loss, spare capacity, and lamp dimming on motor starting would have to be accounted for in determining the actual wire size and overcurrent protection to use for the job. The circuiting shown would be safe but perhaps not as efficient or effective as it could be.

The feeder conductors that will carry both the lighting-load current and the motor-load current must have sufficient capacity for the total load. The basic design steps are as follows:

Step 1. The total load is the sum of the motor load and the lighting load:

Motor load = 65 A + 40 A + 14 A + 14 A + (0.25 × 65 A) = 149 A per phase

Lighting load = 120 A × 1.25 = 150 A per phase

Total load = 149 A + 150 A = 299 A per phase

Note that, because the lighting load of 120 A per phase is a "continuous" load, **NEC** Section 220-10(b) requires that feeder conductors carrying that load must have ampacity at least equal to 125 percent of the load current. The multiplier 1.25 used above satisfies the requirement. But, only the phase conductors of the feeder are subject to that 125 percent factor; the neutral conductor of the feeder requires only an ampacity of 120 A. The 125 percent continuous-load factor is not applied to the feeder neutral because the neutral does not connect to the terminal of a switch, CB, or other device for which heating would be a problem under continuous load.

Step 2. **NEC** Table 310-16 shows that a combined power and lighting load of 299 A can be served by the following copper conductors:

500-kcmil TW (320 A)

350-kcmil THW (310 A)

Table 310-16 shows that this load can be served by the following aluminum or copper-clad aluminum conductors:

700-kcmil TW (310 A)

500-kcmil THW, RHH, or THHN (310 A)

Step 3. The protective device for a feeder supplying a combined motor and lighting load may not have a rating greater than the sum of the maximum rating of the motor feeder protective device and the lighting load; these are obtained as follows:

1. The maximum rating of the motor feeder protective device is the rating or setting of the largest branch-circuit device of any motor of the group being served plus the sum of the full-load currents of the other motors; this is

200 A (50-hp motor) + 40 A + 14 A + 14 A = 268 A

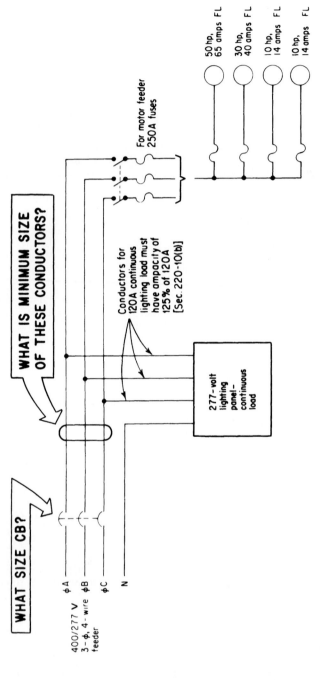

FIGURE 5.6 Conductors and protection for a power and lighting feeder must be sized to satisfy both loads.

MOTOR FEEDERS 403

This calls for a maximum standard rating of 250 A for the motor feeder fuses (the nearest standard fuse rating that does not exceed the maximum permitted value of 268 A).

2. For the lighting load, 120 A × 1.25 = 150 A.

A CB for the combined load must be rated at 268 A + 150 A = 418 A maximum. This calls for a 400-A CB (the nearest standard rating that does not exceed the 418-A maximum).

<u>AGAIN</u>: The **NEC** *does not* have a provision which permits the use of the next higher size, rating, or setting of protective device for a motor feeder when the calculated maximum rating does not correspond to a standard size of device.

Example. Figure 5.7 shows a layout in which a main feeder is run to a point where it is tapped (such as at an auxiliary gutter) by two subfeeders—one for a lighting load and one for a motor load. The main switch for the combined feeder might be in a switchboard or panelboard, or it could be a main service switch (disconnect and protection at a service entrance). The conductors for the combined feeder and the fuses in the 200-A combined feeder switch are sized on the basis of the following loads: The motor load is four 3-phase, 230-V, squirrel-cage induction motors, designed for 40°C temperature rise, marked with code letter H, started across the line, and rated as follows:

- One 10-hp motor
- One 7½-hp motor
- Two 1½-hp motors

The lighting load is a 20-kw, single-phase, 115-V load.

Step 1. The average full-load motor currents (from **NEC** Table 430-150) are:

10-hp motor: 28 A
7½-hp motor: 22 A
1½-hp motors: 5.2 A each

Step 2. The conductors for the motor-load feeder must be selected to have an ampacity based on the following:

10 hp: 1.25 × 28 A	35 A
7½ hp	22 A
1½ hp: 2 × 5.2 A	10.4 A
Total	67.4 A

The required ampacity of 67.4 A could be provided by one of the following:

Three No. 4 TW in 1-in conduit
Three No. 4 THW in 1-in conduit
Three No. 4 RHH (without covering) in 1-in conduit
Three No. 4 THHN in 1-in conduit

Step 3. The motor feeder protection (Section 430-62), using circuit breakers for the branch circuits but fuses in the feeder, is sized as follows:

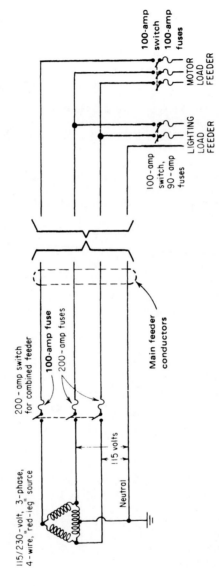

FIGURE 5.7 Sizing of conductors and protection must be based on actual phase loadings.

MOTOR FEEDERS 405

10 hp: 28 A × 2.5   70 A (rating of branch CB)
7½ hp   22 A
1½ hp: 2 × 5.2 A   10.4 A
Total   102.4 A

The maximum size of the feeder fuses is 102.4 A. Therefore, a 100-A switch with 100-A non-time-delay fuses or smaller time-delay fuses would be used.

NOTE: **NEC** Section 430-62 does not permit the motor-feeder fuses to be upsized to the next standard rating above 102.4 A, or 110 A. The fuses must be selected at 100 A, the next *lower* rating.

Step 4. For the lighting load, the full-load current would be

$$\frac{20 \times 1000 \text{ W}}{230 \text{ V}} = 87 \text{ A}$$

A 100-A switch with two 90-A fuses would be used for a lighting load that is not continuous—that is, one where the full load is not on for 3 hours or more. [If the 87-A load were continuous, a fused switch would be selected so that load was not over 80 percent of the fuse rating. This means that the minimum fuse size would be 87 A × 1.25 = 108.75 A, so a 110-A fuse would be used for each phase, in a 200-A switch (per NEMA standards).]

Thus, three No. 2 TW, THW, RHH, or THHN in 1¼-in conduit, or in 1-in conduit (for THHN), would be used. (For a continuous 87-A load, where the fuse size would be increased to 110 A, the feeder conductors would have to be rated for at least 110 A as are No. 1 TW and No. 2 THW, RHH, or THHN.)

Step 5. Steps 1 to 4 cover the sizing of the individual feeders—one for the motor load and one for the lighting load. A single main feeder which supplies the combined motor and lighting load will have the 3-phase motor load on each of its three phase conductors and the lighting load on only two of its phase conductors and a neutral. A 3-phase, 4-wire, "red-leg" circuit would be designed as follows:

Step 6. The two phase conductors which supply both the motor load and a non-continuous lighting load must have a minimum capacity of

Motor load   67.4 A
Lighting load   87 A
Total   154.4 A

Each of these two conductors must be at least No. 3/0 TW (165 A) or No. 2/0 THW (175 A) or RHH or THHN (185 A). [If the lighting load were continuous, a value of 108.75 A (125 percent of 87 A) would be used instead of 87 A in the previous calculation.]

Step 7. The third phase conductor of the main combined feeder serves only the motor load and must have a rating of at least 67.4 A, which calls for No. 4 TW, THW, RHH, or THHN as noted in step 2.

Step 8. Under the assumption that the 20-kW single-phase lighting load is all 115-V loads, equally balanced from hot legs to neutral, the neutral must be rated for the maximum unbalance; this is 87 A, the same as the load on each hot leg. This neutral conductor would then be a No. 2, TW, THW, RHH, or THHN, as shown in step 4.

(If the lighting load were continuous, the neutral could still be sized at 87 A, although the phase legs would require an ampacity at least equal to 1.25 × 87 A in such a case.)

Step 9. Based on the loads on the three hot legs—154, 154, and 67.4 A—a three-pole, 200-A switch is needed for this combined feeder.

Step 10. Overcurrent protection for the two hot legs that carry the motor load and the lighting load is sized according to **NEC** Section 430-63:

Motor load	102.4 A setting
Lighting load	87 A
Combined load	189.4 A

Thus, the two fuses in the 200-A switch protecting the combined load must have a rating not greater than 200 A. (**NEC** Section 430-63 may be interpreted as setting a maximum fuse rating of 190 A—that is, 100 A for the rating of the motor-feeder fuses plus 90 A for the lighting-feeder fuses—thereby requiring the use of the next smaller size fuse, or 175 A.)

Step 11. The fuse which protects the third hot leg—the one serving only the motor load—must be rated not over 100 A.

NOTE: This sizing calculation would not change at all if the feeder constituted the service-entrance conductors to a building with the given load, rather than the feeder to the combined load within the building.

Motor Feeder Circuit Protection Clarifications

The rules governing maximum ratings for motor feeder short-circuit and ground-fault protection are given in Part **E** of Article 430 (Section 430-61 through 430-63). There are two significant details related to sizing feeder protection that have recently been clarified in Section 430-62(a), which covers the rating of feeder protective devices where the load consists of motors only.

The first significant clarification is the additional wording provided in the parentheses of the basic rule in Section 430-62(a). The new phraseology was added to clarify the CMP's intent regarding the maximum rating permitted for feeder protection where the highest-rated branch-circuit protective device is set or rated at some value *less* than that recognized by Table 430-152. Such applications have been the source of controversy for a number of code cycles because the literal wording was ambiguous.

As most are aware, the rule of Section 430-62(a) requires that the maximum rating or setting of the feeder protection be no greater than the sum of the highest-rated branch-circuit protective device plus the full-load current value of all other motors supplied by the feeder. However, in the 1990 and previous editions of the **NEC,** while the first paragraph of Section 430-62 indicated that the ampere value of the highest-rated branch-circuit device was to be based on the maximum given in Table 430-152, the second paragraph referred to the actual value "used." As a result, some believed that it was permissible to use the maximum ampere value permitted by the table and others said the actual rating or setting of the installed device must be used to establish the maximum rating or setting for the feeder protective device.

The problem presented by that lack of clarity is illuminated by the question presented in the Substantiation submitted for Comment No. 11-30 in the 1992 TCD. The relevant portion of that substantiation reads as follows:

Three motors with FLA (full-load amperage) of 40amps are installed each on a separate (branch) circuit with an 80amp breaker. The installer decided not to protect at the maximum 100A permitted (250% FLA for CBs) by 430-52 (and Table 430-152).

The question is: What is the maximum feeder breaker rating permitted? Is it: 80A + 40A + 40A = 160 or 150A breaker or 100A + 40A + 40A = 180 or a 175A breaker? (Fig. 5.8.)

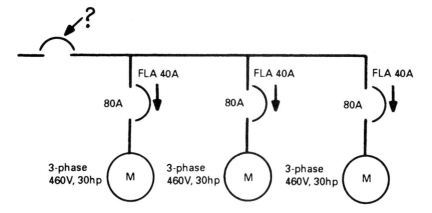

Question: Is the maximum permitted rating for the feeder: (80A + 40A + 40A), or 160A, which would call for a 150A CB?

OR

is it (100A + 40A + 40A), or 180A, which would call for a 175A CB?

FIGURE 5.8 Example submitted to prove clarification of calculation.

As can be seen, if the actual installed rating is used, a 150-A breaker would be the maximum permitted. And, if the table-permitted maximum value is used, then a 175-A breaker would be acceptable.

Although the second paragraph of Section 430-62(a) still refers to the "rating or setting ... used," inclusion in the first paragraph of the phrase "the maximum permitted value for the specific type of protective device shown" is intended to convey that use of the table value is permitted. That is, it is the CMP's intent to allow use of the 100-A rating (250% times 40 A) permitted by Table 430-152 to determine the maximum rating or setting of the feeder protection, even though an 80-A rated CB was actually installed. And, although use of the installed 80-A rating to establish the maximum rating of feeder protective device would not be prohibited because it results in the selection of a lower value, it is *not* required.

Consider a feeder supplying two 50-hp, 3-phase, 230-V motors. Assume that the "next lower standard rating" is adequate to carry the starting current and the branch-circuit fuses are rated as shown in Figure 5.9. The maximum permitted rating of protective device for a feeder supplying two such motors would be:

Max. feeder protection rating = 390 A   (maximum permitted
    by Table 430-152) + 130A

Max. feeder protection rating = 520 A

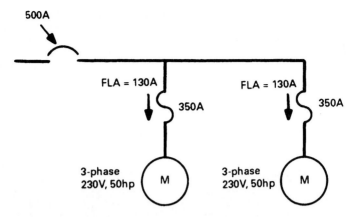

New wording of the basic rule in Sec. 430-62(a) permits use of the maximum value indicated by Table 430-152. In this case, the feeder protective device rating would be:

Max rating = 390A (130 X 3 from Table 430-152) + 130A
           = 520A, or 500A, which is the next lower standard rating.

FIGURE 5.9 Example of **NEC** permitted rating for overtime fuse as feeder protection where one-time fuses are used for branch-circuit protection.

It is worth noting that selection of the next *lower* standard-rated device has long been required for motor feeder protection. Therefore, because it is not possible to get 520-A fuses, 500-A-rated devices would be selected.

The second major change to the procedure for establishing the maximum rating for motor feeder protection is the new exception following Section 430-62(a). The new requirement addresses an inherent inadequacy of the basic rule that results in especially excessive ratings for feeder protection where instantaneous-trip CBs or MSCPs are used on any branch circuit(s) supplied by the feeder.

As we have seen, the rating of a motor branch-circuit protective device is based on the *type* of device to be used. That is, if one-time fuses are used, they must generally be rated at no more than 300 percent of motor full-load current; if an inverse-time CB is used, then the multiplier is 250 percent; for time-delay fuses, the maximum is 175 percent, etc. However, the basic rule for determining the maximum rating or setting for motor feeder short-circuit and ground-fault protection is based on the rating or setting of the largest branch-circuit protective device—without regard for the type of device on either the feeder or the branch. And that can result in undesirably high or low ratings where certain combinations of devices are used.

For example, consider an application using time-delay fuses to protect a feeder supplying two motor branch-circuits protected by one-time fuses. For the ease of explanation, assume the full-load current for one motor is 100 A and the other is 5 A. In accordance with Section 430-52(a) and Table 430-152, the one-time fuses on the branch may be rated at 300 percent of motor full-load current. Therefore, in this example, the branch fuses protecting the 100-A motor should be rated at not more than 300 A (100 A × 3). And, the one-time fuses for the 5-A motor should be rated at 15 A (5 A × 3).

In accordance with Section 430-62(a), the time-delay fuses would also be permitted to be rated at 300 A because the sum of the largest device rating plus the full-load current value of all other motors equals 305 A (300 A + 5 A), which necessitates the use of 300-A fuses—the next lower standard rating. Therefore, the one-time fuses on the branch *and* the time-delay fuses on the feeder would be rated at 300 A.

While the established maximum ratings satisfy the literal wording, the feeder protection is probably rated too high. As we know, time-delay fuses will carry a higher value of current for a longer period of time than one-time fuses of the same ampere rating. Therefore, use of a lower ampere rating for the time-delay fuses on the feeder will assure that the feeder conductors are more closely protected and is probably warranted. The need for lower ratings where time-delay fuses are used is effectively acknowledged by the **NEC,** which logically establishes a lower maximum rating for time-delay fuses (175 percent of motor full-load current) than for one-time fuses (300 percent) where used to provide motor branch-circuit protection.

The inherent deficiency of this **NEC** procedure can be more fully appreciated when one considers reversing the position of the different fuse types in the above example. That is, if time-delay fuses were used on the *branch* circuit and one-time fuses were installed on the *feeder,* the maximum rating for the branch and feeder devices would be as follows:

Per Section 430-52(a) and Table 430-152, the maximum rating for the largest-motor branch-circuit time-delay fuses would be not more than 175A (100 A × 1.75). And, the sum of the highest rating plus the full-load currents of all other motors would also call for a fuse rated at 175 A, the next lower standard rating below 180 A (175 A + 5 A). Therefore, as ridiculous as it sounds, the maximum rating for both the largest-motor branch circuit and the feeder protective device in this case is 175 A, even though we are protecting the same motors and motor circuits as before! However, in this case, the 175-A-rated one-time fuses on the feeder will probably not carry the starting current of the motor and open the circuit on startup. *But,* use of higher-rated one-time fuses on the feeder would be a violation of Section 430-62(a) (Fig. 5.10).

Although the basic rule does *not* generally require any consideration of the types of devices selected when determining maximum ratings for motor feeder protection, the Exception to Section 430-62(a) *does* where an instantaneous-trip breaker or MSCP is used on any branch circuit supplied by the feeder to be protected. That is, where one or more instantaneous-trip CBs or MSCPs are used to provide short-circuit and ground-fault protection on the branch circuits to be supplied, then the maximum rating of the feeder protective device must be calculated with respect to the type of device selected for feeder protection.

As has been discussed, under certain conditions the Exception to Section 430-52(a) would permit instantaneous-trip CBs and MSCPs to be set as high as 13 times the motor full-load current. And, if the basic rule for establishing the maximum permitted rating for feeder protection were applied, the calculated maximum value would be based on the 1300 percent value of the instantaneous-trip CB or MSCP. Because that will result in an unacceptably high rating, the new Exception to Section 430-62(a) now requires that the rating of the instantaneous-trip CB(s) or MSCP(s) on the branch circuit(s) be recalculated at the value permitted by Table 430-152 for the type of device that is to be used as feeder protection. Let's take an example. Consider the application shown in Fig. 5.11.

As can be seen, there are three, 3-phase, 230-V, 50-hp motors protected by instantaneous-trip breakers. According to Table 430-150, the full-load current is to be taken as 130 A.

Although such devices are generally limited to a trip-setting of no more than 700 percent of motor full-load current, here the instantaneous-trip breaker on each

What is the maximum permitted rating
for a time-delay fuse on the feeder?

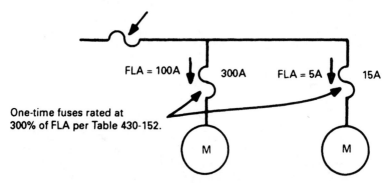

Max rating for feeder protection = 300A + 5A
= 305A, which would result in the selection
of the next lower-rated 300A device...

...*but*, if the device types are reversed, one-time fuses
on the feeder and time-delay fuses on the branches, then...

Max rating feeder protection = 175A + 5A
= 180A, which would result in selection
of next lower-rated 175A rating.

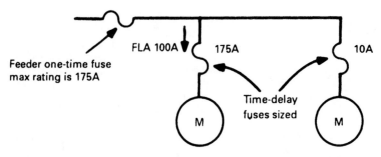

FIGURE 5.10 Comparison of permitted ratings for fuses per Section 430-62(a).

branch is set at 13 times full-load current—or 1690 A—to prevent nuisance-tripping on startup as permitted by the Exception for instantaneous-trip breakers in Section 430-52(a).

The maximum permitted rating for the feeder protective device according to the 1990 and previous editions of the **NEC** would have been the sum of the highest-rated device plus the full-load currents of the other motors, or

MOTOR FEEDERS   411

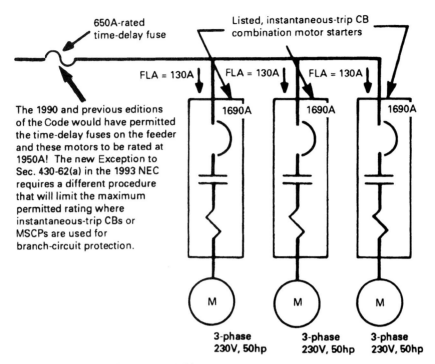

FIGURE 5.11   Rules of Section 430-62(a).

Max. rating of feeder protection = 1690 A + 130 A + 130 A

Max. rating of feeder protection = 1950 A

As shown, the maximum rating would be 1950 A, *regardless* of whether the feeder protective device is a one-time fuse or a time-delay fuse or an inverse-time CB.

The only real difference in the new procedure is that the instantaneous-trip breaker ratings must be correlated with the type of feeder protective device to establish the rating for largest-rated branch-circuit device. Here, one of the 1690-A instantaneous-trip CBs must be rerated based on the percent value that would be permitted by Table 430-152 if the same type of device protecting the feeder were used on that branch. That is, if a one-time fuse is selected for feeder protection, the highest-rated instantaneous-trip CB branch device must be taken as being equal to the maximum value permitted for one-time fuses on a branch circuit. In this case, the highest-rated device must be taken as 390 A (130 A × 3 from Table 430-152 for one-time fuses). The remainder of the calculation remains as before. The full-load currents of the other motors (130 A + 130 A, or 260 A) are added to this adjusted highest rating (390 A), giving a maximum rating of 650 A (260 A + 390 A) for one-time fuses. If time-delay fuses or CBs are used to protect the feeder, the highest-rated device would be taken as equal to 175 or 250 percent, respectively, of the motor full-load current and summed with the full-load currents of the remaining motors.

412 CHAPTER FIVE

What would be the maximum permitted rating of time-delay fuses for a feeder supplying two, 3-phase, 230-V, 50-hp motors where one motor was protected by an instantaneous-trip breaker and the other by a one-time fuse?

As shown in Figure 5.12, the instantaneous-trip breaker is set at 1690 A in accordance with the Exception to Section 430-52(a) and the one-time fuse is rated at 350 A as per Table 430-152 and Section 430-52(a). The adjusted rating of the instantaneous-trip breaker where time-delay fuses are to be used as feeder protection must be taken as being equal to the maximum permitted rating if time-delay fuses were used to protect that branch circuit. In this case, Section 430-52(a) and Table 430-152 would recognize a maximum rating of 227.5 A (130A × 1.75). *But,* because the table permitted a 390-A rating of the one-time fuse to be greater than 227.5 A, the 390-A one-time fuses become the highest-rated device. And, the full-load current of the motor protected by the instantaneous-trip breaker is simply added to the ampere

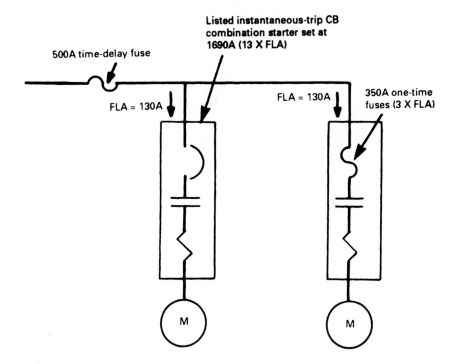

When the rule of Sec. 430-62(a), Ex. is applied, the "adjusted rating" of instantaneous-trip CB becomes 227.5A. Because that "rating" is lower than the 390A maximum permitted rating for the one-time fuse on the other motor, the one-time fuses become the "largest-rated" device and the full-load current of the instantaneous-trip CB protected motor is added to the 390A.

Max rating for feeder time-delay fuses = 390A + 130A
= 520A, or 500A

FIGURE 5.12

rating of the highest-rated branch-circuit device. Therefore, the maximum permitted rating for time-delay fuses on the feeder would be 520 A (390 A + 130 A). And 500-A time-delay fuses—the required next lower rating—would be selected.

What this rule boils down to is this: Instantaneous-trip breakers and MSCPs are never to be taken at their actual settings when establishing the maximum permitted rating of feeder protection. The rating of any instantaneous-trip CB or MSCP must be taken as being equal to the maximum rating permitted on the given branch circuit by Section 430-52 and Table 430-152 for the specific type of protective device to be used on the feeder. And, if the adjusted rating is *not* the highest when compared to the rating of any one-time fuses, time-delay fuses, or CBs used on other branch circuits supplied by the feeder, then the full-load current(s) of the motor(s) protected by an instantaneous-trip breaker or MSCP are simply added to the other full-load currents and summed with the ampere value of the highest-rated fuse or CB.

CHAPTER 6
TRANSFORMER APPLICATIONS (TO 600 V)

The wide availability of economical, efficient, and highly effective dry-type transformers has strongly promoted the use of such transformers to carry electric power, at higher-than-utilization voltages, as close as possible to the center of a layout of utilization equipment—lighting, appliances, and receptacle loads.

Widespread acceptance and application of industrial medium-voltage systems (above 600 V and usually up to 35 kV) have stimulated the use of higher voltages in electrical systems for commercial and industrial buildings. Not too many years ago, 3-wire, single-phase, 120/240-V distribution systems and 3-phase, 4-wire, 120/208-V distribution systems were standards for nonindustrial buildings. Today, although these lower-voltage systems still find wide and efficient application, the 480Y/277-V, 3-phase, 4-wire system is established as the preferred secondary-voltage system for use where substantial motor loads and general area fluorescent or other electric-discharge lighting must be supplied.

Tremendous growth in the size of electrical loads in commercial buildings was responsible for the development of the 480Y system, which has many of the characteristics of industrial-type systems. Air-conditioning loads and businesses and other machine loads increased the ratio of power loads to lighting loads, calling for the type of loadcenter circuit treatments and layouts which are used in industrial plants to economically and efficiently serve heavy motor loads.

Higher-voltage feeders to motor loads and to step-down transformers for lighting and receptacle circuits proved to be the ideal solution. Less copper is needed to distribute the heavy power requirements, and voltage drop and other losses are effectively minimized. The wide availability of extensive lines of lighting equipment which operates on 277-V circuits has also contributed greatly to the success of this system.

Studies have shown that a typical 480Y system can provide savings of more than $40 per kilovoltampere demand over a 120/208-V system handling the same load. Many installations over the past years—in schools, office buildings, industrial plants, shopping centers, and sports arenas—have convincingly proved the economic and operating advantages of the 480/277-V system.

The 480Y system was developed to meet the requirements of commercial building load conditions. Usually, the general lighting is fluorescent and can be served by

277-V circuits. Motors for air-conditioning compressors, circulating fans, elevators, and pumps make up an average load of about 4 VA/ft^2. These motors can be more efficiently and economically supplied at 480 V than at lower voltages, and 480-V motors are less expensive than lower-voltage motors of the same horsepower ratings. The combined power and general lighting loads average between 5 and 15 W/ft^2. Receptacle and miscellaneous loads—desk lamps, local lights, business machines, appliances, water coolers, etc.—average only about 0.5 to 2 W/ft^2.

As can be seen from the above, about 80 percent of a building load may be served directly by 480/277-V feeders. The 120-V circuits may be provided either by using separate 120/208-V substations (fed by medium-voltage feeders of 4160 V, 13.2 kV, etc.) or by using dry-type transformers (480 to 120/240 V or 480-V, 3-phase to 120/208-V) installed locally at the center of each concentration of 120-V loads. The latter method offers greater economy, even when the 120-V loading is as high as 5 W/ft^2.

The design of electrical systems utilizing dry-type transformer applications must include careful consideration of a number of transformer characteristics. These are discussed in the next several paragraphs.

Voltage Rating

The primary and secondary voltage ratings of a transformer are determined very simply from the voltage rating of the supply to the primary winding and the required voltage level on the secondary. Both single- and 3-phase transformers are made with a wide variety of voltage ratings to afford ready application in standard system configurations and even in nonstandard arrangements.

The two basic selection factors are secondary voltage and secondary current. Together they determine the kilovoltampere capacity of the transformer, which is its load-handling ability. The two terms are defined as follows:

- The *rated secondary voltage* of a constant-potential transformer is the voltage at which the transformer secondary is designed to deliver the rated kilovoltampere capacity.

- The *rated secondary current* of a constant-potential transformer is the secondary current obtained by dividing the rated kilovoltampere capacity by the rated secondary voltage.

Transformer selection must first focus on those two basic ratings—depending on the load to be supplied. Then the primary voltage rating of the transformer must be matched to the voltage of the primary feeder circuit to the transformer.

To match actual system conditions, transformers are available that allow some variation from the nominal primary voltage but still produce the required secondary voltages. Figure 6.1 shows some typical transformer primary tap connections for varying input voltages. On general-purpose transformers the primary is commonly provided with a number of taps for input voltages that vary over a range of 15 percent; for instance, two 2½ percent taps above normal voltage and four 2½ percent taps below normal voltage may be provided. On a transformer with a 480-V primary, then, the lowest of these taps would be connected to a primary feeder which operated at 432 V to obtain the rated secondary output of 240/120 V (since 480 V − 4 × 2.5 percent of 480 V = 432 V). If the 432-V primary supply were connected to the 480-V transformer terminals or to any of the other taps, the secondary voltage would be lower than 240/120 V. The taps thus compensate for system voltage drops.

General-purpose transformers have no-load tap changers, requiring that the transformer be deenergized when tap changes are made. Invariably, suitable taps for

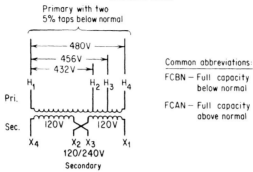

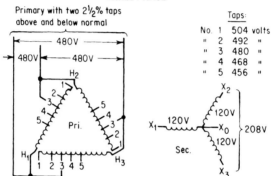

FIGURE 6.1 Transformer primary taps provide for matching the transformer input-voltage rating to the voltage of the supply circuit.

the given primary voltage are set once and left alone, unless some change in load conditions alters the voltage level at the transformer primary. On large transformers serving fluctuating demand loads, automatic tap changers which operate under load can be used on the transformer to maintain a constant voltage level at the transformer secondary terminals or at a remote point on the secondary output feeder.

The selection of transformer taps should be based on the no-load or maximum-voltage conditions at the transformer. Taps above or below the nominal rated voltage should be selected to accommodate anticipated variations in supply voltage and, thereby, provide the required no-load secondary voltage. This requires that voltage drops in secondary circuits be taken into account to arrive at proper voltage levels at utilization devices.

It should be remembered that transformer taps change the voltage transformation by changing the turns ratio of the transformer. Because of this, the kilovoltampere rating of a transformer is not changed when taps are changed.

For a given voltage rating, a transformer winding has a certain current rating, which determines the kilovoltampere rating of the transformer. The current (ampere) rating is a maximum value determined by the impedance of the winding and the capability of the transformer to dissipate heat. The conditions of use must not place a higher current on either the primary or the secondary winding. If the voltage input to a set of primary transformer terminals of given voltage rating is reduced below that voltage rating, the maximum kilovoltampere rating is also reduced, because the current cannot be increased to maintain constant kilovoltamperes. And if an overvoltage is applied, overheating will result and will shorten the life of the transformer.

Figure 6.2 shows the basic current, voltage, and kilovoltampere relationships for single-phase and 3-phase transformers. These relationships are used in design calculations as shown in the following examples.

Example. What are the primary and secondary current ratings of a 50-kVA transformer for stepping 480 V down to 240 V?

First use

$$I_p = \frac{\text{transformer kVA rating} \times 1000}{E_p}$$

$$= \frac{50 \times 1000}{480} = 104 \text{ A}$$

Then

$$I_s = I_p \frac{E_p}{E_s} = 104 \, \frac{480}{240} = 208 \text{ A}$$

Single-phase transformer fundamentals

CIRCUIT CONDITIONS
  E_p = Primary voltage
  I_p = Primary current
  E_s = Secondary voltage
  I_s = Secondary current
  E_{vd} = Voltage drop in circuit conductors
  E_L = Voltage across load
  T_p = Number of turns on primary
  T_s = Number of turns on secondary

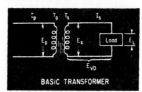
BASIC TRANSFORMER

PRACTICAL RELATIONSHIPS

$\frac{E_p}{E_s} = \frac{T_p}{T_s}$ $T_p \times I_p = T_s \times I_s$ $\frac{I_p}{I_s} = \frac{T_s}{T_p}$ $\frac{I_p}{I_s} = \frac{E_s}{E_p}$ $I_p \times E_p = I_s \times E_s$ Transformer kva rating = $\frac{I_p \times E_p}{1000} = \frac{I_s \times E_s}{1000}$

Three-phase transformers currents and voltages

For any delta connection
(pri. or sec.):

$I_L = \frac{\text{Rated transformer kva} \times 1000}{1.73 \times E_L}$

For any wye connection
(pri. or sec.):

$I_L = \frac{\text{Rated transformer kva} \times 1000}{1.73 \times E_L}$

Where, I_L = Line current (pri. or sec.) of transformer
  I_w = Rated current of each transformer winding
  E_L = Rated voltage (phase-to-phase) of transformer
  E_w = Rated voltage of each transformer winding

NOTE: This data applied to 3-phase transformers and to single-phase transformers connected for 3-phase use.

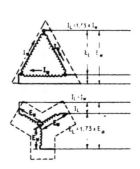

FIGURE 6.2 Transformer circuit design is based on these relationships.

Example. What are the primary and secondary current ratings of a 500-kVA, 3-phase transformer stepping 480-V delta down to 120/208-V wye?

On the delta primary, with $E_L = 480$ V

$$I_L = \frac{500 \times 1000}{1.73 \times 480} = \frac{500{,}000}{830} = 603 \text{ A}$$

On the wye secondary, with $E = 208$ V

$$I_L = \frac{500 \times 1000}{1.73 \times 208} = \frac{500{,}000}{360} = 1388 \text{ A}$$

NOTE: The above formulas do not take transformer losses into consideration and, therefore, provide only approximate determinations. However, because transformers have extremely high efficiencies (e.g., 98 percent), the calculations shown are generally acceptable for practical applications of transformers.

One consideration in selecting a transformer is the "impedance" of the unit, which is indicative of its ability to supply short-circuit currents into faults on the load side of the transformer. The impedance of a transformer is the opposition which the transformer presents to the flow of short-circuit current through it.

Every transformer has an impedance, which is generally expressed as a percentage—the percentage of the normal rated primary voltage which must be applied to the transformer to cause full rated load current to flow in the short-circuited secondary. For instance, if a 480- to 120-V transformer has an impedance of 5 percent, then 5 percent of 480 V, or 24 V, applied to the primary will cause rated load current to flow in the short-circuited secondary. If 5 percent of the primary voltage will cause such a current, then 100 percent of the primary voltage will cause 100/5 (or 20) times the full rated load current to flow through a solid short circuit on the secondary terminals.

From the foregoing, it can be seen that the lower the impedance of a transformer of given kilovoltampere rating, the higher the short-circuit current which it can deliver. Consider two transformers, both rated at 500 kVA. Assume the rated secondary load current is the same for both transformers. If one transformer is rated at 10 percent impedance, it can supply 100/10 (or 10) times the rated secondary current into a short circuit on its secondary terminals. If the other is rated at 2 percent impedance, it can supply 100/2 (or 50) times the rated secondary current into a short circuit on its secondary terminals. Thus, the second transformer can supply 5 times as much short-circuit current as the first, even though they have the same kilovoltampere load-handling ability. Common impedance values for general-purpose transformers are between 3 and 6 percent.

Voltage regulation is another transformer consideration, and impedance is related to it. Voltage regulation is a measure of how the secondary voltage of a transformer varies as the load on the transformer varies from full load to zero, with the primary voltage held constant.

Voltage regulation is expressed as a percentage that is calculated as the no-load voltage minus the full-load voltage divided by the full-load voltage times 100 percent. If a transformer has a no-load secondary voltage of 240 V, and the voltage drops to 220 V when the transformer is fully loaded (supplying the rated current output), then the regulation percentage is

$$\frac{240 - 220}{220} \times 100 \text{ percent} = 9 \text{ percent}$$

This indicates that 9 percent of the secondary voltage is being dropped across the internal impedance of the transformer. It is obvious, then, that the higher the impedance of a transformer, the greater the drop from no-load to full-load voltage and the higher the percentage regulation. It is generally desirable to keep regulation as low as possible, to minimize variations in voltage as load-current demand varies. Typical regulation values are between 2 and 4 percent.

Low regulation becomes very important when the transformer supplies varying load demands for utilization equipment which is sensitive to voltage changes. In cases where the load current on the transformer is relatively constant, however, the transformer taps or the primary supply conditions can be adjusted to provide fixed compensation for the voltage drop in the transformer. The voltage level can be set as required at the utilization devices, and it will not change because the load current is not changing. In such cases, acceptable voltage conditions at the utilization devices can be achieved with transformers of either low or high regulation.

But it should be pointed out that the high-regulation transformer provides a high internal impedance, which reduces the level of possible short circuits on the secondary. As a result, fuses, circuit breakers, and other protective devices need not be rated for as high interrupting capacities as they would need if a low-regulation (and low-impedance) transformer were used.

Kilovoltampere Rating

The rated kilovoltampere output of a transformer is that output which it can deliver for a specified time, at the rated secondary voltage and rated frequency, without exceeding a specified temperature rise based on insulation life and ambient temperature. The output that a transformer can deliver without objectionable deterioration of the insulation may be more or less than the rated output, depending upon ambient temperature and load cycles.

Transformers may be loaded above their kilovoltampere ratings with no sacrifice of life expectancy only in accordance with prescribed temperature testing or short-time loading data from the manufacturer. In addition, the load that can be carried by self-cooled transformers may be increased considerably by the use of fans or other auxiliary cooling equipment. Such an application may be advantageous for short-time peak loading, although the design characteristics of the transformer, including increased voltage regulation, must be considered.

The basic loading conditions, for which the normal life expectancy of a transformer is determined, are:

1. The transformer is continuously loaded at its rated kilovoltamperes and rated voltage.
2. The average temperature of the cooling air during any 24-hour period is 30°C.
3. The temperature of the cooling air at no time exceeds 40°C.

The rated kilovoltamperes, rated voltage, and rated primary and secondary currents are considered to be maximum values. The kilovoltampere rating is based on drawing the rated current at the rated voltage; but, as noted previously, the kilovoltampere handling capacity cannot be kept constant by, say, operating at a lower voltage and a higher current. This capacity is determined by the maximum current the transformer can carry without reaching a dangerous temperature, and that current cannot be exceeded. Moreover, it should be noted that transformers are rated in *kilovoltamperes*, so that their load-handling ability is dependent only upon current and voltage, without regard for the effect of circuit power factor. The power-handling capacity (rated kilowatt load) of a transformer, therefore, is greater for high-power-factor loads than for low-power-factor loads.

A transformer selected for a particular application must have a kilovoltampere rating at least equal to the kilovoltampere rating of the load to be supplied. But modern engineering practice, based on experience with load growth in electrical systems, demands the inclusion of spare capacity in every transformer installation. Of course, this spare capacity will be provided automatically if the feeders and branch circuits supplied by the transformer have spare capacity and the transformer is sized to accommodate that extra capacity.

In selecting transformers to supply motor loads, several general rules are used to determine the required capacity. One such rule says that a transformer should have at least 1 kVA of capacity for each horsepower of motor load. Another guide suggests at least 1¼ kVA per horsepower of motor load for motors rated 5 hp and above. Motors over 50 hp may be satisfactorily served by more or less than 1¼ kVA/hp, depending upon the torque characteristics (NEMA design) of the particular motor, the horsepower rating of the motor, the load characteristics, whether or not reduced-voltage starting is used, and what type of reduced-voltage starter is used. Of course, greater accuracy can be achieved in matching transformer capacity to a number of motors if the nameplate current ratings of the motors are added together and then multiplied by the rated voltage to obtain the required kilovoltamperes. Effective, economical, and reliable operation of large motors may often depend upon selection of a transformer with the correct kilovoltampere rating and internal impedance for the particular conditions of motor starting, duty cycle, and loading. Early engineering assistance from a transformer manufacturer can prevent many problems in motor applications.

Harmonics and the K-Factor

Today's electrical designer has had the rules of design thrown out the window by the explosion of harmonic-producing loads in the workplace. The complex wave form is prevalent, and probably the norm, in many building types.

Transformer Sizing. The kilovoltampere rating of transformers used to be dependent on one factor: load. Assumptions were made on variables such as wave form (e.g., sinusoidal), operating temperature, power factor, load balance, etc. Couple these variables with the calculated load versus the actual load, and the probability of an overloaded transformer resulting in failure and/or shortened life was reasonably low. However, applying these rules today in a high-harmonic environment will be disastrous, resulting in unplanned transformer failures.

Here is a comparison between results calculated using each of two recommended methods that have been proposed for derating transformer kilovoltampere ratings to accommodate harmonic loading. We will compare the Computer and Business Equipment Manufacturer Association (CBEMA) crest factor method and the IEEE methods for derating "off-the-shelf" (non-K-factor-rated) dry-type transformers. Both methods have unique benefits and abilities.

It is worth noting that the concept of transformer derating should not be applied to new construction for two fundamental reasons:

1. The information necessary to execute either of those two methods is simply not available at the design stage.

2. There is some question as to whether a derated, non-K-factor transformer satisfies the rules of **NEC** Sections 110-2 and 110-3. That is, even if derated, a non-K-factor-rated transformer has *not* been evaluated for supplying the derated kilovoltampere to the specific combination of nonlinear loads—either with or without additional linear loads—by a third-party testing lab. Because a standard dry-type (non-K-factor-rated) transformer has not been evaluated to deliver the derated kilovoltampere to the specific load, it may or may *not* be suitable for the application and therefore should not be used.

Although derating is generally not the most desirable approach and is not suited to new construction in existing facilities, it's better than doing nothing; it may be the only economically feasible approach. Even though the accuracy of the derated kilovoltampere for a specific application has not been verified by a third-party test lab, because the derating is based on the actual loading of the existing transformer, the following methods should help to reduce the likelihood of transformer burnout or worse, a fire.

CBEMA Crest Factor. The crest factor method for derating transformers, when harmonic currents are present, was recommended by CBEMA in a 1988 information letter. This method involves determining the crest factor of the load on the transformer and dividing 1.414 by the crest factor.

The crest factor (CF) is the ratio of peak amps to rms amperes. That is, CF = peak amperes/rms amperes. In a pure sine wave, that ratio is equal to the square root of 2, or 1.414. The idea behind the CBEMA crest factor method is to establish a ratio between the actual crest factor of a given load, or combination of loads, and the crest factor of a normal sine wave. That's why the actual crest factor is divided into 1.414 (the value of a "normal" crest factor).

The proper method for determining a load's crest factor requires measurements using a current meter that indicates true rms current and the peak current value of the load. Then, the peak current is divided by the rms current and the crest factor for that load is used to establish a derating factor for the transformer supplying that load as indicated in the following example:

Given. Crest factor of load = 2.0

Find. CBEMA derating value from:

$$\frac{1.414}{\text{actual CF}}$$

In this case:

$$\frac{1.414}{2.0} = \text{CBEMA derating value}$$

$$0.707 = \text{CBEMA Derating Value}$$

That calculated value is then multiplied by the kilovoltampere rating of the transformer to indicate the maximum loading on the transformer. If the transformer is rated 75 kVA, the maximum load should be:

$$75 \text{ kVA} \times 0.707 = 53 \text{ kVA}$$

IEEE Method. The IEEE method is more complex and requires knowing much more information about the harmonic content of the load and also specific electrical characteristics of the transformer being used. The IEEE method is discussed in ANSI/IEEE publication C57.110 1986. This method assesses the impact of the harmonic currents on the operation of the transformer because of I^2R loss, eddy-current loss, and other stray losses. The various characteristics needed to do this calculation are:

- Harmonic content of load, in percent, for the fundamental third, fifth, seventh, ninth, eleventh, thirteenth, and fifteenth harmonic
- Transformer kilovoltampere rating, primary winding resistance, secondary winding resistance, and load loss under rated conditions

A program was written that used the simplified method shown in ANSI/IEEE C57.110-86. The above characteristics are entered into the spreadsheet (Fig. 6.3) and the program calculates the usable kilovoltampere for a specific transformer in accordance with the simplified method.

Comparison of Results. A comparison of the results obtained using both methods of derating is shown in Fig. 6.4, which compares the usable (or maximum) kilovoltampere—determined in accordance with both the CBEMA and IEEE methods—for a 75-kVA transformer with characteristics as shown in Fig. 6.3. The average crest factor and average harmonic current load profiles of field studies were used to establish the CBEMA and IEEE derating values. Then, a comparison percentile was calculated as follows:

$$\% \text{ comparison} = \frac{\text{CBEMA kVA} - \text{C57.110 kVA}}{\text{C57.110 kVA} \times 100}$$

	A	B	C	D	E	F	G	H	I
1	DELTA-WYE TRANSFORMER				h	hs	Ih	Ih2	Ih*Ih2
2									
3	Rating	75	KVA		1	1	1.0000	1.0000	1.0000
4	Primary	480	Volts		3	9	0.5170	0.2673	2.4056
5	Secondary	208			5	25	0.3833	0.1469	3.6730
6	Pri Res	0.4470			7	49	0.1893	0.0358	1.7559
7	Sec Res	0.0243			9	81	0.0570	0.0032	0.2632
8	Load Loss	2600			11	121	0.0317	0.0010	0.1216
9	Pri Fla	90.2			13	169	0.0373	0.0014	0.2351
10	Sec Fla	208.2			15	225	0.0247	0.0006	0.1373
11									
12	Pec-r	334	Watts					1.4563	9.5916
13									
14	Pec-r max	0.762							
15									
16	Imax (pu)	0.541						K factor	6.59
17									
18	Usable KVA	40.6						Pct THD	67.5

FIGURE 6.3 Typical computation of the K-factor.

A negative value indicates that the CBEMA method is more stringent while a positive value indicates that the ANSI/IEEE C57.110 method is more stringent. Please note that out of the 95 case studies, the ANSI/IEEE C57.110 method was more stringent in 68 studies, or 72 percent of the time. Also note that there is as much as a 30 percent difference between the two methods. The total variance between the two methods ranges from −12.93 to +30.97, or a total variance of almost 44 percent.

A look at three case studies in particular demonstrates this variance. Case studies 1, 7, and 29 have an almost identical crest factor of 1.95 to 1.96, yet the percent usable kilovoltampere variance ranges from −8.93 percent to 6.3 percent. The harmonic current load profiles are:

	Current (as % of fundamental)		
Harmonic	Case 1	Case 7	Case 29
Third	20.97	21.60	36.57
Fifth	15.17	15.90	17.37
Seventh	8.37	10.00	5.87
Ninth	4.27	7.40	4.10
Eleventh	1.47	7.00	1.97
Thirteenth	0.73	2.50	1.57
Fifteenth	0.87	4.20	1.13

As can be seen, case study seven has a much higher component of ninth, eleventh, thirteenth, and fifteenth harmonic currents, which results in higher losses in the transformer.

Another two case studies to compare are 10 and 18. Again, the crest factors are almost identical (i.e. 1.98 versus 1.97), but the variance ranges from −12.93 percent to 30.97 percent. The harmonic current load profiles are:

$$\left[\frac{\text{CEMBA KVA} - \text{C57.110 KVA}}{\text{C57.110 KVA}} \times 100\right]$$

		Average Crest			Average Crest
	% Variance	Factor		% Variance	Factor
1. Office Building	−8.93%	1.95	49. Office Building	−0.67%	2.17
2. Office Building	17.45%	2.84	50. Office Building	−0.41%	2.3
3. Office Building	5.36%	2.18	51. Office Building	3.24%	2.19
4. Office Building	8.16%	2.27	52. Office Building	−1.52%	2.18
5. Office Building	11.62%	2.23	53. Office Building	−0.70%	2.09
6. Office Building	6.70%	2.2	54. Office Building	9.05%	2.2
7. Office Building	6.30%	1.96	55. Office Building	−2.50%	2.06
8. Office Building	−5.82%	1.59	56. Office Building	2.94%	2.31
9. Office Building	6.44%	1.65	57. Office Building	3.39%	2.23
10. Office Building	−12.93%	1.98	58. Office Building	4.4%	2.03
11. Office Building	−2.39%	1.59	59. Office Building	10.94%	2.38
12. Office Building	6.59%	2.45	60. Office Building	−3.07%	1.98
13. Office Building	12.72%	1.4	61. Office Building	11.94%	2.36
14. Office Building	4.19%	2.187	62. Office Building	5.87%	2.26
15. Office Building	14.13%	2.51	63. Office Building	1.04%	2.18
16. Office Building	16.97%	2.39	64. Office Building	8.29%	2.39
17. Office Building	3.12%	2.26	65. Office Building	−4.76%	1.69
18. Office Building	30.97%	1.97	66. Office Building	−0.59%	2.09
19. Office Building	27.93%	1.95	67. Office Building	6.80%	2.41
20. Office Building	2.11%	2.30	68. Office Building	7.55%	2.33
21. Office Building	13.53%	2.89	69. Office Building	4.67%	2.37
22. Office Building	4.76%	1.44	70. Office Building	4.4%	2.48
23. Office Building	−8.79%	2.13	71. Office Building	11.41%	2.36
24. Office Building	−7.55%	2.09	72. Office Building	−0.96%	2.06
25. Office Building	3.37%	2.23	73. Office Building	0.21%	2.2
26. Office Building	3.53%	2.14	74. Office Building	−3.16%	2.11
27. Office Building	5.22%	2.27	75. Office Building	6.58%	2.06
28. Office Building	−3.29%	1.98	76. Office Building	0.86%	2.27
29. Office Building	2.34%	1.95	77. Office Building	0.64%	2.26
30. Office Building	3.72%	1.49	78. Office Building	−11.59%	1.84
31. Office Building	−6.13%	1.84	79. Office Building	0.24%	2.3
32. Office Building	−4.68%	1.53	80. Office Building	−12.07%	2.01
33. Office Building	10.37%	2.39	81. Office Building	−12.05%	2.03
34. Office Building	8.08%	2.24	82. Office Building	−10.16%	2.06
35. Office Building	2.67%	2.17	83. Office Building	10.42%	1.43
36. Office Building	6.31%	2.33	84. Office Building	3.30%	1.44
37. Office Building	8.79%	2.31	85. Office Building	2.43%	1.43
38. Office Building	8.41%	2.28	86. Office Building	4.68%	1.407
39. Office Building	11.94%	2.48	87. Government Facility	2.9%	2.22
40. Office Building	5.00%	2.0	88. Government Facility	−8.43%	1.76
41. Office Building	0.38%	2.13	89. Government Facility	0.77%	1.405
42. Office Building	0.99%	2.13	90. Audio Visual Facility	20.71%	1.961
43. Office Building	0.20%	2.14	91. Audio Visual Facility	21.65%	1.532
44. Office Building	2.57%	2.11	92. Industrial Facility	0.12%	1.55
45. Office Building	−1.17%	2.1	93. Industrial Facility	−3.04%	1.82
46. Office Building	6.62%	2.24	94. Industrial Facility	7.90%	1.89
47. Office Building	2.30%	2.09	95. Banking Facility	0.81%	2.1
48. Office Building	3.35%	2.12			

FIGURE 6.4 Comparison of CBEMA and IEEE methods for transformer derating.

	Current (as % of fundamental)	
Harmonic	Case 10	Case 18
Third	17.30	18.00
Fifth	15.30	42.47
Seventh	7.10	17.83
Ninth	2.30	4.20
Eleventh	1.70	4.73
Thirteenth	1.70	2.40
Fifteenth	1.10	2.43

As can be seen, the percentage values for the fifth through fifteenth harmonic currents are substantially higher for case 18 than they are for case 10.

Two important factors stand out when evaluating the comparison between the two methods shown in Fig. 6.4. The first is that loads with substantially different harmonic contents could have the same or similar crest factors. The second is that a small shift in the magnitude of the harmonic currents at the higher frequencies causes the IEEE derating method to become more severe.

Conclusion. The harmonic content of loads can vary substantially as demonstrated by the case studies. However, if the database becomes large enough, some conclusions can be drawn as to the expected harmonic content for derating existing transformers in a given building type. It must be pointed out, however, that nothing is better than knowing the actual harmonic content of a load. And that can only be determined through measurement using the proper techniques and measuring equipment.

The crest factor is a quick tool for determining if harmonics are present. However, care must be used in applying the crest factor for derating transformers since the crest factor *is independent* of harmonic content. Both the CBEMA crest factor method and the IEEE method should be evaluated in each circumstance, and the more severe derating applied. Since the variance can be substantial (40 percent for some case studies) the engineer is gambling if only one method is used.

This has only addressed the derating of *existing* off-the-shelf transformers. The derating is only providing allowance for the additional losses experienced by the transformer because of harmonic currents and the resultant higher operating temperatures. This derating does not affect the magnetic circuit of the transformer. A transformer derated by the methods discussed here could still be driven into saturation by the harmonic currents creating voltage distortion.

There is one other problem associated with derating existing transformers. If it is determined that a transformer requires derating, what does one do with that portion of load that is removed from the transformer? That is, from where does one supply that load? That is a tough one to deal with. It might be possible to find a lightly loaded transformer elsewhere in the building. Or, it may be possible to install another transformer to supply that portion of loading that was removed from the derated transformer. Or, if neither of those options are acceptable, then transformer replacement should be considered.

K-rated transformers that are properly designed for both the higher losses and the effect to the magnetic circuit caused by the harmonic currents can be used and should become the choice of designers and installers. These transformers will eliminate the problem and gamble of derating existing off-the-shelf transformers that were not designed for harmonic currents. And, in an existing installation, where the harmonic content of the load to be supplied is known, the required K-factor can be easily established.

Understanding the K-Factor

The K-factor already plays a significant role in the work of the professional engineer, electrical contractor, building management and maintenance personnel. However, it is not, at present, clearly understood by very many people. The application of the K-factor to electrical distribution systems must be understood first as it relates to the rated device (the transformer in this case), and the load connected to that transformer.

The following is one effort to explain K-factor: what it is and how it is used. We all know that nonlinear loads cause harmonic currents to flow in the circuit conductors. These currents can constitute a significant amount of the total load current and adversely affect other components of the electrical network.

The K-factor is computed by multiplying the square of the individual harmonic current by the square of that current's harmonic number. These products are then summed to produce K-factor. In this case, because we are using percentage values rather than actual current values, the sum of the products of harmonic and current number is divided by the sum of the percentage units squared (Fig. 6.5).

Harmonic	Harmonic Squared	Current % Fund.	Current % Fund. Squared	Product (b)*(d)
(a)	(b)	(c)	(d)	(e)
1	1	1.00	1.00	1.00
3	9	0.70	0.49	4.41
5	25	0.35	0.12	3.06
7	49	0.00	0.00	0.00
9	81	0.00	0.00	0.00
11	121	0.00	0.00	0.00
13	169	0.00	0.00	0.00
			1.61	8.47

Total Harmonic Distortion	78.3%
"K" Factor	5.25

FIGURE 6.5 Typical K-factor computation.

Using the values shown in Fig. 6.5, the K-factor is computed as follows: The harmonic number (order) is shown in column (a) and the harmonic current, as a percentage of the fundamental, for each harmonic order is expressed as a decimal value in column (c). Both values—those from columns (a) and (c)—are squared, and the results are shown in columns (b) and (d), respectively. The values shown in columns (b) and (d) are then multiplied and the product is given in column (e). Next, the individual values in column (e) are added, which in this case equals 8.47. But, as indicated, because percentage values for the harmonic currents are used instead of the actual current values of a specific harmonic order, the total of 8.47 must be divided by the sum of values shown in column (d) to give the K-factor for that particular loading. Therefore, in this case the K-factor equals 8.47/1.61, or 5.25.

| Percent | HARMONIC | | | | | |
of Fundamental	3	5	7	9	11	13
10	1.08	1.24	1.48	1.79	2.19	2.66
20	1.31	1.92	2.85	4.08	5.62	7.46
30	1.66	2.98	4.96	7.61	10.9	14.9
40	2.10	4.31	7.62	12.0	17.6	24.2
50	2.60	5.80	10.6	17.0	25.0	34.6
60	3.12	7.35	13.7	22.2	32.8	45.5
70	3.63	8.89	16.8	27.3	40.5	56.2
80	4.12	10.4	19.7	32.2	47.8	66.6
90	4.58	11.7	22.5	36.8	54.7	76.2
100	5.00	13.0	25.0	41.0	61.0	85.0

FIGURE 6.6 K-factor computation for various harmonics.

The chart in Fig. 6.6 illustrates the computed K-factor for varying individual percentages of harmonic currents at the odd harmonic orders, which are more commonly observed in the real world of electrical distribution. Single-phase nonlinear loads produce large amounts of third harmonic current and moderate amounts of fifth (and sometimes seventh) harmonic currents.

The most interesting aspect shown in the chart (Fig. 6.6) is that the same percentage of fundamental current at higher harmonic orders will produce a higher K-factor. For example, 50 percent of the fundamental as third harmonic produces a K-factor of approximately a K-factor of 2.6, while 50 percent of the fundamental as seventh harmonic produces a K-factor of 10.6. Careful examination of the chart and the accompanying graph (Fig. 6.7), will provide insight into the effect that different values of harmonic current at specific harmonic orders have on K-factor. It becomes immediately clear that higher-order harmonics can produce very high K-factors. For example, 3-phase nonlinear loads, such as six-pulse variable frequency drives, can produce K-factors of 16 and higher.

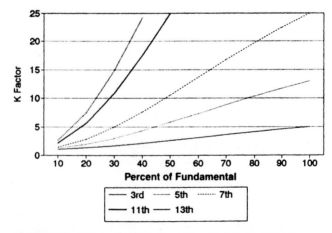

FIGURE 6.7 Effects of different harmonic orders on the K-factor.

Frequency Rating

By far the most common frequency of alternating current used for modern power and lighting electrical systems is 60 Hz. The transformers used on such systems are correspondingly rated for 60 Hz. Where other frequencies are used, care must be taken to ensure the suitability of the transformers.

Number of Phases

Single-phase transformers are used on single-phase systems and on single-phase circuits derived from 3-phase systems. To transform 3-phase voltage from one level to another, either three single-phase transformers or a 3-phase transformer can be used. A 3-phase transformer weighs less and requires less space than three single-phase transformers of the same type, construction, and total kilovoltampere capacity and provides greater ease and economy in installation. But the use of three single-phase transformers has the advantage that failure of one of the units requires that only the one unit be replaced. Failure in a 3-phase transformer requires replacement of the entire device.

Noise Rating

A transformer that is used in a place with a low ambient noise level—school, office, library, church, etc.—requires a low decibel hum rating and/or careful installation to prevent objectionable noise.

Air-cooled, dry-type transformers are generally enclosed in sheet-metal housings. The enclosure may or may not be ventilated. Completely enclosed units are made for use in atmospheres where dust, moisture, chemicals, or other airborne matter might harm exposed coils in a ventilated case. Some units have their core laminations exposed to dissipate heat. Other small-size units have totally enclosed housings filled with epoxy compound, which seals the windings against atmospheric attack and greatly reduces operating hum.

Temperature Rating

Dry-type transformers are designed with various types of insulation on the windings; the rating and loading of a transformer are based on the temperature limits of the particular insulation system. The common systems are as follows:

Class 105 (formerly called "Class A") insulation systems, when properly loaded and applied in an ambient temperature of not over 40°C, will operate at no more than a 55°C average temperature rise on the winding conductors, with an added 10°C of hot-spot temperature gradient. (This gradient is the difference between the average conductor temperature and the temperature of the hottest spot—the point where the greatest possible insulation damage could occur.) The sum—40°C plus 55°C plus 10°C—results in the designation of 105°C for the system. Today's usage limits Class 105 insulation to the smallest single-phase transformers (e.g., up to 0.150 kVA).

Class 150 (formerly called "Class B") insulation systems reach a maximum operating temperature of 150°C—an 80°C rise on the winding, added to a maximum 40°C ambient, plus a 30°C hot-spot temperature gradient. Class 150 transformers are generally available in ratings up to 2 kVA, single phase, for 600-, 480-, or 240-V primary with 240/120-V secondaries.

Class 185 (formerly called "Class F") insulation systems reach a maximum operating temperature of 185°C—a 115°C rise on the windings, added to a maximum 40°C ambient, plus a 30°C hot-spot temperature gradient. Class 185 units are made in single-phase and 3-phase types from 3 through 30 kVA, for 600-, 480-, and 240-V primaries with 240/120-V or 208/120-V secondaries.

Class 220 (formerly called "Class H") insulation systems have a total temperature rise of 220°C—a 150°C temperature rise on the conductors plus a hot-spot temperature gradient of 30°C, plus 40°C ambient temperature. Single-phase and 3-phase transformers of the ventilated indoor type are available in ratings from 15 to 500 kVA for single-phase systems and up to 1000 kVA for 3-phase systems, for primaries up to 600 V. In those ratings, Class H insulation is the type most commonly used in dry-type transformers today.

Transformers must be carefully selected with regard for all the factors that depend upon or are related to the temperature rating, as follows:

Relative loading. Transformers loaded at or very close to their kilovoltampere ratings will run hotter and be more disposed to thermal aging and damage than transformers that are lightly loaded. If, for instance, an 80-kVA demand load is to be fed by a 3-phase, 480- to 208/120-V transformer, a 100-kVA unit would be selected. But, based on this initial loading, the transformer would be only 80 percent loaded.

Duty cycle. A transformer may be operated fully loaded (100 percent of rating) for 24 hours a day, 7 days a week. Or, the same transformer may be loaded at less than 100 percent of its rating for all or part of each 24-hour period and for one or more days every week. The loading level and cycle can vary widely, and such variation should be known and factored into the selection.

Ambient temperature. The average and maximum ambient temperatures at the place where a transformer is to be installed must be determined or estimated and included in the analysis leading to the selection of the transformer.

Efficiency

With so much attention focused on spiraling energy costs today, the relative efficiency of a transformer—its core losses and conductor losses in relation to its kilovoltampere rating—has become an important element in evaluating its life cycle costs (first cost plus operating energy costs over the life of the equipment). Higher losses result in a higher transformer operating-temperature rise on the winding, for a given load.

Normally, transformer insulation is expected to remain effective for 20 to 25 years, if the transformer is loaded to 100 percent of its kilovoltampere rating and operated continuously at its rated voltage, with an average ambient temperature of not more than 30°C for any 24-hour period and a maximum ambient of 40°C, and at any altitude up to 3300 ft. Obviously, these are more severe conditions of operation than those found in the usual transformer application. Most transformers are not loaded to 100 percent of the rated load and operated 24 hours a day, every day, for their operating lives. And most installations are not subjected to a steady ambient temperature of 86°F (30°C) or a maximum of 104°F (40°C).

When transformers are applied at their maximum operating temperatures—that is, when their winding-temperature rise is at its highest value and the overall system is subjected to the maximum temperature—care should be taken to assure that conditions such as excessive ambient temperatures, 100 percent loading, and/or severe duty cycling do not represent a threat to the normal operating life of the equipment. Such conditions as frequent heavy-current motor-starting duty (that subjects a transformer to the exaggerated I^2R effects of repeated high motor inrush currents) and repeated thermal cycling due to off-on loading have been found to be more damaging to insulation than steady-state (continuous) operation.

There is no specific temperature that is critically related to transformer insulation life. That is, there is no single temperature above which insulation will deteriorate quickly and below which deterioration will be slight or negligible. Thus, there is no fixed or correct operating temperature. Heat is always the enemy; insulation deteriorates at just about any temperature, and it is difficult to assign an absolute maximum operating temperature. Operating time is also involved in insulation failure, along with operating temperature—insulation destruction is a function of the product of time and heat. Equal insulation deterioration can result from short-time, high-temperature operation and from long-time, low-temperature operation. Continuous full-load operation and periods of overloading produce high heat that reduces transformer life below the life that would result from lower-temperature operation.

Transformers with Class 220 insulation systems—the most commonly used insulation for the widest range of transformer ratings—are available in three distinct designs, based on temperature of operation. Each design has specific characteristics and must be carefully selected by relating its rating and capabilities to all the requirements and conditions of the particular application. The three types of Class 220 transformers are as follows:

1. *Class 220, 150°C-rise transformer.* A unit of this type operates with the full 150°C temperature rise on the winding conductors. Such a transformer is used at the maximum temperature capability of the overall insulation system. When fully loaded to its kilovoltampere rating and operated continuously in an average ambient temperature not over 30°C (86°F), with the maximum ambient never over 40°C (104°F), such a transformer can be expected to offer what is described as "full economic life." This design of transformer is commonly applied with an 80 percent load limitation under continuous loading, as a conservative measure to afford some degree of spare capacity for future load growth. Of course, such 80 percent loading reduces the production of operating heat and retards thermal aging and ultimate insulation failure. But Class 220, 150°C-rise units are not designed with any continuous overload capability.

2. *Class 220, 115°C-rise transformer.* A unit of this design has an insulation system that is rated for the full 220°C temperature limit and is capable of withstanding a 150°C rise on the windings; however, the actual operating temperature of the windings is limited by design to only 115°C (instead of 150°C) when the transformer is fully and continuously loaded to its kilovoltampere rating. The reduced temperature rise is achieved by using larger conductors, with lower I^2R losses, in the windings; this results in a larger coil, core, and case. Such a unit is larger, weighs more, and costs more than a 150°C-rise unit of the same kilovoltampere rating. Such a transformer runs cooler than a Class 220, 150°C-rise unit and for that reason may be considered to have a built-in "temperature reserve" somewhat similar in effect to the 80 percent load limitation that designers use on Class 220, 150°C-rise units. The reduced operating temperature provides the unit with slower thermal aging—and, therefore, greater life expectancy—than 150°C-rise units under full, continuous loading. The Class 220, 115°C-rise unit, because of its ultimate capability of 150°C rise on the windings, has a continuous overload capability of 15 percent

above its nameplate kilovoltampere rating. And the lower losses of the unit (for a given capacity rating) result in energy savings and reduced operating costs.

3. *Class 220, 80°C-rise transformer.* This unit has an insulation system rated for a maximum operating temperature of 220°C but is designed to provide full, continuous kilovoltampere loading with a temperature rise of only 80°C on the windings. As the next step in reduced operating temperature below the level designed into Class 220, 115°C units, the 80°C-rise unit has even larger winding conductors to further reduce I^2R heat losses. Class 220, 80°C-rise units therefore have an even greater tolerance for the extreme heating effects of full, continuous loading, with greater life expectancy than hotter units under the same operating conditions. Class 220, 80°C units cost more and are generally larger and heavier than either 150°C- or 115°C-rise units. However, such units have a designed continuous overload capability of 30 percent without exceeding the safe operating temperature of the insulation, and they offer even greater reductions in energy losses and operating costs than Class 220, 115°C-rise units.

Class 220, 115 and 80°C-rise transformers are available as UL-listed devices for their capacity ratings at their specific temperature rise—either 115 or 80°C. But it should be noted that such units achieve reduced temperature ratings through their core and coil design, which reduces the temperature rise by using larger conductors with lower resistance (I^2R) losses. They are specifically designed to deliver rated kilovoltamperes at the reduced temperatures; they are not simply 150°C-rise units operated at reduced current loadings. Using an oversized 150°C-rise unit to obtain a lower temperature rise on the conductors will reduce conductor I^2R losses but result in unnecessarily high core losses (heat losses due to hysteresis—"magnetic friction"—in the lamination structure of the core and losses due to eddy currents).

Electric users select 115 or 80°C-rise units where they judge that conservative transformer loading is dictated by heavy, continuous loadings in areas with elevated ambients and excessive temperature is a concern. Or, these higher-priced transformers are selected where extremely high reliability or extra long life is needed.

In applications where transformers are fully and continuously loaded, the higher efficiency (lower losses) of 115 or 80°C-rise units may justify their higher first cost, since savings in electric-energy charges can recoup their added cost in a few years. Cases have been cited in which energy-cost savings over the life of a lower-rise transformer far exceeded its original cost. Of course, detailed analysis and calculations should relate transformer cost, local utility energy rates, required transformer loading and duty cycle, and the ratings and characteristics of the transformer in determining the economic advantages of low-rise units.

Comparisons can be developed between 150°C-rise units and either 115- or 80°C-rise units, as regards first costs and operating costs over the life of the transformers. Data on transformer losses are available and can be combined with utility energy rates and operating data to evaluate the potential economy of using either 115 or 80°C-rise units. And the overload capabilities of the low-rise Class 220 transformers may strongly suggest the use of such units—for the overload capabilities alone or in conjunction with energy savings and greater life expectancy.

Transformer Destruct Curves

As most are aware, overcurrent protective device time-current trip curves are readily available from the fuse or CB manufacturer. But what about the transformer destruct curves? Where do they come from? They must be calculated.

For a number of years there were IEEE/ANSI Standards C57:12:00, C57:12:01, etc., which covered a method for developing a transformer destruct *point* based on the I^2R (i.e., thermal) stress, only. However, years of experience and research indicated that there was a need to be concerned not only for the thermal stresses in transformers under fault conditions but also with the severe effects of mechanical stresses during faults. To eliminate transformer damage from both thermal and mechanical stresses, the IEEE/ANSI Standards were modified and amended to establish a method for developing a curve which covered both the thermal and mechanical stress limits.

Although approved by the Committee in 1982, many engineers are unaware of this change. The following discussion will illuminate the method for developing a transformer destruct curve based on the modified IEEE/ANSI Standards.

Transformer Category.   The magnitude of transformers' thermal and mechanical stresses are related to their kilovoltampere ratings. As a result, in the modified Standards, four different transformer groups or categories were established based on kilovoltampere rating. The four categories, along with thermal and mechanical withstand capabilities and other information necessary to develop the ANSI transformer destruct curve are shown in Fig. 6.8.

As indicated by Note (3) in Fig. 6.8, for Category III and IV transformers, where the "per unit" value of the system impedance (Z_s) is *not* known, the equation shown at right can be used to calculate Z_s.

Max. system voltage (KV)	Available Short circuit MVAsc	Calculation of Z_s
48.3 KV or lower	4300	$Z_s = \dfrac{V_t}{I_{sc}} = \dfrac{(MVA)\text{ base}}{(MVA)\text{ sc}}$
72.5	9800	
121	25100	
145	38200	
169	27900	
242	50200	
550	69300	
800	97000	

As can be seen, the majority of transformers commonly used fall into category "I," which covers single- and 3-phase transformers from 5 to 500 kVA. Because the thermal stress in transformers covered by Category I is considerably greater than the mechanical stress for all values of fault current, only thermal limitations are shown. Therefore, if the selected overcurrent protective device is capable of opening the circuit before the thermal limit is reached, then a Category I transformer can be considered as adequately and properly protected under any fault condition.

As shown in Fig. 6.8, the equation that expresses the thermal limitation (withstand capability) for any Category I transformer is: $I^2t = 1250$, where I is the "per unit" current (see "The Per Unit System," below) whose magnitude is equal to the smaller of the two values shown under Column 5.

Where a transformer larger than 500 kVA (Categories II, III, or IV) is evaluated, then the full ANSI curve—showing both thermal and mechanical stresses—should

Category	Transformer KVA		Thermal Withstand Capability		Mechanical Withstand Capability		Transaction Line (% of Max. Short Circuit Current)
	1 Phase	3 Phase	Equation	I per unit	Equation	I per unit	
I	5 - 25	15 - 75	$I^2t = 1250$	$\frac{1}{Z_t}$ or 40 *	—	—	—
	37.5 - 100	112.5 - 300	$I^2t = 1250$	$\frac{1}{Z_t}$ or 30 *	—	—	—
	167 - 500	500	$I^2t = 1250$	$\frac{1}{Z_t}$ or 25 *	—	—	—
II	501 - 1667	501 - 5,000	$I^2t = 1250$	$\frac{1}{Z_t}$	$I^2t = k$ [1]	$\frac{1}{Z_t}$ [2]	70%
III	1668 - 10,000	5001 - 30,000	$I^2t = 1250$	$\frac{1}{Z_t}$	$I^2t = k$	$\frac{1}{Z_t + Z_s}$ [3]	50%
IV	> 10,000	> 30,000	$I^2t = 1250$	$\frac{1}{Z_t}$	$I^2t = k$	$\frac{1}{Z_t + Z_s}$	50%

Notes:

(1) The constant "k" in the mechanical equation $I^2_1 = k$, has a value indexed at 2 seconds which is dependant on the transformer % impedance. It is determined by the following:

$$k = I^2_1 = \left(\frac{1}{Z_t}\right)^2 \times 2$$

(2) Z_t = per unit transformer impedance on transformer self cooled KVA base.

(3) Z_s = system impedance in per unit based on transformer self cooled rating. If Z_s is not known, use Table 2.

* *Choose smaller value*

FIGURE 6.8 This shows how the four ANSI transformer categories are distinguished and defined. As can be seen, these categories are based on the kilovoltampere rating of a given transformer. Notice that for Category I transformers (5 to 500 kVA) there is no data given under the Mechanical Withstand Column. Because the thermal stress on transformers of those kVA ratings is more significant than the mechanical stresses, one only need consider the thermal stress and assure that the selected overcurrent protective device will open the circuit before the transformer's thermal limit is exceeded.

be used. The transformer's mechanical stress is expressed mathematically as $I^2 t = k$, where I is the per unit current based on the transformers self-cooled kilovoltampere rating (not the elevated kilovoltampere rating that may be possible when the transformer is fan-cooled). And the value for k is determined in accordance with the Notes below Fig. 6.8.

The heading of the last column (far right) of Fig. 6.8 reads "Transition Line (% of Short-Circuit Current)." To develop the individual (thermal and mechanical) stress curves, two points are plotted to establish the thermal stress curve and two points plotted for the mechanical stress curve. Typically, those two curves will have a space between them, and the "transition line" is used to tie the thermal and mechanical curves together. And this "transition" occurs at the lower end (minimum current) of the transformer's mechanical stress limit and at the upper end (maximum current) of the transformer's thermal limit.

ANSI Curve Shift. For the purpose of this discussion, it will be assumed that a fault occurs on the secondary side of the transformer. This fault is then reflected into the transformer's primary side. The magnitude of the fault current depends on: (1) the type of secondary fault (bolted or arcing or line-to-ground, line-to-line, or line-to-line-to-line) and (2) the connection of the transformer's internal windings (delta-delta, delta-wye, etc.) (Figs. 6.9 and 6.10).

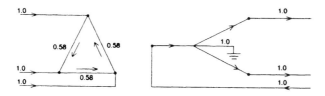

Three Phase Fault *

* Winding and line currents are per unit values with the base current equal to the 3-phase fault current

FIGURE 6.9 Current flow in the transformer primary and secondary windings and conductors during a 3-phase (line-to-line-to-line) bolted secondary fault. Under such a condition, the primary-conductors and primary overcurrent protective device(s) would carry 100 percent of the available primary fault current (i.e., $1.0I_{pu}$).

For a bolted 3-phase (line-to-line-to-line) fault on the secondary of a delta-wye transformer, Fig. 6.9 shows that both the primary and secondary conductors are carrying 1.0 per unit amperes (i.e., 100 percent of the available 3-phase fault current. Thus, the primary feeder fault current multiplied by the transformer's turns ratio equals the secondary fault current. And, where primary-only protection is used [**NEC** Section 450-3(b)(1)], the primary fuse or overcurrent relay must be capable of deenergizing the transformer before damage occurs.

For a line-to-ground fault on the secondary side of a delta-wye transformer, the primary-side circuit conductors will only be at 0.58 "per unit amps" (i.e., 58 percent of the secondary fault current times the transformer's turns ratio.

Based on the fact that the majority of faults are line-to-ground faults, it becomes clear that the primary fuses or overcurrent relays must be rated or set to interrupt a

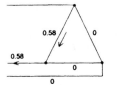
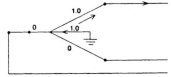

Single Line-to-Ground Fault *

* Winding and line currents are per unit values with the base current equal to the 3-phase fault current

FIGURE 6.10 Current flow in the transformer primary and secondary windings and conductors during a single-phase (line-to-ground) bolted secondary fault. Under such a condition, the primary-conductors and primary overcurrent protective device(s) would carry only 58 percent of the available primary fault current (i.e., $0.58I_{pu}$). To assure that the selected primary overcurrent protective device is capable of protecting the transformer against line-to-ground faults, the current values used to develop the destruct curve must be "shifted" (derated) as shown in Fig. 6.11.

value of current that represents 58 percent of the available primary fault current. Why? Because under line-to-ground faults, the primary conductors will carry less current than they would during a 3-phase (line-to-line-to-line) fault. If the primary fuse or overcurrent relay were rated or set to interrupt 1.0 per unit amperes (i.e., 100 percent of the available primary fault current), the fuses or relays will allow fault current to persist for too long a period of time, and thus, damage the transformer secondary. Consequently, the primary protective devices must be based on 58 percent of the available primary fault current for a delta-wye connected transformer. And the ANSI curves must be adjusted or "shifted" down to 58 percent of the available primary fault current to assure protection of the transformer during any type (line-to-ground, line-to-line, or line-to-line-to-line) of fault condition. A summary of typical transformer connections and the required ANSI "shift" (or derating factor) is shown in Fig. 6.11.

Example of Application. It may seem as if all this is quite complicated. But, really, it is not. The following example shows the steps and the sequence to follow for developing thermal and mechanical stress curves for any Category I, II, III, or IV transformer.

Given. A 2000-kVA, delta-wye connected, solidly grounded, 4160-480Y/277-Vac transformer with a 5.75 percent impedance ($Z = 0.0575$).

Step 1. From Fig. 6.8, select the correct category. A 3-phase 2000-kVA transformer would be a Category II transformer.

Step 2. Find the maximum possible secondary short-circuit current using the per unit system. We select the transformer kilovoltampere as the base kVA, and we select the secondary voltage as the base volts. Therefore,

$$\text{Base kVA} = 2000$$

$$\text{Base volts} = 480$$

TRANSFORMER APPLICATIONS (TO 600 V)

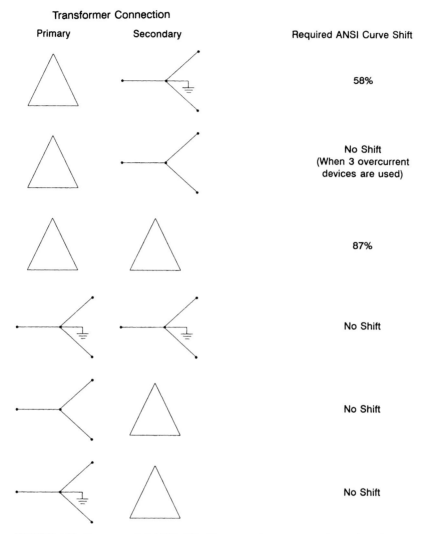

FIGURE 6.11 Recommended ANSI shifts. To assure adequate protection against the most common type of fault—line-to-ground—the designated shift (or derating) factor must be applied to the 3-phase short-circuit value (I_{sc}).

Using the per unit method, we know that

$$\text{Amps}_{pu} = \frac{\text{amps}_{sc}}{\text{Amps}_{base}} \quad \text{or} \quad I_{pu} = \frac{I_{sc}}{I_{base}}$$

This can also be expressed as:

$$I_{pu} \times I_{base} = I_{sc} \tag{1}$$

Where the per unit method is used, Fig. 6.8 tells us that I_{pu} is equal to $1/Z$, or, in this case $1/0.0575$.

And as indicated in the next section, I_{base} (base amps) = base kVA × 1000/Base volts × 1.732, therefore,

$$I_{base} = \frac{2000 \times 1000}{480} \times 1.732$$

$$= \frac{2{,}000{,}000}{831.36}$$

$$= 2405.69,$$

which is rounded up to 2406. (This is also the transformer secondary full-load current.) And I_{base} will always be the full-load current of the transformer if a base kVA and base volts are selected as the point of reference in a per unit system calculation.

Substituting the calculated values for the variables in Equation (1), we see:

$$I_{pu} \times I_{base} = I_{sc}$$

$$\frac{1}{0.0575} \times 2406 = I_{sc}$$

$$41{,}843 = I_{sc}$$

Step 3. What (if any) is the required ANSI shift? From Fig. 6.12, we see it is 58 percent.

Step 4. Determine the transition line. Refer to Fig. 6.8. As stated in the heading at the top of the far righthand column, the Transition Line (TL) value is to be taken as a percentage of short-circuit current available, as indicated for the specific transformer Category. For a Category II transformer,

$$TL = 70\% \times I_{sc}$$

because $I_{sc} = I_{pu} \times I_{base}$, we can say that

$$TL = 0.7 \times I_{pu} \times I_{base}$$

and because $I_{pu} = 1/Z$, we can say that

$$TL = 0.7 \times \frac{1}{Z_t} \times I_{base},$$

or

$$TL = 0.7 \times \frac{1}{0.0575} \times 2406$$

Step 5. Fig. 6.8 indicates that the mechanical stress should be considered and a curve should be plotted. To do that, we must find both the time and ampere value of two points. First, using the equation given in Fig. 6.8 for the mechanical stress—$I^2 t = k$—we determine k.

TRANSFORMER APPLICATIONS (TO 600 V)   439

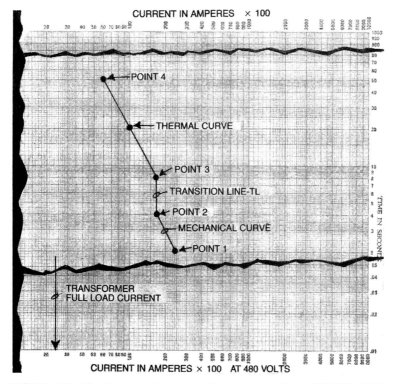

FIGURE 6.12 The thermal and mechanical stress curves for a 2000-kVA, 4160-480Y/277-V ac transformer with a 5.75 percent impedance.

Point 1. Again, in Fig. 6.8, we see that where the per unit method is used, $I_{pu} = 1/Z_t$, and

$$\frac{1}{Z_t} = \frac{1}{0.0575} = 17.39 \text{ A}$$

so

$$I_{pu} = 17.39$$

Next, using the ANSI recommended index of 2 s from Note (1) in Table 6.8, we can determine k:

$$k = I^2 t$$
$$= (17.39 \text{ A})^2 \times 2 \text{ s}$$
$$= 604.8$$

which is rounded up to 605 (this will be used to find the "time" for Point 2).

440 CHAPTER SIX

Now, to determine the line-to-ground fault current at 2 s, we multiply the maximum 3-phase short-circuit current by the shift (or derating) value shown in Fig. 6.11 for a delta-wye transformer.

$$\text{Point 1} = I_{sc} \times \text{shift}$$

$$= 41{,}843 \text{ A} \times 0.58$$

$$= 24{,}269 \text{ A at 2 s}$$

Point 2. Next determine the second point at the TL as follows: First, find I_{pu} at the point of transition. This would be

$$I_{pu} \text{ at transition} = 0.7 \times I_{pu}$$

or

$$I_{pu} \text{ at transition} = 0.7 \times \frac{1}{Z_t}$$

$$= 0.7 \times \frac{1}{0.0575}$$

$$= 12.17 \text{ A}$$

We now know two values for the variables in the equation $k = I^2 t$; therefore, we can find the time of transition as follows:

$$k = 605$$

$$I_{pu} = 12.17$$

at transition: $k = I^2 t$

$$605 = (12.17)^2 \times t$$

$$= 148.1 \times t$$

$$\frac{605}{148.1} = t$$

$$4.08 \text{ s} = t$$

To find the maximum 3-phase short-circuit current at the TL, we multiply the maximum short-circuit current (I_{sc}) times 70 percent (0.7), as indicated by the heading of the far righthand column in Fig. 6.8:

$$\text{TL} = I_{sc} \times 0.7$$

$$= 41{,}843 \times 0.7$$

$$= 29{,}290.1$$

To find the line-to-ground fault current at the transition line, we multiply the 3-phase fault value by the shift (derating) factor shown in Fig. 6.11 for a delta-wye connected transformer (i.e., 58 percent):

$$\text{Point 2} = I_{sc} \times 0.58$$
$$= 29{,}290 \times 0.58$$
$$= 16{,}988 \text{ at } 4.08 \text{ s}$$

Step 6. Find two thermal points using the equation $I^2 t = 1250$ from Fig. 6.8. Because the transition line ties the mechanical and thermal stress curves together at a point where the mechanical stress is at its lower limit and where the thermal stress is at its upper limit, we will first find the time and magnitude of the thermal transition point.

Point 3. Find I_{pu} at the transition line:

$$I_{pu} \text{ at transition} = 0.7 \times I_{pu}$$
$$= 0.7 \times \frac{1}{Z_t}$$
$$= 0.7 \times \frac{1}{0.0575}$$
$$= 12.17$$

Next find the time of transition from the equation:

$$I^2 t = 1250$$
$$(12.17)^2 t = 1250$$
$$148.1 t = 1250$$
$$t = \frac{1250}{148.1} = 8.44 \text{ s}$$

Now, find the maximum 3-phase short-circuit current at the transition line. From the discussion on per unit calculations we see that

$$I_{pu} = \frac{I_{sc}}{I_{base}}$$

$$12.17 = \frac{I_{sc}}{2406}$$

$$12.17 \times 2406 = I_{sc}$$
$$29{,}281 = I_{sc}$$

As indicated by Fig. 6.11, a shift must be applied to this 3-phase value (I_{sc}) to obtain the line-to-ground value at the transition line. Therefore, we multiply I_{sc} times the shift value given in Fig. 6.11 for a delta-wye connected transformer:

$$\text{Point 3} = I_{sc} \times 58\%$$
$$= 29{,}281 \times 0.58$$
$$= 16{,}982 \text{ at } 8.44 \text{ s}$$

Point 4. As recommended by ANSI, assume $I_{pu} = 5$. Determine the time t from the equation $I^2 t = 1250$ as follows:

$$I^2 t = 1250$$
$$(5)^2 t = 1250$$
$$25t = 1250$$
$$t = \frac{1250}{25}$$
$$t = 50 \text{ s}$$

Determine the maximum 3-phase short-circuit current from the equation $I_{pu} = I_{sc}/I_{base}$ as follows:

$$I_{pu} = \frac{I_{sc}}{I_{base}}$$

$$5 = \frac{I_{sc}}{2406}$$

$$5 \times 2406 = I_{sc}$$
$$12{,}030 = I_{sc}$$

Apply shift value (58 percent) to I_{sc} as indicated by Fig. 6.11 to determine line-to-ground fault value:

$$\text{Point 4} = I_{sc} \times \text{shift value}$$
$$= 12{,}030 \times .58$$
$$= 6977 \text{ at } 50 \text{ s}$$

Those values should then be plotted on log-log paper as shown in Fig. 6.12. You will notice that the current values for points 2 and 3 are really the same (the minor difference is due to rounding of some of the numbers); what differs is the time.

For Category I transformers, it is only necessary to plot the thermal destruct curve and compare that with the time-current trip curve of the selected overcurrent protective device. For Category II, III, or IV transformers, both the thermal and mechanical stress curves must be plotted and compared with the protective device's trip curve to assure adequate protection. And, in any case where delta-wye transformers are used, it is best to consider the line-to-ground fault curve since an overcurrent device capable of protecting the transformer against damage from those faults will be more than adequate for transformer protection during line-to-line (phase-to-phase) or line-to-line-to-line (3-phase) bolted faults.

TRANSFORMER APPLICATIONS (TO 600 V)   **443**

When comparing the trip curve with the destruct curve, the destruct-curve values must be greater than (to the right of) the values at which the protective device opens the circuit. If at any point the current value of the selected device's trip curve is greater than (to the right of) the transformer's destruct curve, then another device—either faster acting or a lower-rated—must be used to properly protect the transformer against damage under conditions of short-circuit.

The Per Unit System. The basic advantage of the per unit system is that comparisons between widely differing values can be made with relative ease. It is based on the idea that any two numbers can be essentially "indexed" by using an arbitrarily selected value as the "base" for comparison. This is expressed mathematically as:

$$\text{per unit} = \frac{\text{a number}}{\text{a base number}}$$

Once the differing values are converted to a per unit (pu) value, they can be compared on an "apples to apples" basis. For example, let's compare the efficiencies of two motors, one operating at 4000 V ac and the other at operating at 440 V ac using nominal system voltages as the "base" and the utilization voltage as the "the number."

Motor 1. pu = 4000/4160 = 0.96 pu
Motor 2. pu = 440/480 = 0.92 pu

From that comparison, we can see that the higher voltage motor is more efficient. We can also see that if the voltage drop at the 480-V level is reduced, say to 465 V, the efficiency of the motor operating on the 480 V, nominal, system will be better than the motor operating on the 4160-V system as follows:

Motor 1. pu = 4000/4160 = 0.96153 pu
Motor 2. pu = 465/480 = 0.96875 pu

Perhaps the greatest use for the per unit system is where short-circuit calculations are to be made.

Normally, when making a short-circuit study, all that one need do is add the $R + jX$ values of the various impedances, convert those to an equivalent Z (impedance) value, and then divide the line-to-neutral voltage by the equivalent Z value. However, when there are several voltages in a system, the per unit system is much more convenient.

When using pu values in fault calculations, only two values (i.e. base volts, base amps, base kVA, or base impedance) need be selected because the other values can be calculated from Ohm's Law, as follows:

$$\text{Base amps} = \frac{\text{base volts}}{\text{base ohms}}$$

Then using the selected base values, the per unit value for any part of the electrical system would be:

$$\text{Volts}_{pu} = \frac{\text{volts}}{\text{base volts}}$$

$$\text{Amps}_{pu} = \frac{\text{amps}}{\text{base amps}}$$

$$\text{Ohms}_{pu} = \frac{\text{ohms}}{\text{base ohms}}$$

For short-circuit calculations, it is generally more convenient to select a base kVA (usually that of the largest transformer or generator in the system) and a base voltage (any voltage used in the system that is comfortable to work with). The other quantities (amps and ohms) can be easily calculated:

$$\text{Base amps} = \text{base kVA} \times \frac{1000}{1.73} \times \text{base volts}$$

$$\text{Base ohms} = \frac{\text{base volts}}{1.73} \times \text{base amps}$$

In actual practice, it is best to convert all impedances from ohms directly to base ohms:

$$\text{Ohms}_{pu} = \text{ohms} \times \frac{\text{base kVA}}{\text{base kV}^2} \times 1000$$

The available short-circuit current can be found by adding the per unit impedances from the source to the fault location. Then the actual kVA at the fault location (independent of voltage) would be

$$\text{Actual kVA} = \frac{\text{base kVA}}{Z_{pu(total)}}$$

And the available fault current would be

$$I_{fault} = \frac{\text{actual kVA}}{1.73} \times \text{actual line-to-line volts}$$

Thus, once a fault kilovoltampere is found, coordination curves may be plotted for the voltage at any point within the system.

UL Listing. The *Electrical Construction Materials List* of the UL lists "Transformers—Power." To satisfy **NEC** and OSHA regulations, as well as local code rules on the acceptability of equipment, any transformers of the types and sizes covered by UL listing must be so listed. The use of an unlisted transformer of a type and size covered by UL listing would certainly be considered a violation of the spirit of **NEC** Section 110-2.

UL listing covers "air-cooled" types rated up to 333 kVA for single-phase transformers and up to 1000 kVA for three-phase units (all up to 600 V rating).

Enclosure. The housing of a transformer must be suitable for use where the transformer is installed. Units used outdoors must be approved for such use, and any conditions that prevail at the installed location must be taken into account, as required by **NEC** Section 110-3.

Mounting. Every transformer must be selected so that it is suitable for the manner in which it is intended to be mounted—such as wall-hung, floor-mounted, or ceiling-suspended.

Grounding Provisions. Careful evaluation of bonding and grounding provisions should be an important element of the transformer selection task. Either a built-in jumper conductor or provision for a bonding jumper must be included to afford ready compliance with **NEC** Section 250-26, which covers bonding and grounding of the secondary neutral and transformer case and is discussed below. The choice between internal grounding connection and connection to an external grounding stud on the case should be made with reference to local code rules and inspectors' preferences.

Winding Connections. The winding connections in 3-phase transformers must match primary and secondary electrical requirements on voltage values.

Wye-wye transformer connections should generally be avoided because of the objectionable characteristics of such connections. First, a connection of this type generally requires that the primary neutral point of the transformer windings be connected to the neutral of the source. Without this connection of the primary neutral, unbalanced loading from phases to neutral on the secondary side produces a "floating neutral" on the primary side, with a serious unbalancing of secondary wye voltages and reduced power-handling capacity. And because the feeder neutral to the transformer primary is absolutely necessary to properly supply wye-connected loads on the wye secondary, the source from which the primary feeder originates must be provided with a neutral point, and the extra neutral conductor must be run with the supply circuit. Of course, this objection to wye-wye connection does not apply with balanced wye or delta loading on the secondary.

The second, and major, objection to the general use of wye-wye-connected transformers arises from the character of the harmonic currents generated in such a hookup. Under balanced conditions, a 3-phase system operates with a 120° phase difference between the phases of the fundamental current frequency and the phases of each harmonic frequency of current, except for the third harmonic and multiples of the third harmonic which alternate in phase with each other. With sinusoidal applied voltages, magnetizing currents in iron-core transformers contain appreciable third-harmonic current. But since the vector sum of the currents at the neutral junction in a wye hookup must equal zero, third-harmonic currents and multiples of the third harmonic cannot exist in the primary of a wye-wye connected transformer if the primary neutral point does not have a connection to the neutral of the source supplying the transformer. Without the neutral connection, the third harmonic and its multiples are eliminated.

When the third harmonic of the primary magnetizing current is eliminated by not connecting the primary neutral, the flux wave is distorted so that appreciable third-harmonic voltage is induced in the secondary. In a wye-wye connection the third-harmonic voltages cancel each other between line terminals, but they show up in voltages between line terminals and neutral. These harmonic voltage waves assume a phase relationship with the fundamental voltage wave such that the windings can be subjected to severe overvoltages. Voltage peaks of nearly twice normal amplitude can subject the transformer insulation to destructive magnetic stresses and can even destroy the transformer.

There are basically two ways to eliminate these effects of wye-wye transformer connection. The simplest way is to provide a primary neutral to permit third harmonics in the transformer exciting currents, thereby producing sinusoidal flux and sinusoidal secondary voltages. However, with this solution, the primary harmonic current flow may possibly produce interference on telephone lines. The second method of eliminating unsafe voltage stresses due to harmonic voltages is to use a tertiary winding in each transformer, in a delta connection. The delta tertiary windings short-circuit the third-harmonic voltages, producing the third harmonic of the magnetizing current for sinusoidal flux. Thus, voltage spikes due to wave distortion are virtually eliminated.

The problem of third harmonics occurs in 3-phase, wye-wye hookups of single-phase transformers and in 3-phase transformers with shell-type core and winding construction. However, wye-wye 3-phase transformers with core-type construction produce almost no third-harmonic flux and, therefore, do not produce third-harmonic currents or the overvoltage operating condition.

For given conditions of application, a transformer manufacturer can provide the proper equipment for wye-wye transformation. However, careful attention to voltage levels and the circuit design configuration—whether to use wye or delta connections—can often afford real economy by eliminating the need for special provisions, such as tertiary windings, in transformers.

Transformer Protection

An important element in the design of transformer layouts for electrical distribution systems is the provision of effective overcurrent protection for each transformer. With regard to transformer protection, careful observance of the rules of **NEC** Section 430-3(b) will ensure both compliance with the **NEC** and complete design adequacy for transformers rated up to 600 V on their primaries.

Figure 6.13 shows the basic rule as applied to protection of the dry-type transformers used for power stepdown to 120/240 V, single-phase or to 120/208 V, 3-phase, 4-wire: Overcurrent protection for a dry-type transformer is provided by fuses or CBs rated at not more than 125 percent of the transformer primary full-load current (TPFLC) to protect the circuit conductors that supply the transformer primary. These circuit conductors must have an ampacity that is properly protected by the rating of the overcurrent protection.

In the layout in Fig. 6.13, a CB or a set of fuses rated not over 125 percent of the transformer rated primary full-load current provides all the overcurrent protection required by the **NEC** for the transformer. This overcurrent protection is in the feeder circuit to the transformer, and it is logically placed at the supply end of the feeder so it may also provide the overcurrent protection required for the primary feeder conductors. There is no limit on the distance between the primary protection and the transformer. When the correct maximum rating of transformer protection is selected and installed at any point on the supply side of the transformer (either near to or far from the transformer), the feeder circuit conductors must be sized so that the CB or fuses selected will provide the proper protection as required for the conductors. The ampacity of the feeder conductors must be at least equal to the ampere rating of the CB or fuses unless the exception of Section 240-3(b) is satisfied. That is, when the rating of the overcurrent protection selected is not more than 125 percent of the rated primary current, the primary feeder conductor may have an ampacity such that the overcurrent device is the next higher standard rating.

The rules for protecting a 600-V transformer with a CB or set of fuses in its primary circuit are given in Fig. 6.14. Note that, for transformers with rated primary current of 9 A or more, "the next higher standard" rating of protection may be used, if needed. Figure 6.14 shows the absolute maximum values of protection for smaller transformers. When the factor 1.67 (167 percent) or 3 (300 percent) is used, if the resultant current value is not exactly equal to a standard rating of fuse or CB, then the next *lower* standard rated fuse or CB must be selected.

TRANSFORMER APPLICATIONS (TO 600 V)

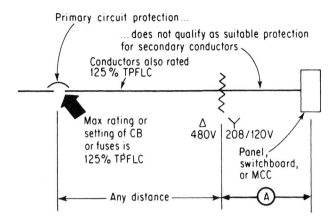

Distance "A" from transformer to panel is limited to 10 or 25 ft, subject to the requirements of Section 240-21 (b) and (d). If overcurrent protection is placed at the transformer secondary connection to protect secondary conductors, the circuit can run any distance to the panel. If the panel fed by the transformer secondary is a lighting panel and requires main protection, the protection must be on the secondary [Section 384-16(d)].

FIGURE 6.13 Most common layout for protection of dry-type 600-V transformers.

It is important to note that transformer primary-circuit protection is not acceptable as suitable protection for the secondary-circuit conductors—even if the secondary conductors have an ampacity equal to the ampacity of the primary conductors times the primary-to-secondary voltage ratio. On 3- and 4-wire transformer secondaries, it is possible that an unbalanced load may greatly exceed the secondary-conductor ampacity that is selected assuming balanced conditions. Because of this, the **NEC** does not permit the protection of secondary conductors with overcurrent devices operating from the primary through a transformer having a 3- or 4-wire secondary. For other than 2-wire-to-2-wire transformers, protection for secondary conductors has to be provided completely separately from any primary-side protection. Section 384-16(d) states that required main protection for a lighting panel on the secondary side of a transformer must be located on the secondary side. However, Section 240-3(i) permits the secondary circuit from a transformer to be protected by means of fuses or a CB in the primary circuit to the transformer—if the transformer has no more than a 2-wire primary circuit and a 2-wire secondary. As shown in Fig. 6.15, the 2-to-1 primary-to-secondary turns ratio of the transformer allows 20-A primary protection to protect against any secondary current in excess of 40 A, thereby protecting, say, secondary No. 8 TW wires rated at 40 A. The protection on the primary (here, 20 A) must not exceed the product of the secondary-conductor ampacity (40 A) multiplied by the secondary-to-primary transformer voltage ratio ($120 \div 240 = 0.5$). Here, $40 \text{ A} \times 0.5 = 20 \text{ A}$, and the rule is satisfied.

An important element in the design of transformer layouts for electrical distribution systems is the provision of effective overcurrent protection for each transformer. With regard to transformer overcurrent protection, careful observance of the rules of

A transformer with rated primary current of *9 amps or more* . . .

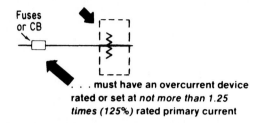

. . . must have an overcurrent device rated or set at *not more than 1.25 times (125%)* rated primary current

NOTE: Where 1.25 times primary current does not correspond to a standard rating of protective device, the next higher standard rating from Section 240-6 is permitted.

A transformer with rated primary current of *less than 9 amps* . . .

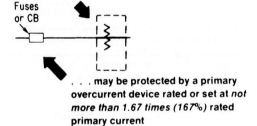

. . . may be protected by a primary overcurrent device rated or set at *not more than 1.67 times (167%)* rated primary current

A transformer with rated primary current of *less than 2 amps* . . .

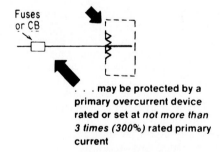

. . . may be protected by a primary overcurrent device rated or set at *not more than 3 times (300%)* rated primary current

FIGURE 6.14 Basic rules on protecting a transformer with an overcurrent device on the primary.

TRANSFORMER APPLICATIONS (TO 600 V)   **449**

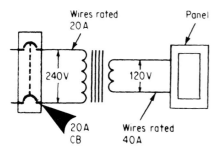

FIGURE 6.15   This kind of protection of secondary conductors is permitted.

NEC Section 450-3(b) and coordination of the protective device trip-curve with transformer inrush and damage points will ensure compliance with the **NEC** and complete adequacy for transformers rated up to 600 V on their primaries.

Coordination

In addition to determining the proper rating or setting for the transformer overcurrent protective device, coordination of the protective device time-current trip-curve with the transformer inrush and damage points is necessary to prevent nuisance tripping and assure that the transformer is adequately protected.

A 125 percent rated primary protective device is generally considered adequate to accommodate transformer inrush if the device can carry a current value of 12 times TPFLC for 0.1 s without opening the circuit. Some transformer manufacturers, however, have indicated that in addition to coordination at 12 times TPFLC for 0.1 s, the protective device trip-curve should be evaluated at 25 times the TPFLC for 0.01 s and at 3 times for 10 s.

Because transformer inrush current may reach (and even exceed) a value of 25 times the TPFLC in the first half cycle as the core is magnetized, this inrush point should be considered when selecting the primary overcurrent protective device. If the primary overcurrent protective device is a fuse, the selected fuse should be such that the minimum melting-current at 0.01 s is greater than 25 times the TPFLC. If the primary device is a CB, the selected CB should be such that its magnetic unlatching-current is greater than 25 times TPFLC for 0.01 s. Additionally, because loads supplied by the transformer may have current requirements of several times full-load

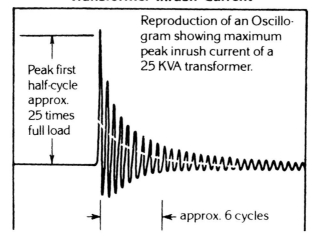

current (e.g., motor inrush), the primary protective device should be capable of carrying 3 times the TPFLC for 10 s without opening the circuit.

Coordination of the protective device trip characteristics with the transformer damage points is necessary to ensure that the selected protective device will interrupt the circuit before the current can reach a value that will cause a significant and irreversible degradation of the transformer insulating material. Because such stresses will adversely affect the service life of the transformer, good design practice dictates careful consideration of transformer damage points, as well as coordination with the recommended inrush clearance points.

When thermal-magnetic circuit breakers are sized at 125 percent of rated TPFLC for transformer primary-only overcurrent protection, coordination with transformer inrush and damage points may be difficult to achieve.

For example, assume that a transformer is rated at 112.5 kVA; what would be the maximum rating of protective device required by the **NEC** for primary-only protection? As covered in Section 450-3(b)(1), the maximum rating of the transformer primary protective device must be 1.25 times the TPFLC. For a 3-phase transformer, TPFLC is determined from the transformer kilovoltampere rating as follows:

$$\text{TPFLC} = \text{kVA} \times \frac{1{,}000}{1.73} \times \text{phase-to-phase voltage}$$

$$= 112.5 \times \frac{1{,}000}{1.73} \times 480$$

$$= \frac{112{,}500}{830}$$

$$= 135.5 \text{ A}$$

Per Section 450-3(b)(1), the maximum rating of the primary overcurrent protective device must be:

$$1.25 \times \text{TPFLC} = \text{maximum rating}$$

$$1.25 \times 135.5 \text{ A} = 169.375 \text{ A}$$

For nonadjustable CBs and fuses, Exception No. 1 to Section 450-3(b)(1) recognizes use of the next higher standard rating of protective device, as covered in Section 240-6, for transformers whose TPFLC is 9 A or more where the "calculated" ampere rating for the protective device does not correspond to a standard rating of protective device. Therefore, the maximum rating of fuse or nonadjustable CB (adjustable-type CBs *must* be *set* at *no more than* 125 percent of the TPFLC) must be 175 A (next standard size) to protect the 112.5-kVA transformer where only primary protection is provided.

Figure 6.16 shows the trip-curve for a typical 175-A molded-case CB with plots of the inrush clearance points and damage points for a 480-V ac to 208Y/120-V ac dry-type transformer.

Generally, the magnetic operation of a standard molded-case thermal-magnetic circuit breaker occurs at approximately 10 times (maximum) its trip rating. Although this curve reveals that this breaker would not open the circuit at 12 times TPLFC for 0.1 s, notice that the thermal magnetic circuit breaker's instantaneous

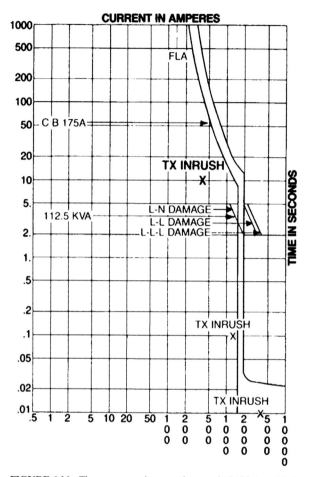

FIGURE 6.16 Time-current trip-curve for a typical 175-A molded-case circuit breaker with transformer inrush-clearance points plotted at 25 times TPFLC for 0.01 s, 12 times for 0.1 s, and 3 times for 10 s for a 112.5 kVA, 3-phase, 480-V ac to 208Y/120-V ac transformer. Transformer damage points are also shown.

pickup at 1750 A is well to the left of the first cycle inrush clearance point of 25 times TPFLC for 0.01 s (3,387 A). We can see graphically that worst-case transformer inrush could cause nuisance trip of the circuit breaker.

To accommodate this worst-case inrush condition, the circuit breaker would have to be up-sized to 250 percent of TPFLC to allow for magnetization of the transformer without nuisance tripping of the circuit breaker. In addition, to comply with the other requirement of Section 450-3(b)(2), secondary protection rated or set at not more than 125 percent of the TSFLC must be provided. And the primary conductors would have to be up-sized to assure compliance with Section 240-3.

Figure 6.16 also plots three sets of time-current damage points for the 112.5-kVA 3-phase dry-type transformer with a 480-V ac delta primary and a 120/208 grounded-wye secondary. The left-side set of damage points describe the combinations of time and current which, during a line-to-neutral fault, could cause damage to the 112.5-kVA transformer. Moving to the right, the next line represents the line-to-line damage points, and the final line, the 3-phase damage points. Note that the total clearing curve of the thermal-magnetic circuit breaker falls to the right of the line-to-neutral transformer damage points. We can see graphically that should a line-to-neutral fault of 1200 to 1749 A occur, damage to the transformer may result before the circuit breaker clears.

Figure 6.17 plots the same 112.5-kVA transformer inrush and damage points as well as the characteristics of an RK-5, 175-A, dual-element, time-delay fuse. Note that the minimum melt values of the 175-A fuse fall well to the right of the transformer inrush points and that the total clearing values give adequate coordination with line-to-line and line-to-line-to-line transformer damage points. However, the transformer could still be damaged by line-to-neutral faults.

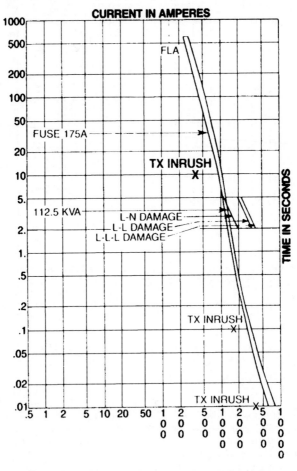

FIGURE 6.17 Time-current curve for a 175-A, RK-5 dual-element time-delay fuse with the 112.5-kVA transformer's inrush clearance and damage points plotted.

Figure 6.18 shows the trip characteristics of an RK-5, 150-A, dual-element, time-delay fuse, sized smaller than the *maximum* allowed by Section 450-3(b). Notice that coordination with first cycle inrush is maintained while coordination with all damage points is adequate.

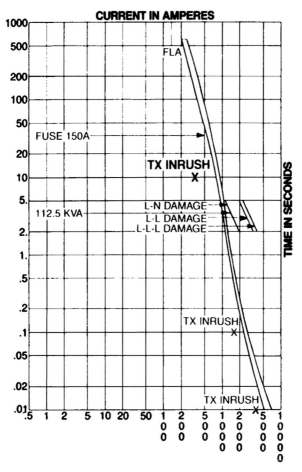

FIGURE 6.18 Time-current curve for a 150-A, RK-5 dual element time-delay fuse with the 112.5-kVA transformer's inrush clearance and damage points plotted.

It is worth noting that, although Section 450-3(b) establishes "maximum" values for transformer overcurrent protective devices, it is permissible to use a device that is rated *less than* the maximum provided all other **NEC** rules, particularly Section 220-10, are satisfied. That is, if the protective device that provides transformer primary protection is to supply a *continuous* load, then the protective device must be rated such that the continuous load would not be greater than 80 percent of the breaker rating (or the breaker must be rated at 125 percent of the continuous current). In this example, if all load supplied from this transformer were general lighting in an office building or industrial facility, which is considered to be a "continuous load," the rating of the protective device must not be less than 125 percent times the TPFLC, or 169.375 A. If the total noncontinuous primary load *plus* 1.25 times the continuous primary load is 150 A or less, then use of the 150-A fuse would satisfy the rule of Section 220-10(b).

The **NEC** contains provisions considered necessary for safety, and per Section 90-1(c), it is not intended as a design guide. We can conclude from our calculations and observations that when sizing thermal-magnetic circuit breakers for transformer primary-only overcurrent protection at the maximum rating permitted by Section 450-3(b) of the **NEC,** coordination of the circuit breaker with transformer first-cycle inrush may be difficult to achieve, and nuisance trips upon transformer energization may occur. If a larger circuit breaker is used, transformer secondary overcurrent protection must be provided. In addition, good design practice may require sizing RK-5 dual-element time-delay fuses smaller than the **NEC** maximums in order to achieve adequate primary-only fused transformer protection.

When the foregoing overcurrent-protection rules are observed, the transformer itself is properly protected, and the primary feeder conductors, sized to correspond, are provided with the required protection. But the secondary side must be evaluated separately and independently for any transformer with a 3- or 4-wire secondary. When a transformer is provided with primary-side overcurrent protection, a whole range of design and installation possibilities are available for a secondary arrangement that satisfies the **NEC.** The basic approach is to provide the required overcurrent protection for the secondary conductors right at the transformer—as with a fused switch or CB attached to the transformer enclosure, as shown in Fig. 6.19. Or, 10- or 25-ft taps may be made, as covered under "Feeder taps" in Chapter 4 of this handbook.

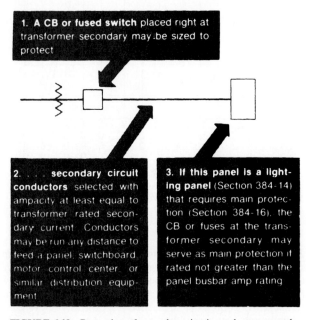

FIGURE 6.19 Protection of secondary-circuit conductors must be independent of primary protection.

In designing transformer circuits, the rules of **NEC** Section 450-3 on overcurrent protection for transformers can be coordinated with Section 240-21(d); which provides special rules for tap conductors used with transformers. This exception would be used mainly where the primary overcurrent devices are rated at 250 percent of rated transformer primary current, according to Section 450-3(b)(2), or where a secondary tap would exceed 10 ft. (Refer to the discussion of "Feeder taps.")

Where secondary feeder taps do not exceed 10 ft in length, the requirements of Exception Section 240-21(b) could apply as in the case of any other feeder tap, with no restriction on the size of the feeder overcurrent device ahead of the tap. In applying the tap rules in parts (b) and (d), the requirements of Section 450-3 on transformer overcurrent protection must always be satisfied.

Another acceptable way to protect a 600-V transformer is described in **NEC** Section 450-3(b)(2). In this method, the transformer primary may be fed from a circuit which has overcurrent protection (and circuit conductors) rated up to 250 percent (instead of 125 percent, as above) of the rated primary current; but, in such cases, there must be a protective device on the secondary side of the transformer, and that device must be rated or set at not more than 125 percent of the transformer's rated secondary current, as shown in Fig. 6.20. This secondary protective device must be located right at the transformer secondary terminals or not more than the length of a 10- or 25-ft tap away from the transformer, and the rules on tap conductors must be fully satisfied.

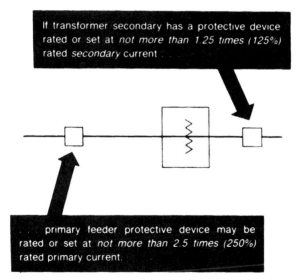

FIGURE 6.20 Use of secondary protection permits higher-rated primary protection and circuit.

The secondary protective device rated at not over 125 percent of the rated secondary current may readily be incorporated as part of other required provisions on the secondary side of the transformer, such as protection for a secondary feeder from the transformer to a panel or switchboard or a motor control center fed from the switchboard. And a single secondary protective device rated not over 125 percent of the rated secondary current may serve as required panelboard main protection as well as the required transformer secondary protection, as shown at the bottom of Fig. 6.21.

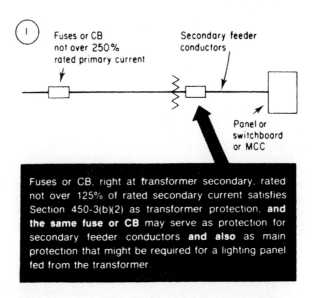

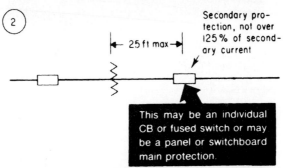

NOTE: In both cases, primary and secondary feeder conductors must be sized to be properly protected by the fuses or CB in both the primary and secondary circuit, which will give them more than adequate ampacity for the transformer full-load current.

FIGURE 6.21 Secondary protection rated at 125 percent of the secondary current may be placed at one of several points.

The use of a transformer circuit with primary protection rated up to 250 percent of the rated primary current offers the opportunity to avoid situations in which a set of primary fuses or CB rated at only 125 percent would cause nuisance tripping or opening of the circuit on transformer inrush current. But the use of 250 percent rated primary protection has a more widely applicable advantage: It makes possible the

TRANSFORMER APPLICATIONS (TO 600 V) 457

feeding of two or more transformers from the same primary feeder. The number of transformers that might be used in any case would depend on the amount of continuous load on all the transformers. But, in all such cases, the primary protection must be rated not more than 250 percent of any one transformer, if they are all the same size, or 250 percent of the smallest transformer, if they are of different sizes. And for each transformer fed, there must be a set of fuses or CB on the secondary side rated at not more than 125 percent of the rated secondary current, as shown in Fig. 6.22.

Figure 6.23 shows an application of 250 percent primary protection to a feeder supplying three transformers (such as at the bottom of Fig. 6.22). This example indicates how the rules of **NEC** Section 450-3(b) must be carefully related to Section 240-21 and to other **NEC** rules:

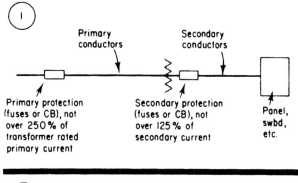

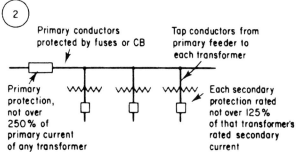

NOTE: Each set of tap conductors from primary feeder to each transformer may be same size as primary feeder conductors **OR** may be smaller than primary conductors if sized in accordance with Section 240-21, —which permits a 25-ft tap from a primary feeder to be made up of both primary and secondary tap conductors. The 25-ft tap may have any part of its length on the primary or secondary but must not be longer than 25 ft and must terminate in a single CB or set of fuses.

FIGURE 6.22 Transformer primary protection rated at 250 percent protects one or more transformers.

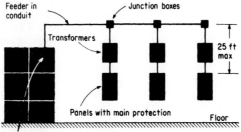

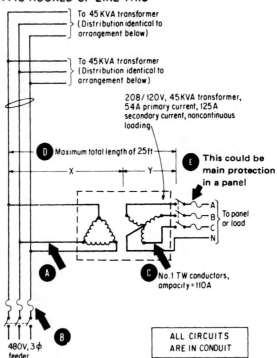

FIGURE 6.23 This layout of feeder and transformer taps provides economical application with **NEC** compliance.

Section 240-21(d) covers the use of a 25-ft unprotected tap from feeder conductors, with a transformer inserted in the 25-ft tap. This rule does not eliminate the need for secondary protection; it sets a special condition for placement of the secondary protective device. It applies to both single- and 3-phase transformer feeder taps.

Figure 6.23 shows a feeder supplying three 45-kVA transformers; each transformer is fed through a 25-ft feeder tap that conforms to Section 240-21(d). Although each transformer has a rated primary current of 54 A at full load, the demand load on each transformer primary was calculated to be 41 A, based on secondary loading. Thus, No. 1 THW copper feeder conductors were considered adequate for the total noncontinuous demand load of the three transformer primaries, which is equal to 3 × 41 A, or 123 A. In the following step-by-step analysis of this system, the letters refer to the circled letters on the sketch:

A. The primary circuit tap conductors are No. 6 TW, rated at 55 A, which gives them "an ampacity at least ½ that of the conductors or overcurrent protection from which they are tapped," because these conductors are tapped from the feeder conductors protected at 125 A. The use of No. 6 TW conductors is permissible for the 41-A primary current.

B. The 125-A fuses in the feeder switch properly protect the No. 1 THW feeder conductors, which are rated at 130 A and feed the taps for the three transformers.

C. According to Section 240-21(d)(2), the conductors supplied by the transformer secondary must have "an ampacity that, when multiplied by the ratio of the secondary-to-primary voltage, is at least ⅓ the ampacity of the overcurrent protection from which the primary conductors are tapped." The secondary-to-primary voltage ratio of the transformer is

$$\frac{208 \text{ V}}{480 \text{ V}} = 0.433$$

(Note that phase-to-phase voltages must be used to determine this ratio.) Then, for the secondary conductors, according to (d)(2),

Minimum ampacity × 0.433 = ⅓ × 125 A = 41.67 A

and

$$\text{Minimum ampacity} = \frac{41.67 \text{ A}}{0.433} = 96 \text{ A}$$

The No. 1 TW secondary conductors, rated at 110 A, are above the 96-A minimum and are, therefore, satisfactory.

D. The total length of the unprotected tap—i.e., the primary conductor length plus the secondary conductor length ($x + y$) for any circuit leg—must not be greater than 25 ft.

E. The secondary tap conductors from the transformer must terminate in a single CB or set of fuses that will limit the load on those conductors to their rated ampacity from Table 310-16. Note that there is no exception to that requirement, and the next higher standard device rating *may not be used* if the conductor ampacity does not correspond to the rating of a standard device.

The overcurrent protection required at E, the load end of the 25-ft tap conductors, must not be rated higher than the ampacity of the No. 1 TW conductors:

Maximum rating of fuses or CB at E = 110 A

But a 100-A main would satisfactorily protect the 96-A secondary load.

NOTE: The overcurrent protective device required at E could be the main protective device required for a lighting and appliance panel fed from the transformer. *Watch out for this trap!!*

The foregoing calculation shows how unprotected taps may be made from feeder conductors by satisfying Section 240-21(d). However, the rules of Section 240-21 are concerned with the *protection of conductors only*. Consideration must now be given to transformer protection, as follows:

1. Note that Section 240-21 makes no reference to transformer protection. But Section 450-3 calls for the protection of transformers, and no exception is made for the conditions of Section 240-21(d).
2. It is clear that each transformer shown in Fig. 6.23 is *not* protected by a primary-side overcurrent device rated at not more than 125 percent of the rated primary current (54 A), as required by Section 450-3(b)(1), because 1.25×54 A $= 68$ A.
3. But Section 450-3(b)(2) does offer a way to provide the required protection. The 110-A protection at E is secondary protection rated not over 125 percent of the rated secondary current (1.25×125 A secondary current $= 156$ A). With that secondary protection, a primary feeder overcurrent device rated not more than 250 percent of the rated primary current would satisfy Section 450-3(b)(2). That would call for fuses in the feeder switch (or a CB), at B in the diagram, rated at not over

$$2.5 \times 54 \text{ A (primary current)} = 135 \text{ A}$$

Because the fuses in the feeder switch are rated at 125 A (which is not in excess of 250 percent of the transformer primary rated current), those fuses satisfy Section 450-3(b)(2).

In addition to the two basic methods described above for protecting transformers, Section 450-3(b)(2) provides for protection with built-in thermal overload protection, as shown in Fig. 6.24.

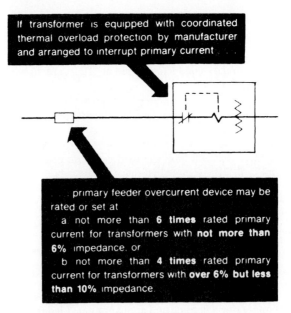

FIGURE 6.24 "Built-in" overload protection is an alternative to the other methods shown for transformer protection.

TRANSFORMER APPLICATIONS (TO 600 V)   461

NOTE: Refer back to "Feeder taps" (in Chapter 4) for additional discussion of the relationships among tap conductors, conductor protection, and transformer protection.

Typical Applications

The following examples are based on the foregoing design considerations and procedures.

Example. In Fig. 6.25, the feeder to the 3-phase dry-type transformer supplies 480-V power for step-down to 120/208 V for a lighting panel with a continuous load of 30 A on each phase. What are the required minimum sizes of conductors and protective devices?

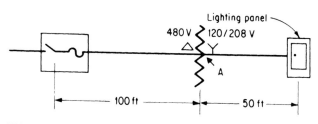

FIGURE 6.25 Sizing of conductors and overcurrent protection must satisfy all transformer-circuit conditions.

NOTE: Knowing these minimums under conditions of continuous use, the designer can effectively provide spare capacity without depriving the conductors and the transformer of the **NEC**-required levels of protection.
Solution. Use the following calculations:

Step 1. The kilovoltampere rating of the 30-A, 120/208-V load on the panel is

$$\frac{30 \times 208 \times 1.73}{1000} = 10.8 \text{ kVA}$$

That load will require at least a 15-kVA transformer, assuming a larger one is not needed for even greater possible load growth.

Step 2. Assuming eventual use of the entire transformer capacity, the rated primary current is computed as

$$\frac{15{,}000}{480 \times 1.732} = \frac{15{,}000}{831} = 18 \text{ A}$$

That value of primary current would require conductors with a minimum current-carrying capacity of 20 A: No. 12 copper or No. 10 aluminum. Because the footnote to **NEC** Table 310-16 says that No. 14 copper wire must not have a load-current rating over 15 A, the 18-A current prohibits the use of No. 14 conductors.

Step 3. **NEC** Section 450-3(b)(1) says that primary overcurrent protection for the transformer must be rated or set at not more than 125 percent of the rated primary current. In this example,

$$1.25 \times 18\text{ A} = 22.5\text{ A}$$

Because Exception No. 1 of Section 450-3(b)(1) says that the next higher rated device may be used here, this would call for a 25-A, 3-pole CB, or 25-A fuses. This transformer overcurrent protection may be installed at the supply end of the primary circuit feeder.

Step 4. Section 220-10(b) says that the feeder overcurrent device must be rated at least 125 percent of the continuous load. Since

$$1.25 \times 18\text{ A} = 22.5\text{ A}$$

the 25-A CB protection computed in step 3 above for an 18-A full-load transformer current would not violate the limitation set by Section 220-10(b).

Step 5. The primary-circuit conductors to the transformer are also required by Section 220-10(b) to have an ampacity of not less than 125 percent of 18 A, or 22.5 A. This would call for No. 12 copper or No. 10 aluminum circuit conductors. From **NEC** Table 310-16, No. 12 TW copper wire has an ampacity of 25 A, and No. 10 aluminum has that same ampacity. And No. 12 copper has a load rating of 20 A (from the footnote to Table 310-16), which exceeds the 18-A load current.

Step 6. The rated secondary current of a 15-kVA transformer would be

$$\frac{15{,}000}{208 \times 1.732} = 41.7\text{ A}$$

Section 220-10(b) requires that the secondary-circuit conductors from the transformer to the panel have an ampacity of at least 125 percent of that value, or 52 A. To carry this load, the following secondary conductors could be used in conduit:

Copper. No. 6 TW (55 A), THW (65 A), or THHN (75 A)
Aluminum. No. 4 TW (55 A), THW (65 A), or THHN (75 A)

Step 7. But the secondary conductors cannot be protected "through" the transformer. Section 240-3(i) permits the protection of transformer secondary conductors by overcurrent devices placed on the primary side of the transformer only for single-phase transformers with 2-wire primary and 2-wire secondary circuits. That is not the case here. Therefore, protection must be provided for the secondary feeder conductors which run 50 ft from the transformer to the panel. Such protection must be located right at the transformer—that is, "at the point where the conductor to be protected receives its supply," as covered in Section 240-21. Of course, secondary conductors from the transformer can be used as 10- or 25-ft tap conductors provided they fully comply with Section 240-2(b), (c), and (d). But where the length from the transformer to a panel is over 25 ft, as in this case, separate overcurrent protection must be inserted to protect the secondary conductors at their prescribed capacity.

A 60-A CB or 60-A fuses would protect any of the secondary conductors listed in step 6 above in accordance with its ampacity. An enclosure containing this 60-A protection could be placed right on the transformer case. Or, it could be placed so that 10-ft tap conductors or 25-ft tap conductors could reach it.

NOTE: The protection for the secondary conductors must never be placed further away from the transformer than the distance at which 25-ft tap conductors could be used. Overcurrent protection must be placed at *A* in Fig. 6.25, and the protection for the secondary conductors may also provide required protection for the lighting panel supplied—if the rating of the protection is not higher than the mains rating of the panel.

Thus, in this example, a 60-A device, for instance, could protect, say, both No. 6 THW secondary conductors and a 60-A lighting panel fed from the transformer.

Example. The circuit components in Fig. 6.26 are sized as follows:

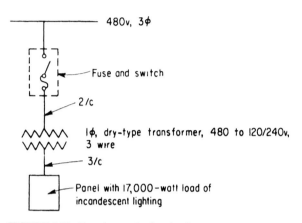

FIGURE 6.26 Transformer feeder circuits may be sized for the load capacity or for the full capacity of the transformer selected.

Step 1. The lighting load will be served by 15-A, 2-wire circuits. Because this lighting load will be operating continuously to supply fixtures in a store, each circuit must not be loaded to more than 80 percent of its rating. At 15 A and 120 V, each circuit has a maximum capacity of 15×120, or 1800 W. The 80 percent limit gives 1800 W $\times 0.8 = 1440$ W. The required number of circuits is determined as follows:

$$\text{Number of circuits} = \frac{\text{total load in watts}}{\text{watts per circuit}} = \frac{17{,}000}{1440} = 12 \text{ circuits}$$

Step 2. The feeder to the 120/240-V panel must have sufficient capacity for the 17-kW load, which operates continuously. The required feeder conductor ampere loading is found as follows:

$$I = \frac{\text{load in watts}}{2 \times \text{phase-to-neutral voltage}} = \frac{17{,}000}{240} = 71 \text{ A}$$

Although the feeder circuit is described as being rated at 120/240 V, these are high values of voltage that usually will not occur during normal operation. It is better to use 115/230 V as more likely and typical of the voltage conditions that will exist. And, since the current is determined by dividing wattage by voltage, it is safer to use the lower voltage value, which yields a higher current value to be used in sizing the feeder conductors. This will provide a higher-rated conductor and reduce the chance of the circuit being overloaded. Thus,

$$I = \frac{17{,}000}{230} = 74\text{ A}$$

The demand factors which may be applied to an overall feeder load, as covered in **NEC** Section 220-11, are not applicable because this is a store building and all the lights will be operating simultaneously and constantly.

Step 3. But, because the load on the feeder is continuous (operates for longer than 3 hours), the value of 74 A must be multiplied by 125 percent to satisfy **NEC** Section 220-10(b) for continuous loading. The feeder must be sized for

$$1.25 \times 74\text{ A} = 93\text{ A}$$

To meet that ampacity requirement, the secondary feeder could be made up of three No. 2 TW copper conductors (rated at 95 A); or No. 2 THW, XHHW, or THHN copper conductors could be used.

Step 4. According to **NEC** Section 384-13, the busbars of a panel must have a current rating not lower than the minimum feeder capacity required for the load. This means that, in this case, the 120/240 V panel must have a rating of at least 93 A—the load on the feeder. A 100-A panel would satisfy this requirement.

Step 5. Based on a total load of 17 kW, the smallest standard size of transformer which could be used to continuously supply the load would be rated at 25 kVA. If the load were not continuously at the maximum load value, it might be possible to use a 15-kVA transformer and allow it to operate overloaded during peak load times. But a 25-kVA transformer is properly required for a continuous 17-kW load.

Step 6. To obtain data for sizing the transformer's overcurrent protection, its primary and secondary full-load currents should be calculated using the formula for a single-phase transformer:

$$\text{Primary current} = \frac{\text{transformer kVA rating} \times 1000}{\text{primary voltage}} = \frac{25 \times 1000}{480} = 52\text{ A}$$

Its full-load secondary current is found with the same formula, applied to the secondary:

$$\text{Secondary current} = \frac{\text{transformer kVA rating} \times 1000}{\text{secondary voltage}} = \frac{25{,}000}{240} = 104\text{ A}$$

Step 7. Under the load conditions given, the load current on the transformer secondary will be the 74 A of load current drawn by the lighting panel. The load current drawn by the transformer primary is found from the formula which relates the currents and voltages on the primary and secondary sides of the transformer:

$$\frac{I_p}{I_s} = \frac{E_s}{E_p}$$

where I_p = primary current, A
 I_s = secondary current, A
 E_s = secondary voltage, V
 E_p = primary voltage, V

For a secondary current of 74 A, the primary current is calculated as

$$I_p = I_s \frac{E_s}{E_p} = 74 \frac{240}{480} = 37 \text{ A}$$

To handle a continuous load of 37 A, the primary circuit conductors must have an ampacity of

$$1.25 \times 37 \text{ A} = 46 \text{ A}$$

Two No. 6 TW copper conductors (rated at 55 A) could be used, as could No. 6 THW, THHN, XHHW, or RHH conductors.

But, because a 25-kVA transformer has been selected, good design practice would call for sizing the entire feeder for the capacity of the transformer—even though the present loading is lower than that capacity.

Step 8. In an application of this sort, the transformer must be protected by a primary CB or fuses in a switch, rated or set at not more than 125 percent of the transformer rated primary current. The rated primary current is computed as 25,000 VA ÷ 480 V = 52 A, and 1.25 × 52 A = 65 A. That rating of protection would not properly protect No. 6 TW conductors (55 A) but would protect No. 6 THW (65 A) or No. 6 THHN or XHHW (75 A) conductors. However, to obtain the full output of the transformer for future use, primary conductors rated at 1.25 × 52 A, or 65 A, would have to be used. And because equipment rated up to 100 A is required by UL to be wired with 60°C conductors (TW)—or with higher-temperature conductors that are sized as if they were TW conductors—the primary circuit would require at least No. 4 TW (70 A) or THW, THHN, or XHHW conductors.

Step 9. If the primary feeder is sized to allow the full 52 A of rated primary current for a 25-kVA transformer, the secondary conductors must be resized at 25,000 VA ÷ 240 V × 1.25, or 130 A, to allow for future load increase up to the transformer's rated 25-kVA capacity. That calls for No. 1 THW copper conductors (130 A) or No. 2/0 TW copper conductors (145 A). If THHN or XHHW conductors are used, they must be sized as if they were 75°C conductors (THW)—which means No. 1 copper must be used.

Step 10. The 100-A lighting panel must then be equipped with 100-A main protection. That protection could be placed right at the transformer secondary, or it could be placed at the load end of unprotected tap conductors that run no more than 25 ft from the transformer secondary to the main 100-A CB or fused switch that serves as main protection in the panel.

Example. Figure 6.27 shows a single-phase transformer with a 2-wire primary circuit and a 2-wire secondary, that is used to feed ungrounded circuits for a hospital operating room.

Step 1. In such a circuit, the secondary conductors may be protected by the overcurrent device protecting the primary circuit to the transformer if the size of the conductors and the rating of the protective device are calculated on the basis of the transformer turns ratio.

Step 2. In the hookup shown, the 30-A, two-pole CB protecting the primary circuit to the transformer is acceptable as protection for the transformer if the transformer

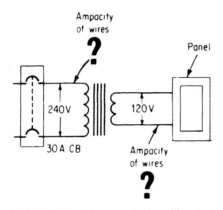

FIGURE 6.27 Primary protection will protect feeder-circuit conductors.

rated primary current is more than 20 A [since 1.25 × 21 A = 26 A, and a 30-A CB is the next higher standard rating of protective device, as permitted by **NEC** Section 450-3(b)(1)]. The minimum size of copper conductors acceptable for use as the primary circuit to the transformer is No. 10 TW. The minimum acceptable size of secondary-circuit conductor is No. 6 TW copper (55 A), for which the 30-A primary CB acts as a 60-A protective device through the 2-to-1 current step-up from primary to secondary. And such 60-A protection is the "next higher standard rating" of CB above 55 A, as permitted by Section 240-3(b).

Step 3. If the transformer shown in the sketch were rated at 5 kVA, it would have a rated primary current (at 240 V) of 5000 VA ÷ 240 V, or 20.8 A. The maximum rating of overcurrent device for such a transformer would be 125 percent of 20.8 A, or 26 A. Because Exception No. 1 of Section 450-3(b)(1) permits the next higher standard rating of protective device from Section 240-6 when 1.25 times the rated primary current does not correspond to a standard rating, the maximum standard rating of CB acceptable as protection for this transformer is 30 A, which is the next higher standard size of CB above 26 A.

Example. Calculations of the various loads supplied by the 208/120-V switchboard shown in Fig. 6.28 reveal that the switchboard would have a noncontinuous demand load of 762 A per phase.

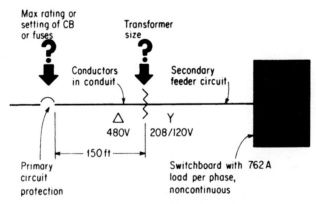

FIGURE 6.28 Switchboard loading dictates transformer size, which then dictates conductor and protection ratings.

Step 1. The minimum transformer kilovoltampere rating that can handle that load is determined with the formula

$$\text{kVA rating} = \text{line-to-line voltage} \times \text{current per phase} \times \frac{1.732}{1000}$$

$$= 208 \text{ V} \times 762 \text{ A} \times \frac{1.732}{1000} = 274 \text{ kVA}$$

Step 2. The nearest standard rating of transformer, which must be used to handle the load, is 300 kVA.

Step 3. A transformer with a 300-kVA rating has a rated primary current of 300 × 1000 ÷ (480 × 1.732), or 361 A.

Step 4. The CB or fuses used to protect the primary feeder circuit to the transformer must have a rating or setting of not more than 125 percent of the transformer rated primary current if the primary-circuit protection is to qualify by itself as the required overcurrent protection for the transformer.

Step 5. Therefore, the primary overcurrent device must have a rating or setting not over 1.25 × 361 A, or 451 A.

Step 6. The maximum standard rating of protective device permitted by Exception No. 1 to **NEC** Section 450-3(b) is 450 A. (Actually, Exception No. 1 would permit the use of 500-A protection as the next higher rating above 451 A, as calculated in step 6 above.)

Step 7. Based on the use of primary overcurrent protection of the rating established in step 6 above, the primary-circuit conductors may be of the same ampacity or of such ampacity that the primary overcurrent device in step 6 is of the "next higher standard rating or setting" above the ampacity rating of the conductors. For this circuit, that condition is achieved if each phase leg of the primary circuit has an ampacity of over 400 A; either a single conductor per phase leg of that ampacity may be used, or two conductors per leg may be used, with each conductor rated over 200 A.

Step 8. If the primary circuit to the transformer is made up of two parallel sets of three conductors, with all six conductors in the same conduit, each conductor could be a 300-kcmil THHN or XHHW aluminum conductor (in a dry location), which has a normal rating of 255 A (from **NEC** Table 310-16); but these conductors must be derated to 80 percent of their ampacity (0.8 × 255 A, or 204 A), as required by Note 8 to Table 316-16, because there are more than three current-carrying conductors in the conduit.

Step 9. With two parallel sets of conductors as described in step 8 above, the ampacity of each phase leg of the primary circuit is 2 × 204 A, or 408 A, and the overcurrent protection sized in step 6 is the next higher rating above the ampacity of the circuit phase legs.

Step 10. The minimum size of conduit required for the six 300-kcmil THHN or XHHW circuit conductors of step 8 above is 3 in.

<u>NOTE</u>: Although 90°C conductors are being used (THHN or XHHW), they will be operating at only 361 A when the transformer is loaded to its capacity. Because 361 A is lower than the 368-A rating of two 300-kcmil THWs in parallel (2 × 230 A × 0.8 = 368 A), the equipment terminal lugs will be within the 75°C operating limit that UL places on conductor terminal operations.

Step 11. If the load on the switchboard fed from the transformer secondary were continuous instead of noncontinuous, the calculations given above for sizing

the primary-circuit overcurrent protection and the primary conductors would have to be evaluated in light of **NEC** Section 220-10(b), which requires feeder conductors and overcurrent protection to have an ampacity at least equal to 125 percent of the load current. The 274-kVA load on the transformer draws a primary current equal to 274 × 1000 ÷ (480 × 1.732), or 330 A, of primary current. Then 1.25 × 330 A = 412 A. The primary-circuit conductors selected in step 9 above have a derated capacity of 408 A per phase, which is not quite up to the required 412-A level. The 300-kcmil conductors are, therefore, not fully adequate for the continuous load. If 350-kcmil THHN aluminum conductors (280 A) are used instead of the 300-kcmil conductors, their derated ampacity of 2 × 280 A × 0.8 = 448 A will satisfy the need for an ampacity of 412 A. And the 450-A rating of the primary protection is acceptable because it is greater than 125 percent of the continuous load of 330 A.

Step 12. The above calculations cover the primary protection, primary-circuit makeup, and transformer protection. The conductors and other components on the secondary side of the transformer must be sized separately, on the basis of other, independent calculations; the primary protection does not affect the required secondary protection or the sizing of the secondary feeder conductors.

Step 13. The secondary conductors from the transformer to the switchboard must be sized on the basis of the demand load on the panel. For the noncontinuous load indicated, the secondary conductors must have an ampacity of not less than 762 A.

Step 14. The secondary-circuit conductors are selected on the basis of the load on the conductors, the distance of the switchboard from the transformer, and consideration of the need for overcurrent protection for the conductors.

Step 15. If the transformer is located adjacent to the switchboard, as shown at the top of Fig. 6.29, so that secondary conductors only 9 ft long can be used in conduit to connect the transformer secondary terminals to the main lugs in the switchboard, those conductors need only have an ampacity that is not less than the main current rating of the switchboard; in addition, they qualify as 10-ft tap conductors and may be used without overcurrent protection.

Step 16. If the secondary conductors are run in conduit from the transformer to the switchboard located 13 ft away, as at the bottom of Fig. 6.29, they may be run without overcurrent protection at the transformer secondary to a single CB or set of fuses in the switchboard, provided that the conductors have an ampacity of at least 762 A (for a noncontinuous load), they are not more than 25 ft long, and the protective device in which they terminate protects them at their rated ampacity.

Step 17. If the secondary conductors are run in conduit to the switchboard located 50 ft away from the transformer, they must be supplied with overcurrent protection at the transformer (their supply end); this overcurrent protection must protect the secondary conductors at their rated ampacity, which must be not less than 762 A.

Step 18. If the load on the switchboard consisted of 500 A of continuous load and 262 A of noncontinuous load, the rating of the overcurrent protection at the transformer in step 17 above would have to be not less than 125 percent of 500 A plus 262 A, or 887 A. And the current rating of the switchboard would then have to be at least 1000 A.

Step 19. The condition described in step 18 would call for a standard secondary overcurrent device (CB or set of fuses) rated not below 1000 A. Then the secondary-circuit conductors would have to have at least that ampacity and not a lower ampacity, because the overcurrent device would be rated over 800 A.

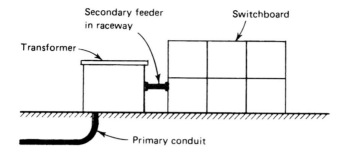

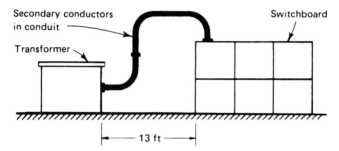

FIGURE 6.29 Distance between a transformer and the equipment it feeds can alter calculations.

Transformer Grounding

With the great popularity of distribution systems using transformer loadcenters, particularly 480/277-V systems, effective grounding of transformer secondaries has become an important consideration in modern electrical-system design. In a 480/277-V system, 120-V circuits for incandescent lighting and receptacle outlets may be provided either by using single-phase transformers to step 480 V down to 240/120-V, single-phase secondaries or by using 3-phase transformers to step the 480-V, 3-phase primary supply down to 208/120-V, 3-phase, 4-wire secondaries. The ac systems supplied from the secondaries of such transformers are considered to be "separately derived ac wiring systems" and are subject to the grounding requirements of **NEC** Section 250-26. The same rules on grounding and bonding apply to the output circuit of an engine generator used for emergency or standby power supply and to that of a generator used for on-site supply of the normal power to a building or facility.

Details on the grounding of transformer secondaries are as follows:

1. Any system that operates at over 50 V but not more than 150 V to ground must be grounded; that is, it must be operated with one of its circuit conductors connected to a grounding electrode.

2. This requires the grounding of secondaries of dry-type transformers serving 208/120-V, 3-phase or 240/120-V, single-phase circuits for lighting and appliance

outlets and receptacles at loadcenters throughout a building, as shown in Fig. 6.30. And 480/277-V, wye-connected secondaries must also be grounded if indoor lighting is to be supplied—with loads connected from line to neutral and/or from line to line. Electric-discharge lighting that is supplied by 480-V ballasts indoors *must* be fed from a 480/277-V transformer secondary with a grounding connection to the neutral point to provide a voltage that is not over 300 V to ground, as required by **NEC** Section 210-6.

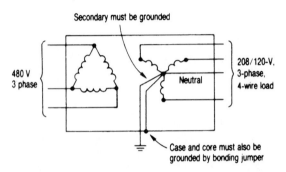

FIGURE 6.30 Transformer secondary neutral must be both grounded and bonded to the case.

3. All rules applying to both system and equipment grounding must be satisfied in such installations.

The wiring-system conductor to be grounded in compliance with items 1 and 2 above depends upon the type of system. In 3-wire, single-phase systems, the midpoint of the transformer winding—the point from which the system neutral is derived—is grounded. In grounded 3-phase wiring systems (either 3-wire or 4-wire systems), the neutral point of the wye-connected transformer or generator is the point connected to ground. In delta-connected transformer hookups, grounding can be effected by grounding one of the three phase legs, by grounding a center-tap point on one of the transformer windings (as in the 3-phase, 4-wire "red-leg" delta system), or by using a special grounding transformer which establishes a neutral point of a wye connection which is grounded.

The steps involved in designing transformer secondary connections are as follows (Fig. 6.31).

Step 1. A bonding jumper must be installed between the transformer secondary neutral terminal and the metal case of the transformer. The size of this bonding conductor is obtained from **NEC** Table 250-94, based on the size of the transformer secondary phase conductors (as if they were service-entrance conductors) and is the same as the size of the required grounding-electrode conductor. For cases where the transformer secondary-circuit conductors are larger than 1100-kcmil copper or 1750-kcmil aluminum per phase leg, the size of the bonding jumper must not be less than 12½ percent of the cross-sectional area of the secondary phase leg.

Example. A 75-kVA transformer has a 208/120-V, 3-phase, 4-wire secondary, with current of

$$\frac{75{,}000}{208 \times 1.732} = 209 \text{ A}$$

STEP 1 – BONDING JUMPER

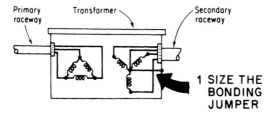

STEP 2 – GROUNDING ELECTRODE CONDUCTOR

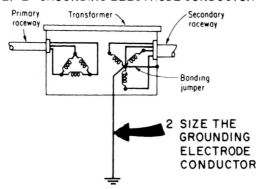

STEP 3 – GROUNDING ELECTRODE

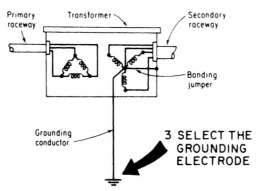

FIGURE 6.31 These design steps provide for transformer secondary connections.

If No. 4/0 THW copper conductors (with a 230-A rating) were used for the secondary phase legs, then the size of the required bonding jumper would be obtained from Table 250-94 as if 4/0 service conductors were being used. The table shows that 4/0 copper service conductors require a minimum of No. 2 copper or No. 1/0 aluminum for a grounding-electrode conductor. The bonding jumper would have to be of either of those two sizes.

If the transformer were a 500-kVA unit with a 120/208-V secondary, its rated secondary current would be

$$\frac{500 \times 1000}{1.732 \times 208} = 1388 \text{ A}$$

If say, THW aluminum conductors were used, then each secondary phase leg would be made up of four 700-kcmil aluminum conductors in parallel (each 700-kcmil THW aluminum is rated at 375 A, and four are rated at 4 × 375 A, or 1500 A, which suits the 1388-A load). Then, because 4 × 700 kcmil equals 2800 kcmil per phase leg, which is in excess of 1750 kcmil, Section 250-79(c) would require the bonding jumper from the case to the neutral terminal to be at least equal to 12½ percent of 2800 kcmil (0.125 × 2800 kcmil), or 350-kcmil aluminum.

Step 2. A grounding-electrode conductor must be installed from the transformer secondary neutral terminal to a suitable grounding electrode. This grounding conductor is sized the same as the required bonding jumper in the above example. That is, this grounding-electrode conductor is sized according to Table 250-94 as if it were a grounding-electrode conductor for a service with service-entrance conductors equal in size to the phase conductors used on the transformer secondary side. But, this grounding-electrode conductor does not have to be larger than 3/0 copper or 250-kcmil aluminum when the transformer secondary phase legs are over 1100-kcmil copper or 1750-kcmil aluminum.

Example. For the 75-kVA transformer in step 1, the grounding-electrode conductor must not be smaller than the required minimum size shown in Table 250-94 for 4/0 phase legs, which makes it the same size as the bonding jumper—that is, No. 2 copper or No. 1/0 aluminum. For the 500-kVA transformer, the grounding-electrode conductor is sized directly from Table 250-94, which requires 3/0 copper of 250-kcmil aluminum where the phase legs are over 1100-kcmil copper or 1750-kcmil aluminum. Note that the grounding-electrode conductor for this 500-kVA transformer is smaller than the bonding jumper sized in step 1.

In the 1978 **NEC,** this rule called for the bonding and grounding of a grounded transformer secondary or generator output (for example, 208/120-V wye) "at the source" of the separately derived system. That phrase was frequently interpreted very rigidly to mean "only at the transformer itself." In the present **NEC,** the last sentence of Sections 250-26(a) and (b) permits the bonding and grounding connections to be made either right at the transformer or generator or at the first disconnect or overcurrent device fed from the transformer or generator, as in Fig. 6.32.

The last sentence of Sections 250-26(a) and (b) does, however, say that the bonding and grounding "shall be made at the source"—which appears to mean right at the transformer or generator, and not any other point—where the transformer supplies a system that "has no disconnecting means or overcurrent device." A local transformer that has no disconnect or overcurrent devices on its secondary is one that supplies only a single circuit and has overcurrent protection on its primary, such as a control transformer to supply motor starter coils. But in such applications, the transformer does not supply a separately derived "system" in the usual sense that a "system" consists of more than one circuit. However, for a transformer supplying only one circuit, any required bonding and grounding of a secondary grounded conductor—such as the grounded leg of a 2-wire, 120-V control circuit—would normally have to be done right at the transformer.

Another interpretation that might be put on that last phrase is that it refers to a hookup in which the transformer secondary feeds main lugs only in a panelboard,

TRANSFORMER APPLICATIONS (TO 600 V)   473

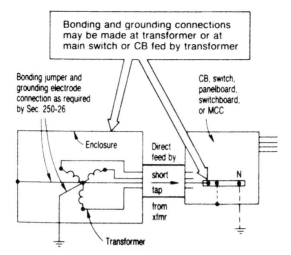

FIGURE 6.32 Secondary neutral grounding and bonding may be accomplished at either of two places.

switchboard, or motor control center. In such cases, the absence of a main CB or fused switch would mean that there is no disconnect means or overcurrent device for the overall "system" fed by the transformer, even though there are disconnects and overcurrent protection for the individual "circuits" that make up the "system." That interpretation would require bonding and grounding of the secondary right at the source (the transformer itself).

Step 3. The grounding-electrode conductor, installed and sized as in step 2 above, must be properly connected to a grounding electrode that must be "as near as practicable to and preferably in the same area as the grounding conductor connection to the system." That is, the grounding electrode must be as near as possible to the transformer itself. In order of preference, the grounding electrode must be one of the following:

1. The nearest available structural steel of the building, provided it is established that such building steel is effectively grounded
2. The nearest available water pipe, provided it is effectively grounded

Section 250-112 clarifies the term "effectively grounded" by noting that the grounding connection to a grounding electrode must "assure a permanent and effective ground." Thus, when a nearby cold-water pipe is used as a grounding electrode, it would appear to be necessary to bond around any unions or valves that might be opened and thereby might break the piping connection to the earth (Fig. 6.33). And Section 250-81(a) requires a bonding jumper to be used around all indoor water meters to ensure continuity to earth or through the interior water-pipe system. Such bonding jumpers must be at least the same size as the grounding conductor from the transformer to the water pipe and other electrodes.

At least 10 ft of the metal water pipe must be buried in earth outside the building for the water-pipe system to qualify as a grounding electrode for the transformer secondary neutral. However, even where there is not 10 ft of metal pipe in the earth,

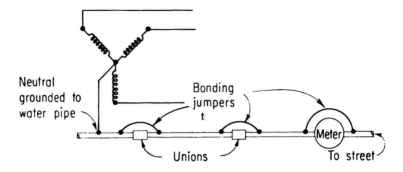

NOTE: The bonding shown here are clearly required by the NE Code does not specifically clarify this point, although literal wording appears to require bonding.

FIGURE 6.33 Reliability and conductivity of grounding-electrode connections must be assured.

there must always be a connection between an interior metal water-pipe system and the service entrance grounded conductor (the neutral of the system that feeds the primary of transformers in the building). The grounding connection for the service neutral or other system grounded conductor must be made at the service. Where a metallic water-pipe system in a building is fed from a nonmetallic underground piping system or has less than 10 ft of metal pipe underground, the water pipe is not a grounding electrode, and the service neutral or other service grounded conductor must have a connection to a ground rod or other electrode in addition to the connection to the interior metal water-pipe system.

Note that Section 250-26 accepts an "effectively grounded" metal water pipe as a suitable grounding electrode for a "separately derived system" such as a transformer secondary. But, Section 250-81 *does not* permit the use of a water-pipe electrode (with at least 10 ft in the ground) as the *sole* grounding electrode for the service to a building. Any such water-pipe electrode must be supplemented by bonding to at least one other grounding electrode—such as grounded building steel, reinforcing bars in the footing or foundation, or a driven ground rod. (See "Grounding Electrodes" in Chapter 7 of this handbook.)

Figure 6.34 shows the various elements involved in grounding and bonding at a transformer. Transformer housings must be grounded by connection to grounded cable armor or metal raceway, by use of a grounding conductor run with the primary circuit conductors (either a bare conductor or a conductor with green-colored covering) or by a separate equipment grounding conductor. Such an effective equipment grounding conductor (metal raceway or a separate conductor) is necessary to provide for fault-current return in case an energized conductor of a grounded primary supply circuit (such as a 3-phase, 480-V primary circuit derived from a grounded-wye, 480/277-V system) faults to the metal transformer housing. Even though the secondary neutral is bonded to the case and grounded to building steel, a low-impedance ground-return path must be provided for primary circuit faults in the metal housing to clear a primary protective device—as required by **NEC** Section 250-57(b). But the clearing of ground faults on the *secondary* side of the transformer is facilitated by the bonding that Section 250-26 requires between the secondary neutral and the metal case of the transformer, as shown in Fig. 6.35.

TRANSFORMER APPLICATIONS (TO 600 V) 475

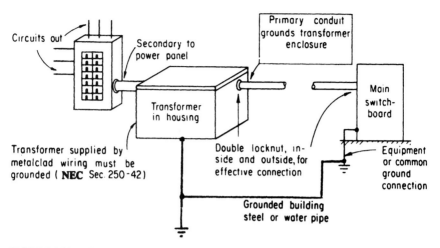

FIGURE 6.34 All the elements of transformer grounding and bonding.

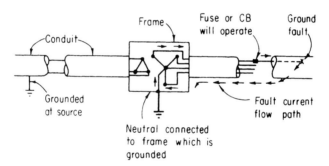

FIGURE 6.35 Secondary neutral bonding provides for fault clearing on secondary faults but not on primary faults.

Figure 6.36 shows an important detail for effective grounding of a transformer case and secondary neutral. A common technique for protecting a bare or insulated system grounding conductor (one which grounds the wiring system and equipment cases) and for protecting equipment-only grounding conductors makes use of a metal conduit sleeve, run open or installed in concrete. In all such cases, the **NEC** says that "metallic enclosures for grounding conductors shall be electrically continuous from the point of attachment to cabinets or equipment to the grounding electrode, and shall be securely fastened to the ground clamp or fitting." This means that the grounding conductor must be connected to its protective conduit at both ends so that any current which might flow over the conductor will also have the conduit as a parallel path.

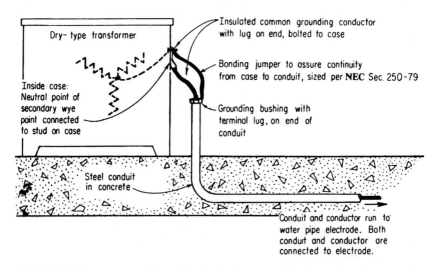

FIGURE 6.36 Protective metallic conduit sleeving for a grounding-electrode conductor *must* be bonded to the conductor at both ends.

The necessity for making a grounding conductor electrically parallel with its protective conduit applies to any grounding conductor. If the protective conduit were arranged so that the conductor and conduit were not acting as parallel conductors—such as would be the case in Fig. 6.36 if there were no bonding jumper from the conduit bushing to the conductor lug—the presence of magnetic metal conduit (steel) would serve to greatly increase the inductive reactance of the grounding conductor in limiting the flow of current to ground. The steel conduit would act as the core of a "choke" to restrict current flow.

Where building steel or a metal water pipe is not available for the grounding of in-building transformers (or generators), other electrodes may be used, based on **NEC** Section 250-81 or 250-83.

Figure 6.37 shows techniques of transformer grounding that have been used in the past but are no longer acceptable.

As shown in Fig. 6.38, the Exception to Section 250-26(b) exempts small control and signaling transformers from the basic requirement for a grounding-electrode conductor run from the bonded secondary grounded conductor (such as a neutral) to a grounding electrode (nearby building steel or a water pipe). A Class 1, 2, or 3 remote-control or signaling transformer that is rated not over 1000 VA simply has to have a grounded secondary conductor bonded to the metal case of the transformer; no grounding-electrode conductor is needed—provided that the metal transformer case itself is properly grounded by grounded metal raceway supplying its primary or by means of a suitable equipment grounding conductor that ties the case back to the grounding electrode for the primary system. The Exception to Section 250-26(a) permits the use of a No. 14 copper conductor to bond the grounded leg of the transformer secondary to the transformer frame, leaving the supply conduit to the transformer to provide the path to ground back to the main service ground, but depending on the connection between neutral and frame to provide an effective return for clearing faults, as shown. Transformer housings must be grounded by connection to grounded cable armor or metal raceway or by use of a grounding conductor run with circuit conductors (either a bare conductor or a conductor with green-colored covering).

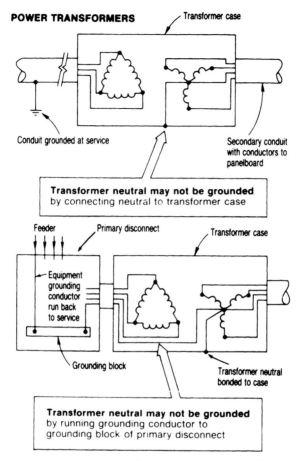

FIGURE 6.37 These grounding methods are *not* acceptable.

Because the rule on bonding jumpers for the secondary neutral point of a transformer refers to Section 250-79(c) and therefore ties into Table 250-94, the smallest size that may be used is No. 8 copper, as shown in that table. But for small transformers (such as those used for Class 1, 2, or 3 remote-control or signaling circuits) a bonding jumper that large is not necessary and is not suited to termination provisions. For that reason, the Exception to Section 250-26(a) permits the bonding jumper for such transformers rated not over 1000 VA to be smaller than No. 8 wire. The jumper simply has to be at least the same size as the secondary phase legs but in no case smaller than No. 14 copper or No. 12 aluminum.

Other Design Examples

In Fig. 6.39, for a feeder of a 480/277-V, 3-phase, 4-wire system, a 140-kVA demand load of incandescent lighting and receptacle circuits must be fed from a transformer

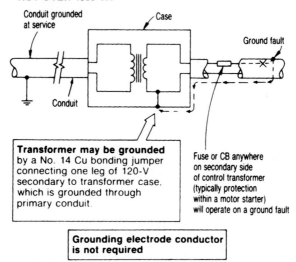

FIGURE 6.38 Grounding-electrode connection is *not* required here.

which will feed two 120/208-V, 3-phase, 4-wire panelboards for the required branch circuits to the loads:

1. What size 3-phase, dry-type transformer must be used for the total 120/208-V load?
2. What size primary and secondary conductors must be used to carry the transformer current?
3. What size bonding jumper and grounding-electrode conductors must be used at the transformer?
4. What ratings and arrangements of primary and secondary overcurrent protection are required?

These questions are answered in the next several paragraphs, for both noncontinuous and continuous loading.

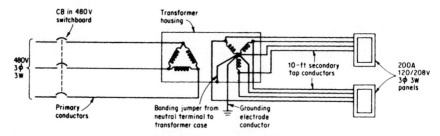

FIGURE 6.39 Sizing of all circuit components must be carefully coordinated for all conditions.

TRANSFORMER APPLICATIONS (TO 600 V) 479

Transformer. The total demand load to be supplied from the 120/208-V, 3-phase, 4-wire secondary is 140 kVA, and the next highest standard size of dry-type transformer is 150 kVA. If the extra 10 kVA of capacity is sufficient to accommodate anticipated load growth, a 150-kVA transformer would do the job.

Conductors. The full-load primary current for a 150-kVA transformer with a 480-V, 3-phase, 3-wire delta primary hookup is determined using the standard formula:

$$\text{Rated primary current} = \frac{\text{rated transformer kVA} \times 1000}{1.73 \times \text{rated primary phase-to-phase voltage}}$$

$$= \frac{150 \times 1000}{1.73 \times 480} = \frac{150{,}000}{831} = 180 \text{ A}$$

This means that the ampacity of the primary-circuit conductors from the CB in the 480-V switchboard to the transformer must be at least 180 A to provide full use of the 150-kVA rating of the transformer.

NOTE: If the load fed by the circuit is continuous, a 125 percent multiplier must be used in sizing all circuit conductors, as explained later.

Conductors are selected with **NEC** Table 310-16. If the primary conductors are to be copper and in a raceway, then No. 4/0 Type TW conductors (rated 195 A) or No. 3/0 Type THW (rated 200 A) or No. 3/0 Type THHN or XHHW (rated 225 A) appear to be acceptable. All these conductors would be operating within the ampacity and heating limitation for 75°C (THW) conductors. If aluminum conductors are to be used, Table 310-16 shows that three No. 4/0 TW in 2-in conduit or three No. 3/0 THW in 2-in conduit or three No. 3/0 THHN or XHHW in 1½-in conduit would be acceptable.

As shown in Fig. 6.39, the primary circuit run to the transformer from the 480-V switchboard is long, but the two panels fed by the transformer are located immediately adjacent to the transformer, and each is fed by tap conductors not over 10 ft long from the transformer to each panel.

We know that the rated secondary current of the transformer is

$$\text{Rated secondary current} = \frac{\text{rated transformer kVA} \times 1000}{1.73 \times \text{rated secondary phase-to-phase voltage}}$$

$$= \frac{150 \times 1000}{1.73 \times 208} = \frac{150{,}000}{360} = 417 \text{ A}$$

But, in this case, the transformer capacity will be divided equally between the two 200-A panels; i.e., a set of secondary conductors rated 200 A will be used for the 10-ft tap to each panel (assuming a separate raceway to each panel).

From **NEC** Table 310-16, each 10-ft feeder tap could be made up of four of any of the following copper conductors:

250-kcmil Type TW (rated 215 A)

No. 3/0 Type THW (rated 200 A)

No. 3/0 Type THHN or XHHW (rated 225 A)

NOTE: The 225-A rating of the THHN or XHHW and the 215-A rating of the 250-kcmil TW make the full secondary capacity of the transformer available, so that two conductors of any of these types, when paralleled, would have a current-carrying capacity greater than 417 A, the full secondary current rating. If a single set of secondary conductors had been taken from the transformer to supply, say, a single panel, the conductors would have had to be rated at least 417 A (as are 600-kcmil Type THW, rated 420 A). However, in this example the secondary capacity is divided between two panels.

Taps from the transformer, if not more than 10 ft long, are acceptable, just as taps from a feeder with no change in voltage level would be. It is only necessary that all design factors be accounted for:

- The primary-circuit conductors to the transformer must be sized and protected properly.
- The transformer must be protected.
- The tap conductors must be sized and installed in accordance with Section 240-21(b):

1. The tap conductors must not be over 10 ft long.
2. The ampacity of the tap conductors must not be less than the combined computed loads on the panel feed and not less than the rating of the panel mains or not less than the rating of fuses or a CB in which they terminate.
3. The tap conductors must be enclosed in a raceway from the transformer case to each panel.

Protection of the panel mains must also be considered. If either of these panels is a "lighting and appliance" panel (more than 10 percent of its CB poles or fuses rated 30 A or less, with a neutral terminal block provided in the panel), each such panel must have main protection as described in Section 384-16. Basically, this means that a lighting and appliance panel must be protected by a main CB or fused switch with a rating not greater than the rating of the panel mains (i.e., not over 200 A).

Grounding. The bonding jumper and grounding-electrode conductor required by **NEC** Section 250-26 for the transformer would be sized as follows:

The bonding jumper used to connect the system neutral of the transformer to the metal housing of the transformer must be sized in accordance with Section 250-79(c), which says that Table 250-94 must be used just as in sizing the grounding-electrode conductor that runs to a water piper building steel.

The required size of the bonding jumper, either copper or aluminum, is selected by referring to Table 250-94 and using the size of the transformer secondary conductors as if they were service-entrance conductors. In this example, with the secondary circuit made up of two sets of, say, No. 3/0 copper THW, it is necessary to convert the two No. 3/0 conductors to an equivalent single copper conductor with a cross section equal to that of the two No. 3/0 THWs.

NOTE: In **NEC** Table 250-94, the phrase "or Equivalent for Parallel Conductors" in the heading of the left-hand column must be taken as meaning equivalent in cross-sectional area, not in ampacity. And the type of insulation on the service-entrance conductors is of no concern in this calculation. Physical size of conductor material—effective cross section per phase—is the sole determinant of the required size of grounding-electrode conductor (and bonding jumper).

The cross-sectional area of a single No. 3/0 stranded conductor is 167,800 cmil (from Table 8 of **NEC** Chapter 9). The single conductor "equivalent" to two No. 3/0 conductors would have a cross-sectional area of $2 \times 167,800$ or 335,600 cmil. From Table 8, the single conductor having cross-sectional area closest to this figure is the 350 kcmil, which has a cross-sectional area of 350,000 cmil.

Table 250-94 shows that, for 350-kcmil service conductors, a minimum of No. 2 copper or No. 1/0 aluminum must be used for a grounding-electrode conductor. For the transformer in this example, then, the bonding jumper must not be smaller than No. 2 copper or No. 1/0 aluminum.

The grounding-electrode conductor used to connect the transformer neutral (and the bonded equipment enclosure) to the grounding electrode must be sized according to Section 250-26, which says in effect that the grounding-electrode conductor must be sized the same as the bonding jumper, using Table 250-94 as described above. Thus, for this transformer application, the grounding-electrode conductor must not be smaller than No. 2 copper or No. 1/0 aluminum (again, from Table 250-94).

IMPORTANT: The size of a grounding-electrode conductor and bonding jumper for a dry-type transformer application is based on the conductor material and cross-sectional area of the secondary phase conductors. Thus, the use of higher-temperature-rated secondary phase conductors to obtain the required capacity in a smaller conductor (say THW instead of TW) will result in a smaller required grounding-electrode conductor and bonding jumper.

In this example, if TW conductors had been used for the secondary phase conductors rather than THW, they would have had to be at least 250-kcmil conductors. Two per phase would have been equivalent in size to one 500 kcmil, which would require a minimum of No. 1/0 copper or No. 3/0 aluminum for the grounding-electrode conductor and transformer bonding jumper (Table 250-94).

When transformer secondary conductors are selected at 125 percent of the "continuous" load current, they are still larger than those required for noncontinuous operation—and thereby often require larger grounding-electrode conductors and bonding jumpers.

If the 150-kVA transformer of this example had been used to supply, say, one 400-A panel with a single set of conductors, each secondary phase conductor might have been a single 600-kcmil THW, rated at 420 A. The minimum required size of grounding-electrode conductor and bonding jumper would then have been No. 1/0 copper or No. 3/0 aluminum.

Overcurrent Protection. Overcurrent protection for this 150-kVA dry-type transformer must conform to **NEC** Section 450-3(b). According to subparagraph (1) of that section, a dry-type transformer must be protected by an overcurrent device rated or set at not more than 125 percent of the rated primary current. This device may be right at the transformer primary or at the supply end of the circuit that feeds the transformer primary.

In this example, say, three No. 3/0 THW conductors are used for the primary supply circuit to carry the 180 A of rated primary current (assuming future full use of the transformer). These conductors must then be protected at not more than their 200-A ampacity. A 200-A CB or 200-A fuses would provide such protection.

Because Section 450-3(b)(1) sets a basic limit of 125 percent of the primary current on the rating of the transformer protective device, a CB or fuses used to protect this transformer must be rated at not more than 1.25×180 A or 225 A. Any rating or setting below this value would be acceptable according to Section 450-3(b)(1), so that a 200-A CB or 200-A fuse could be used.

Could these No. 3/0 THW primary conductors be used, then, for the circuit to the transformer, with a 200-A CB or 200-A fuses in a 200-A switch for both transformer and circuit protection? Before this question can be answered, other rules must be accounted for. If the 120/208-V load fed by the transformer is a 180-A "continuous" load (the full load operates continuously for 3 hours or more), the following must be considered.

1. When a feeder supplies a continuous load, the rating of the overcurrent device protecting the feeder must be not less than 125 percent of the continuous load [**NEC** Section 220-10(b)].
2. If the CB or fused switch feeding the transformer is located in a panelboard, the load on either one of them (180 A, under full-load conditions) must not exceed 80 percent of the rating of the CB or fuses [**NEC** Section 384-16(c)].
3. A fused switch used for other than a motor circuit must not be loaded in excess of 80 percent of the rating of the fuses in the switch (UL requirements). And a general UL rule limits the continuous loading on any molded-case circuit to not over 80 percent of the CB rating.

If the full load on the transformer is, say, lighting that will operate for 3-hour periods or longer, or if it is assumed that the load will be continuous, the foregoing rules would prohibit the use of a 200-A CB or fuses for this transformer circuit. Certainly the circuit to the transformer is a feeder which must have protection complying with item 1 above. Then, the use of a 200-A CB or 200-A fuses is not acceptable.

If the transformer's rated primary current of 180 A must not exceed 80 percent of the rating of the overcurrent protection (that is, the rating of the device is not less than 125 percent of 180 A), then the minimum acceptable rating of overcurrent protection is 180 A × 1.25 = 225 A. Thus, two conditions must be met:

- The primary circuit protection device (CB or fuses) must not be rated below 225 A [Section 220-10(b)].
- The same protective device, if it is also to provide protection for the transformer, must not be rated above 225 A.

But, when 225-A overcurrent protection is substituted for the initial selection of a 200-A CB or 200-A fuses, the size of the primary circuit conductors must be increased from No. 3/0 THW, because the 225-A protection does not protect these 200-A conductors within their allowable ampacity. It is important to note that Section 220-10(b) does *not* also require the circuit conductors to be rated not less than 180 A × 1.25, or 225 A, to correspond with standard fuse or CB ratings. But, the conductor size must be properly protected by the 225-A device, which virtually requires that the conductor be rated at 125 percent.

Primary-circuit conductors which are acceptable for use with 225-A protection must be selected from Table 310-16. The conductors could be as follows:

Copper. 300-kcmil TW (rated 240 A), No. 4/0 THW (rated 230 A), or 4/0 THHN or XHHW (rated 260 A)

Aluminum. 400-kcmil TW (rated 225 A), 300-kcmil THW (rated 230 A), or 300-kcmil THHN or XHHW (rated 255 A)

NOTE: For other than continuous loads, where conductors have a current rating that does not correspond to a standard fuse or CB rating, the next larger size of fuse or CB may be used. However, for continuous loading, Section 220-10(b) does not provide that type of exception to the rule that the conductors be rated at 125 percent of the load.

But, for transformer protection, it must be understood that exceptions to Section 450-3(b) do give permission to use the next larger standard overcurrent device rating when the transformer full-load current does not correspond to a standard rating of CB or fuse. However, in some cases, the next smaller size must be used to avoid exceeding the maximum value. (Permission for use of the "next higher size" is also given in Section 430-52, covering motor branch-circuit protection.)

Example. In Fig. 6.40, a 500-kVA transformer is used to step 480-V, 3-phase power down to 208-120 V for incandescent lighting, electric heating and receptacle outlets. No motor loads are fed by the secondary. The total load on the transformer is 422 kVA, of which 70 percent is continuous.

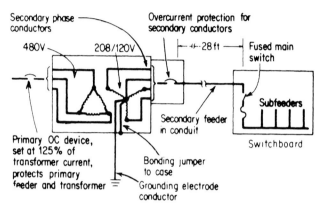

FIGURE 6.40 Loading, conductor sizing, and the relative locations of equipment influence design calculations.

Step 1. The secondary current rating of the transformer is determined from the formula

$$I = \text{kVA rating} \times \frac{1000}{1.732 \times 208}$$

Step 2. The rated secondary current is

$$\frac{500 \times 1000}{1.732 \times 208} = 1388 \text{ A}$$

Step 3. To use the full secondary capacity of the transformer, a circuit makeup of four 700-kcmil THW aluminum conductors per phase leg (each rated at 375 A) would be acceptable for the load (4 × 375 A = 1500 A). Of course, to maintain that load ampacity, each set of three phases and a neutral must be installed in a separate conduit.

Step 4. Based on the use of four 700-kcmil aluminum conductors per phase, the bonding jumper must be sized for a phase-leg size of 4 × 700 kcmil, or 2800 kcmil. The minimum acceptable size of aluminum bonding jumper would be 12½ percent of 2800 kcmil, or 350 kcmil, which would require the use of a standard 350-kcmil jumper. If a copper bonding jumper were used, it would have to be sized at 12½ percent of 2000 kcmil [four 500-kcmil THW copper conductors, which have a total ampacity of 4 × 380 A, or 1520 A, and comprise the copper equivalent of four 700-kcmil THW aluminum conductors per phase, as required by **NEC** Section 250-79(c)]. That would call for a 250-kcmil copper jumper.

Step 5. The minimum acceptable size of grounding-electrode conductor to connect this transformer neutral to a nearby grounded structural-steel column is 3/0 copper or 250-kcmil aluminum.

Step 6. The secondary conductors from the transformer to the switchboard are provided with overcurrent protection right at the transformer, and the protective device must have a rating or setting that is at least equal to the total load because by definition "ampacity" *is* a continuous rating. But, to facilitate selection of the overcurrent device, the conductor is also taken at 125 percent of the continuous load plus the noncontinuous load. The total load on the secondary is equal to 422 kVA ÷ (208 V × 1.732), or 1171 A. The continuous part of that load is 70 percent of 1171 A, or 820 A. The secondary conductors and their protection must be rated for at least 1.25 × 820 A + 351 A, or 1376 A.

Step 7. With such loading, the circuit breaker protecting the secondary would have to be at least a 1400-A standard size (the next rating above 1376 A). The 1500-A rating of the secondary circuit of four 700-kcmil THW aluminum conductors fully satisfies the need for an ampacity of at least 1376 A. And the 1400-A CB is acceptable protection for conductors that have an ampacity of 1500 A.

NOTE: Under those conditions the entire circuit and transformer are loaded to within 24 A of the maximum permissible load on a 1400-A circuit breaker that is *not* UL-listed for continuous operation at full-load rating (1400 A – 1376 A). Up to 24 A of noncontinuous load may be added. If it is added, the load on the transformer will then be 1171 A + 24 A, or 1195 A. But the additional transformer capacity of 193 A (1388 A rated less 1195 A load) could not be utilized. This example makes a strong case for use of 100 percent continuous-rated circuit breakers or 100 percent fusible switching equipment, in which case a CB or fused switch rated at 1200 A could be used for the 1171-A secondary load, with conductors having an ampacity of at least 1200 A.

Step 8. If the switchboard were located 5 ft closer to the transformer, there would be no need for overcurrent protection at the transformer, provided the secondary conductors were tap conductors not over 25 ft long, terminating in a single main CB or fused switch in the switchboard with protection rated at not more than the ampacity of the conductors.

Autotransformers

Most buildings with 480/277-V electrical systems obtain 208/120-V power by means of 2-winding (or insulating) transformers. The same function can be obtained from autotransformers. Figure 6.41 shows a comparison of the 2-winding hookup versus the autotransformer hookup.

In a large, modern commercial, institutional, or industrial building, the usual practice is to locate small step-down transformers—generally in the 15- to 45-kVA range, with 480-V delta primary and 208/120-V wye secondary windings—at regular intervals to provide power for convenience outlets and miscellaneous small equipment. And, quite often, relatively large amounts of 208/120-V power may be required for larger load concentrations such as computer equipment and kitchen loads. In such cases, transformers rated up to 500 kVA are commonly used.

The following comparison of one manufacturer's 225-kVA dual-winding transformers and autotransformers indicates the relative efficiency, weight, and sound-

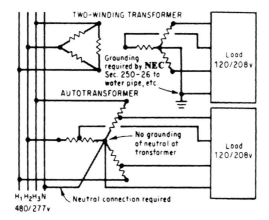

FIGURE 6.41 Autotransformer may be used to derive a 120/208-V, 3-phase, 4-wire system from a 480/277-V system.

level advantages of the autotransformer for the common 480/277- to 208/120-V transformation.

	Dual-winding transformer	Autotransformer
Full-load loss, W	5800	3200 (55%)
Sound level, db	47	44 (94%)
Weight, lb	1750	1050 (60%)

NEC Section 210-9, headed "Circuits Derived from Autotransformers," says "branch circuits shall not be supplied by autotransformers" (transformers "in which a part of the winding is common to both primary and secondary circuits") unless the system supplied has an identified grounded conductor which is solidly connected to a similar identified grounded conductor of the system supplying the autotransformer. In effect, this section requires that a grounded neutral conductor be included in the supply circuit to the primary side of a 480/277- to 208/120-V autotransformer. And the secondary side of the autotransformer must also have a grounded neutral which is connected to the primary grounded neutral.

The addition of this fourth conductor for the primary supply to an autotransformer is not a significant disadvantage in the usual layout of 480/277-V systems. Dry-type transformers for stepping down to 208/120 V are located very close to (commonly within a few feet of) the 480/277-V panel. The neutral feeder conductor must be brought to this panel to provide for the 277-V grounded circuits for fluorescent lighting. Then the only additional requirement for supplying an autotransformer instead of a 2-winding transformer is the short run of neutral conductor from the 480/277-V panel to the autotransformer.

Although the 480-V primary of a 2-winding transformer (normally a delta-wye connection) requires only connections of the three-phase conductors (no neutral is needed), an **NEC** rule requires a service-type grounding connection to be made at each 2-winding transformer (as described previously for 2-winding transformers, which are "separately derived systems" as covered by **NEC** Section 250-26). The definite grounding connection required by the **NEC** for all 2-winding units adds labor and material costs when 2-winding transformers are used. These added grounding costs for 2-winding units are generally greater than the cost of neutral connection for an autotransformer primary.

In specifying autotransformers to derive 4-wire, 208/120-V systems from 480-V feeders, it is very desirable to require three-legged core-type units. This construction leads to reduced third-harmonic currents and is more tolerant of unbalanced phase loads. Autotransformers are also lighter and, therefore, easier and less expensive to install. Since they are more efficient, they require less ventilation. They are also significantly quieter in the larger sizes.

Figure 6.42 shows two other applications for autotransformers. The top diagram shows how a 110-V system for lighting may be derived from a 220-V system by means of an autotransformer. The 220-V system may be either a single-phase system or one leg of a 3-phase system. In the case illustrated, the "supplied" system has the required grounded wire solidly connected to a grounded wire of the "supplying" system: a 220-V single-phase system with one conductor grounded.

Exception No. 1 of **NEC** Section 210-9 permits the use of an autotransformer in existing installations for an individual branch circuit without connection to a similar identified grounded conductor where it is used for transforming from 208 to 240 V

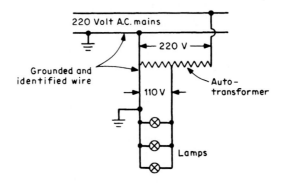

Autotransformer used to derive a two-wire 110-V system for lighting from a 220-V power system.

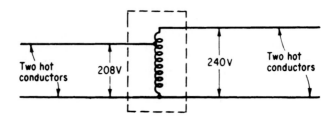

Autotransformers without grounded conductors are recognized.

FIGURE 6.42 Two common uses of autotransformers, mainly on branch circuits.

or vice versa, as shown at the bottom of Fig. 6.42. Typical applications are concerned with cooking equipment, heaters, motors, and air-conditioning equipment. For such applications, transformers are commonly used. This is a long-established practice for voltage ranges where a hazard is not considered to exist.

Buck or *boost* transformers are designed for use on single- or 3-phase circuits to supply 12/24- or 16/32-V secondaries with a 120/240-V primary. When connected as autotransformers, they will handle kilovoltampere loads that are large in comparison with their physical size and relative cost.

Grounding Autotransformers. In recent years, autotransformers have been used increasingly to convert existing ungrounded 480-V industrial distribution systems to grounded operation. There are two basic ways to convert an ungrounded system to a grounded type:

First, one of the three-phase legs of the 480-V delta can be intentionally connected to a grounding-electrode conductor that is then run to a suitable grounding electrode. Such grounding gives the two ungrounded phases (A and B) a voltage of 480 V to ground. The system then operates as a grounded system, so that a ground fault (phase to conduit or other enclosure) on the secondary can cause fault-current flow that opens a circuit protective device to clear the faulted circuit.

But corner grounding of a delta system does not give the lowest possible phase-to-ground voltage. In fact, the voltage to ground of a corner-grounded delta system is the same as it is for an ungrounded delta system, because the voltage to ground for ungrounded circuits is defined as the greatest voltage between the given conductor and any other conductor of the circuit. Thus, the voltage to ground for an ungrounded delta system is the maximum voltage between any two conductors, on the assumption that an accidental ground on any one phase puts the other two phases at full line-to-line voltage above ground.

In recognition of increasing emphasis on the safety of grounded systems over ungrounded ones, **NEC** Section 450-4 covers the use of zig-zag grounding autotransformers to convert 3-phase, 3-wire, ungrounded delta systems to grounded wye systems. Such grounding of a 480-V delta system, therefore, lowers the voltage to ground from 480 V (when ungrounded) to 277 V (the phase-to-grounded-neutral voltage) when the system is converted to a wye system as shown in Fig. 6.43.

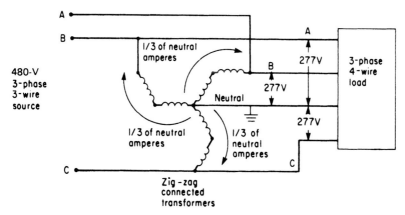

FIGURE 6.43 Zig-zag grounding autotransformer can be used to change voltage to ground from 480 to 277 V.

The *zig-zag* grounding autotransformer gets its name from the angular phase differences among the six windings that are divided among the three legs of the transformer's laminated magnetic-core assembly. The actual hookup of the six windings is an interconnection of two wye configurations, with specific polarities and locations for each winding. Just as a wye or delta transformer hookup has a graphic representation that looks like the letter *Y* or the Greek letter Δ, so a zig-zag grounding autotransformer is represented as two wye hookups with pairs of windings in series but phase-displaced, as in Fig. 6.44.

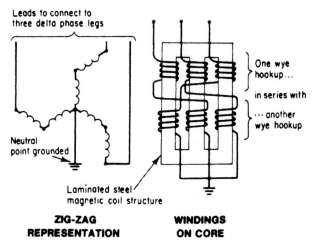

FIGURE 6.44 In zig-zag connection, a path is provided for the flow of fault current or neutral current.

With no ground fault on any leg of the 3-phase system, current flow in the transformer windings is balanced, because equal impedances are connected across each pair of phase legs. The net impedance of the transformer under balanced conditions is very high, so that only a low level of magnetizing current flows through the windings. But when a ground fault develops on one leg of the 3-phase system, the transformer windings assume a very low impedance in the fault path, permitting a large fault current to flow and operate the circuit protective device—just as it would on a conventional grounded-neutral wye system, as shown in Fig. 6.45.

Because the kilovoltampere rating of a grounding autotransformer is based on short-time fault current, the selection of such a transformer is much different from the sizing of a conventional 2-winding transformer to supply a load. Careful consultation with a manufacturer's sales engineer should precede any decisions about the use of these transformers.

NEC Section 450-4 points out that a grounding autotransformer may be used to provide a neutral reference for grounding purposes or for the purpose of converting a 3-phase, 3-wire delta system to a 3-phase, 4-wire, grounded wye system. In the latter case, a neutral conductor can be taken from the transformer to supply loads connected phase to neutral—such as 277-V loads on a 480-V delta system that is converted to a 480Y/277-V system.

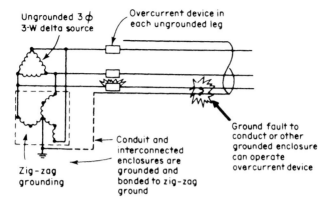

FIGURE 6.45 Zig-zag autotransformer provides the same automatic fault clearing as obtained on a solidly grounded wye system.

Such transformers must have a continuous rating and a continuous neutral current rating. The phase current in a grounding autotransformer is one-third the neutral current, as shown in Fig. 6.43.

The same **NEC** rule requires the use of a three-pole CB rated at 125 percent of the transformer phase current. The requirement for "common-trip" in the overcurrent device excludes the conventional use of fuses in a switch as overcurrent protection. A three-pole CB prevents single-phase opening of the circuit.

CHAPTER 7
SERVICES

The service for any building consists of the conductors and equipment used to deliver electric energy from the utility supply lines (or from an on-site generator) to the distribution system of the building (or other premises being supplied). Service may be brought to a building either overhead or underground—from a utility pole line or from an underground transformer vault. Figure 7.1 shows the various component parts of services.

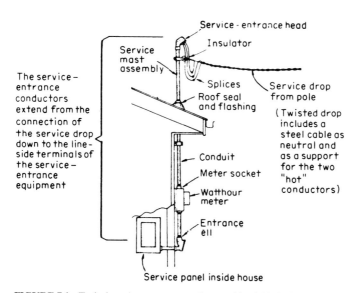

FIGURE 7.1 Typical service arrangement for a residential building.

The word "service" includes all the materials and equipment involved with the transfer of electric power from the utility distribution line to the electrical wiring system of the premises being supplied. Although service layouts vary widely, depending upon the voltage and ampere rating, the type of premises being served, and the type of equipment selected to do the job, every service generally consists of *service-drop* conductors (for overload service from a utility pole line) or *service-lateral* conductors (for an underground service from either an overhead or underground utility system), along with metering equipment, some type of switch or circuit-breaker control, overcurrent protection, and related enclosures and hardware. A typical layout of "service" for a one-family house breaks down as shown in Fig. 7.1.

That part of the electrical system which directly connects to the utility supply line is referred to as the *service*. Depending upon the type of utility line serving the building, there are two basic types of service—overhead and underground.

The "service point" is the point where the utility connects to the user's equipment. Both overhead drops and underground laterals may or may *not* belong to the utility. Generally speaking, if the conductors that connect to a service entrance conductor are sized by the utility, then they belong to the utility. Usually, the overhead drop *is* sized by the utility, whereas the lateral conductors are about 50/50 utility/user where either the drop or the lateral is owned by the user, they must be sized per Sections 230-23 and 230-31, respectively. (See **NEC** Article 100, "DEFINITIONS".)

The overhead service has been the most commonly used type. In a typical example, the utility supply line is run on wood poles along the street property line or back-lot line of the building, and a cable connection is made overhead from the utility line to a bracket installed high up on the building (Fig. 7.2). This line of wood poles also carries the telephone lines, and the poles are generally called "telephone" poles.

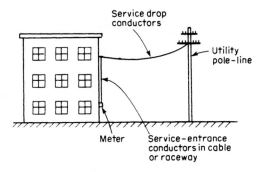

FIGURE 7.2 Basic overhead service.

The aerial cable that runs overhead from the utility lines to the bracket on the outside wall of the building is called the *service drop*. This cable is usually installed by the utility line worker. At the bracket which terminates the service drop, conductors are then spliced to the service entrance conductors to carry power down to the electric meter and into the building.

In the underground service, the conductors that run from the utility line to the building are carried underground. Such an underground run to a building may be tapped from either an overhead utility pole line or an underground utility distribution system. Figure 7.3 compares overhead with underground services. Although underground utility services tapped from a pole line at the property line have been used for many years to eliminate the unsightliness of overhead wires coming to a building, the use of underground service tapped from an underground utility system has only recently started to gain widespread usage in residential areas. This latter technique is called *URD*, for underground residential distribution.

SERVICES

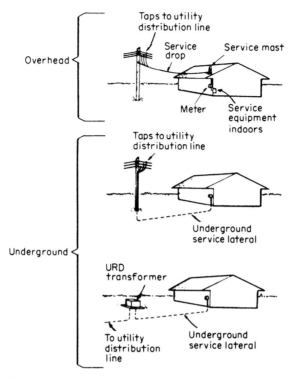

FIGURE 7.3 Service may be made overhead *or* underground.

As noted above, when a building is supplied by an overhead drop, an installation of conductors must be made on the outside of the building to pick up power from the drop conductors and carry it into the meter enclosure and service entrance equipment (switch, CB, panelboard, or switchboard) for the building. On underground services, the supply conductors are also brought into the meter enclosure on the building and then run into the service equipment installed, usually, within the building. "Service conductors" is a general term that covers all the conductors used to connect the utility supply circuit or transformer to the service equipment of the premises served. This term includes *service drop* conductors, *service lateral* (underground service) conductors, and *service entrance* conductors. In an overhead distribution system, the service conductors begin at the line pole where connection is made. If a primary line is extended to transformers installed outdoors on private property, the service conductors to the building proper begin at the secondary terminals of the transformers. Where the supply is from an underground distribution system, the service conductors begin at the point of connection to the underground street mains. In every case the service conductors terminate at the service equipment.

Design Procedure

The design of the service entrance for any premises can be reduced to a procedure involving five basic steps:

1. Calculate the required current rating and power capacity of the service feeder.
2. Lay out the elements of the service connection to the source of power supply.
3. Provide a disconnect means and effective overcurrent protection.
4. Consider an emergency supply source.
5. Consult with the supplying electric utility company and carefully observe all the detailed regulations set out in their published standards. Figure 7.4 shows part of the table of contents of a typical utility company booklet covering such requirements.

NEC Article 230 covers services.

Although the **NEC** states that a building or other premises must be supplied by "only one service"—that is, by only one set of service drop or service lateral (underground) conductors—exceptions are made. If a separate service is required for fire pumps or for emergency lighting and power purposes, more than one set of service drop or lateral conductors may be used to supply the building. In the case of apartment houses, shopping centers, and other multiple-occupancy buildings, two or more sets of service entrance conductors may be tapped from a single set of service drop or lateral conductors, or two or more subsets of service entrance conductors may be tapped from a single set of main service entrance conductors.

Two or more services (drops or laterals) to one building are permitted when the total demand load of all the feeders is more than 2000 A (600 V) or is more than the capacity of one drop (because the premises' load requirements are greater than the capacity of the utility's largest drop) or by special permission (Fig. 7.5). Cases in which separate light and power services are run to a single building, or separate services are provided to water heaters or electric space heaters to allow the use of different rate schedules, are also exceptions to the general rule requiring a single service (Fig. 7.6). And if a single building is so large, special permission can be obtained for additional services.

An important **NEC** rule says that for any installation where more than one service is permitted to supply one building, a "permanent plaque or directory" must be mounted or placed "at each service drop or lateral or at each service-equipment location" to advise personnel that there are other services to the premises and to tell where they are and what building parts they supply.

As discussed thoroughly later, when a single service is brought to a building or structure, the usual requirement of electrical codes is that the service disconnect means must consist of not more than six normal disconnect switches or circuit breakers. Some local codes mandate the use of a single disconnect for a service. There are many **NEC** rules that modify the basic arrangement of service disconnects—as described below.

In its simplest form, a service to a building would consist of a single service drop from a pole line or a single underground service (a lateral) supplying a single set of service entrance conductors, which are connected to a single service disconnect (a circuit breaker or a fused switch). A minimal variation on that form is a single drop or lateral supplying a single set of service entrance conductors, which are run to a location just outside or inside the building, where two or more service disconnects (up to a maximum of six switches or CBs) are tapped from the SE (service entrance) conductors, with the taps made within a panelboard or switchboard housing the switches or CBs or with taps made in an auxiliary gutter and nippled into individual enclosures for the two to six service disconnects.

However, the **NEC** permits more than one set of SE conductors to be tapped from a single drop or lateral. Where a single service drop supplies a building which uses

INTRODUCTION

This Booklet is published for the benefit of our customers, architects, engineers and contractors to provide a convenient reference. However, design or construction should not be undertaken until complete information is obtained from us. Such information and assistance is available from our Energy Management Services or Electric Operations Departments. See pages 6 thru 17 for location of our offices.

We supply electricity subject to Rules and Regulations, Terms and Conditions, policies and procedures, rate schedules, and industry standards —all of which are made a part of these requirements. Because these change from time to time, they are not included in this booklet, but are available upon request from our offices.

Legal restrictions, changes in the art, judgment and safety require this booklet to be revised from time to time, and we reserve the right to make such revisions. Our present schedule calls for publishing any such revisions in 1983.

We endeavor to supply electricity adequately and reliably. We do not guarantee a continuous supply and do not assume liability for direct or consequential loss or damage to persons or property due to the supply delivered, or as a result of any interruption or variation in the supply.

Failure to comply with our requirements, applicable codes, or orders of an enforcement authority can result in our refusal to energize the service or in the disconnection of an existing service.

Section 5—Your Service Facilities
A. Service Location 28
B. Service Equipment 28
C. Service Entrance Conductors 28
D. Pole-Mounted Service Equipment and Metering
   (Special Installation) 29
E. Identification 29

Section 6—Meter Installation
A. General 30
B. Standard Meter Installation 30
C. Meter Locations 31
D. Meter Equipment Mounting and Supports 31
E. Grounding 32
F. Cover Plates 32
G. Meter and Equipment Seals 32
H. Self-Contained Single Phase Meter Installations 33
I. Self-Contained Three Phase Meter Installations 34
J. Instrument Transformer Meter Installations 35

Section 7—Your Utilization Equipment
A. General 38
B. Motor Installations 38
C. Motor Starting Current 38
D. Motor Protective Devices 39
E. Power Factor 39
F. System Disturbances 39

Section 8—Your Alternate Electric Energy Sources
A. Non-Parallel Generation (Standby or Emergency) 40
B. Parallel Generation 40
C. Uninterruptible Power Supply (UPS) 40

Section 9—Illustrations
Figure 1—Temporary Electric Service—Direct Buried System 42
Figure 2—Direct-Buried Service 44
Figure 3—Conduit Service (optional) 45
Figure 4—Temporary Service From Overhead System 46
Figure 5—Overhead Service 48
Figure 6—Overhead Service Entrance Facilities 49
Figure 7 Service Mast 50

FIGURE 7.4 Excerpts from a utility regulations booklet.

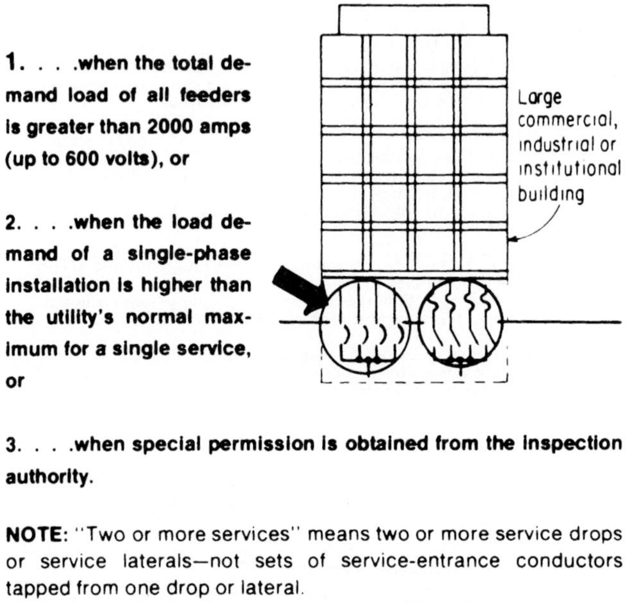

1. . . .when the total demand load of all feeders is greater than 2000 amps (up to 600 volts), or
2. . . .when the load demand of a single-phase installation is higher than the utility's normal maximum for a single service, or
3. . . .when special permission is obtained from the inspection authority.
Large commercial, industrial or institutional building
NOTE: "Two or more services" means two or more service drops or service laterals—not sets of service-entrance conductors tapped from one drop or lateral.

FIGURE 7.5 Under certain conditions, a building may be supplied by more than a single service.

two to six service disconnecting means, each in a separate enclosure and each feeding a separate load, a separate set of service-entrance conductors may supply each of the disconnects (see **NEC** Section 230-40, Exception No. 2). Thus, the two to six sets of service entrance conductors are tapped from the single drop. Figure 7.7 shows how two to six separate sets of service-entrance conductors may be supplied by a single service drop for either single- or multiple-occupancy buildings. Disconnects can be of same or different ratings, and each set of service entrance conductors can be installed using any approved wiring method. This sketch shows some possible combinations that may be used. And it should be noted that this **NEC** permission to use up to six sets of service entrance conductors applies to single-occupancy buildings as well as to multiple-occupancy buildings.

230-volt, 3-phase
3-wire service
115/230-volt,
1-phase,
3-wire service

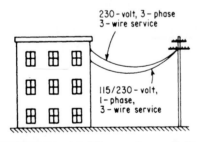

FIGURE 7.6 A building may be supplied by two services when the services are for different purposes—e.g., a 3-phase power service for motors and a single-phase lighting service.

NEC Section 230-2 provides the basic rule requiring that any "building or other structure" be supplied by "only one service" (that is, by a single drop or a single lateral). Exception No. 7 adds an important qualification of that rule but applies only to those cases where service entrance layouts with two to six service disconnects are to be fed by underground service (lateral) conductors and are installed in separate individual enclosures at one location, with each disconnect supplying a separate

For apartment houses and other multiple-occupancy buildings

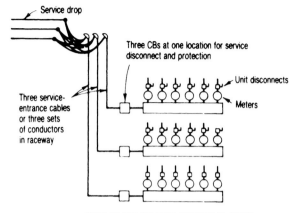

THIS IS OK BY THE PRESENT NEC

For single-occupancy buildings such as factories, schools and stores

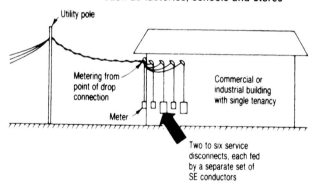

FIGURE 7.7 More than one set of service-entrance conductors (up to six sets) may be tapped from a single drop.

load. As described in Section 230-40, Exception No. 2, such a service equipment layout may have a separate set of service entrance conductors run to "each or several" of the two to six enclosures. Exception No. 7 to Section 230-2 notes that where a separate set of underground conductors of size 1/0 or larger is run to each or several of the two to six service disconnects, the several sets of underground conductors are considered to be one service (that is, one service lateral) even though they are run as separate circuits, that is, connected together at their supply end (at the transformer on the pole or in the pad-mount enclosure or vault) but not connected together at their load ends. The several sets of conductors are taken to be "one service" in the meaning of Section 230-2, although they actually function as separate circuits (Fig. 7.8).

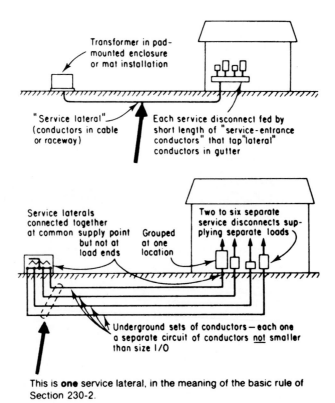

FIGURE 7.8 In both these cases, the building is supplied by a "single" service lateral.

Although Section 230-2 applies to "service-entrance conductors" and service equipment layouts fed by either a "service drop" (overhead service) or a "service lateral" (underground service), Exception No. 7 is addressed specifically and only to service lateral conductors (as indicated by the word "underground") because of the need for clarification based on the **NEC** definitions of "service drop," "service lateral," "service-entrance conductors, overhead system," and "service-entrance conductors, underground system." (Refer to these definitions in the **NEC** to clearly understand the intent of Exception No. 7 and its relation to Section 230-40, Exception No. 2.)

The matter involves these separate but related considerations:

1. Because a "service lateral" may (and usually does) run directly from a transformer on a pole or in a pad-mount enclosure to gutter taps where short tap conductors feed the terminals of the service disconnects, most layouts of that type literally do not have any "service-entrance conductors" that would be subject to the application permitted by Section 230-40, Exception No. 2—other than the short lengths of tap conductors in the gutter or box where splices are made to the lateral conductors.

2. Because Section 230-40, Exception No. 2 refers only to sets of "service-entrance conductors" as being acceptable for individual supply circuits tapped from one

drop or lateral to feed the separate service disconnects, that rule clearly does not apply to "service lateral" conductors which, by definition, are not "service-entrance conductors." So there is no permission in Section 230-40, Exception No. 2 to split up "service lateral" capacity. And the basic rule of Section 230-2 has the clear, direct requirement that a building or structure be supplied through only one lateral for any underground service. That is, either a service lateral must be a single circuit of one set of conductors, or, if circuit capacity requires multiple conductors per phase leg, then the lateral must be made up of sets of conductors in parallel—connected together at both the supply and load ends—in order to constitute a single circuit (that is, one lateral).

3. Exception No. 7 permits "laterals" to be subdivided into separate, nonparallel sets of conductors in the way that Section 230-40, Exception No. 2 permits such use for "service-entrance conductors"—but only for conductors of size No. 1/0 and larger and only where each separate set of lateral conductors (each separate lateral circuit) supplies one or several of the two to six service disconnects.

Section 230-2, Exception No. 7 recognizes the importance of subdividing the total service capacity among a number of sets of smaller conductors rather than as a single parallel circuit (that is, a number of sets of conductors connected together at both their supply and load ends). The single parallel circuit would have much lower impedance and would, therefore, require a higher short-circuit interrupting rating in the service equipment. The higher impedance of each separate set of lateral conductors (not connected together at their load ends) would limit short-circuit current and reduce short-circuit duty at the service equipment, permitting the use of lower interrupting-capacity-rated equipment reducing the destructive capability of any faults at the service equipment.

Current Rating of Service Feeders

The basic requirement of the **NEC,** dictated by reason and logic, is that the service conductors have adequate capacity for the total load current that might possibly be drawn simultaneously. Unless tests and experience (such as demand-meter studies) indicate that something less than 100 percent of all load currents will flow simultaneously or unless control provisions (such as sequence switches or interlocked contactors) assure a maximum demand of less than the total connected load, it is imperative that the conductors have adequate current rating for the entire system load.

The sizing of service entrance conductors usually involves the same type of step-by-step procedure as set forth previously for sizing feeders. A set of service entrance conductors is considered to be a feeder and must be sized just as if it were a feeder (Fig. 7.9). In general, the service entrance feeder conductors must have a minimum current-carrying capacity sufficient to handle the total lighting and power load served. Where the **NEC** gives demand factors that may be used (Tables 220-11, 220-13, 220-18, 220-19, and 220-20) or where it allows the use of other acceptable demand factors based on sound engineering determination of a less-than-100 percent demand requirement (as in Exception No. 1 of Sections 220-15 and 220-17), the minimum required ampacity for service entrance conductors may be reduced from the total of the connected loads being supplied.

From the analysis and calculations given in the section of this handbook on feeder circuits, a total power and lighting demand load can be developed for use in sizing service entrance conductors. Of course, where two or more sets of service entrance conductors are used (as shown in Fig. 7.7), the basic conductor sizing procedure must be applied to each separate set of service entrance feeder conductors.

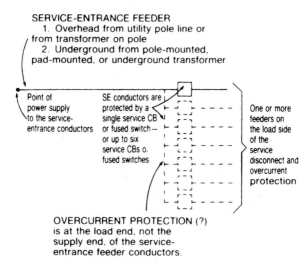

FIGURE 7.9 Evaluation and application of **NEC** rules must be based on an understanding that service entrance conductors constitute a "feeder" but depart from the usual feeder requirements.

When a total load (consisting of both initial and spare capacity for the future) has been established for the service entrance conductors, the required current-carrying capacity is easily determined by dividing the total load in kilovoltamperes (or in kilowatts with proper correction for the power factor of the load) by the voltage of the service. But, it should be carefully noted that **NEC** Section 230-42 establishes certain minimum capacities for service entrance conductors. These are for small-size service entrances, and the minimum sizes of conductors must be used regardless of the actual loads supplied. A 100-A service conductor ampacity is the **NEC** mandatory minimum if the system supplied is a one-family dwelling with six or more 2-wire branch circuits (or the equivalent of that for multiwire circuits) or a one-family dwelling with an initial computed load of 10,000 W. Now that three 20-A small-appliance branch circuits are required in a single-family dwelling, the average new home will need a 100-A, 3-wire service, because even without electric cooking, heating, drying, or water-heating appliances, more than five 2-wire branch circuits will be installed. And **NEC** Sections 220-30 through 220-41 offer "optional calculations" for sizing service entrance conductors for dwelling units, schools, and farm buildings—where experience dictates the use of reasonable demand factors.

From the required current rating of the conductors, the required size of the service entrance phase conductors is determined. Sizing of the service neutral is the same as for feeders. Although suitably insulated conductors must be used for the phase conductors of service entrance feeders, the **NEC** does permit the use of bare grounded conductors (such as neutrals).

For service-lateral conductors (underground service), an individual grounded conductor (such as a grounded neutral) of aluminum without insulation or covering may not be used in an underground raceway. A bare *copper* neutral may be used in a raceway, in a cable assembly, or even directly buried in soil where local experience establishes that soil conditions do not attack copper. An aluminum grounded conductor of an underground service lateral may be without individual insulation "when part of a cable assembly identified for underground use," where the cable is directly buried or run in a raceway. Of course, a lateral made up of individual insulated phase legs and an insulated neutral is acceptable in underground conduit or raceway.

Except for the use of a bare neutral (as permitted), all service entrance conductors must be insulated. Again, a bare individual aluminum grounded conductor (grounded neutral or grounded phase leg) may not be used in a raceway or for direct burial. Any such bare conductor is acceptable only when enclosed in an "identified" jacketed cable assembly—whether installed directly on the side of a building, structure, or pole or installed within a raceway on a building structure.

NOTE: Any nonmetallic cable used for underground installation—either directly buried or in a raceway—must be "Type USE" cable. "Type SE" cable is recognized by UL only for above-ground installation.

An extremely important element of service design is that of fault consideration. Service busway and other service conductor arrangements must be sized and designed to ensure their safe application with the service disconnect and protection. That is, service conductors must be capable of withstanding the let-through thermal and magnetic stresses on a fault, as determined by the service overcurrent device.

Just as the sizing of a feeder involves adding the continuous and noncontinuous current loads for lighting, heating, motors, and/or other power equipment and then applying any applicable demand factors, the service entrance conductors are sized for the entire connected load in a building or other premises. The following simple example shows the procedure.

Example. Assume a 120,000-ft^2 store building has the following loads:

252 kVA of continuous lighting

73 kVA demand load for receptacles

116 kVA of electric heating

346 kVA of motor loads

38 kVA of appliances

The general approach to the sizing of service entrance conductors might be as follows:

Step 1. Although the conductors must have sufficient capacity for the 252 kVA of continuous lighting, the **NEC** requires a minimum feeder capacity of 3 W/ft^2 for general lighting—for a total of $120,000 \times 3 = 360,000$ W. Assuming the use of high-power-factor lighting equipment, which can be taken as 360 kVA, **NEC** Table 220-11 requires it be taken at 100 percent demand.

Step 2. The 73-kVA demand load for receptacles was calculated from **NEC** Section 220-13. Refer to "Receptacle Loads" in Chapter 2 of this handbook.

Step 3. The 116 kVA of electric space heating is a 100 percent demand figure, based on connected load and control provisions.

Step 4. The 346-kVA motor load is the sum of 125 percent of the load of the largest motor in the building and the demand loads of all the other motors (in accordance with **NEC** Sections 430-24, 430-26, and 430-33, second paragraph).

Step 5. The 38 kVA of appliance load is the total load of all appliances that could be operating simultaneously.

The service entrance conductors would thus seem to be required to have sufficient capacity for a total of

$$360 + 73 + 116 + 346 + 38 = 933 \text{ kVA}$$

On a 480Y/277-V system, the minimum required current capacity is most effectively calculated using a voltage value of 460 V, which is a multiple of 115 and 230 V. Using the lower value of voltage yields a higher value of current, with greater assurance of adequacy in the conductor size. This gives

$$\frac{933,000}{460 \times 1.732} = 1171 \text{ A}$$

This result indicates that service entrance conductors with 1171-A usable capacity would be adequate for the initial demand load.

But immediately, the critical need to add even minimal spare capacity would dictate the use of conductors rated for at least 1200 continuous amperes (which is the lowest standard **NEC** rating of overcurrent protection above 1171 A). And it is always necessary to focus on "usable" ampacity, because **NEC** Section 220-10(b) requires overcurrent devices for "feeder" conductors (and, therefore, of devices for overcurrent service entrance "feeder" conductors) to have a rating or setting at least equal to "the noncontinuous load plus 125 percent of the continuous load" if the service protective device is not a circuit breaker listed for continuous loading at 100 percent of its rating or if the device is not a switch-and-fuse assembly listed for 100 percent continuous load. Sizing the service conductor on the same basis as the overcurrent device simplifies the procedure. And Note 8 to **NEC** Table 310-16, if applicable here, requires ampacity derating for continuous or noncontinuous loads where more than three current-carrying conductors are installed in a raceway or in a cable or are directly buried in the earth.

In this example, where service conductors are being sized for a store, the minimum calculated lighting load of 360 kVA must be taken at 125 percent of its value because it is continuous (operating steadily for over 3 hours). And any parts of the other loads that might operate continuously must also be taken at 125 percent of their values. Ventilation fans or air conditioning that would run for 3 hours or more would be included. It might be determined, for instance, that 20 percent of the nonlighting load is continuous. Then, of the 573 kVA of nonlighting load, 114.6 kVA (0.2 × 573 kVA) would be taken at 125 percent and the balance at 100 percent.

Still another typical factor that would have to be considered is the effect of "noncoincident" loads on the required service conductor capacity. If it can be determined that none of the 116-kVA heating load would operate at the same time as the air-conditioning load, then either the heating load or the air-conditioning load, whichever is smaller, may be omitted from the total demand load, because capacity available for the larger of the two loads would more than satisfy the current draw of the smaller load. Of course, all conditions must be accounted for, such as simultaneous operation of the air conditioning and electric-resistance reheat coils in air-conditioning ducts.

If the air-conditioning load is greater than the 116-kVA electric space-heating load in the store building, then the nonheating part of the nonlighting load is 73 kVA + 346 kVA + 38 kVA, or a total of 457 kVA, of which, say, 20 percent is continuous. Then, the sizing of the service conductors would proceed as follows:

Lighting load: 360 kVA × 1.25	450 kVA
Nonlighting load: 457 kVA × 0.2 × 1.25	114 kVA
Nonlighting load: 457 kVA × 0.8 × 1.00	366 kVA
Total demand	930 kVA

To handle that demand load, the conductor ampacity would have to be

$$\frac{930,000}{460 \times 1.732} = 1167 \text{ A}$$

which indicates that service conductors with an ampacity of 1167 A would be adequate for the initial design load; but, again, there would be no spare capacity for load growth. And note that 1167 A of ampacity from Table 310-16 would be adequate only if the makeup of the circuit did not require derating because more than three current-carrying conductors were installed in a single raceway or in a cable or were directly buried.

If the 1167-A circuit is made up of conductors installed in conduit, multiple conductors will be used for the phase legs and the neutral, with three phases and a neutral in each of the multiple conduits. If the major part of the load on a service feeder is electric-discharge lighting, such as fluorescent, metal-halide, or high-pressure sodium, then the neutral conductor of the feeder must be counted as a current-carrying conductor. In such a case, the three-phase conductors and the neutral conductor would have to be taken as four current-carrying conductors in each of the conduits making up the multiple 1167-A circuit, requiring the application of the rule of Note 8 to **NEC** Table 310-16, which calls for an ampacity-derating factor of 80 percent to be applied to all of the conductors in the service conduits.

If that 80 percent ampacity-reduction factor must be applied to the conductors making up the phase legs for the above 1167-A service feeder, then the conductor selected from Table 310-16 to be used in multiple for each phase leg must have an ampacity value such that its derated ampacity, multiplied by the number of conductors making up the phase leg, equals the required ampacity of 1167 A. For instance, if the service is made up of four conduits, with each one containing a neutral conductor plus three 500-kcmil THW copper conductors for phases A, B, and C in that conduit, the derating factor of 80 percent would have to be applied to the ampacity of 380 A shown in Table 310-16 for a 500-kcmil THW copper conductor. With each service feeder phase leg made up of four 500-kcmil THW copper conductors (one in each of the four conduits), then the ampacity of each phase leg would be

4 (conductors) × 380 A (per conductor) × 0.8 (derating factor) = 1216 A

This derated ampacity of 1216 A is the basic ampacity of the conductors under those conditions of application. It fully satisfies the previous calculated requirement for an ampacity of at least 1167 A in the service feeder. In addition, protection of that feeder by a 1200-A single main circuit breaker or 1200-A fuses in a 1200-A switch would also satisfy the rule of Section 220-10(b) that calls for overcurrent protection to have a rating not less than the value of 1167 A.

On a 480Y/277-V, 3-phase, 4-wire service to the example store building, a neutral conductor will be brought in to provide for 277-V lighting and other 277-V loads, so that each set of conductors of the parallel circuit will consist of four conductors: three phase legs and a neutral. But that neutral will be providing for current flow to a load that is only 252 kVA, out of a much larger total. Although the electric-discharge lighting load will be carrying significant third-harmonic current, even under balanced phase-loading conditions, the lighting is not a major portion of the load and the neutral is not required by Note 10(c) of **NEC** Table 310-16 to be counted as a current-carrying conductor. As a consequence, if the circuit is made up of multiple sets of four conductors in each conduit, Note 8 of Table 310-16 does not require derating of the ampacity of each of the four conductors, and conductors for the above calculated 1167 A may be selected directly from Table 310-16. But if there is to be a large number of nonlinear (computers, faxes, copiers, etc.) loads, serious consideration should be given to counting the neutral in such calculations.

The value of 1167 A of ampacity for the service conductor is, of course, based solely on providing adequate rating for the initial load. But spare capacity must be added. Although economics always constrains the decision to provide spare capacity for unforeseen future loads, experience consistently confirms that such provision is not only prudent but economical in the long run. Loads start getting added the day the premises is occupied, and they keep piling up with the passage of time. Many designers argue that spare capacity must always be at least 25 percent of the connected demand load in commercial and institutional buildings and at least 10 percent in industrial buildings, where loads are better accounted for and future changes are more readily anticipated.

In the store building, the original connected load is 825 kVA (252 + 73 + 116 + 346 + 38); adding a full 25 percent would give 825 × 1.25, or 1031 kVA. Dividing 1,031,000 VA by 460 × 1.732 yields an ampacity of 1294 A in the service conductors. The 1294-A value is the minimum load rating for conductors whose ampacity is required by Section 220-10(b) to be equal to 125 percent of the load current for continuous operation. That would call for a minimum ampacity of 1.25 × 1294 A, or 1617 A. And even the minimum value of 1167 A for service conductors to the initial load is based on a spare capacity of 108 kVA (360 kVA of calculated general lighting minus the actual connected load of 252 kVA).

Service conductors with an ampacity of 1600 A would constitute an economically realistic basis of effective design for present and future loads, to ensure the full capacity of 1294 A, with 1600 A being the next larger standard rating of protective device above 1200 A. But it also becomes important to make sure that the conductors selected actually have that ampacity.

Assuming use of aluminum service entrance conductors in conduit, each phase leg of the circuit could be made up of four 700-kcmil THHN conductors, each with an ampacity of 420 A (Table 310-16), for a total of 1680 A per phase leg, with each of four conduits containing a phase-A conductor, a phase-B conductor, a phase-C conductor, and a neutral conductor (which would be sized differently than the phase legs and would be smaller). Because electric-discharge lighting does not constitute the *major* portion of the load on the service conductors (only 252 kVA out of the total), Note 10(c) to Table 310-16 does not require that the neutral be counted as a current-carrying conductor in each conduit, and derating for conduit fill (Note 8) is not required. Dividing the 1294-A design load among the four conductors per phase leg of the service feeder shows that each 700-kcmil THHN aluminum conductor is loaded to 1294 ÷ 4, or 324 A. Because that value is less than the 375-A ampacity of 700-kcmil THW (75°C) aluminum conductors, the THHN (90°C) conductors are not being used in excess of the 75°C ampacity of 700-kcmil aluminum conductors. They would, therefore, satisfy the UL (and **NEC**) requirement that conductor connections to equipment terminals (the one or more service CBs or switches) must not operate above the ampacities of 75°C conductors for equipment rated over 100 A (or above 60°C ampacities up to a 100-A equipment rating) unless the equipment is marked otherwise. To satisfy that UL rule, the maximum load must not exceed 4 × 375 A, or 1500 A.

As will be discussed later, **NEC** Section 220-10(b) is also a factor in the selection of the one or more circuit breakers or fused switches that are required for disconnect and overcurrent protection of service entrance conductors. Section 220-10(b) requires a rating of 125 percent of the load current for overcurrent protection, applied to that part of the load current that is continuous (operates steadily for 3 hours or longer). But the Exception to that rule permits continuous operation with conductors and overcurrent protection rated at 100 percent (instead of 125 percent) *if* the CB or fused switch is UL-listed for full-load continuous operation. In the foregoing analysis, the use of service conductors rated at 1200 A and protected by a CB or fused switch rated for continuous loading to 100 percent of its rating could be considered an effective and economical alternative in supplying spare capacity.

Although the foregoing might seem to be an involved and almost laborious evaluation of required service conductor ampacity, there is really no alternative to that type of procedure if all the related factors are to be effectively included. Any other approach quickly becomes a guess and includes the risk of being inadequate, dangerous, and costly.

Figure 7.10 shows a service and distribution panel for a small building, where sizing of the service conductors and overcurrent protection would follow a typical feeder-sizing approach:

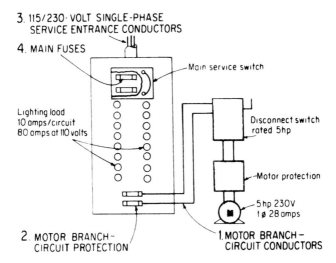

FIGURE 7.10 Sizing of a small service follows the numbered steps to obtain the required ampacity and protection.

1. *Size of motor branch-circuit conductors.* 125 percent of 28 A = 35 A. This requires No. 8 TW, THW, RHH, or THHN copper or No. 6 TW, THW, RHH, or THHN aluminum.

2. *Size of motor branch-circuit fuses (using one-time, non-time-delay fuses).* 300 percent of 28 A = 84 A. This requires a maximum fuse size of 90 A. Smaller fuses, such as the time-delay type, may be used.

3. *Size of service entrance conductors.* 125 percent of 28 A + 80 A (lighting load) = 115 A. Conductors for the hot legs must be No. 1/0 TW or No. 2 THW, RHH, or

THHN copper. (Note that if the lighting load is continuous, the SE conductors must be sized for 125 percent of 28 A + 125 percent of 80 A.)

4. *Size of main fuses.* 90 A (from item 2 above) + 80 A = 170 A. This requires a maximum fuse size of 175 A, in a 200-A switch. Again, smaller fuses may and should be used where possible to improve the overload protection on the circuit conductors.

Residential Service Entrance Conductors. The sizing of service entrance conductors for a residential building—an apartment house, condominium, or 2- or 4-family house—would follow the feeder-sizing procedures that apply to other types of buildings. And that same general procedure may be followed in sizing service entrance conductors for a one-family house or an apartment in a multifamily dwelling.

Example. A house has two floors with outside dimensions 30 by 25 ft. For both floors, that gives a total living area of 30 ft $\times$ 25 ft $\times$ 2 = 1500 ft^2. The house has oil-fired heating and hot water and only one large electric appliance, a 12-kW electric range. What is the minimum required size of circuit conductors?

Solution. Using the standard **NEC** method for feeder sizing (as covered in **NEC** Sections 220-10 through 220-21), the loads are totaled as follows:

General lighting:	
1500 ft^2 $\times$ 3 W/ft^2	4500 W
Two kitchen-appliance circuits: 1500 W per circuit	3000 W
One laundry circuit	1500 W
Total	9000 W

Then the demand factors are applied:

3000 W at 100%	3000 W
6000 (9000 − 3000) W at 35%	2100 W
Basic service entrance load	5100 W

The feeder demand load for the 12-kW range must be added to the 5100 W. As stated in **NEC** Section 220-19, the range feeder demand load is selected from Table 220-19. In this case, it is 8 kW (column A, one appliance):

Basic load	5100 W
Range feeder capacity	8000 W
Total feeder load	13,100 W

The minimum required ampacity for the ungrounded service entrance conductors of a 230/115-V, 3-wire, single-phase service is readily found by dividing the total feeder load by 230 V:

$$\frac{13,000 \text{ W}}{230 \text{ V}} = 56.96 \quad \text{or 57 A}$$

Although a load current of that rating could be readily supplied by No. 4 copper TW, THW, THHN, or XHHW conductors, an important provision of **NEC** Section 230-41(b)(2) comes into play here. Where the initial computed total feeder load is 10 kW or more (and it is 13.1 kW here) for a single-family dwelling, the ungrounded conductors of a service feeder must be rated at least 100 A; that is, the minimum capacity of the service entrance hot legs is 100 A.

That would call for a minimum of one of the following (using 60°C ampacity):

- No. 1 TW copper conductors (110 A)
- No. 1 THW copper conductors (130 A)
- No. 1 THHN copper conductors (150 A)
- No. 1 XHHW copper conductors (150 A)
- No. 1/0 TW, THW, THHN, or XHHW aluminum conductors (100 to 125 A)

The above example represents the **NEC** minimum sizing of service entrance conductors in accordance with the steps indicated in Sections 220-10 through 220-21. The calculation accounts only for the initial loads and includes no capacity for load growth. A simplified calculation procedure for residential services can be based on the same **NEC** capacity requirements but can include a measure of extra capacity to cover the virtually inevitable growth in connected loads. Such a design procedure is as follows:

1. Calculate the general lighting load and the general-purpose receptacle outlet load—for those outlets served by 20-A, 115-V, general-purpose circuits. Although each circuit has a full capacity of 20×115, or 2300 W, the circuit will be considered as being rated at only 2000 W for purposes of this calculation, thereby allowing some spare capacity.

For each 500 ft^2 of floor area in the house (excluding porches, garages, unused spaces, unfinished areas, etc.), allow one circuit. Allow an extra circuit for any part of 500 ft^2 left over.

Multiply the required number of circuits by 2000 W to get the total load in watts. (The same result can be obtained by multiplying the total floor area in square feet by 4 W/ft^2.)

2. Find the total circuit capacity—in watts—to be allowed for the appliance load in the kitchen, dining room, pantry, laundry, and utility area that is to be served by 20-A, 115-V, small-appliance circuits. This total capacity can be obtained by multiplying the number of such circuits by 2000 W. If the exact number of such circuits is not known, a 6000-W load can be assumed (two kitchen appliance circuits and one laundry circuit).

3. Take 3000 W of the sum of the amounts of steps 1 and 2 at 100 percent demand.

4. Add to this figure 35 percent (demand) of the remainder of the sum of the amounts of steps 1 and 2. The result is the capacity that must be provided in the service entrance conductors to supply the general lighting and receptacle loads.

5. Add 8000 W for an electric range (rated not over 12 kW). (Refer to **NEC** Table 220-19.) If the electric cooking appliances consist of a built-in oven and range top, the **NEC** must be consulted to get the proper demand load. The **NEC** allows the nameplate ratings of a counter-mounted cooking unit and not more than two wall-mounted ovens to be added together and treated as a single range of that total rating.

6. Add together the wattage ratings of all appliances to be served by individual circuits not previously accounted for in the calculation. A demand factor of 75 percent may be applied to the sum of the nameplate ratings of these appliances—excluding cooking units, clothes dryers, air-conditioning units, and any electric space-heating load.

If both electric heating and air conditioning are to be used in the house, the rating (in watts) of only the larger of these two connected loads need be used in this total—since the two loads will not be used simultaneously.

7. Add together the following: the general lighting and general-purpose receptacle load (from step 4), the electric-range demand load (from step 5), and the appliance/air-conditioning/heating loads (from step 6).

8. Divide the total number of watts obtained in step 7 by 230 V (for 230/115-V, 3-wire, single-phase service) to get the required ampere rating of the service conductors.

Optional Service Entrance Calculation

The **NEC** provides an optional method for sizing residential service entrance conductors in Section 220-30. That optional calculation may be used:

1. Only for a one-family residence or an apartment in a multifamily dwelling, or other "dwelling unit," that is

2. Served by a 115/230- or 120/208-V, 3-wire, 100-A or larger service, and

3. Where the total load is supplied by one set of service-entrance conductors.

The method recognizes the greater diversity attainable in large-capacity installations. It therefore permits a smaller size of service entrance conductors for such installations than would be permitted by using the load calculations of Sections 220-10 through 220-21.

A typical example of the optional calculation of Section 220-30, based on Table 220-30 (shown in Fig. 7.11), is as follows.

Example. A 1500-ft^2 house (excluding unoccupied basement, unfinished attic, and open porches) contains the following specific electric appliances:

12-kW range

2.5-kW water heater

1.2-kW dishwasher

9 kW of electric heat (in five rooms)

5-kW clothes dryer

6-A, 230-V air-conditioning unit

There is recognition in Section 220-21 that if "it is unlikely that two dissimilar loads will be in use simultaneously," then it is permissible to omit the smaller of the two in calculating the required capacity of feeder or service entrance conductors. In Section 220-30, that concept is spelled out to require adding only the larger of either an air-conditioning load or the connected load of four or more separately controlled electric space-heating units. For the residence considered here, these loads would be as follows:

SERVICES 509

Largest one of the following five selections.
(1) 100 percent of the nameplate rating(s) of the air conditioning and cooling, including heat pump compressors.
(2) 100 percent of the nameplate ratings of electric thermal storage and other heating systems where the usual load is expected to be continuous at the full nameplate value. Systems qualifying under this selection shall not be figured under any other selection in this table.
(3) 65 percent of the nameplate rating(s) of the central electric space heating, including integral supplemental heating in heat pumps.
(4) 65 percent of the nameplate rating(s) of electric space heating if less than four separately controlled units.
(5) 40 percent of the nameplate rating(s) of electric space heating of four or more separately controlled units.
Plus: 100 percent of the first 10 kVA of all other load.
40 percent of the remainder of all other load.

The loads identified in Table 220-30 as "other load" and as "remainder of other load" shall include the following:

(1) 1500 volt-amperes for each 2-wire, 20-ampere small appliance branch circuit and each laundry branch circuit specified in Section 220-16.

(2) 3 volt-amperes per square foot (0.093 sq m) for general lighting and general-use receptacles.

(3) The nameplate rating of all appliances that are fastened in place, permanently connected, or located to be on a specific circuit, ranges, wall-mounted ovens, counter-mounted cooking units, clothes dryers, and water heaters.

(4) The nameplate ampere or kVA rating of all motors and of all low-power-factor loads.

FIGURE 7.11 Optional calculation for residential service conductors is based on this table.

Air conditioning: 6 A × 230 V = 1.38 kW

Heating (five separate units): 9.00 kW

It is permissible, therefore, to disregard the air-conditioning load in using the optional method.

The heating and other loads (except air conditioning) must be totaled in accordance with Section 220-30:

1. 1500 W for each of two small-appliance circuits (2-wire, 20-A) required by Section 220-4(b)(1): 3000 W

2. Laundry branch circuit (3-wire, 20-A): 1500 W

3. 3 W/ft^2 of floor area for general lighting and general-use receptacles (3 × 1500 ft^2): 4500 W

4. Nameplate rating of fixed appliances:

Range. 12,000 W

"4 or more" separately controlled heating units: 9000 W

Water heater. 2500 W
Dishwasher. 1200 W
Clothes dryer. 5000 W

5. Total: 38,700 W

The first three load categories given in Table 220-30 are not applicable here. "Air conditioning" has already been excluded as a load because the heating load is greater. There is no "central" electric space heating; and there are not "less than four" separately controlled electric space-heating units.

The total load of 38,700 W, as summed above, is classified as "all other load," as referred to on line 6 of **NEC** Table 220-30. The last two references in Table 220-30 constitute all the "optional" calculations in this particular example:

1. Take 10 kW of "all other load" at 100 percent demand to obtain 10,000 W.
2. Take the "remainder of other load" at 40 percent demand factor to obtain 0.4 (38,700 − 10,000) = 11,480 W.
3. Add the results of steps 1 and 2 to obtain 10,000 W + 11,480 W = 21,480 W.

At 230 V, single phase, the minimum ampacity of each service hot leg would then have to be

$$\frac{21,480 \text{ W}}{230 \text{ V}} = 93.39 \quad \text{or 93 A}$$

But Section 230-42(b)(2) requires a minimum conductor rating when the demand load is 10 kW or more. Thus,

$$\text{Minimum service conductor rating} = 100 \text{ A}$$

The neutral service entrance conductor size is calculated in accordance with Section 220-22. The 230-V loads have no relation to the required neutral capacity. The water heater, clothes dryer, and electric space-heating units operate at 230 V, 2-wire and have no neutrals. By considering only those loads served by a circuit with a neutral conductor and determining their maximum unbalance, the minimum required size of neutral conductor can be determined.

When a 3-wire, 230/115-V circuit serves a total load that is balanced from each hot leg to neutral—that is, half the total load is connected from one hot leg to neutral, and the other half of the total load from the other hot leg to neutral—the condition of maximum unbalance occurs when all the load fed by one hot leg is operating and all the load fed by the other hot leg is deenergized. Under that condition, the neutral current and hot-leg current are equal to half the total load wattage divided by 115 V (half the voltage between hot legs). But that current is exactly the same as the current that results from dividing the total load (connected hot leg to hot leg) by 230 V (which is twice the voltage from hot leg to neutral). Because of this relationship, it is easy to determine the neutral-current load by simply calculating the hot-leg current load—the total load from hot leg to hot leg divided by 230 V.

In this example, calculation of the neutral current load begins with the following steps, in which the components of the neutral load are summed:

1. Multiply 1500 ft^2 by 3 W/ft^2 to obtain 4500 W.
2. Add three small-appliance circuits (two kitchen, one laundry) at 1500 W each to obtain 4500 W.

3. Add the results of steps 1 and 2 to obtain the total lighting and small-appliance load as 9000 W.
4. Take 3000 W of that value at 100 percent demand factor to obtain 3000 W.
5. Take the balance of the load at 35 percent demand factor to obtain 0.35 (9000 − 3000) = 2100 W.
6. Add the results of steps 4 and 5 to obtain 5100 W.

The remainder of the calculation is performed with reference to Fig. 7.12.

Assuming an even balance of this 5100-W load on the two hot legs, under maximum unbalance the corresponding neutral current will be the same as the total load (5100 W) divided by 230 V:

$$\frac{5100 \text{ W}}{230 \text{ V}} = 22.17 \text{ A}$$

The neutral unbalanced current for the range load can be taken as equal to the 8000-W range demand load multiplied by the 70 percent demand factor permitted by Section 220-22 and then divided by 230 V:

$$\frac{8000 \times 0.7}{230} = \frac{5600}{230} = 24.34 \text{ A}$$

The neutral current load that is added by the 115-V, 1200-W dishwasher must also be included:

$$\frac{1200 \text{ W}}{115 \text{ V}} = 10.43 \text{ A}$$

The minimum required neutral capacity is, therefore, the sum 22.17 A + 24.34 A + 10.43 A = 56.94 A, or, after rounding up, 57 A. From **NEC** Table 310-16, the minimum neutral conductors are

Copper. No. 4 TW, THW, THHN, or XHHW

Aluminum. No. 3 TW, THW, THHN, or XHHW

And the 75 or 90°C conductors must be used at the ampacity of 60°C conductors, as required by the UL *Electrical Construction Materials Directory.*

NOTE: The above calculation of the minimum required capacity of the neutral conductor differs from the calculation and results shown in Example No. 2(a) in **NEC** Chapter 9. There, the 1200-W dishwasher load is added as a 230-V load to the range load and general lighting and receptacle load. To include this 115-V load as a 230-V load (and then to divide the total by 230 V, as shown) does not accurately represent the neutral load that the 115-V, 1200-W dishwasher will produce. In fact, it yields exactly half the neutral load that the dishwasher represents. The optional calculation method of Section 220-30 does indicate, in part (3), that fixed appliances be added at nameplate load, and it does not differentiate between 115- and 230-V devices. It simply totals all loads and then applies the 100 and 40 percent demand factors as indicated. That method clearly is based on well-founded data about load diversity and is aimed at obtaining a reasonable size of service hot legs. But, calculation of the feeder neutral in accordance with Section 220-22 is aimed at determining the maximum unbalanced current to which the service neutral might be subjected.

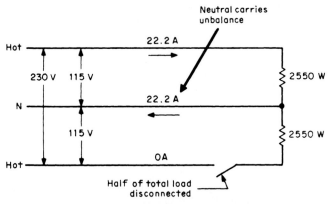

Neutral current for lighting and receptacles.

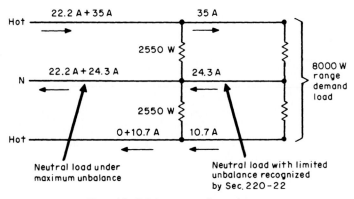

Neutral for lighting, receptacles, and range.

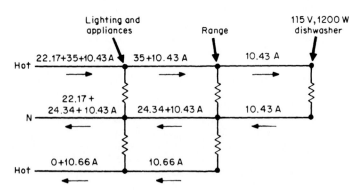

Neutral load = 22.17 + 24.34 + 10.43 = 56.94 amps

Neutral current for total load.

FIGURE 7.12 These loads determine the required neutral-conductor ampacity.

Although there is only a small difference between the **NEC** value of 51.7 A and the value of 56.9 A determined here, precise calculation should be made to ensure the adequacy of conductor ampacities. The difference of 5.2 A actually changes the required minimum size of neutral conductor from No. 6 up to No. 4 for copper, and from No. 4 up to No. 3 for aluminum. A load like a dishwasher, which draws current for a considerable period of time and is not used for just a few minutes like a toaster, should be factored into the calculation with an eye toward ensuring adequate capacity for conductors.

Another example of the application of Section 220-30 to the sizing of residential service conductors shows the very heavy loading that is permitted on a 100-A service.

Example.   The following loads apply to a 1500-ft^2 house with all-electric utilization:

- 1500 W for laundry receptacle circuit: 1500 W
- 1500 W for each of two (minimum of two) required kitchen-appliance circuits: 3000 W
- 1500 ft^2 at 3 W/ft^2 for general lighting and receptacles: 4500 W
- 14 kW of electric space heating from more than four separately controlled units: 14,000 W
- 12-kW electric range: 12,000 W
- 3-kW water heater: 3000 W
- 5-kW clothes dryer: 4500 W
- 3-kW load of unit air conditioners (because this load is smaller than the space-heating load and will not be operated simultaneously with it, no load need be added): 0 W
- Total: 42,500 W
  Solution.   The calculation is:

First 10 kW at 100 percent	10,000 W
Remainder at 40 percent = 32,500 W × 0.4	13,000 W
Total demand load	23,000 W

Size of service = 23,000 W ÷ 230 V = 100-A service

Under certain load conditions, the calculated required service capacity may be substantially less than 100 A. In such cases, however, 100 A is the minimum size service which can be used.

Service Entrance Wiring Methods

A number of specific design requirements apply to the layout and installation of raceways containing service entrance conductors:

1. Service entrance conductors may be run in rigid metal conduit, intermediate metal conduit, EMT, rigid nonmetallic conduit, or wireway. Busway or cable bus may also be used; and service entrance conductors may be run as service entrance cable, Type MC cable (metal-clad cable of the interlocked-armor or corrugated type), or Type MI cable. Cables that are suitable for use as service entrance conductors may be run in a cable tray.

The list of acceptable wiring methods for running service entrance conductors also includes flexible metallic conduit (Greenfield) or liquid-tight flexible metallic conduit. Such raceways are if a bonding jumper is used to assure ground continuity between the ends of the flex, they may be used to enclose service entrance conductors, such as to provide flexible routing around a pipe or other obstruction (see Fig. 7.13).

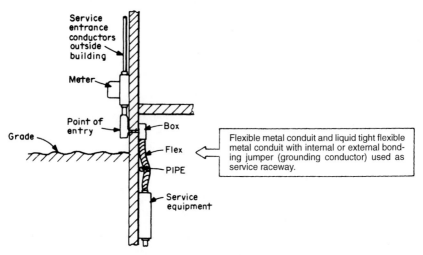

FIGURE 7.13 Flexible conduit may be used as a service raceway.

2. A very important design requirement for service entrances calls for the service overcurrent protection (and the service entrance disconnect) to be located right at the point where service entrance conductors enter a building, either inside or outside the building. But for those cases where it may be difficult or impossible to satisfy that requirement directly, a standard technique is available. Conductors in conduit or duct enclosed by concrete or brick not less than 2 in thick are considered to be outside the building, even though they are actually run within the building. Figure 7.14 shows how a service conduit was encased within a building so that the conductors are considered as entering the building right at the service protection and disconnect, where the conductors emerge from the concrete, to satisfy the **NEC** rule requiring the service disconnect to be as close as possible to the point where the service entrance conductors enter the building. In the typical application, forms are hung around the service conduit and then filled with concrete to form the required case.

3. The need to prevent moisture from entering energized equipment applies to underground service conduits (Fig. 7.15). Where service raceways are required to be sealed—as where they enter a building from underground—the sealing compound used must be marked on its container or elsewhere as suitable for safe and effective use with the particular cable insulation, with the type of shielding used, and with any other components it contacts. Some sealants attack certain insulations, semiconducting shielding layers, etc.

4. Service conductors—whether directly buried cables, conductors in metal conduit, conductors in nonmetallic conduit, or conductors in EMT—must satisfy all the rules of **NEC** Section 300-5 for protection against physical damage.

SERVICES 515

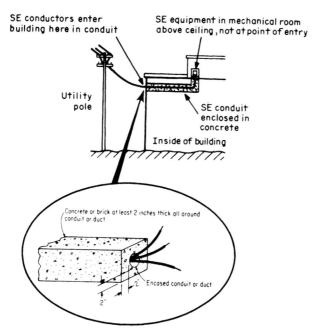

FIGURE 7.14 Conduit encased in concrete is considered to be outside the building.

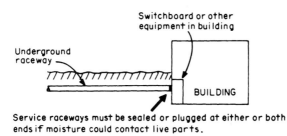

Service raceways must be sealed or plugged at either or both ends if moisture could contact live parts.

FIGURE 7.15 Underground service conduit may have to be sealed to keep out water.

5. Service-entrance conductors in EMT or rigid conduit must be made rain-tight, using rain-tight raceway fittings, and must be equipped with a drain hole in the service ell at the bottom of the run or must be otherwise provided with a means for draining off condensation.

6. When rigid metal conduit, IMC, or EMT is used for a service, the raceway must be provided with a service head (or weather head). Figure 7.16 shows details of a service-head installation.

Service heads must be located above the service drop attachment. Although this arrangement alone will not always prevent water from entering service raceways and equipment, such an arrangement will solve most water-entrance problems. But

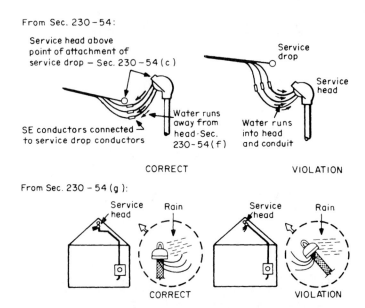

FIGURE 7.16 Service head position must exclude water from the raceway.

it is permissible for a service head to be located not more than 24 in from the service drop termination where it is impractical for the service head to be located above the service-drop termination. In such cases a mechanical connector is required at the lowest point in the drip loop to prevent siphoning.

Connections or conductor arrangements, both at the pole and at the service, must be made so that water will not enter connections and siphon under head pressure into service raceways or equipment. Where the pole connection is higher than the connection of the service entrance conductors at the service head, the connections at the pole should be made in such a manner that moisture will not enter the conductor. In many cases the connection at the utility pole is higher than the connection at the building. Any stranded service-drop conductors act as a hose, and, at the service head, moisture is forced up through the service entrance conductors by this head pressure and down into the meter or service equipment.

7. Figure 7.17 shows the difference between copper and aluminum when used for a bare, grounded, underground service entrance conductor. For service-lateral conductors (underground service), an individual grounded conductor (such as a grounded neutral) of aluminum without insulation or covering may be used either directly buried or in a raceway underground where "identified" for the use. A bare copper or aluminum neutral may be used in a raceway, in a cable assembly, or even directly buried in soil where local experience establishes that soil conditions do not attack copper.

Service entrance conductors must be insulated, except for a bare grounded conductor, such as a grounded neutral. But, again, a bare individual aluminum grounded conductor (grounded neutral or grounded phase leg) may be used in raceway or for direct burial only where identified for the use. Any such bare conductor is acceptable only when enclosed in an "identified," jacketed cable assembly such as Type SE (above ground), which may be used outdoors, with or without conduit enclosure. Underground, Type USE must be used (Fig. 7.18).

INSULATED PHASE CONDUCTORS and a bare copper neutral for an underground service lateral in buried raceway. Note: A bare aluminum or copper-clad aluminum neutral could be used here provided it is "identified" for underground use in raceway.

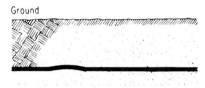

BARE COPPER NEUTRAL in a direct-buried cable assembly with moisture- and fungus-resistant outer covering. Note: A bare aluminum or copper-clad aluminum could be used like this, where within a cable assembly.

TYPE USE PHASE CONDUCTORS and a bare copper neutral directly buried where soil conditions are suitable for the bare copper. Note: a bare aluminum may be used if identified for direct burial.

FIGURE 7.17 Underground, a bare aluminum conductor may not be directly buried or located in a raceway.

Service Disconnect

Service entrance conductors must be equipped with a readily accessible means of disconnecting the conductors from their source of supply.

An extremely important **NEC** rule requires that switches and circuit breakers used as service entrance disconnecting means must be approved for use as service equipment. The switch or CB must be listed and labeled by the UL as suitable for service entrance use. Manufacturers' catalogs should be checked on this point.

The disconnect means for service entrance conductors may consist of not more than six switches or six circuit breakers, in a common enclosure or individual enclosures, located either inside or outside the building wall, as close as possible to the point at which the conductors enter the building (Fig. 7.19). The intent of this rule [**NEC** Section 230-72(a)] is to limit to six operations of the hand the disconnection of all conductors of the service. The limitation is applicable to cases where all the service disconnecting means are located in one place or "grouped" as required in Section 230-72.

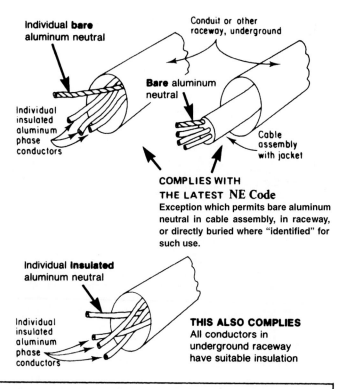

FIGURE 7.18 Bare aluminum service conductor may be used underground only when part of a jacketed, listed cable assembly.

The first sentence of Section 230-72(a), which refers to "two to six disconnects" for *each* service, ties directly to the requirements of Section 230-2. It is the intent of this basic rule that where a multiple-occupancy building (such as an apartment building, store building, or office building that is not over two floors high) is provided with more than one set of service entrance conductors tapped from a single drop or lateral, each of those sets of service entrance conductors may have up to six switches or circuit breakers to serve as the service disconnect means for that set. The rule does recognize that six disconnects for each set of service entrance conductors at a building with, say, 10 sets of SE conductors tapped from a drop or lateral does result in a total of 60 (6×10) disconnect devices for completely isolating the building's electrical system from the utility supply. And the same permission for use of up to six disconnect devices applies to "separate" services for fire pumps, emergency lighting, etc., which are recognized in Section 230-2 as being separate services for specific purposes.

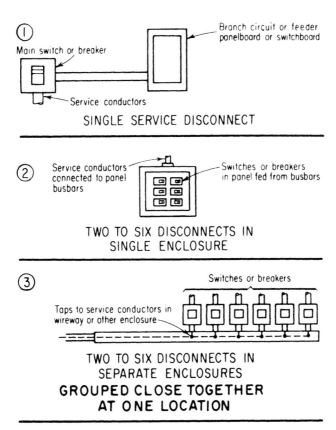

FIGURE 7.19 Service disconnect arrangement may take any one of these three forms.

For any type of occupancy, a power panel (not a lighting and appliance panel, as described in Section 384-14) containing up to six switches or circuit breakers may be used as the service disconnect. A lighting and appliance panel used as service equipment for the renovation of an existing service in an individual residential occupancy (but not for new installations) may have up to six main breakers or fused switches. However, a "lighting and appliance" panel used as service equipment for a new building of any type must have not more than two main devices—with the sum of their ratings not greater than the panel bus rating.

The first sentence of Section 230-71(a) and that of Section 230-72(a) note that from one to six switches (or circuit breakers) may serve as the service disconnecting means for each service to a building. For example, if a single-occupancy building has a 3-phase power service and a separate single-phase lighting service, each such service may have up to six disconnects (Fig. 7.20). Where the two sets of service equipment are not located adjacent to each other, a clear, conspicuous, and legible plaque or directory must be installed at each service equipment location indicating where the other service equipment is—as required by the second paragraph of Section 230-2. This is a very important safety consideration, to ensure that the building will be disconnected from *all* supply conductors in the case of an emergency such as a fire. Design of the service must always fully account for conditions of emergency, panic, or chaos.

FIGURE 7.20 Each separate service may have up to six disconnect devices.

Single-pole switches or circuit breakers equipped with handle ties may be used in groups as "single" disconnects for multiwire circuits, with fuses in the switch poles or the CB trip mechanism simultaneously providing overcurrent protection for the service (Fig. 7.21). Of course, multipole switches and circuit breakers may also be used as single disconnects. The requirements of the **NEC** are satisfied if all service entrance conductors can be disconnected with no more than six operations of the hand—regardless of whether each hand motion operates a single-pole unit, a multipole unit, or a group of single-pole units with "handle ties" or a "master handle" controlled by a single hand motion. Of course, a single main device for service disconnect and overcurrent protection—such as a single main CB or fused switch—gives better protection to the service conductors, is often required by local codes, and is a preferred design concept.

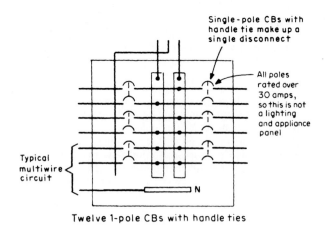

FIGURE 7.21 This arrangement constitutes *six* disconnect devices.

Section 384-16(a) requires a higher degree of overcurrent protection for lighting and appliance branch-circuit panelboards when used as service equipment or on the load side of service equipment. Each such panelboard must be protected on the supply side by not more than two main circuit breakers or two sets of fuses having a combined rating not greater than that of the panelboard. But, as noted before, up to six main protective devices may be used in a lighting and appliance branch-circuit panelboard where such a panelboard is used as service entrance equipment for the renovation of an existing installation (but not for new jobs) in an individual residential occupancy. (Refer to the discussions in Chapter 2 of this handbook.) It should be noted that these rules concern only a lighting and appliance branch-circuit panelboard, which is defined as a panelboard having more than 10 percent of its overcurrent devices rated 30 A or less, for which neutral connections are provided. Other types of panelboards used as service equipment may still utilize the concept of six main disconnect devices fed by the supply conductors to the panelboard.

Grouping of Disconnects. In a service disconnect arrangement of more than one disconnect—such as where two to six disconnect switches or CBs are used—all the disconnects making up the service equipment "for each service" must be grouped and not spread out at different locations, as required by **NEC** Section 230-72. The basic idea is that anyone operating the two to six disconnects must be able to do so while standing at one location. Service conductors must be readily disconnected from all loads at one place. And each of the individual disconnects must have lettering or a sign to specify what load it supplies, as shown in Fig. 7.22. The two to six service disconnects that are permitted for each service, or for each set of service entrance conductors in a multiple-occupancy building, must be grouped. But the individual groups of two to six breakers or switches do not have to be located together, and each grouping of two to six disconnects may be within a unit occupancy—such as an apartment—of the building.

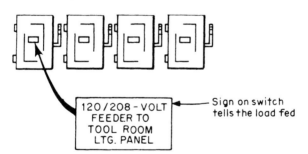

FIGURE 7.22   Each service disconnect must be "marked" to indicate the load served.

Special or emergency service equipment does not have to be grouped with the normal service equipment. It should also be noted that **NEC** Section 700-12(d) requires emergency services to be widely separated from other services, to prevent failure of both due to a single fault. And Section 230-72(b) makes it mandatory to install emergency disconnect devices where they will not be disabled or affected by any fault or violent electrical failure in the normal service equipment. If a service disconnect for emergency and exit lighting is installed very close to a normal service switchboard, an equipment burndown or fire near the main switchboard might knock out the emergency circuit.

Each separate service permitted for fire pumps or for emergency service may be equipped with up to six disconnects, just as the normal service (or any service) may have up to six service entrance disconnects. But, again, the disconnect or disconnects for a fire-pump or emergency service must be remote from the normal service disconnects, as shown in Fig. 7.23.

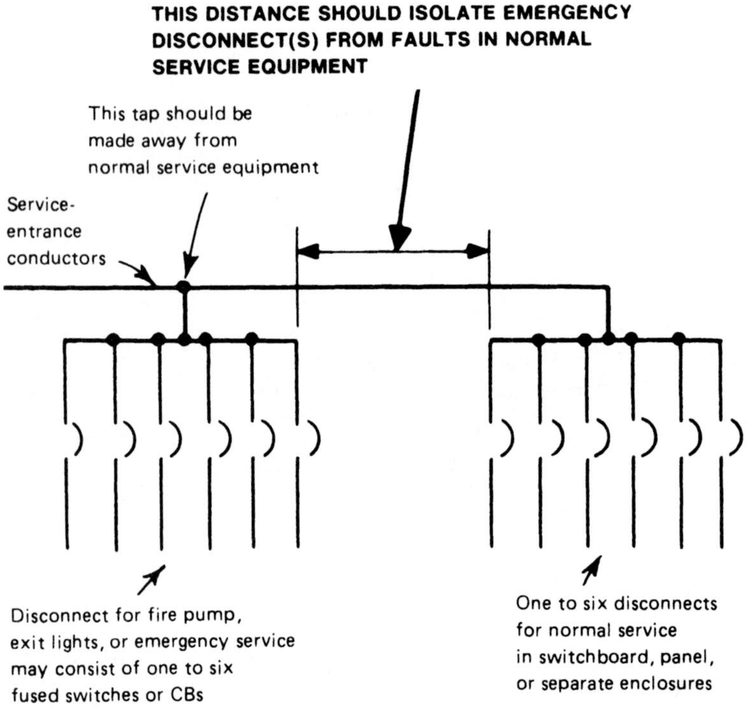

FIGURE 7.23 Emergency disconnects must be isolated from the effects of destructive faults in normal service entrance disconnects.

Another exception to the general **NEC** rule on the grouping of service disconnects is made for those cases where one of the two to six disconnects is for a circuit to a water pump that is counted on for firefighting purposes. This exception permits (but *does not require*) one of the two to six service disconnects to be located remote from the other disconnecting means that are grouped in accordance with the basic rule—provided that the remote disconnect is used only to supply a water pump that is also intended to provide fire protection. In a residence or other building that gets its water supply from a well, spring, or lake, a remote disconnect for the water pump affords improved reliability of the water supply for fire suppression in the event that fire or other faults disable the normal service equipment. Remote location, and marking on the disconnect, will help to distinguish the water-pump disconnect from the other normal service disconnects, minimizing the chance that firefighters will unknowingly open the pump circuit when they routinely open service disconnects during a fire. This exception ties into the rule of Section 230-72(b), which requires (not simply permits) remote installation of a fire-pump disconnect switch that is permitted to be tapped ahead of the one to six switches or CBs that constitute the normal service disconnecting means (see Exception No. 5 to Section 230-82). The exception provides remote installation of a normal service disconnect when it is used for the same purpose (water pump used for firefighting) as the emergency service disconnect (fire pump) covered in Section 230-72(b). In both cases, remote installation of the pump disconnect isolates the critically important pump circuit from interruption or shutdown due to fire, arcing-fault burndown, or any other fault that might knock out the main (normal) service disconnects.

A wide variety of layouts can be made to satisfy the **NEC** permission for remote installation of a disconnect switch or CB serving as a normal service disconnect (one of a maximum of six) supplying a water pump. Figure 7.24 shows three typical arrangements that would provide the isolated fire-pump disconnect.

Disconnect Location. The disconnecting means required for every set of service entrance conductors must be located at a readily accessible point nearest to the point at which the service conductors enter the building, on either the inside or the outside of the building (Fig. 7.25). The service disconnect switch (or circuit breaker) is generally placed on the inside of the building as near as possible to the point at which the conductors enter.

Although the **NEC** does not set maximum distance from the point of conductor entry to the service disconnect, various inspection agencies do set limits on this distance. For instance, service cable may not run within the building more than 18 in from its point of entry to the point at which it enters the disconnect. Or, service conductors in conduit must enter the disconnect within 10 ft of the point of entry. Or, as one agency requires, the disconnect must be within 10 ft of the point of entry, but overcurrent protection must be provided for the conductors right at the point at which they emerge from the wall into the building. The concern is to minimize the very real and proven potential hazard of having unprotected service conductors within the building. Faults in such unprotected service conductors must burn themselves clear, and such application has caused fires and fatalities.

Special design considerations apply to service disconnects for multiple-occupancy buildings, such as apartment houses, condominiums, town houses, office buildings, and shopping centers. The disconnect means for each unit in a multiple-occupancy building must be accessible to occupants of that unit. For instance, the disconnect means for deenergizing the circuits in an apartment of an apartment building must be in the apartment (such as in a panel), in an accessible place in the hall, or in a place in the basement or outdoors where it can be reached.

Multiple-occupancy buildings having individual occupancy above the second floor must have service equipment grouped in a common accessible place, and the disconnect means may consist of not more than six switches (or six CBs) as in Fig. 7.26. Although specific provisions on that application have been deleted from the present **NEC,** such practice was recognized and required in previous editions, is not objectionable under the present **NEC,** and must be understood to be required by the present **NEC** because no proposal was made to delete the requirement. That there was no intention to remove the rule can be determined from the following sequence:

1. The above-described provision for multiple-occupancy buildings with individual occupancy above the second floor has been deleted from Section 230-72(d) and Exception No. 4 of Section 230-90. In the discussions and comments published by

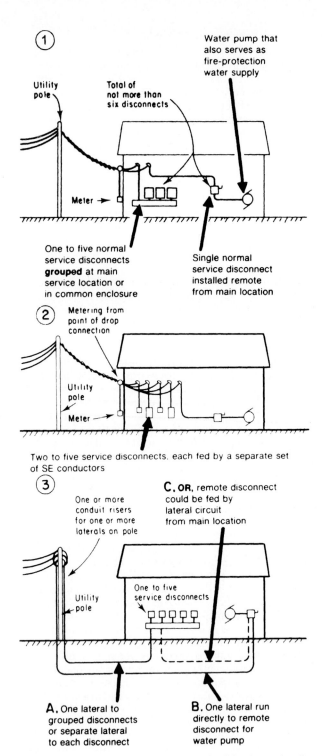

FIGURE 7.24 Remote installation is permitted for the service disconnect on a fire-pump circuit.

SERVICES

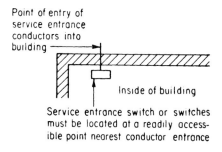

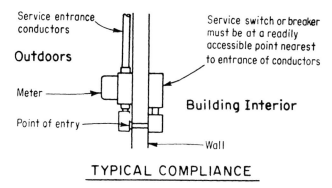

FIGURE 7.25 Service disconnect must be placed at the point where the service entrance conductors enter the building.

the **NEC** code-making panel on the change in these sections, a clear explanation is not given of the intent of the deletion. The original proposal that resulted in the deletion of all but the first sentence of Section 230-72(d) sought a new exception as follows: "Exception: Subject to serving agency approval and where Section 230-2, Exception No. 5, conditions are met, more than one service location shall be permitted. The area served by each service shall be clearly defined."

2. The substantiation presented in favor of the addition of that rule made clear that the intent was to give an exception to the rule of the second sentence of the section, which said, "A multiple occupancy building having individual occupancy above the second floor shall have service equipment grouped in a common accessible location." Note that the use of the word "a" indicates *one* location. The exception sought recognition of grouping of service equipment in two or more common accessible locations for those multiple-occupancy buildings that have individual occupancy above the second floor but cover such a large property area that the utility company must provide two or more service drops or laterals spaced along the length of the building, as provided by Exception No. 5 of Section 230-2. Thus, the use of a common location for the one to six disconnects fed by each drop or lateral required an exception to the rule requiring "a" (that is, one) common location for "not more than six switches or six circuit breakers" for any multiple-occupancy building with individual occupancy above the second floor.

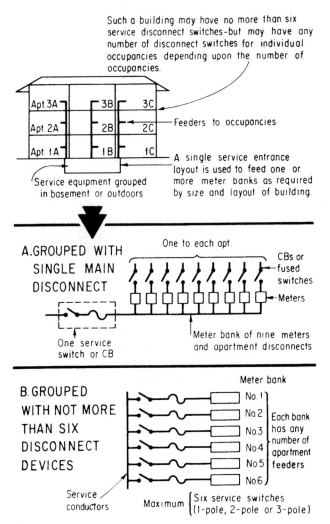

FIGURE 7.26 These basic rules must guide the design of services for multiple-occupancy buildings with individual occupancy above the second floor.

3. Because the code-making panel voted to accept the idea, such a service layout is now acceptable to the **NEC**, as shown in Fig. 7.27. But deletion of the last three sentences that appear in Section 230-72(d) of the 1978 **NEC** cannot be taken to mean that there has been a change in the intent of the rules for locating service disconnects for multiple-occupancy buildings.

The phrase "individual occupancy" means any space such as an office or apartment that is independent of any other occupancy in the building. Generally, each such space is supplied through a separate meter. Each apartment intended for use as

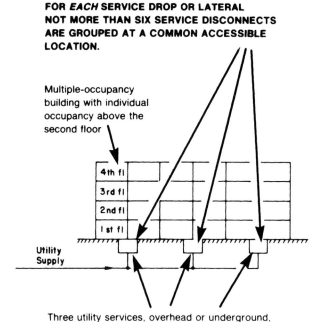

FIGURE 7.27 Buildings covering large areas may have more than one service but must have service equipment grouped for each service.

living quarters by one family is an individual occupancy. Each apartment might be supplied through a separate meter, or all might be supplied through one meter.

Multiple-occupancy buildings that do not have individual occupancy above the second floor may have service conductors run to each occupancy. The service disconnecting means in each occupancy may then consist of not more than six switches (or six CBs) as shown in Fig. 7.28. Example B in that illustration could be a hotel with stores on the lower two floors. The hotel owner would represent an "occupant" who has occupancy on the lower floors and is the only occupant of the upper floors. Such a building would not have "individual" occupancy above the second floor.

Although specific permission for the use of separate sets of service conductors to each occupancy of a multiple-occupancy building that does not have individual occupancy above the second floor was deleted from Section 230-72 in the 1981 **NEC**, there is no indication that such practice is objectionable. Some members of the **NEC** code-making panel warned that the word "deletions" must not be construed as revoking permission to run separate sets of service entrance conductors to an individual occupancy above the second floor—a practice that has long been allowed by this **NEC** section. It was stated that:

> Separate service to individual occupants of multiple occupancy buildings has been permitted since 1933 with a good record of safety. The multiple services serve to reduce

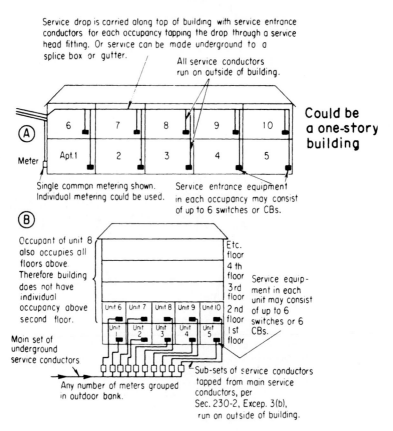

FIGURE 7.28 Buildings without individual occupancy above the second floor may have any number of sets of service entrance conductors and disconnects.

available fault current at the service location, and provide the consumer's ready access to the service equipment.

Three panel members objected to the deletion as unclear and confusing—and possibly interpretable as permitting a separate set of service entrance conductors to be run to individual unit occupancies in buildings with individual occupancy above the second floor.

Because no objection or complaint was ever raised about the rules of Section 230-72(d) as they apply to buildings without individual occupancy above the second floor, it must be assumed that running service entrance conductors to individual occupancies in such buildings—which has long been permitted—is still completely acceptable.

In buildings with occupancy above the second floor, it was never acceptable to the **NEC** to run separate sets of service entrance conductors to individual occupancies, and such practice was not the letter or intent of any proposals for the 1981 **NEC**. In addition, such practice would be extremely poor design and completely grotesque in, say, a 10-story apartment house.

It is clear that the only intended change is that described in Fig. 7.27 to permit grouping of disconnects at more than "a" common location for multiple-occupancy buildings over two floors high. Where there is individual occupancy above the second floor, the disconnecting means must be located in one or more commonly accessible places and must not consist of more than six normal service disconnects at each location, as explained in Figs. 7.26 and 7.27.

It should be noted that the requirement for access to the disconnect means for each occupant would be modified where the building was under the management of a building superintendent or the equivalent and where electrical service and maintenance were furnished. [See the Exception to Section 240-24(b).]

To comply with the requirement that the disconnect be located where conductors enter a building, the conductors should either be run on the outside of the building to each occupancy or, if run inside the building, be encased in 2 in of concrete or masonry in accordance with Section 230-6. In either case the service equipment should be located "nearest to the entrance of the conductors inside the building," and each occupant should have "access to his disconnecting means," which is satisfied by CBs or switches in the occupant's premises.

Any desired number of sets of service entrance conductors may be tapped from the service drop or lateral, or two or more subsets of service entrance conductors may be tapped from a single set of main service conductors.

Type of Disconnect. Any switch or CB used for service disconnect must be manually operable. In addition to manual operation, a switch may have provision for electrical operation—such as for remote control of the switch—provided it can be manually operated to the open or off position (Fig. 7.29). But an electrically operated switch does not have to provide manual closing. An electrically operated breaker with a mechanical trip button that will open the breaker even if the supply power is dead is suitable for use as a service disconnect. The manually operated trip button assures that the breaker "can be opened by hand." **NEC** Section 240-80 requires circuit breakers to be capable of being closed and opened by hand. To provide manual closing of electrically operated circuit breakers, manufacturers provide emergency manual handles as standard accessories. Thus, such breaker mechanisms can be both closed and opened manually if operating power is not available.

Local requirements on the use of electrically operated service disconnects should be carefully included in service entrance design involving that type of equipment. If a switch can be opened and closed without exposing the operator to contact with live parts, it is an externally operable switch, even though access to the switch handle requires opening the door of a cabinet. Electrically operated switches and circuit breakers are required to be externally operable only to the open position; they are not required to be externally operable by hand to the closed position.

Disconnect Rating. In general, the service disconnecting means must have a rating that is not below 60 A; this rule is applicable to both fusible switches and CBs. However, 100 A is the minimum rating for a single switch or CB service disconnect for any "one-family dwelling" with an initial load of 10 kW or more, or where the initial installation contains more than five 2-wire branch circuits. It should be noted that, because the definition of a one-family dwelling is given in **NEC** Article 100 as "a building consisting solely of one dwelling unit," that rule applies to one-family houses only; it does not apply to apartments or similar dwelling units that are in two-family or multifamily dwellings.

If the demand of a total connected load, as calculated according to **NEC** Article 220, is 10 kW or more, a 100-A service disconnect, as well as 100-A service entrance

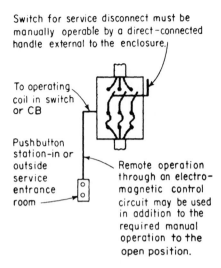

FIGURE 7.29 Service disconnect *must* be manually operable but may also be electric-power operable.

conductors, must be used. Any one-family house with an electric range rated 8¾ kW or more must have a disconnect (or service equipment) rated at 100 A; such a range provides a demand load 8 kW, the two required 20-A kitchen-appliance circuits provide a demand load of 3 kW at 100 percent demand (from Table 220-11), and the sum of the two loads exceeds the 10-kW level at which the minimum 100-A service is required.

If 100-A service is used, the demand load may be as high as 23 kW. By using the optional service calculations of Table 220-30, a 23-kW demand load is obtained from a connected load of 42.5 kW. This shows the effect of diversity on large-capacity installations.

Where two to six disconnects are used for a service instead of a single main service disconnect switch or CB, another basic **NEC** rule applies. Figure 7.30 shows that the total rating of the multiple service disconnects (two to six) must be at least equal to the rating of the single service disconnect that would be required if a single disconnect were used instead of multiple disconnects. It should be noted that the sum of the CB ratings in Fig. 7.30 is well above 400 A, but it does comply with the rule of **NEC** Section 230-80, even though the 400-A service entrance conductors could be heavily overloaded. Exception No. 3 of Section 230-90(a) clearly exempts this type of layout from the need to protect the conductors at their rated ampacity, as required in the basic rule. The 400-A rating of the service entrance conductors must be based on the minimum calculations set forth in Article 220 and is considered by the **NEC** to be safe because it is adequate for the maximum sum of the demand loads fed by the five disconnects shown in the layout.

The continuous current rating of a service (or another) disconnect must be sufficient for full-load current, and its momentary current rating must be such that the device can withstand possible short-circuit currents for the length of time it takes for the circuit to be cleared. The interrupting capacity of a disconnect must be sufficient

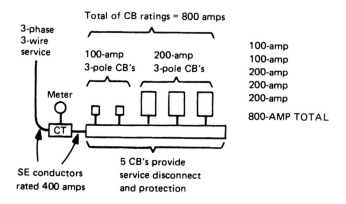

From Art. 220 (Secs. 220-10 through 220-21), calculation of demand load indicated that a single disconnect for this service must be rated at least 400 amps. The rating of multiple disconnects must total at least that value.

FIGURE 7.30 Sum of the ratings of multiple service disconnects must *at least* be equal to the minimum required rating of a single disconnect.

for any possible conditions of operation. Requirements for interrupting loads and overloads must be related to the time-current characteristics of service protection devices. On primary circuits, the disconnect should usually be able to break full-load current up to its own rating. On secondary circuits, a switch may need an interrupting capability ranging anywhere from its own current rating up to more than 10 times its current rating.

Multibuilding Premises. In a group of buildings or other structures under single management, disconnect means must be provided for each building. Each building or other structure in the group must always be provided with a readily accessible means, installed either inside or outside the building or other structure, at a location nearest the point of entrance of the conductors, for disconnecting all ungrounded conductors from the supply (Fig. 7.31). The disconnect means required for each outbuilding must be suitable for use as service equipment. The **NEC** rule on the provision of disconnect means for other buildings is usually interpreted to permit the disconnect for each building or structure to be the same kind as permitted for a service disconnect—that is, up to six switches or CBs.

The wording of that **NEC** rule (in Section 225-8) has often caused confusion in applying the rule. Some authorities had allowed use of the feeder switch in the main building as the only disconnect for each feeder to the outlying buildings, provided the switches in the main building were accessible to the occupants of the outlying buildings. But in the present **NEC,** the effect of the last sentence of Section 225-8(b) is to clearly require that the disconnect for any outbuilding be located within or just outside each building "nearest the point of entrance of the service-entrance conductors."

Overcurrent Protection

Each ungrounded service entrance conductor must be protected by a short-circuit overcurrent device in series with the conductor. Although there are many situations

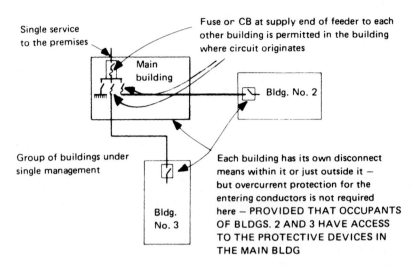

FIGURE 7.31 Each separate building must have its own disconnect for the supply conductors to it.

where, of necessity, the rating or setting of the overcurrent device will be higher than the service-conductor ampacity, the basic rule starts with the concept that the overcurrent device must have a rating or setting that is not higher than the allowable current capacity of the conductor. The idea is to ensure that the overcurrent protection required in the service entrance equipment protects the service entrance conductors from "overload." Because the service overcurrent protection devices are in the service disconnect equipment, which is installed at the load end of the service entrance conductors, these devices cannot protect the service entrance conductors from a "fault" that occurs in the service entrance conductors on the line side of the protection; but these devices can protect them from overload if they are of the proper rating. Conductors on the load side of the service equipment are considered as feeders or branch circuits and are required by the **NEC** to be protected as described in Articles 210, 215, and 240.

Each ungrounded service entrance conductor must be protected by an overcurrent device in series with the conductor. Figure 7.32 shows that such protection is always required where a single circuit breaker or fused switch is used for service disconnect and protection. The overcurrent device must have a rating or setting not higher than the allowable current capacity of the conductor, with the exceptions noted in **NEC** Section 230-90:

Exception No. 1. If the service supplies one motor in addition to some other load (such as lighting and heating), the overcurrent device may be rated or set in accordance with the protection required for a branch circuit supplying the one motor (Section 430-52) plus the other load, as shown in Fig. 7.33 and calculated as follows:

1. *Size of motor branch-circuit fuses.* 300 percent of 28 A, or 84 A. This requires a maximum fuse size of 90 A. Smaller fuses, such as the time-delay type, may be used.
2. *Size of service-entrance conductors.* Must be adequate for a load of 125 percent of 28 A plus 80 A (lighting load), or 115 A.

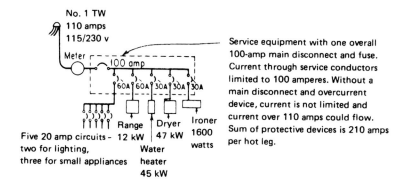

FIGURE 7.32 Single main service protective device must normally protect service entrance conductors at their ampacity.

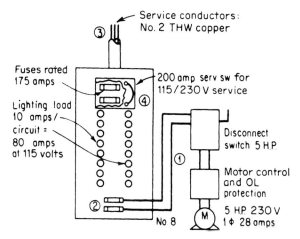

FIGURE 7.33 Motor load permits the sizing of service entrance protection above the conductor ampacity.

3. *Size of main fuses.* 90 A (from item 1 above) plus 80 A, or 170 A. This requires a maximum fuse size of 175 A. Again, smaller fuses may and should be used where possible to improve the overload protection on the circuit conductors.

The use of 175-A fuses where the calculation calls for 170 A conforms to Exception No. 2 of Section 230-90—the next higher standard rating of fuse (Section 240-6). For motor branch circuits and feeders, Articles 220 and 430 permit the use of overcurrent devices having ratings or settings higher than the capacities of the conductors. Article 230 makes similar provisions for services where the service supplies a motor load or a combination of both motors and other loads.

If the service supplies two or more motors as well as some other load, then the overcurrent protection must be rated in accordance with the required protection for a feeder supplying several motors plus the other load (Section 430-63). Or, if the service supplies only a multimotor load (with no other load fed), then Section 430-62 sets the maximum permitted rating of service overcurrent protection.

Exception No. 3. Not more than six CBs or six sets of fuses may serve as overcurrent protection for the service entrance conductors, even though the sum of the ratings of the overcurrent devices is in excess of the ampacity of the service conductors supplying the devices—as illustrated in Fig. 7.34. The grouping of single-pole CBs as multipole devices, as permitted for disconnect means, may also apply to overcurrent protection. And a "set" of fuses is all the fuses required to protect the ungrounded service entrance conductors.

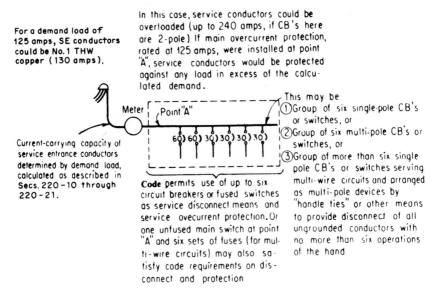

FIGURE 7.34 With two to six service entrance protective devices, conductors could be overloaded.

This exception ties into Section 230-80. Service conductors may be sized for the total maximum demand load, with permitted demand factors from Table 220-11 applied. Then each of the two to six feeders fed by the service entrance conductors is also sized from Article 220, based on the load fed by each feeder. When those feeders are given overcurrent protection in accordance with their ampacities, it is frequently found that the sum of the ratings of the overcurrent devices is greater than the ampacity of the service entrance conductors, which were sized by applying the applicable demand factors to the total connected load of all the feeders. Exception No. 3 recognizes that possibility as acceptable, even though it departs from the rule in the first sentence of Section 230-90(a). The assumption is that if the demand load for the service entrance conductors is correctly calculated, there will be no overloading of those conductors because the diversity of feeder loads (some loads on and some off) will be adequate to limit the load on the service entrance conductors.

Suppose the load of a building, computed in accordance with Article 220, is 255 A, and conductors are selected at that ampacity (250-kcmil THW copper is rated at 255 A). Under Section 240-3(b), 300-A fuses or a 300-A CB may be considered as overcurrent protection of the proper size for service conductors rated between 255 and 300 A if a single service disconnect is used. If the load were separated in such a manner that six 70-A CBs could be used instead of a single service disconnect means, then the total rating of the CBs would be 6 × 70 A, or 420 A, which is greater than the ampacity of the service entrance conductors. And that would be acceptable to the **NEC**, although heavy overload is possible.

Figure 7.35 shows a very important design consideration. The **NEC** makes it mandatory to oversize the protection for the fire-pump circuit. That rule is intended to prevent opening of the fire-pump circuit on any overload up to and including stalling or even seizing of the pump motor. Because the conductors are "outside the building," operating overload is no hazard; and, under fire conditions, the pump must have no prohibition on its operation. It is better to lose the motor than to attempt to protect it against overload when it is needed.

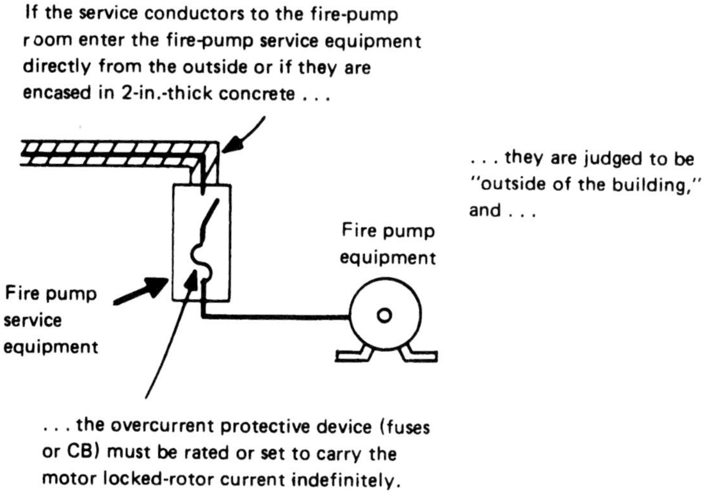

FIGURE 7.35 Overcurrent protection for a fire pump must permit overload up to locked-rotor current.

The basic design considerations on overcurrent protection for conductors feeding from one building to other buildings are covered in Fig. 7.31, which also illustrates the requirements on disconnect means for each feeder to a separate building. The rule of **NEC** Section 225-8(b) Exception No. 1 permits the overcurrent protection for a feeder from one building to another to be at the supply end of the circuit, in the building where the circuit originates, provided it is accessible to the occupants of the building being fed. Because a switch or CB must be used as the disconnect in each building being fed, a CB would constitute protection, and the use of fuses in a fusible switch would also eliminate concern about access to protection in the building where the feeder originates.

Ground-Fault Protection

NEC Section 230-95 requires ground-fault protection equipment to be provided for each service disconnecting means rated 1000 A or more in a solidly grounded-wye electrical service that operates with its ungrounded legs at more than 150 V to ground (Fig. 7.36). This applies to the rating of the disconnect and not to the rating of the overcurrent devices or to the capacity of the service entrance conductors.

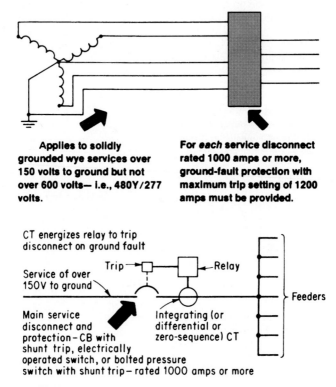

FIGURE 7.36 The basic condition that requires GFP at a service.

The wording of the **NEC** rules makes clear that service ground-fault protection (GFP) is required only for grounded-wye systems that have voltages over 150 V to ground. In effect, that means the rule applies only to 480/277-V grounded-wye systems and not to 208/120-V systems or any other commonly used systems. And GFP is not required on any systems operating over 600 V, although it is generally good design practice to include such protection.

A detailed discussion of ground-fault protection is given in Chapter 4 of this handbook. As noted there, the basic rules on ground-fault protection at a service entrance are as follows:

1. In a typical service GFP hookup, a ground-fault current of 1200 A or more must cause the disconnect to open all ungrounded conductors. Thus the maximum permitted GFP pickup setting is 1200 A, although the device may be set lower.
2. The rule requiring GFP for any service disconnect rated 1000 A or more (on 480/277-V services) specifies a maximum time delay of 1 s for ground-fault currents of 3000 A or more. The maximum permitted setting of a service GFP hookup is 1200 A, but the time-current trip characteristic of the relay must ensure opening of the disconnect in not more than 1 s for any ground-fault current of 3000 A or more. This is intended to establish a specific level of protection in GFP equipment by setting a maximum limit of I^2t for fault energy.
3. Selective coordination between GFP and conventional protective devices (fuses and CBs) on service and feeder circuits is a very clear and specific task as a result of the wording of Section 230-95(a) that calls for a maximum time delay of 1 s at any ground-fault current of 3000 A or more.

Additional considerations for service GFP are as follows:

1. In applying the rule of Section 230-95, the rating of any service disconnect means is to be determined as shown in Fig. 7.37.

FUSED SWITCH (bolted pressure switch, service protector, etc.)

Rating of switch is taken as . . . . . . the amp rating of the largest fuse that can be installed in the switch fuseholders

EXAMPLE

If 900-amp fuses are used in this service switch, ground-fault protection would be required, because the switch can take fuses rated 1200 amps—which is above the 1000-amp level at which GFP becomes mandatory.

CIRCUIT BREAKER

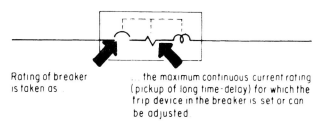

Rating of breaker is taken as . . . . . . the maximum continuous current rating (pickup of long time-delay) for which the trip device in the breaker is set or can be adjusted.

Example: GFP would be required for a service CB with, say, an 800-amp trip setting if the CB had a trip device that can be adjusted to 1000 amps or more.

FIGURE 7.37 Service-disconnect rating must be carefully determined.

2. Because the rule on required service GFP applies to the rating of each service disconnect, in many instances GFP would be required if a single service main disconnect were used but not if the service subdivision option of using up to six service entrance disconnects were taken, as shown in Fig. 7.38.

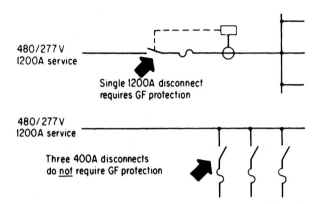

FIGURE 7.38 Subdividing the service disconnect to eliminate any 1000-A (or higher) disconnects evades mandatory GFP at the service.

3. Continuous industrial process operations are exempted from the GFP rules of Section 230-95(a) where the electrical system is under the supervision of qualified persons who will effect orderly shutdown of the system and thereby avoid the hazards, greater than ground fault itself, that would result from the nonorderly, automatic interruption that GFP would produce in the supply to such critical continuous operations. The GFP requirements are not applicable where a nonorderly shutdown will introduce additional or increased hazards. The idea behind that is to provide maximum protection against service outage for industrial processes. With highly trained personnel at such locations, design and maintenance of the electrical system can often accomplish safety objectives more readily without GFP on the service. Electrical design can include provisions to minimize any danger to personnel resulting from loss of process power while also protecting electrical equipment.

4. Another exception in Section 230-95 excludes fire-pump service disconnects from the basic rule that requires ground-fault protection on any service disconnect rated 1000 A or more on a grounded-wye 480/277-V system. Because fire pumps are required by Exception No. 4 of Section 230-90 to have overcurrent protection devices large enough to permit locked-rotor current of the pump motor to flow without interruption, larger fire pumps (100 hp and more) would have disconnects rated 1000 A or more. Without the exception, those fire-pump disconnects would be subject to the basic rule and would have to be equipped with ground-fault protection. GFP protection on any fire pump is objectionable for the same reason that leads to Exception No. 4 of Section 230-90, which permits locked-rotor current to flow continuously. The intent is to give the pump motor every chance to operate during a fire—to prevent opening of the motor circuit on an overload up to and including stalling or seizing of the shaft or bearings. For the same reason, Section 430-31 exempts fire pumps from the need for

overload protection, and Exception No. 4 to Section 430-72 requires overcurrent protection to be omitted from the control circuit of a starter for a fire pump. And it should be noted that Exception No. 2 in Section 230-95 says that ground-fault protection "shall not apply" to fire-pump motors—which appears to make omission of GFP mandatory, as shown in Fig. 7.39.

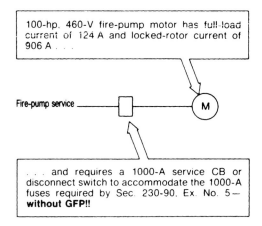

FIGURE 7.39 Fire-pump service disconnects of 1000 A or more must not be equipped with ground-fault protection.

5. Obviously, the selection of ground-fault equipment for a given installation merits detailed study. The option of subdividing services, so that up to six service entrance disconnects are fed by one service drop or lateral, should be evaluated. A 4000-A service, for example, could be divided using five 800-A disconnecting means, and in such cases GFP would not be required.

6. One very important consideration that should be studied is the potential desensitizing of ground-fault sensing hookups when an emergency generator and transfer switch are provided in conjunction with the normal service to a building. The problem arises in those cases where a solid neutral connection from the normal service is made to the neutral of the generator through a three-pole transfer switch. With the neutral both grounded at the normal service and bonded to the generator frame, ground-fault current on the load side of the transfer switch can return over two paths, one of which will escape detection by the GFP sensor, as shown in Fig. 7.40. Such a hookup may also cause nuisance tripping of the GFP due to normal neutral current. Under normal (nonfault) conditions, the neutral current due to normal load unbalance on the phase legs can divide at the common neutral connection in the transfer switch, with some current flowing toward the generator and returning to the service main on the conduit—and indicating falsely that a ground fault exists and causing nuisance tripping of the GFP. A four-pole neutral-switched transfer switch will eliminate this problem and ensure proper and effective operation of the GFP hookup, as shown in Fig. 7.41.

7. Although item 6 above is concerned with the desensitizing of service GFP where connection is made from an emergency generator through an automatic transfer

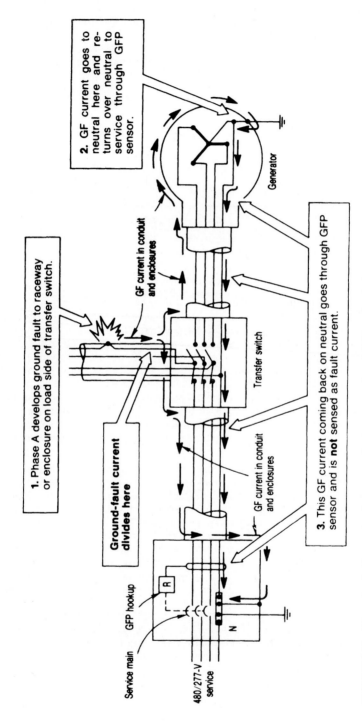

FIGURE 7.40 Alternative fault-current paths can cause improper operation of GFP sensing devices.

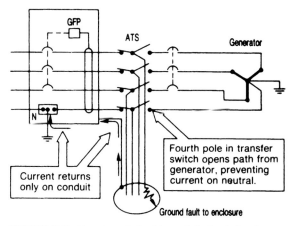

FIGURE 7.41 Transfer switch with a switched pole for the neutral prevents the occurrence of two return paths for neutral current.

switch, the whole question of using a GFP hookup on the generator-output disconnect means must be considered. Because a disconnect for the generator output constitutes a "service disconnect," such a unit rated 1000 A or more would be required by the wording of Section 230-95 to have GFP. But, recognizing that the reliability and continuity of an emergency supply is a design consideration that should be left to the judgment of the designer, **NEC** Section 700-26 states that the emergency generator disconnect does not require ground-fault protection. This rule clarifies the relationship between Section 230-95 and disconnect means for emergency generators. But, an indication must be provided. To afford the highest reliability and continuity for an emergency power supply, the **NEC** does not require GFP on any disconnect for an emergency generator, although it may be used if desired (Fig. 7.42).

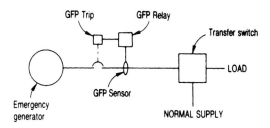

NOTE: On the output of "optional standby" generator, a disconnect rated 1000 A or more on a 480/277-V grounded wye system *must* be equipped with GFP.

FIGURE 7.42 Ground-fault protection is optional for the disconnect means on the output of an emergency or legally required standby generator.

VERY IMPORTANT: Because there have been many reports of improper and/or unsafe operation (or failure to operate) of ground-fault protective hookups, Section 230-95(c) requires (a mandatory rule) that every GFP hookup be "performance tested when first installed." This rule requires that such testing be done according to "approved instructions . . . provided with the equipment." A written record must be made of the test and made available to the inspection authority.

Service Grounding

The basic grounding concepts are presented under "System and Equipment Grounding" in Chapter 4 of this handbook. Every service entrance must have a grounding-electrode connection—that is, grounding of an intentionally grounded circuit conductor (such as a neutral conductor) and interconnection and grounding of the metal enclosures of the electrical system, or simply grounding of the metal enclosures where the system does not include an intentionally grounded circuit conductor. In all such cases, a major design consideration is the selection and layout of conductors and electrodes to provide the grounding connection to earth.

A grounding-electrode arrangement is required at the service entrance of a premises or in a building or other structure fed from a service in another building or other structure, as described under "Grounding at Separate Buildings" in Chapter 4. But the methods and requirements involved here do not apply to the grounding of a separately derived system, such as a local step-down transformer, which is covered under "Transformer Grounding" in Chapter 6.

To satisfy the **NEC,** it is now necessary to provide a suitable "grounding electrode system" and not simply a "grounding electrode" as required by previous **NEC** editions. Several considerations are involved in this design task.

Until the 1978 **NEC,** the "water-pipe" electrode was the premier electrode for service grounding, and "other electrodes" or "made electrodes" were acceptable only "where a water system (electrode) . . . is not available." If a metal water pipe to a building had at least 10 ft of its length buried in the ground, that pipe had to be used as the grounding electrode, and no other electrode was required. The underground water pipe was the preferred electrode and the best electrode.

Of all the electrodes previously and still recognized by the **NEC,** the water pipe is the least acceptable to the present **NEC** and the only one that may never be used by itself as the sole electrode. It must always be supplemented by at least one "additional" grounding electrode. Any of the other grounding electrodes recognized by the **NEC** is acceptable as the sole grounding electrode, by itself. In addition, where other electrodes are connected to the grounded-conductor terminal within the service equipment through the water pipe, all connections—including those to the grounded conductor—must be made within 5 ft of the water pipe's point of entrance.

Consider a typical water supply of 12-in-diameter metal pipe running, say, 400 ft underground to a building with a 4000-A service. According to Section 250-81(a), that water pipe, connected by a 3/0 copper conductor to the bonded service-equipment neutral, must be used as a grounding electrode; but it may not serve as the only grounding electrode. It must be supplemented by one of the other electrodes from Section 250-81 or 250-83. So the installation can be made acceptable by, say, running a No. 6 copper grounding-electrode conductor from the bonded service neutral to an 8-ft, ½-in-diameter ground rod. Although that seems like using a mouse to help an elephant pull a load, it is the literal requirement. And if the same building did not have 10 ft of metal water pipe in the ground, the 8-ft ground rod would be entirely acceptable as the only electrode.

The basic **NEC** rule on grounding-electrode systems requires that all or any of the electrodes specified in Sections 250-81(a) to (d), if available on the premises, must be bonded together to form a "grounding electrode system":

1. If there is at least a 10-ft length of underground metal water pipe, a grounding-electrode conductor must be connected to the water pipe.
2. If, in addition, the building has a metal frame that is "effectively grounded" (see Article 100 Definitions), the frame must be bonded to the water pipe—or vice versa, because the rules do not spell out where actual connections are to be made for grounding-electrode conductors and bonding jumpers.
3. Then, if there is at least a total of 20 ft of one or more ½-in-diameter steel reinforcing bars or rods embedded in the concrete footing or foundation, a bonding connection must be made from one of the other electrodes to one of the rebars—and, obviously, that has to be done before concrete is poured for the footing or foundation.

When two or more grounding electrodes are to be combined into a "grounding electrode system," the bonding jumper between pairs of electrodes must not be smaller than the size of grounding-electrode conductor indicated in Table 250-94 for the particular size of the largest phase leg of the service feeder, up to No. 3/0 copper, which is shown as the maximum required size of copper grounding-electrode conductor in Table 250-94.

The "unspliced" grounding-electrode conductor at the service may be connected to whichever one of the interbonded electrodes provides the most convenient and effective point of connection.

The grounding-electrode conductor must be sized as if it were the grounding-electrode conductor for whichever one of the interbonded electrodes requires the largest grounding-electrode conductor. If, for instance, a grounding-electrode system consists of a metal underground water-pipe electrode bonded to a driven ground rod, the grounding-electrode conductor must be sized from Table 250-94; and on, say, a 2000-A service, it would have to be No. 3/0 copper or 250-kcmil aluminum, whether it is connected to the water-pipe electrode (which would require that size of grounding-electrode conductor) or to the driven ground rod (which requires only a No. 6 copper or No. 4 aluminum grounding-electrode conductor if it is used by itself as a grounding electrode). And the bonding jumper between the water-pipe electrode and the ground rod would also have to be of that size.

The "20 feet of bare copper conductor" referred to in Sections 250-81(c) and (d) is going to be "available" at a building only if the 20-ft arrangement has been specified by the electrical designer, because it is clearly and only a grounding electrode; the other electrodes mentioned in Sections 250-81(a) to (c) (including the rebars) are specified by persons other than the electrical designer, and they may be "available." If a building has all or some of the electrodes described, they must all be bonded together and used as a single electrode system. If it has none, then any one of the electrodes described in Section 250-83 may be used for service grounding.

An electrode (such as a driven ground rod) that supplements an underground water-pipe electrode may be "bonded" to any one of several points in the service arrangement. It may be "bonded" (1) to the grounding-electrode conductor or (2) to the grounded service conductor (grounded neutral), such as by connection to the neutral block or bus in the service panel or switchboard or in a current-transformer cabinet, meter socket, or other enclosure on the supply side of the service disconnect and (3) to a grounded metal service raceway. A bonding jumper must always be used around any water meter within a building and any place where piping on both sides of the meter is to be grounded.

The several requirements set by **NEC** rules and the conditions established for application of the rules can best be understood through a step-by-step approach relative to a typical installation:

Consider a building fed by an underground water-pipe system. At least 10 ft of metal water pipe is buried in the earth ahead of the point at which the metal pipe enters the building. As shown in Fig. 7.43, such a buried pipe is a grounding electrode, and connection must be made to the underground pipe with a grounding-electrode conductor sized from Table 250-94 and run from the grounding point in the service equipment. But now, a number of other factors must be accounted for, as follows:

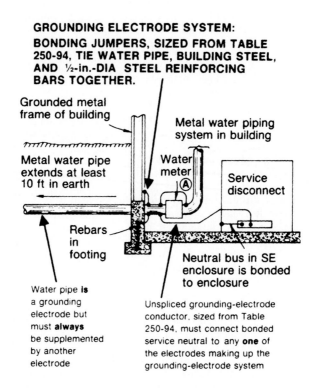

FIGURE 7.43 Water pipe, metal building frame, and reinforcing in footing must be bonded into an "electrode system."

1. Even though the water pipe is a suitable grounding electrode, at least one more grounding electrode must be provided and must be bonded to the water-pipe electrode. A water pipe, by itself, is not an adequate grounding electrode and must be supplemented by at least one other electrode to provide a "grounding electrode system."

2. The additional electrode may be:
- The metal frame of the building, provided the frame is effectively grounded (embedded in earth and/or in buried concrete).
- Or a concrete-encased electrode within, and near the bottom of, a concrete foundation or footing in direct contact with the earth. The electrode must consist of at least 20 ft of one or more steel reinforcing bars or rods of not less than ½-in diameter, or it must consist of at least 20 ft of bare solid copper conductor not smaller than No. 4 AWG. Because the building footing or foundation in Fig. 7.43 contains such steel reinforcing, connection of a bonding conductor has to be made to the steel, and the conductor must be brought out for connection to the water pipe, the building steel, or the bonded service neutral.
- Or a "ground ring encircling the building or structure," buried directly in the earth at least 2½ ft down. The ground ring must be "at least 20 feet" of bare No. 2 or larger copper conductor. (In most cases, the conductor will have to be considerably longer than 20 ft to "encircle" the building or structure.)
- Or underground bare metal gas piping or other metal underground piping or tanks.
- Or a buried 8-ft ground rod or pipe or a plate electrode. An example of the bonding of a supplemental grounding electrode is shown in Fig. 7.44. And an interesting point about the bonding of the ground rod to the neutral bus in the service entrance enclosure relates to the size of the bonding jumper. Because the first paragraph of Section 250-81 requires that all the electrodes that make up a grounding-electrode system be "bonded together" by a jumper sized from Table 250-94, and because the last sentence of Section 250-81(a) requires the supplemental electrode to be "bonded" to the service neutral or other specified point, it seems clear that the conductor shown connecting the ground rod to the neutral bus in the sketch is a "bonding jumper" and must be the same size as the grounding-electrode conductor (from Table 250-94). Because that conductor is not described as a "grounding electrode conductor," use of No. 6 copper or No. 4 aluminum—as recognized by Exception No. 1(a) of Section 250-94—would appear to be a violation. That concept is supported by the last sentence of the first paragraph of Section 250-81.

Section 250-81(c) states that the encased grounding electrode may be 20 ft of bare No. 4 copper—and it may be stranded. It was required to be "solid" copper conductor in the 1978 **NEC**.

3. According to the wording of Section 250-81, any of the four types of grounding electrodes mentioned in parts (a) to (d) of that section, if present, must be bonded together. Note that the rule does not state that any of those electrodes must be provided. But if any or all of them *are* present, they must be bonded together to form a "grounding electrode system." And, where a water-pipe electrode, as described, is present, any one of the other three electrodes in Section 250-81 may be used as the required "additional electrode."

In the case of a building fed by a nonmetallic underground piping system or one that does not have 10 ft of metal pipe underground, the water-pipe system is not a grounding electrode; however, the interior metal water-pipe system must be bonded to the service grounding. But when the water pipe is not a grounding electrode, another type of electrode must be provided to accomplish the service grounding. Any one of the other three electrodes of Section 250-81 may be used as the required grounding electrode. For instance, if the metal frame of the building is effectively grounded, a grounding-electrode conductor, sized from Table 250-94 and run from

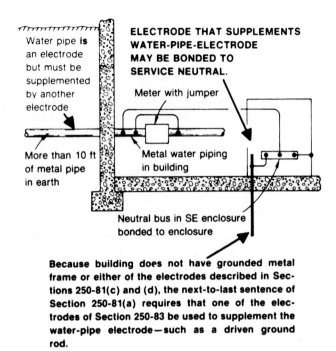

FIGURE 7.44 An additional electrode must always be used to supplement a water-pipe electrode.

the bonded service neutral or ground terminal to the building frame, may satisfy the **NEC,** as shown in Fig. 7.45. Where none of the four electrodes described in Section 250-81 is present, one of the electrodes from Section 250-83 must be used (e.g., a ground rod, as shown in Fig. 7.46). Note, again, that any type of electrode other than a water-pipe electrode may be used by itself as the sole electrode.

As shown in Fig. 4.46, the wording of Exception No. 1 of Section 250-94 would permit the use of a No. 6 copper or No. 4 aluminum grounding-electrode conductor. Because the ground rod is the only grounding electrode, there is no relationship to the rules given in the last sentence of Section 250-81 and the last sentence of Section 250-81(a), which seem to require a larger "bonding jumper" in cases where a ground rod is a supplemental electrode, as shown in Fig. 7.44. Actually, it is more consistent with long-time **NEC** practice, based on tests, that the ground rod shown in Fig. 7.44 be connected to the neutral terminal by a conductor that is not required to be larger than No. 6 copper or No. 4 aluminum.

The four illustrations shown in this discussion are only typical examples of the many specific ways in which **NEC** rules on grounding electrodes may be applied.

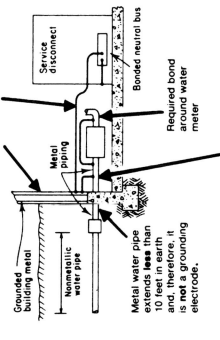

FIGURE 7.45 Building's metal frame may be the sole grounding electrode.

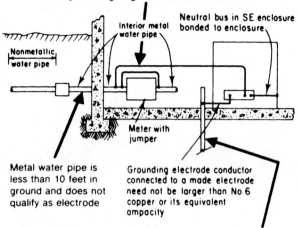

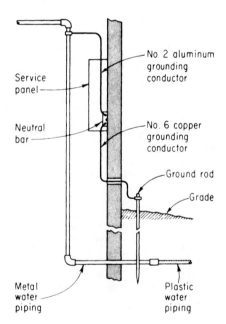

FIGURE 7.46 For many systems, a driven ground rod may be the sole grounding electrode serving a building without other available electrodes.

As noted previously, "continuity of the grounding path or the bonding connection to interior piping shall not rely on water meters." A bonding jumper always must be used around a water meter. This is required because of the chance of loss of grounding if the water meter is removed or replaced with a nonmetallic water meter. The bonding jumper around a meter must be at least the same size as the grounding-electrode conductor required for the given size of service conductors.

The concrete-encased electrode that is described in Section 250-81(c), known as the "Ufer system," has particular merit in new construction, where the bare copper conductor or steel reinforcing bar or rod can be readily installed in a foundation or footing form before concrete is poured. Electrodes of this type using a bare copper conductor have been installed as far back as 1940, and tests have proved the system to be highly effective (Fig. 7.47).

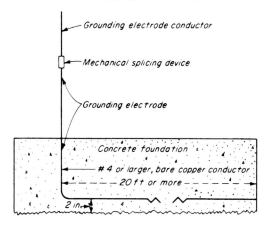

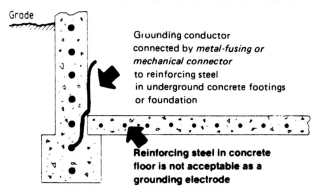

FIGURE 7.47 The so-called "Ufer" system grounding electrode is especially effective.

The intent of the words "bottom of a concrete foundation" is to completely encase the electrode within the concrete, in the footing near the bottom. The footing must be in direct contact with the earth, which means that dry gravel or polyethylene sheets between the footing and the earth are not permitted.

It may be advisable to provide additional corrosion protection in the form of plastic tubing or sheath at the point where the grounding electrode leaves the concrete foundation.

For concrete-encased steel reinforcing bar or rod systems used as grounding electrodes in underground footings or foundations, welded-type connections (metal-fusing methods) may be used for connections that are encased in concrete. Compression or other types of mechanical connectors may also be used.

"Made" Electrodes. As a general rule, if a water-pipe system or other approved electrode is not available, a driven rod or pipe is used as the electrode. A rod or pipe driven into the ground does not always provide as low a ground resistance as is desirable, particularly where the soil is very dry. In some cases where several buildings are supplied, grounding at each building reduces the ground resistance.

Where it is necessary to bury more than one pipe or rod to lower the resistance to ground, they must be placed at least 6 ft apart. If they were placed closer together, there would be little improvement.

Where two driven or buried electrodes are used to ground two different systems that should be kept entirely separate from one another (such as a grounding electrode for a wiring system for light and power and a grounding electrode for a lightning rod), care must be taken to guard against the conditions of low resistance between the two electrodes and high resistance from each electrode to ground. If two driven rods or pipes are located 6 ft apart, the resistance between the two is sufficiently high and cannot be greatly increased by increasing the space. Thus, at least 6 ft of spacing is required between electrodes serving different systems.

Whenever a ground rod is used, it must be driven straight down into the earth, with at least 8 ft of its length in the ground (in contact with soil). If rock bottom is hit before the rod is 8 ft into the earth, it is permissible to drive it into the ground at an angle—not over 45° from the vertical—to ensure that at least 8 ft of its length is in the ground. However, if rock bottom is so shallow that it is not possible to get 8 ft of the rod in the earth at a 45° angle, then it is necessary to dig a 2½-ft-deep trench and lay the rod horizontally in the trench. Figure 7.48 shows these rules.

Another **NEC** requirement calls for the upper end of the rod to be flush with or below the ground level—unless the above-ground end and the conductor clamp are protected either by locating them in a place where damage is unlikely or by placing a metal, wood, or plastic box or enclosure over the end.

This two-part rule was added to the **NEC** because it had become common practice to use an 8-ft ground rod driven, say, 6½ ft into the ground with the grounding-electrode conductor clamped to the top of the rod and run to the building. Not only is the connection subject to damage or disconnection by lawnmowers or vehicles, but also the length of unprotected, unsupported conductor from the rod to the building is a tripping hazard. The rule says, in effect, *bury everything or protect it!*

Of course, a buried conductor-clamp assembly that is flush with or below grade must be resistant to rusting or corrosion that might affect its integrity, and it must be UL-listed as suitable for direct earth burial. Such clamps are available.

In addition to ground rods, plate electrodes are a form of "made" (manufactured) electrodes. Such electrodes are listed in the UL *Electrical Construction Materials Directory* under the heading "Grounding and Bonding Equipment," which also covers bonding devices, ground clamps, grounding and bonding bushings, ground rods,

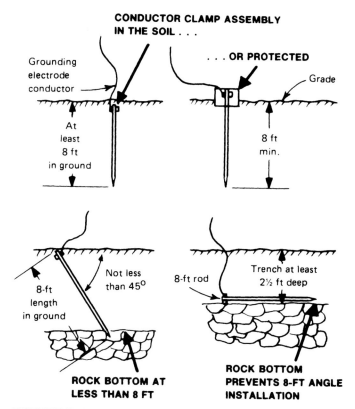

FIGURE 7.48 In a ground-rod installation, there must be at least 8 ft of rod in contact with soil.

armored grounding wire, protector grounding wire, grounding wedges, ground clips for securing the ground wire to an outlet box, water-meter shunts, and similar equipment. Only listed devices are acceptable for use. And listed equipment is suitable only for use with copper, unless it is marked "AL" and "CU."

When a made electrode is used for grounding, care must be taken to ensure that it provides an effective low-resistance path to the earth (or ground). A basic **NEC** rule calls for a made electrode to have a resistance to ground of 25 Ω or less. In any case where a single made electrode (rod, pipe, or plate) shows a resistance to ground over 25 Ω, the **NEC** simply requires that one additional made electrode be used in parallel; but there is then no need to make any measurement or add more electrodes or to be further concerned about the resistance to ground. In previous **NEC** editions, this rule implied that additional electrodes had to be used in parallel with the first one until a resistance of 25 Ω or less was obtained. Now, once the second electrode is added, it does not matter what the resistance to ground is, and there is no need for more electrodes, as shown in Fig. 7.49. Of course, good design practice and the specifications of many engineering organizations generally require a maximum resistance of 25 Ω.

When more than one made electrode is used for grounding, there must be at least 6-ft spacing between any pair of made electrodes. That minimum spacing is essential

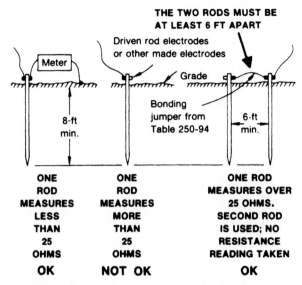

FIGURE 7.49 Resistance to earth of a "made" electrode must be measured.

where more than one ground rod, pipe, or plate is connected to a single grounding-electrode conductor and the resistance of a single grounding electrode is over 25 Ω to ground. Even greater spacing is better for rods longer than 8 ft. Separation of the rods reduces their combined resistance to ground.

Insofar as made electrodes are concerned, wide variation in resistance is to be expected; the present requirements of the **NEC** concerning the use of such electrodes do not provide for a system that is in any way comparable to that which can be expected where good underground metallic piping can be utilized.

It is recognized that some types of soil may cause a high rate of corrosion, resulting in a need for periodic replacement of grounding electrodes. Also, intimate contact between two dissimilar metals such as iron and copper, under wet conditions, can lead to electrolytic corrosion.

When a cross occurs between a high-tension conductor and one of the low-tension secondary conductors, the electrode may be called upon to conduct a heavy current into the earth. The voltage drop in the ground connection, including the conductor leading to the electrode and the earth immediately around the electrode, will be equal to the current multiplied by the resistance. This results in a difference in potential between the grounded conductor of the wiring system and the ground. It is therefore important that the resistance be as low as practicable.

Where made electrodes are used for the grounding of interior wiring systems, resistance tests should be conducted on a sufficient number of electrodes to determine the conditions prevailing in each locality. The tests should be repeated several times a year to determine whether conditions have changed because of corrosion of the electrodes or drying out of the soil.

Grounding Connections.   (Refer to the discussion of "Ground Connections" in Chapter 4.)

In addition to connection of the grounding-electrode conductor to the neutral terminal or the equipment grounding terminal as described in Chapter 4 above, another concern is bonding of the neutral terminal to the equipment grounding terminal at the service entrance.

The equipment grounding conductor at a service (such as the ground bus or terminal in the service equipment enclosure, or the enclosure itself) must be connected to the system grounded conductor (the neutral or grounded phase leg). The equipment ground and the neutral or other ground leg must be bonded together *on the supply side* of the service disconnecting means—which means either within or ahead of the enclosure for the service equipment, as shown in Fig. 7.50.

The ground bus or the enclosure must be simply bonded to the grounding-electrode conductor within or ahead of the service disconnect for an ungrounded system.

As shown in Fig. 7.51, some switchboard sections or interiors include neutral busbars that are factory-bonded to the switchboard enclosure and marked "Suitable for use only as service equipment." They may not be used as subdistribution switchboards; i.e., they may not be used on the load side of the service, except (1) where they are used, with the inspector's permission, as the first disconnecting means fed by a transformer secondary or a generator and where the bonded neutral satisfies Section 250-26(a) for a separately derived system or (2) where they are used as the main disconnect for a building that is fed from another building (as described under "Grounding at Separate Buildings" in Chapter 4 of this handbook).

After service enclosures are bonded to equipment grounding buses and to the grounded neutral (or grounded phase leg) of a grounded system, a common connection to ground must be made for the interbonded components. All the bonded components—the service equipment enclosure, the grounded neutral or grounded phase leg, and any equipment grounding conductors that come into the service enclosure—must be connected to a common grounding electrode with a single grounding-electrode conductor. A common grounding-electrode conductor must be run from the common point so obtained to the grounding electrode, as shown in Fig. 7.52.

Connection of the system neutral to the switchboard frame or ground bus within the switchboard provides the lowest impedance for the equipment ground return to the neutral. The main bonding jumper that bonds the service enclosure and equipment grounding conductors (which may be either conductors or conduit, EMT, etc.) to the grounded conductor of the system must be installed within the service equipment or within a service conductor enclosure on the line side of the service, as shown in Fig. 7.53. And it should be noted that in a service panel, equipment grounding conductors for load-side circuits may be connected to the neutral block, and there is no need for an equipment grounding terminal bar or block.

If one grounding conductor is used to ground the neutral to a water pipe or other grounding electrode, and a separate grounding conductor is used to ground the switchboard frame and housing to the water pipe or other electrode—without the neutral and the frame being connected together in the switchboard—then the length and impedance of the ground path are increased. The proven hazard is that the impedance of the fault-current path may then limit the fault current to a level too low to operate the overcurrent devices "protecting" the faulted circuit.

Note that a number of grounding electrodes that are bonded together are considered to be one grounding electrode.

The electrode or electrodes used to ground the neutral or other grounded conductor of an ac system must also be used to ground the entire system of interconnected raceways, boxes, and enclosures. The single common grounding-electrode conductor then connects to the single grounding electrode or electrode system and thereby grounds the bonded point of the system and equipment grounds.

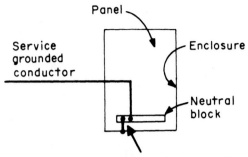

Bonding is the insertion of a bonding screw into the panel neutral block to connect the block to the panel enclosure, or it is use of a bonding jumper from the neutral block to an equipment grounding block that is connected to the enclosure.

NOTE: Bonding—the connection of the neutral terminal to the enclosure or to the ground terminal that is, itself, connected to the enclosure—might also be done in an individual switch or CB enclosure.

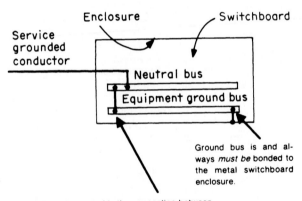

Ground bus is and always *must be* bonded to the metal switchboard enclosure.

Bonding of the neutral is the connection between the neutral bus and the equipment grounding bus or between the neutral bus and the metal enclosure itself.

FIGURE 7.50 Equipment ground terminal must be "bonded" to the grounded conductor at the service equipment.

In any building housing livestock, all piping systems, metal stanchions, drinking troughs, and other metalwork with which animals might come in contact must be bonded together and to the grounding electrode used to ground the wiring system in the building.

Grounding to Neutral. Connection between a grounded neutral (or grounded phase leg) and equipment enclosures is permitted by the **NEC** for the purpose of

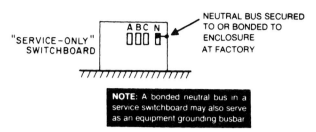

FIGURE 7.51 Bonding of the service neutral may be accomplished by the equipment manufacturer.

grounding the enclosures to the grounded circuit conductor; however, such general permission is limited to the supply side of service equipment and the supply side of a main disconnect for a separate building. Figure 7.54 shows such applications. In part (*a*), the grounded service neutral is bonded to the meter housing by means of the bonded neutral terminal lug in the socket—and the housing is grounded by this connection to the grounded neutral, which itself is grounded at the service equipment as well as at the utility-transformer secondary supplying the service. In part (*b*), the service equipment enclosure is grounded by connection (bonding) to the grounded neutral—which itself is grounded at the meter socket and at the supply transformer. These same types of grounding connections may be made for current-transformer cabinets, auxiliary gutters, and other enclosures on the line side of the service entrance disconnect means, including the enclosure for the service disconnect. (In some areas, utilities and inspection departments will not permit the arrangement shown in Fig. 7.54 because the connecting lug in the meter housing is not always accessible for inspection and testing purposes.)

Aside from the above-described connections, the **NEC** generally prohibits connection between a grounded neutral and equipment enclosures on the load side of the service. So aside from the few specific exceptions, bonding between any system grounded conductor, neutral or phase leg, and equipment enclosures is prohibited on the load side of the service, as indicated in Fig. 7.55. The use of a neutral to ground a panelboard or other equipment on the load side of service equipment would be extremely hazardous if the neutral became loosened or disconnected. In such cases any line-to-neutral load would energize all metal components connected to the neutral, creating a dangerous potential above ground. Hence, the prohibition of such practice.

Although the **NEC** prohibits neutral bonding on the load side of the service, it clearly requires such bonding at the service entrance. And the exceptions to the prohibition of load-side neutral bonding to enclosures are few and very specific:

- In a system, when voltage is stepped down by a transformer, a grounding connection must be made to the secondary neutral of a 208/120- or 240/120-V system, even though it is on the load side of the service.
- When a circuit is run from one building to another, it may be necessary or simply permissible to connect the system "grounded" conductor to a grounding electrode at the other building, as covered under "Grounding at Separate Buildings" in Chapter 4 of this handbook.
- The **NEC** permits grounding of meter enclosures to the grounded circuit conductor (generally, the grounded neutral) on the load side of the service disconnect if

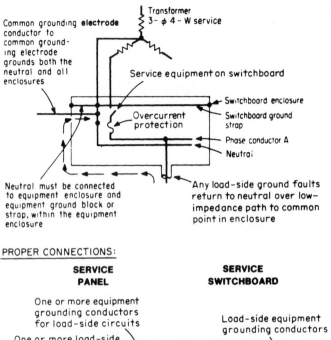

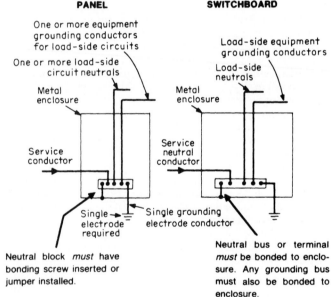

FIGURE 7.52 "Common" grounding-electrode conductor must be used for the service *grounded* conductor (neutral) and the equipment ground terminals.

the meter enclosures are located near the service disconnect and the service is not equipped with ground-fault protection, as shown in Fig. 7.56. There is no definition for the word "near," but it can be taken to mean in the same room or general area. This rule applies, of course, to multioccupancy buildings (apartments, office buildings, etc.) with individual tenant metering.

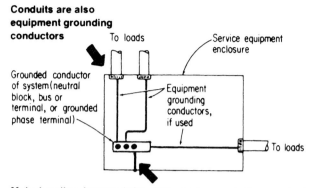

FIGURE 7.53 Main service bonding jumper must be within the service equipment.

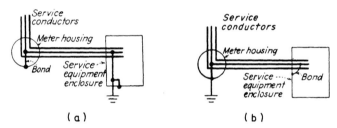

FIGURE 7.54 Grounded circuit conductor may be used to ground equipment housings on the line (supply) side of the service disconnect.

If a meter bank is on the upper floor of a building, as in a high-rise apartment house, or otherwise away from the service disconnect, the meter enclosures would not satisfy the requirement that they be "near" the service disconnect. In such cases, the enclosures must not be grounded to the neutral. And, if the service has ground-fault protection, meter enclosures on the load side must not be connected to the neutral, even if they are "near" the service disconnect.

Unused-Neutral Service. Whenever a service is derived from a grounded-neutral system, the grounded neutral conductor must be brought into the service entrance equipment, even if the grounded conductor is not needed for the load supplied by the service. This is required to provide a low-impedance ground-fault-current return path to the neutral to ensure operation of the overcurrent device, for the safety of personnel and property (Fig. 7.57).

In such cases, the neutral functions strictly as an equipment grounding conductor, to provide a closed-circuit path back to the transformer for automatic circuit opening in the event of a phase-to-ground fault anywhere on the load side of the service equipment. If the neutral were not brought in as shown in Fig. 7.57, fault current would be forced to take the high-impedance path through the earth in flowing from

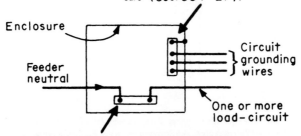

FIGURE 7.55 Neutral must *not* generally be bonded to equipment enclosures on the load side of the service.

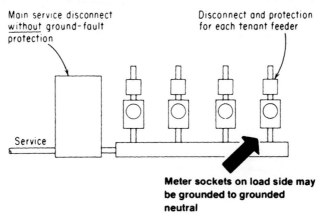

FIGURE 7.56 Grounding of meter enclosures to the grounded conductor (neutral) on the load side of the service is limited.

the service grounding electrode to the transformer ground. Not enough current would flow back to the transformer neutral to cause the protective device to operate. (Only one phase leg is shown in these diagrams to simplify the concept. The other two phase legs have the same relation to the neutral.)

The same requirements apply to separate power and light services derived from a common 3-phase, 4-wire, grounded "red-leg" delta system. The neutral from the outdoor center-tapped transformer winding must be brought into the 3-phase power-service equipment as well as into the lighting service, even though the neutral will not be used for power loads. (This is shown in Fig. 7.58.) Such an unused neutral

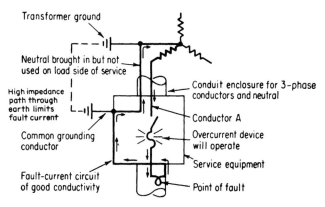

FIGURE 7.57 Grounded neutral must be brought into service entrance equipment even if it is not used for any circuits to loads.

must be at least equal in size to the required minimum size of grounding-electrode conductor specified in **NEC** Table 250-94 for the size of the phase conductors. In addition, if the phase legs associated with that neutral are larger than 1100 kcmil copper, the grounded neutral must not be smaller than 12½ percent of the area of the largest phase conductor. An example of this requirement is shown in Fig. 7.59.

The final part of the rule covers cases where the service phase conductors are paralleled, with two or more conductors in parallel per phase leg and neutral, and it requires that the size of the grounded neutral must be calculated on the equivalent area for parallel conductors. If a calculated size of neutral (at least 12½ percent of the phase-leg cross section) is to be divided among two or more conduits, and if dividing the calculated size by the number of conduits being used calls for a neutral conductor smaller than 1/0 in each conduit, the FPN calls attention to Section 310-4, which gives No. 1/0 as the minimum size of conductor that may be used in parallel in multiple con-

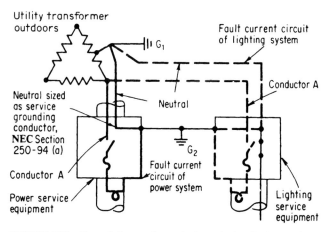

FIGURE 7.58 Grounded neutral must be brought into both sets of service equipment.

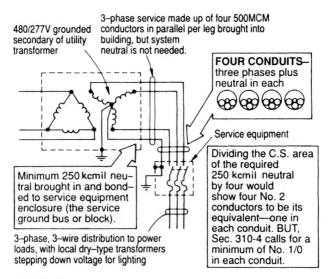

FIGURE 7.59 Grounded service conductor must *always* be brought in. [Sec. 250-23.]

duits. For that reason, each neutral would have to be at least a No. 1/0, even though the calculated size might be, say, No. 1 or No. 2 or some other size smaller than No. 1/0. But, the **NEC** rule does permit subdividing the required minimum 12½ percent grounded (neutral) conductor size by the total number of conduits used in a parallel run, thereby permitting a multiple makeup using a smaller neutral in each pipe.

As shown in Figure 7.59, the minimum required size for the grounded neutral conductor run from the supply transformer to the service is based on the size of the service phase conductors. In this case, the overall size of the service phase conductors is 4 × 500 kcmil per phase leg, or 2000 kcmil. Because that is larger than 1100 kcmil, it is not permitted to simply use Table 250-94 in sizing the neutral. Instead, 2000 kcmil must be multiplied by 12½ percent. Then 2000 kcmil × 0.125 equals 250 kcmil—the minimum permitted size of the neutral conductor run from the transformer to the service equipment. It is **NEC** intent to permit the required 250-kcmil-sized neutral to be divided by the number of conduits. From **NEC** Table 8 in Chapter 9, it can be seen that four No. 2 conductors, each with a cross-sectional area of 66,360 circular mils, would approximate the area of one 250 kcmil (250,000 circular mils divided by 4 = 62,500 circular mils). But, because No. 1/0 is the smallest conductor that is permitted by Section 310-4 to be used in parallel for a circuit of this type, it would be necessary to use a No. 1/0 copper conductor in each of the four conduits, along with the phase legs.

Pole-Type Service. As described at the beginning of this section, the basic approach to service grounding is to bond the neutral of a grounded service to the service equipment enclosure (as in a service panel or a service switchboard) and then run a grounding-electrode conductor from the service neutral bus or block to the common grounding electrode (or electrode system). But **NEC** Section 250-23(a) states a very important requirement concerning connection of the grounding-electrode conductor when a building is fed by service conductors from a meter on a yard pole.

The basic rule of Section 250-23 is that the grounding-electrode conductor required for the grounding of both the grounded service conductor (usually a grounded neutral) and the metal enclosure of the service equipment must be connected to the grounded service conductor within or on the supply side of the service disconnect. But the rule further requires that the grounding-electrode conductor be connected at some point between the load end of the service drop or lateral and the service equipment.

As a result of that requirement, if a service is fed to a building from a meter enclosure on a pole or other structure some distance away (as is commonly done on farm properties), and an overhead or underground run of service conductors is made to the service disconnect in the building, the grounding-electrode conductor will not satisfy the **NEC** if it is connected to the neutral in the meter enclosure at the pole; it must be connected at the load end of the underground or overhead service conductors. And, the connection should preferably be made within the service-disconnect enclosure.

This rule on grounding connection is shown in Fig. 7.60. If, instead of an underground lateral, an overhead run were made to the building from the pole, the overhead line would be a "service drop." The rule would likewise require the grounding connection to be at the load end of the service drop. If a fused switch or CB were installed as service disconnect and protection at the load side of the meter on the pole, then that would establish the service at that point, and the grounding-electrode connection to the bonded neutral terminal would be required at that point. The circuit from that point to the building would be a feeder and not service conductors. But, electrical safety and effective operation would require that an equipment grounding conductor be run with the feeder circuit conductors to ground the interconnected system of conduits and metal equipment enclosures along with metal piping systems and building steel within the building. Or, if an equipment grounding conductor were not in the circuit from the pole to the building, the neutral could be bonded to the main-disconnect enclosure in the building, and a grounding-electrode connection could be made at that point also.

Grounding Multiple Service Entrance Disconnects. When the service disconnect means (one to six circuit breakers of fused switches) is contained in a single enclosure—as with an individual main service switch or CB or a service panelboard or switchboard—a single common grounding-electrode conductor is run from the bonded neutral bus or block (of a grounded system) or from the equipment ground terminal (of an ungrounded system) to the grounding electrode or electrode system. But, a different approach has to be followed where the service disconnect means consists of two to six switches or circuit breakers, each in a separate enclosure. Two to six service disconnects might be fed by taps from an auxiliary gutter that is supplied by a single set of service entrance conductors, or a separate set of service entrance conductors may be run to each of the two to six switches or CBs, with the sets of service entrance conductors run down from a single service drop to the building (or fed by a single underground lateral). With such layouts of multiple service disconnects, the question arises, "Where should the grounding-electrode conductor be connected?"

Where a single service drop supplies a building using two to six service disconnecting means, each in a separate enclosure and each feeding a separate load, service grounding may be provided by a separate grounding-electrode conductor run from each disconnect to the water pipe or other grounding electrode, or a single grounding-electrode conductor may be used with a tap connected from each separate disconnect to the single grounding-electrode conductor (Fig. 7.61). Previous editions of

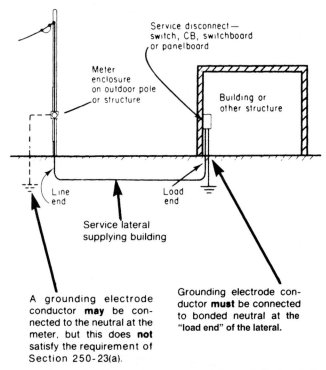

FIGURE 7.60 Pole-type service must be grounded at the load end of a service drop or lateral to a building.

the **NEC** required either that a separate grounding-electrode conductor be run from each enclosure to the grounding electrode or that a single, unspliced conductor be looped from enclosure to enclosure. A single grounding-electrode conductor used to ground all the service disconnects had to be without splice—run from one enclosure to the other and then to the water pipe or other grounding electrode. Exception No. 2 of Section 250-91(a) now says that splices may be made in the grounding-electrode conductor to tap into each of the service disconnect enclosures.

Exception No. 2 also covers the sizing of a main grounding-electrode conductor and taps from it to provide system grounding in each separate service enclosure where two to six service disconnects are used in separate enclosures. The wording of the exception requires the main grounding-electrode conductor to be sized from Table 250-94 but does not state the basis on which that table is to be used. The main grounding-electrode conductor might be sized for the sum of the cross-sectional areas of all the conductors connected to one hot leg of the service drop. Or, the main grounding-electrode conductor could be sized from Table 250-94 for the largest phase-leg conductor of one of the two to six sets of service entrance conductors. Or, the main grounding-electrode conductor could be sized from Table 250-94 on the basis of the size of one hot leg of a service of sufficient size to handle the demand load fed by the two to six service disconnects.

The wording of the rule does, however, make clear that the size of the grounding-electrode tap to each separate enclosure may be determined from Table 250-94 on

WHERE THE LAYOUT IS LIKE THIS . . .

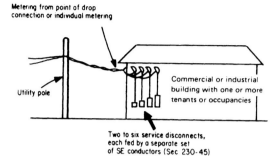

GROUNDING MAY BE DONE LIKE THIS . . .

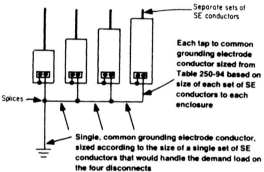

FIGURE 7.61 A choice of method is available for grounding multiple service disconnects in separate enclosures.

the basis of the largest service hot leg serving each enclosure, as shown in Fig. 7.62. Although that illustration shows an overhead service to the layout, the two to six service disconnects could be fed by individual sets of underground conductors, making up a "single" service lateral as permitted by Exception No. 7 of Section 230-2.

Service "Bonding." To ensure the electrical continuity of the grounding circuit, "bonding" (special precautions to ensure a permanent, low-resistance connection) is required at all conduit connections in the service equipment and where any nonconductive coating exists which might impair such continuity. This includes bonding at connections between service raceways, service cable armor, and all service-equipment enclosures containing service entrance conductors, including meter fittings, boxes, and the like.

The need for effective grounding and bonding of service equipment arises from the electrical characteristics of utility-supply circuits. In the common arrangement, service conductors are run to a building, and the service overcurrent protection is placed near the point of entry of the conductors into the building, at the *load end* of the conductors. With such a layout, the service conductors are not protected against ground faults or shorts occurring on the supply side of the service overcurrent protection. Generally, the only protection for the service conductors is on the primary side of the utility's distribution transformer. By providing "bonded" connections (connecting with special attention to reliable conductivity), any ground fault in a metal service raceway or other metal enclosure for service entrance conductors is given the greatest chance of burning itself clear—since there is not effective overcurrent protection ahead of those conductors to provide opening of the circuit on such fault currents. For any contact between an energized service conductor and grounded service raceway, fittings, or enclosures, bonding provides discharge of the fault current to the system grounding electrode—burning the fault clear. This condition is shown in Fig. 7.62. Any fault to an enclosure of a hot service conductor of a grounded electrical system must find a firm, continuous, low-impedance path to ground if there is to be sufficient current flow either to operate the primary protective device or to burn the fault clear quickly. This means that all enclosures containing the service conductors—service raceway, cable armor, boxes, fittings, cabinets—must be effectively bonded together; that is, they must have low impedance through themselves and must be securely connected to each other to provide a continuous path of sufficient conductivity to the conductor that makes the connection to ground, as shown in Fig. 7.63.

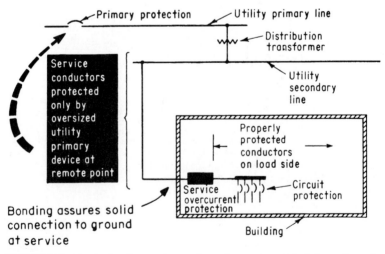

FIGURE 7.62 Service bonding must ensure burn-clear on ground faults in service enclosures.

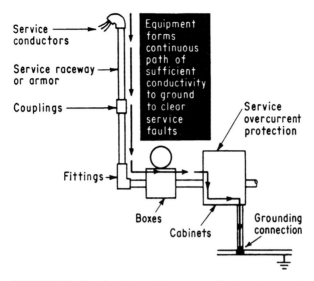

FIGURE 7.63 Bonding is intended to ensure a low-impedance path through all service enclosures.

Good engineering practice has long recognized that the conductivity of any equipment path should be at least equivalent to 25 percent of the conductivity of any phase conductor with which the ground path will act as a circuit conductor on a ground fault. Or, put another way, without reference to insulation or temperature rise, the impedance of the ground path must not be greater than 4 times the impedance of any phase conductor with which it is associated. This figure should always be used as a specific guide in evaluating the suitability of raceways or conductors as a grounding path.

In ungrounded electrical systems, the same careful attention should be paid to the matter of bonding together the non-current-carrying metal parts of all enclosures containing service conductors. Such a low-impedance ground path will quickly and surely ground any hot conductor which might accidentally become common with the enclosure system.

NEC Section 230-63 requires that service raceways, metal sheaths of service cables, metering enclosures, and cabinets for service disconnect and protection be grounded. The **NEC** requires that flexible metal conduit used as part of a run of service raceway be bonded around (Fig. 7.64).

As indicated in Fig. 7.65, service equipment that must have bonded connections includes (1) service raceway, cable trays, cable sheath, and cable armor; (2) all service-equipment enclosures containing service entrance conductors, including meter fittings, boxes, etc., interposed in the service raceway or armor; and (3) conduit or armor that encloses a grounding-electrode conductor that runs to and is connected to the grounding electrode or system of electrodes.

Bonded terminations must be used at the ends of conduit (rigid metal conduit, IMC, or EMT) or cable armor that encloses a grounding-electrode conductor. That means that conduit or cable armor must be connected with a bonding locknut or bonding bushing (with a bonding jumper around unpunched concentric or eccentric rings left in any sheet-metal knockout) or must be connected to a threaded hub or

566 CHAPTER SEVEN

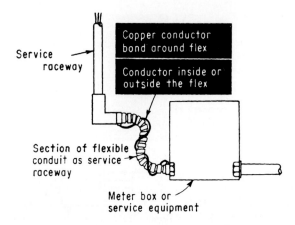

THIS IS NOW A VIOLATION !

FIGURE 7.64 Flex may be used as service raceway but only if bonded.

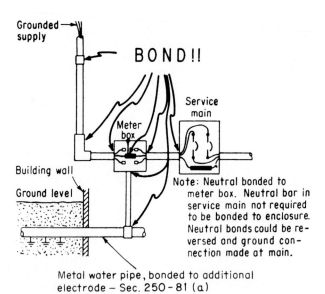

FIGURE 7.65 "Bonding" makes use of prescribed fittings and methods to connect enclosures containing service entrance conductors.

boss. A metal enclosure for a grounding-electrode conductor must be electrically in parallel with the grounding-electrode conductor. As a result, a metal conduit or EMT enclosing a grounding-electrode conductor forms part of that grounding-electrode conductor. In fact, such conduit or EMT has a much lower impedance than its enclosed grounding-electrode conductor (owing to the relation of magnetic fields) and is, therefore, even more important than the enclosed conductor in providing an effective path for current to the grounding electrode.

Figure 7.66 shows the "skin effect" of current flow over a conduit in parallel with a conductor run through it. This clearly shows the conduit itself to be a more important conductor than the actual conductor. In Fig. 7.66, the steel conduit acts as an iron core to greatly increase the inductive reactance of the conductor. This choke action raises the impedance of the conductor to such a level that only 3 A flows through the conductor while the balance of 97 A flows through the conduit. This division of current between the conduit and the conductor points up to the importance of ensuring that couplings and connectors are tight throughout every conduit system and in every metal raceway system and all metal cable jacketing. In particular, it emphasizes the need to bond both ends of any raceway used to protect a grounding conductor run to a water pipe or other grounding electrode. Such conduit protection must be securely connected to the ground electrode and to the equipment enclosure in which the grounding conductor originates. If such a conduit is left open, lightning and other electric discharges to earth through the grounding conductor will find a high-impedance path. High conductivity in the conduit system is also important for effective equipment grounding, even when a specific equipment grounding conductor is used in the conduit. Of course, nonmagnetic conduit—such as aluminum—would have a different effect, but it should also be in parallel with the enclosed conductor.

Ensuring the continuity of the raceway or armor for a grounding conductor reduces the impedance of the ground path, compared with what the impedance would be if the raceway or armor had poor connections or opens. Effective bonding of the raceway or armor minimizes the dc resistance of the ground path and reduces the overall impedance, which includes the choke action due to the presence of magnetic material (steel conduit or armor), that is, the increased inductive reactance of the circuit. Because of that, the "bonding" connection of such raceway to service enclosure must be made as shown in Fig. 7.67.

Bonding Techniques. A variety of techniques may be used to provide the very important (and **NEC** required) bonded connections of service-conductor enclosures:

Where rigid metal conduit or IMC is the service raceway, threaded or threadless couplings can be used to couple sections of conduit together. Conduit may be connected to a meter socket by connecting a threaded conduit end to a threaded hub or boss on the socket housing, where the housing is so constructed; with a locknut and bonding bushing; with a locknut outside and a bonding wedge or bonding locknut and a standard metal or completely insulating bushing inside; or sometimes with a locknut and standard bushing where the socket enclosure is bonded to the grounded service conductor. Conduit can be connected to knockouts in sheet-metal enclosures with a bonding locknut. A bonding locknut is a recognized device for bonding a service-conduit nipple to a meter socket when the knockout is clean (no rings left in the enclosure wall) or is cut on the job. With plastic bushing permitted, this is the most economical of the several methods for making a bonded conduit termination.

The bonded termination of rigid metal conduit or IMC may also be accomplished with a bonding wedge or a bonding bushing, where no knockout rings remain around the opening through which the conduit enters. Where a knockout ring does

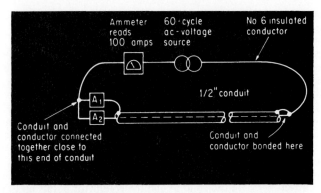

Ammeter A₁ (indicates amount of current in conduit) = 97 amps

Ammeter A₂ (indicates amount of current in conductor) = 3 amps

FIGURE 7.66 Grounding-electrode conductor must be bonded to both ends of any protective steel raceway sleeve for this reason.

remain around the conduit entry hole, a bonding bushing or wedge with a jumper wire must be used to ensure a path of continuity from the conduit to the enclosure.

Figure 7.68 summarizes the various acceptable bonding techniques. It should be noted that the use of the common locknut-and-bushing type of connection is not allowed. Neither is the use of double locknuts—one inside, one outside—and a bushing, although this technique is permitted on the load side of the service equipment. The special methods set forth for "bonded" connections are designed to prevent poor connections and the loosening of connections due to vibration. This minimizes the possibility of arcing and the damage which might result when a service conductor faults to the grounded equipment.

Similar provisions are used to ensure the continuity of the ground path when EMT is the service raceway. EMT is coupled with threadless devices—of the compression, indenter, or set-screw type, using rain-tight couplings outdoors. Standard connectors are used to connect EMT to enclosures. Bonding between the EMT fittings and the enclosures is accomplished in the same way as with rigid conduit. And the fittings used with service-cable armor must ensure the same degree of continuity for the ground path.

Common sense and experience have dictated modern field practice in making raceway and armored service-cable connections to service cabinets. The top of Fig. 7.69 shows how a bonding wedge is used on existing connections at services or for raceway connections on the load side of the service—such as are required by **NEC** Section 501-16(b) for Class I hazardous locations. A bonding bushing, with provision for connecting a bonding jumper, is the common bonding device for new service installations where some of the concentric or eccentric "doughnuts" (knockout rings) are left in the wall of the enclosure, so a bonding jumper is required. Great care must be exercised to ensure that each and every bushing, locknut, or other fitting is used in the way in which it is intended to best perform the bonding function.

Bonding Jumpers. The sizing of a bonding jumper within the service-equipment enclosure or on the line or supply side of that enclosure is based on **NEC** Table 250-94, from which the jumper size is selected as if it were a grounding-electrode con-

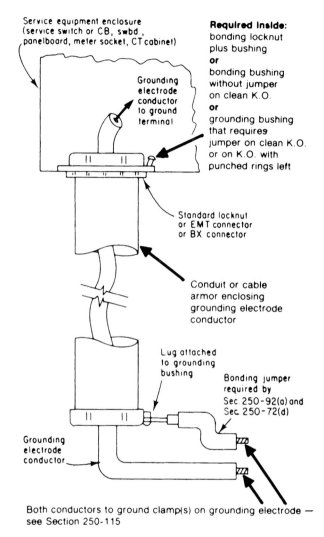

FIGURE 7.67 Every grounding-conductor conduit sleeve must be "bonded" at both ends.

ductor for the particular size of service entrance conductors involved. Figure 7.70 shows examples of the sizing of bonding jumpers. In part (*a*), the bonding bushing and jumper are used to comply with **NEC** Section 250-72. In Table 250-94, with 500-kcmil copper as the "largest service-entrance conductor," the minimum permitted size of grounding-electrode conductor is No. 1/0 copper (or No. 3/0 aluminum). That, therefore, is the minimum permitted size of the required bonding jumper.

In part (*b*), since each service phase leg is made up of two 500-kcmil copper conductors in parallel, the left-hand column heading in Table 250-94 requires computation of the "equivalent for parallel conductors." The phase leg is thus taken at 2×500

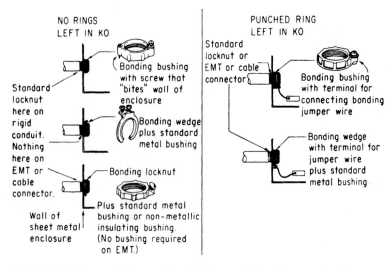

FIGURE 7.68 Recognized ways to "bond" wiring methods to sheet-metal enclosures.

kcmil, or 1000 kcmil, which is the physical equivalent of the makeup. Then Table 250-94 requires a minimum bonding-jumper size of No. 2/0 copper (or No. 4/0 aluminum).

In the hookup of service bonding jumpers shown in Fig. 7.71, the jumper between the neutral bus and the equipment ground bus is a "main bonding jumper," and the minimum required size of this jumper for this installation is determined by calculating the size of one service phase leg. With three 500 kcmil per phase, that works out to 1500 kcmil copper per phase. Because that value is in excess of 1100 kcmil copper, as noted in **NEC** Section 250-79(c), the minimum size of the main bonding jumper is 12½ percent of the phase-leg cross-sectional area. Then,

$$0.125 \times 1500 \text{ kcmil} = 187.5 \text{ kcmil}$$

Table 8 in **NEC** Chapter 9 shows that the smallest conductor with at least that cross-sectional area is No. 4/0, with an area of 211,600 kcmil. Note that a No. 3/0 has a cross-sectional area of only 167.8 kcmil. Thus, No. 4/0 copper with any type of insulation would serve as the bonding jumper.

In Fig. 7.71, the jumper shown running from one conduit bushing to the other and then to the equipment ground bus is an "equipment bonding jumper." It is sized in the same way as a main bonding jumper (as above). In this case, therefore, the equipment bonding jumper would have to be at least No. 4/0 copper.

If each of the three parallel 4-in conduits in Fig. 7.71 had a separate bonding jumper connecting it individually to the equipment ground bus, the size of the separate bonding jumpers could not be less than the size of the grounding-electrode conductor for a service the size of the phase conductor used in each conduit. According to Table 250-94, a 500-kcmil copper service calls for at least a No. 1/0 grounding-electrode conductor. Therefore, a bonding jumper run from the bushing lug on each conduit to the ground bus must be at least No. 1/0 copper (or 3/0 aluminum) as shown in Fig. 7.72.

Bonding of Piping. Within a building or other structure, any available "interior metal water piping system" must be bonded to the service-equipment enclosure, to

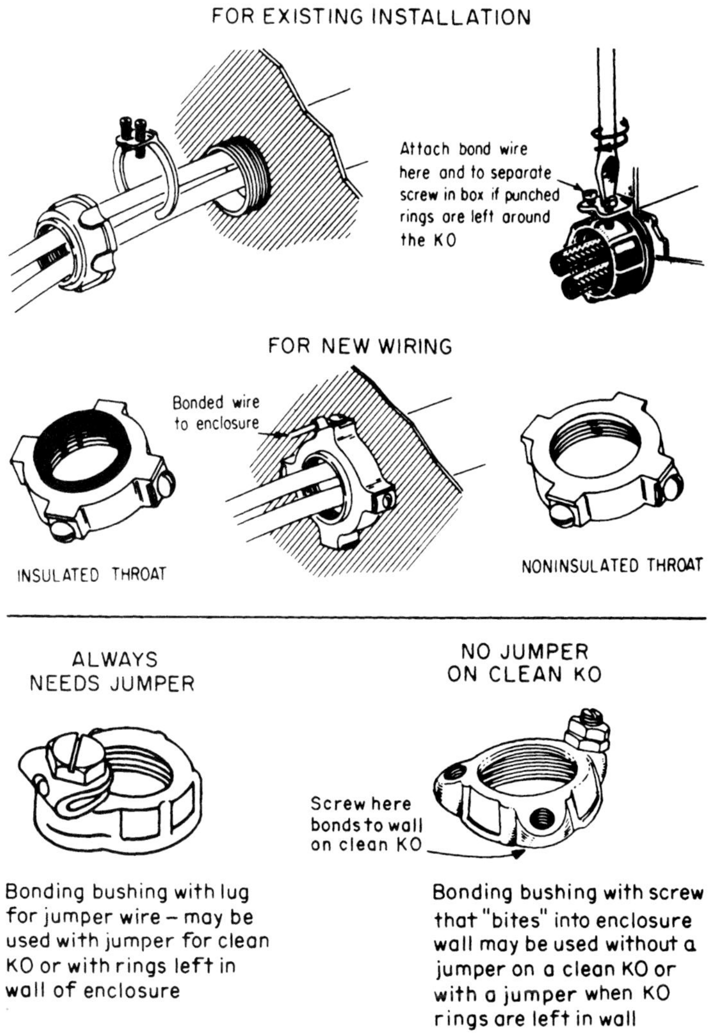

SERVICES

FOR EXISTING INSTALLATION

Attach bond wire here and to separate screw in box if punched rings are left around the KO

FOR NEW WIRING

Bonded wire to enclosure

INSULATED THROAT

NONINSULATED THROAT

ALWAYS NEEDS JUMPER

NO JUMPER ON CLEAN KO

Screw here bonds to wall on clean KO

Bonding bushing with lug for jumper wire – may be used with jumper for clean KO or with rings left in wall of enclosure

Bonding bushing with screw that "bites" into enclosure wall may be used without a jumper on a clean KO or with a jumper when KO rings are left in wall

FIGURE 7.69 Bonding bushings and similar fittings must be used in the intended manner.

the grounded conductor (usually a neutral) at the service, to the grounding-electrode conductor, or to the one or more grounding electrodes as used. This rule applies where the metal water-pipe system does or does not have 10 ft of metal pipe buried in the earth and is, therefore, not a grounding electrode. In all cases, the water-pipe system must be bonded to the service grounding arrangement. And the bonding jumper used to connect the interior water piping to, say, the grounded neutral bus or terminal (or to the ground bus or terminal) must be sized from Table 250-94, based on the size of the service conductors. The jumper is sized from that table because that is the table that would have been used if the water piping had had 10 ft

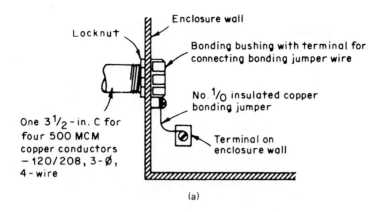

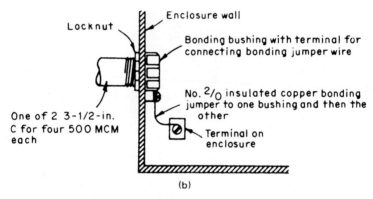

NOTE: Bushing with jumper is acceptable bonding for a clean KO or one with punched rings still in place.

FIGURE 7.70 Bonding jumpers at services must be sized carefully.

buried under the ground, making it suitable as a grounding electrode. Note that the "bonding jumper" is sized from **NEC** Table 250-94 [and not from Section 250-79(c)], which means it never has to be larger than No. 3/0 copper or 250-kcmil aluminum.

A bonding connection must also be made from "other" (than water) metal piping systems that "may become energized"—such as those for process liquids or fluids—to the grounded neutral, the service ground terminal, the grounding-electrode conductor, or the grounding electrodes. But, for these other piping systems, the bonding jumper is sized from **NEC** Table 250-95, using the rating of the overcurrent device of the circuit that may energize the piping.

Understanding and application of the foregoing rule of **NEC** Section 250-80 hinge on the reference to metal piping "which may become energized." What does that phrase mean? Is it not true that any metal piping "may" become energized? Or does the rule apply only to metal piping that conductively connects to metal enclosures of electrical equipment (such as pump motors, solenoid valves, pressure switches, etc.) in

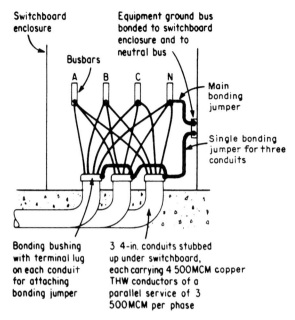

FIGURE 7.71 Service bonding jumpers must be sized and installed in an acceptable manner.

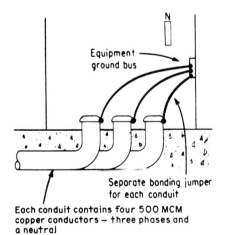

FIGURE 7.72 Separate bonding jumper may be used for each of parallel conduits.

which failure of electrical insulation would put a potential on the metal piping system? It does seem that the latter is what the rule means. Where a particular circuit poses the threat of energizing a piping system, the equipment grounding conductor for that circuit (which could be the conduit or other raceway enclosing the circuit) may be used as the means of bonding the piping back to the service ground point. That has the effect of saying, for instance, that the equipment grounding conductor for a circuit to a solenoid valve in a pipe may also ground the piping, as shown in Fig. 7.73.

EMERGENCY SUPPLY

Where the interruption of electric power supply to a building would result in panic, hazard to life or property, or major loss of production, provision should be made for an emergency supply of power, to

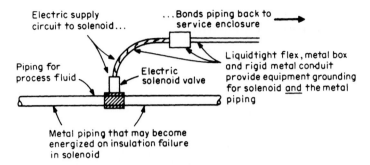

FIGURE 7.73 Metal process piping may be grounded by the grounding conductor of the circuit that would energize the piping under fault conditions.

be used in the event of failure of the normal supply. Such provision may be made by designing an emergency system supplied by

1. Storage batteries,
2. An emergency generator set,
3. Separate emergency service, or
4. A subservice tapped ahead of the disconnect means for the normal service conductors

Detailed requirements and recommendations for emergency electrical systems are given in **NEC** Article 700. But, all the regulations of that article apply to the designated "circuits, systems, and equipment" only when the systems or circuits are required by law and classified as emergency provisions by federal, state, municipal, or other code or by a governmental authority. The **NEC** itself does not require emergency light, power, or exit signs. The **NEC** does exclude from all the rules any emergency circuits, systems, or equipment that are installed on a premises but are not legally mandated for the premises. Of course, any emergency systems that are provided at the option of the designer (or the client) must necessarily conform to all the other **NEC** regulations that apply to the work.

The placement or location of exit lights is not regulated by the **NEC** but is covered in the Life Safety Code, NFPA No. 101 (formerly Building Exits Code). But, where exit lights are required by law, the **NEC** considers them to be part of the emergency system. The **NEC** indicates how the installation will be made, but not where the emergency lighting is required, except as specified in Part **C** of Article 517 for essential electrical systems in health-care facilities.

Prior to OSHA, there was no universal requirement that all buildings or all places of employment have exit signs. The **NEC** does not require them. And, although the NFPA Life Safety Code does include rules on emergency lighting and exit signs, that code was enforced where state or local government bodies required it—that is, in some areas and for some specific types of occupancies. As a result, there are existing occupancies which do not have the exit signs now required by federal law.

OSHA makes clear that every building must have a means of egress—a continuous, unobstructed way for occupants to get out of a building in case of fire or other emergency, consisting of horizontal and vertical ways as required. Egress from all parts of a building or structure must be provided at all times during which the building is occupied. The law substantially states:

Every exit shall be clearly visible or the route to reach it shall be conspicuously indicated in such a manner that every occupant of every building or structure who is physically and mentally capable will readily know the direction of escape from any point, and each path of escape, in its entirety, shall be so arranged or marked that the way to a place of safety outside is unmistakable.

Then it says:

Exit marking. (1) Exits shall be marked by a readily visible sign. Access to exits shall be marked by readily visible signs in all cases where the exit or way to reach it is not immediately visible to the occupants.

(2) Any door, passage, or stairway which is neither an exit nor a way of exit access, and which is so located or arranged as to be likely to be mistaken for an exit, shall be identified by a sign reading "Not An Exit" or similar designation, or shall be identified by a sign indicating its actual character, such as "To Basement," "Storeroom," "Linen Closet," or the like.

(3) Every required sign designating an exit or way of exit access shall be so located and of such size, color, and design as to be readily visible.

Note that marking must be supplied for the way to the exit as well as for the exit itself. Further,

(5) A sign reading "Exit," or similar designation, with an arrow indicating the direction, shall be placed in every location where the direction of travel to reach the nearest exit is not immediately apparent.

(6) Every exit sign shall be suitably illuminated by a reliable light source giving a value of not less than 5 footcandles on the illuminated surface.

A lot of discussion has been generated by that last rule. Note that an exit sign does not have to be internally illuminated, although it may be. And this lighting is required on "every exit sign," which means signs over exit doors and exit signs indicating the direction of travel.

The phrase "reliable light source" raises questions as to its meaning. Just what is reliable? Does this mean that the light source must operate if the utility supply to a building fails? And is there a difference between required applications in new buildings and those in existing buildings?

Because the OSHA regulations on exit signs do not require emergency power for lighting such signs, the light units that illuminate exit signs may be supplied from normal (nonemergency) circuits. OSHA does not make it mandatory to supply such circuits from a tap ahead of the service main or from batteries or an emergency generator. And this applies to new buildings as well as existing buildings. As far as OSHA is concerned, **NEC** Article 700, on emergency systems, does not apply to circuits for exit-sign lighting. Of course, OSHA does not object to the extra reliability such arrangements give. But the **NEC** does classify exit lights as emergency equipment.

Emergency Equipment

Where emergency lighting and/or power is mandated by law, the **NEC** requires the use of emergency equipment that is third-party certified—that is, listed for such application by UL or another qualified testing laboratory. Under the heading

"Emergency Lighting and Power Equipment," the UL *Electrical Construction Materials Directory* states:

> This listing covers battery-powered emergency lighting and power equipment, for use in ordinary indoor locations in accordance with Article 700 of the **National Electrical Code**. The lighting circuit ratings do not exceed 250 volts for tungsten lamps or 277 volts ac for electric discharge lamps. Other ratings may be included (motor loads, inductive loads, resistance loads, etc.) to 600 volts. This listing covers unit equipment, automatic battery charging and control equipment, inverters, central station battery systems, distribution panels, exit lights, and remote lamp assemblies, but not lighting fixtures. The investigation of emergency equipment includes the determination of their suitability of transferring operation from normal supply circuit to an immediately available emergency supply circuit.

Wherever emergency equipment is used, good design practice, as well as the **NEC**, requires that it be carefully selected to have adequate capacity for its intended performance. It is important that power be available for the necessary supply to exit lights and emergency and egress lighting, as well as to operate such equipment as elevators and other equipment connected to the emergency system. In hospitals, there may be a need for an emergency power supply for lighting in operating rooms and for such equipment as inhalators, iron lungs, and incubators. An emergency lighting system in a theater or other place of public assembly must have adequate capacity for exit signs, the chief purpose of which is to indicate the location of the exits, and lighting equipment commonly called "emergency lights," the purpose of which is to provide sufficient illumination in the auditorium, corridors, lobbies, passageways, stairways, and fire escapes to enable persons to leave the building safely.

Any switch or other control device that transfers emergency loads from the normal power source of a system to the emergency power supply must operate automatically on loss of the normal supply, and the transfer must be made in not more than 10 s. Transfer equipment must also be automatic for legally required standby systems, as covered in **NEC** Section 701-7.

To ensure proper control of emergency light and power, audible and visual signal devices must be included in the system to monitor and indicate the operating condition of the system and all its components. To be effective, the signal devices should be located in some room where an attendant is on duty. Lamps may readily be used as signals to indicate the position of an automatic switching device. An audible signal in any place of public assembly should not be so located or of such a character that it will cause a general alarm.

The standard signal equipment furnished by one battery manufacturer with its 60-cell battery for emergency lighting includes an indicating lamp which is lighted when the charger is operating at the high rate; in addition, a voltmeter marked in three colored sections indicates (1) that the battery is not being charged or is discharging into the emergency system or (2) that the battery is being trickle-charged or (3) that the battery is being charged at the high rate. This last indication duplicates the indication given by the lamp.

Engine-driven generators (diesel, gasoline, or gas) are commonly used to provide an alternative source of emergency or standby power when normal utility power fails. Gas-turbine generators also are used. The first step in selecting an on-site generator is to consider applicable requirements of the **NEC,** which differ depending on whether the generating set is to function as an emergency system, a standby power system, or a power source in a health-care facility such as a hospital. For example, an internal-combustion-type engine-generator set selected for use under Article 700 must be provided with automatic starting and automatic load transfer, with enough on-site fuel to power the full demand load for at least 2 hours. The engine driving an emergency generator must not be dependent on a public water supply for its cooling. That means a roof tank or other on-site water supply, with its pumps connected to the emergency source, must be used (Fig. 7.74).

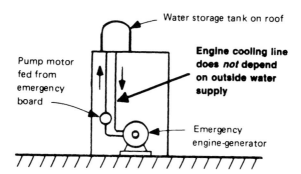

FIGURE 7.74 Engine cooling must be ensured for a water-cooled emergency generator set to provide continuous operation.

A utility gas supply may be used as the sole supply of fuel to an emergency generator—at the discretion of the local inspector—where a simultaneous outage of both electric power and gas supply is highly unlikely.

Where dual utility supply is used to minimize the chance of a total electrical outage in a building, two separate services brought to different locations in the building are always preferable; these services should at least receive their supply from separate transformers where this is practicable. In some localities, municipal ordinances require either two services from independent sources of supply or auxiliary supply for emergency lighting from a storage battery or a generator driven by a steam turbine, internal-combustion engine or other prime mover. Figure 7.75 shows the separate-service type of emergency supply.

As shown in Fig. 7.76, an emergency energy supply derived from a tap ahead of the normal main service disconnect provides protection only against failure of the main service equipment and is of no help on utility outage. In that diagram, the tap ahead of the main could feed directly to the emergency panel, without the automatic transfer switch.

Typical wall-hanging battery-pack emergency lighting units are shown in Fig. 7.77. The 1971 **NEC** accepted only fixed wiring as the means of connection of emergency light units. The present **NEC** recognizes permanent wiring connection or cord-and-plug connection to a receptacle. But even though the unit equipment is allowed to be hooked up with flexible cord-and-plug connections, it is necessary that the unit equipment be permanently fixed in place.

Individual battery-pack unit equipment provides emergency illumination only for the area in which it is installed; therefore, it is not necessary to carry a circuit back to the service equipment to feed the unit. The branch circuit feeding the normal lighting in the area to be served is the same circuit that should supply the unit equipment. A battery-pack emergency unit must provide lighting for at least 1½ hours and must not be connected on the load side of a local wall switch that controls the supply to the unit or the receptacle into which the emergency unit is plugged. Such an arrangement exposes the emergency unit to accidental energizing of the lamps and draining of the battery supply.

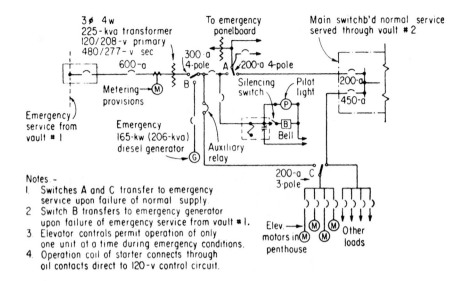

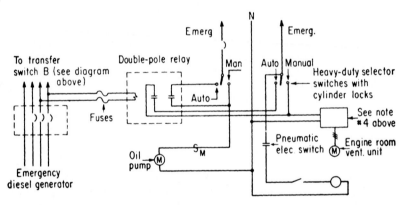

FIGURE 7.75 Two separate services constitute a form of power supply that reduces the chance of an outage.

For the transfer of emergency lighting from the normal source to the emergency source, if a single emergency system is installed, a transfer switch must be provided which will automatically transfer the emergency system to the other source in case of failure of the source of supply on which the system is operating (Fig. 7.78). Where the two sources of supply are two services, the single emergency system may normally operate on either source (Fig. 7.79). Where the two sources of supply are one service and a storage battery, or one service and a generator set, the single emergency system would normally be operated on the service, with the battery or generator used only as a reserve in case of failure of the service.

The use of ground-fault protection on an emergency-service disconnect means that it is rated at 1000 A or more on a 480/277-V grounded-wye supply is made optional by the **NEC**. To afford the highest reliability and continuity for an emergency power supply, GFP is not required on any 1000-A or larger disconnect for an emergency generator, although it may be used if desired.

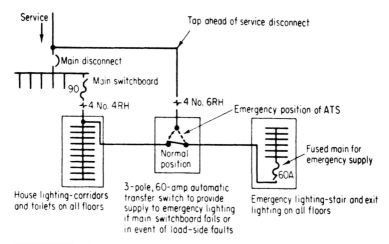

FIGURE 7.76 Tap ahead of the service main disconnect protects against only internal failures.

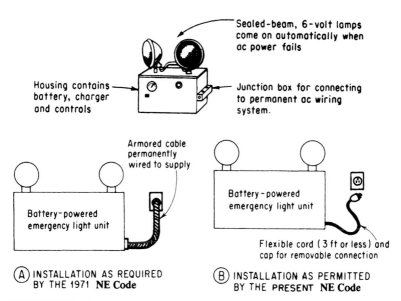

FIGURE 7.77 Battery-pack emergency light units provide simple, direct application to local areas.

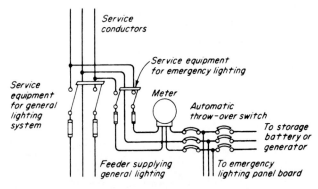

FIGURE 7.78 Automatic transfer must be provided to switch emergency loads from the normal supply to a generator.

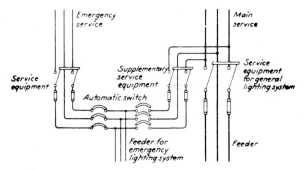

FIGURE 7.79 Emergency lighting may be supplied from a separate emergency service.

Figure 7.80 shows two other emergency power-supply arrangements that were used in industrial buildings to satisfy the particular requirements of the building's operations.

Emergency Wiring. The **NEC** requires that the wiring for emergency systems be kept entirely independent of the wiring used for normal lighting and thus that it be in separate raceways, cables, and boxes. This requirement ensures that faults which may occur on the normal wiring will not affect the emergency-system wiring.

For transfer switches, an exception to this rule is intended to permit normal supply conductors to be brought into the transfer-switch enclosure; these conductors would then be the only ones within the transfer-switch enclosure which were not part of the emergency system. Where two separate sources supply an emergency or exit lighting fixture, both supplies are, of course, permitted to enter the fixture and its junction box.

Figure 7.81 shows two applications that violate **NEC** rules on the hookup and wiring of emergency circuits. The top application is a violation because the wiring to the emergency light (or to an exit light, which is classed as an emergency light) is run in the same raceway and boxes as the wiring to the decorative floodlight. Emergency

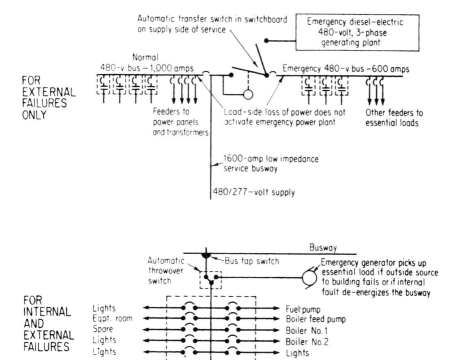

FIGURE 7.80 Emergency-supply hookups can be made to provide different types of protection.

wiring may not be run in enclosures with wiring for nonemergency circuits. The bottom application is a violation because the water cooler is not a piece of emergency equipment, and the **NEC** prohibits emergency circuits from supplying appliances or loads other than those prescribed for truly emergency power and/or lighting.

Standby Systems

NEC Article 701 covers standby power systems that are required by law. Legally required standby power systems are systems that are required and so classified by municipal, state, federal, or other codes or by any government agency having jurisdiction. These systems are intended to supply power automatically to selected loads (other than those classed as emergency systems) in the event of failure of the normal power source.

Legally required standby power systems are typically installed to serve such loads as heating and refrigeration systems, communication systems, ventilation and smoke-removal systems, sewage-disposal systems, lighting, and industrial processes that, when stopped during a power outage, could create hazards or hamper rescue or firefighting operations.

NEC Article 701 covers the circuits and equipment for such systems that are permanently installed in their entirety, including power source.

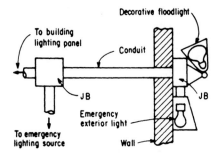

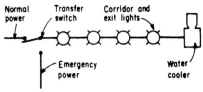

FIGURE 7.81 Emergency wiring must not be located in enclosures for nonemergency circuits, and emergency circuits must not supply nonemergency equipment.

Article 702 covers "optional" standby systems. Life safety is not the purpose of optional standby systems. Instead, optional standby systems are intended to protect private business or property where life safety does not depend on the performance of the system. Optional standby systems (other than those classed as emergency or legally required standby systems) are intended to supply on-site generated power to selected loads, either automatically or manually.

Optional standby systems are typically installed to provide an alternative source of electric power for such facilities as industrial and commercial buildings, farms, and residences, to serve such loads as heating and refrigeration systems, data-processing and communications systems, and industrial processes that, when stopped during a power outage, could cause discomfort, serious interruption of the process, or damage to the product or process.

Because of the constant expansion in electrical applications in all kinds of buildings, the use of standby power sources is growing at an accelerating rate. Continuity of service has become increasingly important with the widespread development of computers and intricate automatic production processes. More thought is being given to—and more money is being spent on—the provision of on-site power sources to back up or supplement purchased utility power to ensure the needed continuity as well as provide for public safety in the event of utility failure.

The most fundamental application of standby power is the portable alternator, used for standby power where the electric-utility supply is not sufficiently reliable or is subject to frequent outages (Fig. 7.82). An optional "standby" power load may be manually or automatically transferred from the normal supply to the standby generator. Any generator used as an "emergency" source must always have provision for

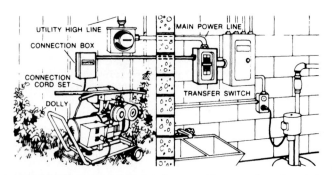

FIGURE 7.82 Optional standby generator with a manual transfer switch is commonly used.

automatic transfer of the load from normal supply to the emergency generator. And automatic transfer must always be used if a standby power system is required by law.

Uninterruptible Power Supply. Any standby power source and system must be capable of fully serving its demand load. This requirement can generally be satisfied relatively easily by generators, but the task can be complex for uninterruptible power supply (UPS) systems. The UPS is either a motor-generator assembly or an all-solid-state power-conversion system designed to protect computers and other critical loads from blackouts, brownouts, and transients. It is usually connected in the feeder supplying the load, with bypass provisions to permit the load to be fed directly. Figure 7.83 shows a typical basic layout for a solid-state UPS system. Such a

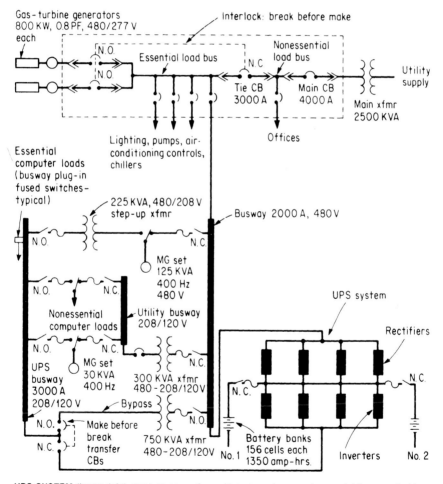

UPS SYSTEM (lower right) receives power from either normal source (upper right) or gas-turbine generators (upper left). From UPS, regulated power is delivered to 3000-amp bus feeding essential computer loads. Key components of UPS system are rectifiers, inverters, batteries and transfer CBs.

FIGURE 7.83 Uninterruptible power supply (so-called "UPS system") is a common form of optional standby supply for data processing.

system utilizes a variety of power sources to ensure continuous power. When the utility supply fails, the normal ac input to the UPS rectifiers ceases. The battery is the power source while the back-up engine-generator is being started. When the generator is running properly, it is brought onto the line through its transfer switch, and the 2000-A busway is then fed from the generator and again feeds ac power to the UPS rectifiers.

The required capacity of a UPS system must be carefully calculated. Data-processing installations normally require medium-to-large 3-phase UPS systems ranging from 37.5 to over 2000 kVA. Some typical ratings available from UPS manufacturers are 37.5 kVA/30 kW, 75 kVA/67.5 kW, 125 kVA/112.5 kW, 200 kVA/180 kW, 300 kVA/270 kW, 400 kVA/360 kW, and 500 kVA/450 kW. Larger systems are configured by paralleling two or more of these standard sizes.

The necessary rating is chosen based on the size of the critical load. If a power profile itemizing the power requirements is not available from the computer manufacturer, the load may be measured using a kilowattmeter and a power-factor meter. Since UPS modules are both apparent-power (kilovoltampere) and real-power (kilowatt) limited, a system should be specified with both kilovoltampere and kilowatt ratings. The required kilovoltampere rating is obtained by dividing the actual-load kilowatts by the actual-load power factor. For example, an actual 170-kW load with an actual 0.85 power factor would require a 200-kVA UPS. The system should be specified as 200 kVA/170 kW, and the standard 200 kVA/180 kW UPS module would be selected for the application.

CHAPTER 8
EQUIPMENT SELECTION AND LAYOUT

Once the voltage, current ratings and other electrical characteristics of feeder and service entrance circuits have been established, the design is completed by determining the actual physical makeup of the circuits and overall system. This means selecting the physical characteristics of all switching, control, protection, and transformation equipment—along with their housings and interconnecting components. And for current flow between system equipment units, either a raceway system or a cable system must be selected.

Because circuit control and overcurrent protection are required for every electrical installation, small or large and simple or complex, such equipment is generally centralized in concentrated load areas in the form of switchboards and/or panelboards. For smaller systems, one or more panelboards will usually be sufficient. For larger installations, such as industrial plants, commercial buildings, and office and apartment buildings, main switchboards receive the total incoming power and then distribute smaller blocks of power to a number of panelboards conveniently located near connected electrical loads.

Modern distribution centers for lighting and appliance feeders include dead-front panelboards and dead-front switchboards. Feeders are tapped through fused switches or circuit breakers from the buses in such distribution centers.

Distribution centers frequently are supplied by the service conductors to buildings and contain the main service switches. This is the most common arrangement in commercial and institutional buildings, although subfeeder distribution centers are sometimes supplied by feeders from the main distribution center (the service panel or switchboard). In many industrial buildings, feeder distribution centers may be supplied from transformers which step high distribution voltages to utilization levels.

Switchboards

Switchboards are assemblies of switches, circuit breakers, fuses, metering equipment, relays and/or other forms of equipment which control or record the electric current distributed throughout a building or area. Widths may range from 3 ft to more than 100 ft. Currently, most large electrical installations, particularly where loads exceed 1200 A, use factory-assembled switchboards for electrical control and distribution purposes. The major reasons for selecting this type of equipment over an installation of separate fused-switch, breaker, and gutter enclosures are overall reduced labor costs and a better quality design. Where the width (sometimes referred to as the length) of a switchboard is more than 40 in, the unit usually is constructed in sections to simplify shipping and installation.

Switchgear are selected according to voltage rating, ampere rating, number of phases, and short-circuit rating. Circuit-breaker switchgear may be of the stationary type, in which the breakers are bolted to the bus and frame, or of the draw-out type, in which the breakers are mounted on a slide-out mechanism for easy removal and maintenance and for disconnecting the breaker from the bus. Switchboards may have provisions for instrument transformers and metering. Feeders from the switchboard may be carried directly to lighting or power branch-circuit panelboards, to subfeeder distribution switchboards or panelboards, to motor control centers, or to individual motor loads.

Good switchboard design provides extra capacity and spare overcurrent devices (or space for them) so that future loads can be added or shifted with minimum cost and inconvenience.

Most switchboards contain no provision for utility meters. Generally an incoming service board contains space for current transformers; from these, metering circuits are run to a separate wall location provided for metering equipment. It is prudent to check the serving utility's metering requirements before switchboard designs are finalized, to determine how much board space is needed for current transformers and/or meters or how much adjacent wall space will be required.

Enclosed, metal-clad switchboards are almost universally used for low- and medium-voltage systems, with all electrical components receiving maximum protection within grounded metal enclosures. Only the operating handles, levers, or pushbuttons are accessible from the outside.

Boards of this type commonly are freestanding units rated from 800 to 6000 A at 600 V or less. For most installations, they contain the service disconnecting means and overcurrent protective devices, together with any current transformers for metering or system ground-fault protection.

While the section in the **NEC** outlining minimum standards for switchboards is relatively short, there are a great many options involving maintenance and operating convenience, extra safety features, overload capability, etc. The NEMA standards publication *Switchboards* is useful in understanding construction and specification details along with test and performance specs.

Operating convenience, ease of maintenance, and proximity to areas with concentrated loads are major considerations in locating switchboards. Proximity to loads may become a problem when the switchboard includes, or constitutes, the service equipment. The location of the utility supply determines where the main switchboard must be placed. However, service raceways can be run within a building in a 2-in concrete or brick envelope and still be considered as being "outside" the building up to the point where the concrete-encased raceway emerges from the concrete at the switchboard or panelboard.

If a switchboard is to be installed in other than dry areas, or outdoors, it must be of weatherproof construction. In any hazardous location, the equipment must be of a type approved for use in the relevant Class and Division of hazardous area as indicated in the **NEC**. It is usually safer and much less costly to locate switchboards outside such areas.

The place of installation of a switchboard must provide the minimum safe working clearances shown in Fig. 8.1. In addition, there must be work space at least 30 in wide in front of electrical equipment and sufficient "elbow room" in front of such equipment as column-type panelboards and single enclosed switches (e.g., 12 in wide) to permit the equipment to be operated or maintained safely.

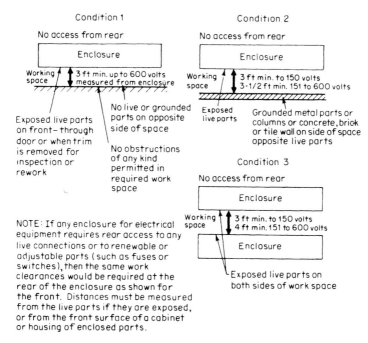

FIGURE 8.1 These minimum clearances must be provided for work spaces around switchboards.

The clearances shown in Fig. 8.1 must be measured from the live parts if exposed or from the enclosure containing live parts. The width measurement is to be made parallel with the face of the equipment.

To prevent situations in which personnel might be trapped in the working space around burning or arcing electrical equipment, two "entrances" or directions of access must be provided to the working space around switchboards and control panels rated 1200 A or more and over 6 ft wide. At each end of the working space of such equipment, an entranceway or access route at least 24 in wide must be provided. Because personnel have been and can be trapped in work spaces when fire erupts between them and the only exit route from the space, design engineers should routinely provide two exit paths in their drawings and specs. Although two doors are not required into an electrical equipment room, it may be necessary to provide two doors to obtain the required two entrances to the work space—especially where the switchboard or control panel is located in tight quarters that do not afford a 24-in-wide exit path at each end of the work space.

At the top, Fig. 8.2 shows one way to provide two paths for entering or leaving the defined work space. In that sketch, placing the switchboard with its front to the larger area of the room and/or other layouts would be acceptable. It is only necessary to have a sure means of exit from the defined work space. If the space in front of the equipment is deeper than the required depth of work space, then a person could simply move back out of the work space at any point along the length of the equipment.

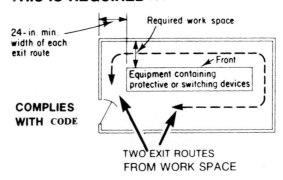

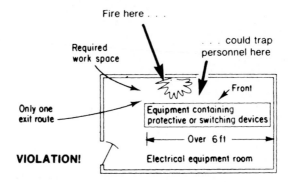

FIGURE 8.2 Positioning of a switchboard or motor control center must assure safe, ready exit from the work space.

Figure 8.2 shows a layout that must be avoided. With sufficient space available in the room, the layout of equipment over 6 ft long and rated 1200 A or more with only one exit route from the required work space would most likely be a violation of the NEC rule. A door at the right end of the work space would eliminate the violation.

Figure 8.3 shows minimum design requirements for headroom and lighting in work spaces around switchboards and other electrical equipment. Although lighting is required for the safety of personnel in work spaces, the kind of lighting (incandescent, fluorescent, mercury vapor), the minimum footcandle level, and such details as the position and mounting of lighting equipment are left to the designer and/or installer, with the inspector the final judge of acceptability.

A minimum headroom of 6¼ ft is required in working spaces around electrical equipment rated up to 600 V. That requirement applies to "service equipment, switchboards, panelboards, or motor control centers."

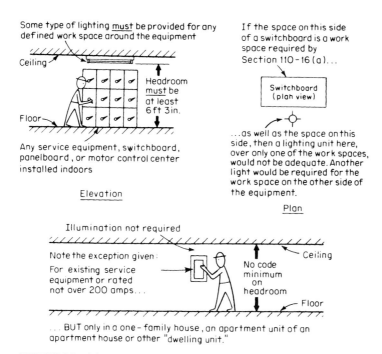

FIGURE 8.3 Adequate headroom and illumination must be provided at work spaces around electrical equipment.

The design drawings and specifications for an electrical system must include the requirement that all disconnect devices (that is, all switches and circuit breakers) for load devices and for circuits be clearly and permanently marked to show the purposes of the disconnects. This is a "must" that, under OSHA, applies to all existing electrical systems, no matter how old, and to all new, modernized, expanded, or altered electrical systems. This marking requirement has been widely neglected in the past. Panelboard circuit directories must be fully and clearly filled out. And all such marking on equipment must be in painted lettering or other substantial identification (Fig. 8.4).

Effective identification of all disconnect devices is a critically important safety matter. When a switch or CB has to be opened to deenergize a circuit quickly—as when dictated by a threat of injury to personnel—it is absolutely necessary to be able to identify quickly and positively the disconnect for the circuit or equipment that constitutes the hazard. Painted labeling or embossed identification plates affixed to enclosures would comply with the requirement that disconnects be "legibly marked" and that the "marking shall be of sufficient durability." Ideally, the marking should tell exactly what piece of equipment is controlled by each disconnect (switch or CB) and where the controlled equipment is located.

Figure 8.5 shows the single-line arrangement of an effectively designed switchboard for a large modern office building. The design of the complete distribution system provided for separation of the lighting and power loads, and skillful loading of the lighting feeders produced the uniform condition in which only 400-A CBs are used in the lighting section. The single size of CB assures ready and effective main-

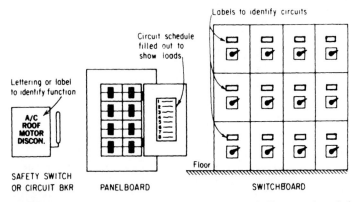

FIGURE 8.4 Clear identification of the purpose of each disconnect is a vital safety matter.

tenance of the assembly. In all phases of electrical design, the kind of simplicity shown in this switchboard layout is a mark of excellence—and especially in large, high-capacity systems. Note the emergency lighting supply tapped from the line side of one of the power CBs. Both voltage levels are available for connection to essential power and lighting loads, which can be kept operating when the lighting switchboard fails. Of course, this actually provides power in only limited emergencies; it would not cover the loss of utility power.

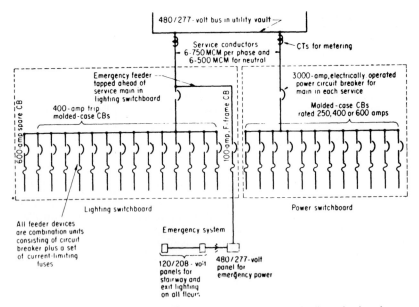

FIGURE 8.5 Switchboard design and layout should strive for simple, direct circuit makeup.

Panelboards

Panelboards comprise the major item of distribution equipment between main or subdistribution points and individual connected loads. Almost every electrical installation involves one or more panelboards for lighting and receptacle circuits. A single panel permits the distribution of electric power from a feeder to as many as 42 lighting and appliance branch circuits, each with overcurrent protection and disconnect.

The **NEC** makes it mandatory that panelboard assemblies be installed in metal cabinets. A metal trim covers the gutter space and circuit connections, leaving only the switch or circuit-breaker handles accessible to personnel. All electrical panelboard parts can be replaced or serviced from the front, after trim removal—an important feature when panels are either surface-mounted or installed flush in walls or partitions. Special colors and metallic finishes are available for the exposed trim on flush-mounted panelboards that are located in conspicuous positions in esthetically designed areas.

A wide variety of panelboards is available to meet specific load and control requirements and installation conditions. Though somewhat similar electrically, the panels fall into five categories:

- *Service-equipment panelboards* are available for loads up to 1600 A. They contain six or fewer fused switches, fused pullouts, or circuit breakers connected to the incoming mains. They may also have split buses, one set supplying branch-circuit overcurrent devices installed in the same enclosure.
- *Feeder distribution panels* generally contain circuit overcurrent devices rated at more than 50 A to protect subfeeders that run from this panel to smaller branch-circuit panelboards throughout the area it serves.

  An interesting application of a CB distribution panelboard for individually metered subfeeds to unit occupancies in an apartment house is shown in Fig. 8.6. Five meters were mounted along each of the three horizontal gutters connecting the panel to the vertical trough. The three neutrals were installed as shown to provide the potential taps. Hot legs for the subfeeds were then carried through the gutters to supply the current coils in the meters. All subfeed runs were made from the pullbox at the top.
- *Lighting and appliance branch-circuit panelboards* have more than 10 percent of their overcurrent devices rated at 30 A or less, and neutral connections are provided for these. Generally, these panels have bus ratings up to 225 A at 600 V or less. However, larger bus ampacities must sometimes be selected to meet load requirements.
- *Power distribution panels,* similar to the feeder distribution type, have buses normally rated up to 1200 A at 600 V or less. They contain control and overcurrent devices sized to match connected motor or other power-circuit loads. Generally, the devices are 3-phase.
- *Special panelboards* containing relays and contactors can be obtained and installed where remote control of specific equipment is specified. Some panelboards contain branch-circuit snap switches rated 30 A or less. These boards must have overcurrent protection rated no higher than 200 A on the line side of the switches.

Any panel installed in a wet or damp location, in a dust-laden area, outdoors, or in any hazardous area must be of the type approved for use in such a location.

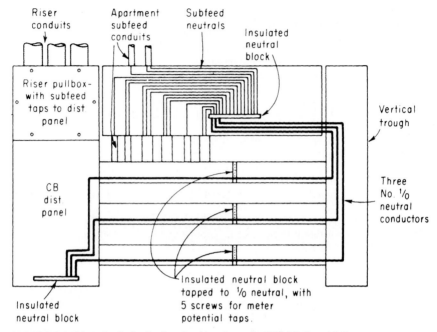

FIGURE 8.6 Meter-bank distribution circuiting demands skillful design techniques.

NEMA standards list three panelboard enclosures: Type 1, indoor—general purpose; Type 3R, outdoor—rainproof and sleet-resistant; and Type 12, for indoor industrial use (dust-, drip-, and oil-tight).

If only qualified persons are to have access to the installed panelboards, then panel trims with locking-type doors should be specified.

A few precautions, if observed by designers in developing plans and specifications, can simplify panelboard installation, improve workmanship, and produce a neater job:

- Coordinate panelboard cabinet installation with the raceway system, and maintain proper alignment.
- Align floor-slab conduit stubups with openings in the bottoms of cabinets, particularly with surface-mounted panels. Maintain alignment.
- Ensure that surface-mounted panelboard enclosures are securely fastened to walls or other structural surfaces. The addition of supplementary pedestals or "legs" (structural steel or pipe sections with floor flanges) will increase rigidity and help distribute the weight of large, heavy panelboards.
- Make sure the gutter space around an overcurrent-device panel assembly is sufficiently wide to permit conductor installation in accordance with Table 373-6. If more space is necessary, use an oversized enclosure, supplementary wiring troughs, or auxiliary gutters.
- Where a number of panelboard enclosures are to be mounted side by side, the addition of a suitably sized auxiliary gutter, wiring trough, or pullbox above and/or

below the grouped cabinets will facilitate cable pulling and circuit installation. In many cases, it will eliminate the need for oversized cabinets to accommodate the required conductors.

Switching Devices

Many types of switches are used for on-off control of loads and for no-load circuit opening. The selection of the switch best suited to a specific task involves a number of considerations:

- The electrical rating and characteristics of the switch must be proper for the circuit function to be performed. The switch must have the right number of poles and the proper voltage and current ratings to carry the load impressed on it. UL-listed enclosed switches are rated up to 4000 A, 500 hp, and 600 V, and nonfusible switches are tested to provide continuous operation (periods over 3 hours) at marked rated loads.
- Enclosed switches rated 800 or 1200 A at more than 250 V are available in two classes. One is for general use and may be used as a disconnect up to its rating; the other is for isolating use only and must be so marked. Any enclosed switch rated over 1200 A must be marked "For isolating use only—do not open under load." (Bolted-pressure switches and CB-switches are rated for load-break above 1200 A.)
- For motor loads, horsepower rating must be considered. Enclosed motor circuit switches are listed by UL up to 500 hp and 600 V. A switch that is marked "Motor circuit switch" is intended for use only in motor circuits. Enclosed switches with horsepower ratings in addition to current ratings may be used in motor circuits as well as for general-purpose circuits.

In sizing feeder disconnect means in switchboards or panelboards, care must be exercised to ensure that the ampere rating of each switch or circuit breaker is adequate for the load, for the switching duty to which it will be subjected, and for any load growth to be designed into the system. A first step in such sizing of a switch or circuit breaker is as follows (Fig. 8.7):

Motor load only in power panel. The switch must have an ampere rating at least equal to the required current rating of the feeder conductors. This current rating is equal to 125 percent of the full-load current of the largest motor plus the sum of the full-load currents of the other motors fed from the panel. But a switch must accommodate the fuses which protect the feeder, and these may have a maximum rating equal to the largest rating or setting of branch-circuit protective device in the power panel plus the sum of the currents of the other motors fed from the panel. Or, a CB may have a higher trip setting. This may require a larger device. And if two or more motors fed from the panel must be started simultaneously, the starting-current drain may be so heavy as to require larger feeder conductors and disconnect.

Motor and lighting load in power panel. The switch must have an ampere rating at least equal to that required for the motor load (as described above) plus the current for the lighting and appliance load on the panel.

When fused switches are used, the heat load on the fuses should be considered in sizing the assembly. Because fuses are rated according to their ability to carry current continuously when placed in open air, on a bench, and in a horizontal position, the effect on fuses of heat accumulation within enclosed mountings in safety switches

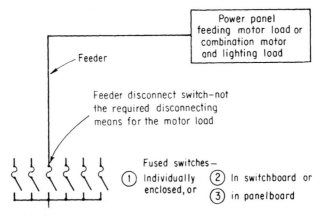

FIGURE 8.7 Fused switches for feeders and subfeeders must be carefully selected.

must be considered. To ensure the continuous operation of loads—minimizing the chance of nuisance fuse blowing—fusible switches must be rated to accommodate fuses which are at least 25 percent higher in rating than the load currents which will be flowing for several hours or longer. Another way of stating this application requirement is that the load current must not exceed 80 percent of the fuse rating, or we could say that the noncontinuous load plus 125 percent of the continuous load must *not* exceed the rating of the overcurrent device rating.

In selecting feeder disconnect and protection equipment, consideration should be given to the provision of spare capacity in this equipment. Depending upon anticipated future requirements and the manner in which extra capacity was included in the feeders, space should be provided in switchboards and/or feeder panelboards for additional switches or circuit breakers or for future installation of larger switching and protection units. And the design of the distribution system must integrate all provisions for future expansion of feeder capacities. This includes the routing of spare feeder conduits, accessibility of the feeder raceway in which capacity has been allowed for additional feeder conductors, and ease of connection in switching and protection assemblies.

Modern switchboards and feeder panelboards are constructed to accommodate all these design provisions for load growth. Care in selecting the proper types of switching and protection assemblies—with buses of substantial capacity for future expansion—is essential to economy of distribution design.

An important consideration in the selection of fusible switches is coordination between the electrical characteristics of the switch and the particular type of fuses used in the switch. With very high short circuits available in the large feeders of modern distribution systems, the fused switches used in panels, switchboards, or individual enclosures must be carefully rated for the following two conditions:

1. *Fault-interruption service.* In addition to being fully rated to make and break load currents up to its full current rating, every switch must have the ability to interrupt ground-fault or short-circuit currents which it might be called upon to break. This is shown in Fig. 8.8 and works as follows:

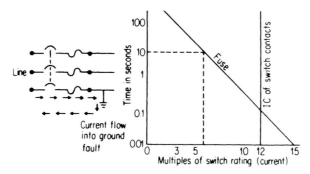

FIGURE 8.8 Switch should be rated to interrupt fault currents that fuses may not quickly clear.

The contact mechanism has a repetitive load-break capability of up to 12 times the continuous-current rating of the contact assembly. This means that the contact mechanism of the switch can safely break any overload up to that value. But the opening time-current line for the fuses shows that at 12 times the contact current rating, the fuse will open in 0.1 s—which is so short a time that the operator could not possibly open the breaker switch while the fault current is flowing. For values of fault current lower than 12 times the switch rating, it would be possible to open the switch before the fuse operated. But, in such cases, the switch contacts would be operating within their rated load-brake range. For values of fault current higher than 12 times the switch rating, the fuse will always operate to open the circuit before anyone could possibly open the switch contact mechanism. This is effective coordination between switch interrupting capacity and speed of fuse operation.

An example of a dangerous application would be the use of the fuses in a switch which had an interrupting capacity of, say, 2 times the contact continuous rating. As shown, a fault current of better than 5 times the contact rating would take 10 s to operate the fuse. If someone opened the switch during that 10 s (which is possible), the switch contacts would be breaking a current higher than that for which they are safely rated—with hazard to life and property as well as to the switch itself.

Switches are coordinated with their current-limiting fuses to assure safe operation. Closing and quickly reopening a switch (without intentional delay) before the fuse clears the fault is called "fuse racing." Coordination between a switch and fuses depends upon the construction and operating characteristics of the switch and the time-current characteristics of the particular fuses used in it. When a switch and its fuses are properly coordinated, the fuses will always operate fast enough to break any current which the switch could not safely open, before any person can operate the switch.

2. *Short-circuit withstandability.* Every switch must be capable of safely withstanding the amount of energy which passes through it from the time a short-circuit fault develops until the fuses in the switch open the fault. Every switch—fusible or nonfusible—must be able to take the thermal and magnetic forces produced by let-through fault current. This ability of a switch is a function of the construction of the switch and the operating speed of the fuses used in it. To be properly applied, a switch must remain operable after any fault that is cleared by the fuses in the switch.

Figure 8.9 shows the coordination between a typical time-delay fuse and a heavy-duty interrupter switch. The melting time versus current curve is compared with two different switch interruption ratings—with ¼ to ½ s taken as the fastest possible time for closing and reopening the switch. Note that an adequate switch interrupting capacity could pose a hazard.

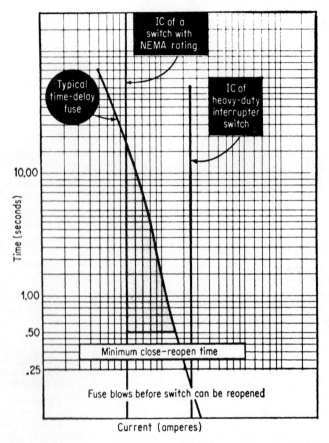

FIGURE 8.9 Heavy-duty switches offer greater safety and reliability under fault conditions.

The above discussion leads to the following factors as considerations in selecting fused switch:

he required fuse interrupting capacity is determined by the short-circuit current ilable from the system at the point of application.

ninimum switch withstandability is determined by the I^2t of fuse let-through t (from the fuse manufacturer).

imum switch interrupting capacity is based on fuse speed and switch time.

It can be seen that switches with high withstandability and high interrupting capacity can be used with slower-operating fuses which have longer melting times and higher let-through currents. But switches with lower withstandability and interrupting capacity must be protected by fuses with lower let-through and faster opening characteristics. The heavy-duty switches, for instance, could be used with Class K-5, RK-5, and K-9 fuses in high-capacity circuits; lighter-duty switches would require the use of Class RK-1, K-1, or J fuses in the same circuits.

Transformer Inrush. Another important consideration in selecting a fusible switch is the effect of inrush current on the operation of the fuse. In selecting an overcurrent protective device (fuse or CB) to protect a transformer, consideration must be given to the time-current inverse-trip characteristics of the protective device to ensure that the inrush current (which flows when the circuit to the transformer primary is closed) will not open the protective device. This is particularly critical where the fuse or CB is set at 125 percent of the transformer primary current [as required in **NEC** Section 450-3(b)(1)] but is not so much of a concern where the primary protective device is set at 250 percent of the primary current [as permitted in **NEC** Section 450-3(a)(1) or 450-3(b)(2)]. To cover the most extreme condition of inrush current—when the circuit is closed at that point in the voltage wave where maximum asymmetry occurs and causes the highest current—tests have indicated that a fuse or CB will operate properly and will not open on inrush if its time-current trip curve will permit a current value equal to 12.1 times the transformer rated primary current to flow for at least 0.1 s. Effective operation therefore depends on the shape of the time-current curve for the particular type of fuse or CB being used.

Example. A 300-kVA, dry-type transformer is used to step a 480-V, 3-phase, 3-wire supply down to a 208/120-V, 3-phase, 4-wire wye secondary. A fuse switch at the supply end of the primary feeder circuit to the transformer will be equipped with fuses rated not over 125 percent of the transformer primary rated full-load current. That level of protection represents the maximum value permitted for primary protection by **NEC** Section 450-3(b)(1). And because the primary-circuit conductors must be protected at their ampacity by the primary protective device (**NEC** Section 240-3), using a device rated at 125 percent provides the economy of using the smallest acceptable size of circuit conductors and the smallest size of switch. The protection is provided as follows:

Step 1. The transformer primary current is equal to 300,000 VA divided by 480 × 1.732, or 361 A.

Step 2. The maximum permitted rating of fuses in switch to protect the transformer is 1.25 × 361 A, or 451 A.

Step 3. In evaluating the possible nuisance tripping of the fuses due to transformer inrush magnetizing current, consideration must be given to the basic rule that the fuse will operate properly, without tripping or inrush, if its time-current characteristic will permit a current equal to 12.1 times the rated primary current to flow for at least 0.1 s. In this case, that value of inrush current works out to be 12.1 × 361 A, or 4368 A.

Step 4. The largest rating of standard fuse that could be used in this application is 450-A fuses. (Or, 500-A fuses might be used as the next larger standard size above 451 A. But use of the larger fuse would call for circuit conductors of such ampacity that the 500-A fuses protect them in accordance with their ampacity, as per **NEC** Section 240-3.)

Step 5. In Fig. 8.10, the time-current plot of the 450-A fuse in the graph at the left shows that the intersection of 0.1 s and 4368 A falls to the right of the fuse melting curve, indicating that the fuse can blow on transformer inrush current.

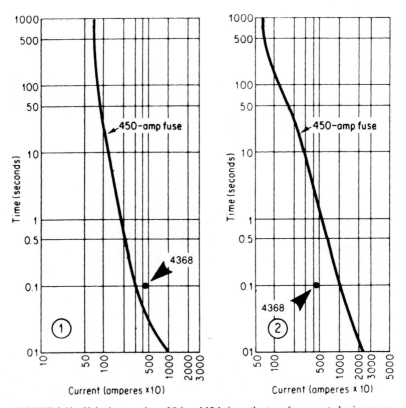

FIGURE 8.10 If the intersection of 0.1 and 12.1 times the transformer rated primary current is to the right of the protective-device curve, inrush current can open the circuit.

Step 6. But, in the graph at the right in Fig. 8.10, which is for a different type of 450-A fuse, with a greater time delay in its operation (a left-to-right slope that is not as steep as that in the curve at left), the intersection of 0.1 s and 4368 A falls to the left of the fuse melting curve—assuring that, on closing of the switch to energize the transformer, the maximum possible inrush current will not blow the fuse.

In evaluating the possibility of nuisance tripping on transformer inrush current as described above, the designer should use the minimum melting curve of a fuse (not the average melting curve or the total clearing-time curve) to ensure that the inrush current $I^2 t$ will not operate even those fuses that happen to be at the low-current side of the manufacturer's calibration tolerance band. And the same evaluation should be performed for circuit-breaker tolerance bands.

Figure 8.11 shows how a dual-element Class RK-1 fuse would be exposed to possible blowing on inrush current to a 150-kVA transformer. The 150-kVA, 480-V,

3-phase transformer primary has a rated full-load current of 150,000 ÷ (480 × 1.732), or 180 A. The maximum rating of the primary fuses that would be used to protect such a transformer is 180 A × 1.25, or 225 A. To ensure that transformer inrush current will not blow the fuses, the fuse time-current curve must be evaluated at the intersection of 0.1 s and 180 A × 12.1, or 2178 A. In Fig. 8.9, that intersection point falls far to the right of the curve for the 200-A Class RK-1 dual-element fuse—far enough to indicate that a 225-A fuse of the same type would be exposed to blowing on inrush.

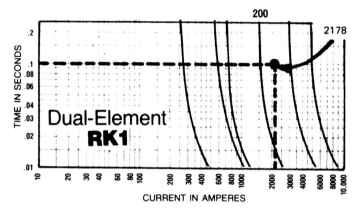

FIGURE 8.11 Inrush current of 2178 A would blow a 200-A fuse and probably a 225-A fuse.

Other Selection Considerations. The following must also be considered:

- Enclosed switches for service use must be UL-marked "Suitable for use as service equipment."
- The construction of the switch assembly must be suited to the anticipated switch usage. Switches used frequently must have sturdy makeup to assure long, trouble-free life.
- Auxiliary provisions—such as fuseholders and electrical operation of the contact mechanism—must be considered.
- Enclosure requirements, including the type of housing required for the place of installation, an interlock between the enclosure cover and operating handle, and provisions for locking the switch in the off position, must be carefully weighed.
- For switches with dual horsepower ratings, the higher rating is based on the use of time-delay fuses in the switch fuseholders to hold in on the inrush current of the higher-horsepower-rated motor.

Safety switches are air-break switches whose current-carrying parts are enclosed in metal cases and which are manually operable by means of external handles. The word "safety" derives from the enclosed form, which reduces the possibility of external contact with energized parts of the switch assembly. Safety switches are referred to as "enclosed switches" in NEMA standards, UL regulations, and the **NEC**.

In practice, general-use and motor-circuit switches are referred to as "safety" switches. However, specific industry standards classify general-use, motor-circuit, and isolating switches as "enclosed switches." NEMA standards list two categories of enclosed switches: general duty and heavy duty.

General-duty devices include door-operated fused switches, fusible pullouts, and enclosed switches designated "general duty." None of these is meant for frequent opening and closing. They do not have quick-make, quick-break contacts and should not be used where more than 10,000 rms symmetrical amperes of short-circuit current is available.

Heavy-duty switches include fused bolted-pressure contact switches and NEMA fused heavy-duty switches. They are designed for use where duty is severe or frequent or where the fault current may exceed 10,000 rms amperes. NEMA fused heavy-duty switches carry ratings of 30 to 1200 A; bolted-pressure switches conforming to UL standards are rated 800 through 4000 A. Both classes of switches are also rated in horsepower.

High-Capacity Switches. A wide range of types and sizes of switches is available for use in modern high-capacity feeder and service entrance applications. In the range from 200 to 1200 A, up to 600 V, service entrance and feeder applications can make use of heavy-duty quick-make, quick-break fusible safety switches. In the range from 600 to 6000 A, either a "fused power circuit device" or a power circuit-breaker-type load interrupter is used.

Manual and electrically operated switches designed to be used with Class L current-limiting fuses rated 601 to 4000 A and 600 V ac are listed by UL as "fused power circuit devices." This category covers bolted-pressure contact switches and high-pressure, butt-type contact switches. Such devices "have been investigated for continuous-duty use at 100 percent of their rating on circuits having available fault currents of 100,000, 150,000 or 200,000 rms symmetrical amperes" as marked. Units suitable for use as service switches are marked "Suitable for use as service equipment." Electrically tripped and/or operated switches of this type that are suitable for ground-fault protection are so marked.

High-capacity secondary-voltage service entrances commonly make use of bolted-pressure contact or butt-contact switches. Typical switches of this type are made in sizes from 600 A up. They are generally manually operated with external handles. Their most common application is in general-purpose, ventilated enclosures, either individually enclosed in an incoming line section or combined within a main secondary switchboard. Such switches are generally used with high-capacity fuses of the current-limiting type.

Modern fused power-circuit devices are rated for breaking load, for withstanding let-through thermal and magnetic stresses during fuse operation on short circuit, and for closing against a short circuit. Typical units are rated for load breaks up to 12 times their nameplate current rating.

With proper fusing, the assembly is capable of fault clearance through the fuses. With the switch closed and fitted with high-interrupting-capacity fuses, faults up to 200,000 rms amperes at 600 V can be cleared promptly and safely without damage to equipment or personnel.

Another type of high-capacity switching assembly is the fusible-CB load interrupter. One form of this assembly consists of a fully coordinated combination of a circuit-breaker mechanism that provides switching and a set of high-capacity current-limiting fuses that provide overload and short-circuit protection. Another form consists of a molded-case or air CB to provide overload protection up to the interrupting duty of the CB contact mechanism, with fuses connected on the load terminals to provide short-circuit protection at current levels above the safe interrupting capacity of the breaker and up to 200,000 A.

EQUIPMENT SELECTION AND LAYOUT

IMPORTANT: Switches, motor starters, panelboards, and other equipment that has been investigated and found suitable for use with specific fuses are marked with the class of fuse intended to be used in the equipment and with an available current rating applicable to that piece of equipment. With the designated fuses installed in equipment that is so marked, the switch or panel or other equipment is suitable for use on circuits that can deliver short-circuit currents up to the current rating of the equipment or the interrupting rating of the fuse—whichever is lower.

NOTE: An interrupting rating on a fuse included in a piece of equipment does not automatically qualify that equipment for use on circuits with available current higher than the current rating of the equipment itself.

Automatic Transfer Switches. Automatic transfer switches are used to automatically transfer load from normal to emergency power upon failure of the normal power supply. An automatic transfer switch (ATS) also automatically returns the load to normal power when the normal source returns.

An ATS can be designed as a single device with a double-throw switch arrangement. Basic types are the mechanically and the magnetically held forms, with either single- or dual-coil operators and interlocking protection. They can also consist of two circuit breakers or magnetic contactors furnished with individual operators and mechanical and electrical interlocking to provide two definite positions: normal closed/emergency open, and normal open/emergency closed. Ratings of automatic transfer switches range up to about 4000 A at 600 V.

Figure 8.12 shows two types of automatic-transfer operating mechanisms. In (a) is shown the single-coil, double-throw contactor with magnetic operation for either transfer, which is mechanically held in either operating position. Relays 1V, 2V, and 3V provide protection against phase failure or voltage drop on one line. Relay SE is the circuit selector, and relay LO supervises emergency. The ATS in Fig. 8.12*b* makes use of two separate contactors, mechanically and electrically interlocked to prevent simultaneous opening or closing. Undervoltage relay UV senses loss of normal power and initiates opening of the normal contactor. The availability of emergency power then closes the emergency contactor. Each contactor has a normally closed auxiliary contact in the coil circuit of the other contactor.

Figure 8.13 shows an ATS making use of two separate circuit breakers that are mechanically and electrically interlocked with a common operator or individual motor operators. Overcurrent (short-circuit) protection is provided for both normal and emergency supply. The control panel may incorporate individual-phase supervisory relays and automatic starting and connection of the emergency generator or other source.

Circuit-breaker transfer switches in the larger ampere sizes are made with motor-driven breaker handles. Again, these are electrically and mechanically interlocked assemblies, with double-source design of the transfer control circuit and with no neutral position in the mechanism. Some use a single motor for transfer; others use two motors.

Depending upon the electrical control circuit, circuit-breaker-type transfer switches may incorporate protection against voltage drop or failure on any one phase or may simply respond to failure of all phases. Full-phase protection is provided by adjustable voltage relays connected line to neutral or line to line.

Although different control-circuit elements are indicated above for each of the three types of automatic transfer switching, any one of them may use any control circuits to provide a desired response of the transfer assembly to load, normal-source, or emergency-supply conditions. And a wide variety of control arrangements is possible for controlling the source of emergency power.

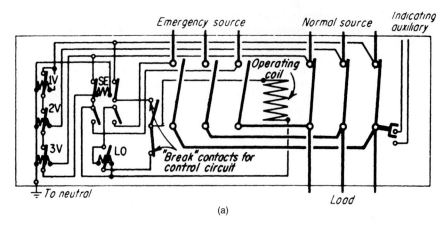

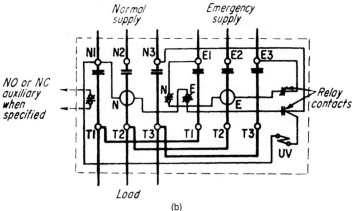

FIGURE 8.12 Automatic transfer switches are made with either one or two operating coils. (a) Circuits shown with normal source energized. (b) Circuits shown without any energy.

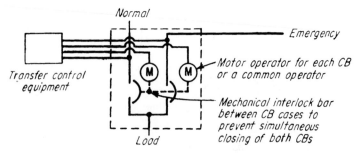

FIGURE 8.13 Two circuit breakers combine to provide an automatic transfer operation.

EQUIPMENT SELECTION AND LAYOUT

NEC Article 700, "Emergency Systems," Article 701, "Legally Required Standby Systems," Article 702, "Optional Standby Systems," and Article 517, "Health Care Facilities," incorporate requirements for the automatic transfer of normal and alternative power sources. However, a specific type is not defined, so any of the above-described ATS assemblies may meet application requirements.

In selecting an automatic transfer switch, the designer should refer to the UL *Electrical Construction Materials List* (*Green Book*) and Standard UL-1008, which covers automatic transfer switches. In the *Green Book*, the information given under "Transfer Switches" also applies and is supplementary to the data given under "Automatic Transfer Switches." This information provides requirements on an approved switch. For example, automatic transfer switches with or without integral overcurrent protection are suitable for continuous use at 100 percent of rated current. Some designs may be marked "Continuous load current not to exceed 80 percent of switch rating," and they should be so applied. Also, listed switches are generally considered to be suitable for all types of loads—motors, tungsten lamps, electric heaters, etc.

Underwriters Laboratories Standard 1008 requires that automatic transfer switches be marked with (1) load classification, (2) voltage sensing on normal supply, (3) voltage and frequency on alternative supply, (4) time-delay features, and (5) closing rating and withstandability rating. The entire standard should be studied and used as a guide in selecting ATSs. Other important selection considerations, most of which are covered in UL Standard 1008, include:

1. The continuous-current rating must be sufficient for the load, in both the normal and emergency positions.

2. The interrupting capacity of the switching units must be sufficient to handle the currents encountered during closing and opening, on either the normal or the emergency supply.

3. The short-circuit rating must be sufficient to withstand let-through fault currents. The UL minimum requirement is 20 times the switch rating but not less than 10,000 A.

4. The voltage, enclosure type, number of poles, whether alternating or direct current, and related factors, such as mounting space and location and installation requirements, must be considered.

5. Electrical and mechanical interlocking must ensure that both switching mechanisms cannot possibly be closed at the same time. Single-throw units have inherent mechanical interlocking. On dual-coil contactor types, each contactor must have normally closed auxiliary contacts in the coil circuit of the other contactor to provide electrical interlocking. Circuit-breaker types should have mechanical as well as electrical interlocking.

6. The protection may be full-phase protection (protection against single-phasing) or feeder-failure protection. Full-phase protection ensures that transfer to the emergency source will occur when the voltage on any phase drops to approximately 70 percent or less. Feeder-failure protection provides for load transfer only on complete feeder failure; it does not protect against partial voltage loss or phase failure.

7. Proper time-delay characteristics should be incorporated in the control circuit to isolate the response of the assembly from transient voltage fluctuations in the normal system. Sensitivity to such fluctuations causes false operation of the switch and may damage the starting circuit of the generator.

Other important selection considerations include: (1) adjustable time delay (on transfer to emergency, or retransfer, and on engine starting); (2) adjustable-voltage sensing relay (for full protection on 3-phase systems, three units are required; on single-phase systems only one relay is used; for accurate voltage sensing, some of these relays utilize circuits incorporating transistors and zener diodes); (3) manual operator; (4) test switch; (5) frequency relays; (6) speed of transfer; (7) pilot lights to indicate the position of the switch.

Motor Centers

A power panelboard is a fused-switch, circuit-breaker, or fuse panelboard from which motor branch circuits originate. It may provide only protection for the branch-circuit conductors, but in most cases the power panel also provides disconnect means for each motor.

A motor control center is a dead-front assembly of cubicles, each of which contains branch-circuit overcurrent protection, motor disconnect means, motor controller, and motor running overcurrent protection. It is a type of switchboard which contains all the protective and control means for the motors supplied from it. The selection and application of power panels and motor control centers should be related to future requirements. Such units must have necessary spare capacity in their buses and must be adaptable to anticipated changes or expansion. The layout of power panels and motor control centers should be related to voltage drop and the lengths of feeders.

Miscellaneous motor loads in the majority of commercial and institutional buildings and in many industrial buildings are usually circuited from power panels to which feeders deliver power. This type of distribution is a standard method of motor circuiting, generally limited to handling a number of small integral- or fractional-horsepower motors located in a relatively small area such as a fan room or pump room.

The feeder to a power panel may be a riser in a multistory commercial or institutional building, run from a basement switchboard. In a one-level industrial area, the feeder to a power panel may be run from a main or loadcenter switchboard or from a loadcenter substation in a high-voltage distribution system. In commercial and industrial buildings, motor feeders may be run from a main switchboard or a loadcenter substation to a motor control center serving a large group of motors in a machine room or in any compact area where the motors are relatively close together and close to the control-center assembly.

In industrial buildings, where a large number of motors are used over a large area, power is generally distributed to the individual motor loads by tapping motor branch circuits from the feeder. As was indicated in the discussion of branch circuits, a feeder may be tapped to motor branch circuits in several ways, by providing various combinations of branch-circuit overcurrent protection and motor control disconnect means. The feeder-tap method of supplying motor branch circuits is, of course, the basis for plug-in busway distribution.

Wiring Methods

The selection and layout of suitable wiring methods to provide for circuiting of electrical systems is a vital design task.

Although busway and interlocked armor cable have been constantly gaining popularity for use in distribution systems, wire in rigid conduit (metal or nonmetallic), intermediate metal conduit, and electrical metallic tubing are still the most used methods. For underground applications, a number of metal-jacketed cables and non-metal-jacketed cables have proved extremely reliable for direct earth burial. And for in-building feeders. Type AC and MC cables offer advantages in many locations. Cable assemblies—such as aluminum-sheathed cable, corrugated-metal-jacketed cables, and prewired plastic conduit—also offer advantages in many types of installations.

For outdoor distribution, a choice has to be made between overhead and underground circuits. Primary and secondary distribution requirements between buildings can usually be served best by underground installations for a number of reasons:

1. Accident hazards are minimized and isolated.
2. The ambient temperature is lowest, and cable capacities are highest.
3. Cables do not get in the way, obstructing the expansion of buildings or interfering with work.
4. The appearance of installations is neater, and architectural and landscape design is not marred by unsightly poles.

The cost of underground installation is not always justified under the following conditions:

1. Where chemical ingredients in the soil constitute a serious corrosion threat to directly buried metal conduits (galvanized steel or aluminum).
2. Where extensive changes to or expansion of buildings or plants would require the removal or alteration of underground feeders.
3. Where proper installation of underground circuits is complicated by such ground conditions as excessive rock.

Conductors. Feeder conductors for primary and secondary circuits should be selected carefully, on the basis of the many factors discussed in previous sections of this book.

Conductor size is determined by:

1. Load current to be supplied
2. Voltage drop and regulation
3. Temperature rise, within the limitations of the insulation
4. Reasonable energy losses
5. Ability to withstand short-circuit heating
6. Spare capacity for load growth

Although copper has been the long-time standard electric conductor material, there has been steady and sometimes rapid growth in the use of aluminum conductors. Often, the material cost savings effected with aluminum conductors can be very substantial. And the light weight of such conductors provides a reduction in the cost of installation labor. Lugs and connectors suitable for aluminum conductors are widely available.

Figure 8.14 presents data on the use of aluminum conductors for large feeders in commercial, institutional, and industrial buildings. The top chart shows feeder make-ups using copper and aluminum conductors, giving equivalents (reading across) on the basis of ampacity and voltage drop. The bottom chart shows how compact-strand aluminum conductors (second column from left) of the same ampacity as copper conductors (first column at left) permit the same number of conductors in a given size of conduit as their copper equivalents—except for the few cases as shown there. [See Note 4 in Part (a) (Tables) of **NEC** Chapter 9.] Allowable current-carrying capacities of both copper and aluminum conductors in a raceway or cable, for direct burial and in free air, are given in **NEC** Tables 310-16 through 310-19. Although copper and aluminum conductors are usually compared on the basis of current rating alone, voltage drop is a better basis for comparison because practical equivalents are thereby indicated. Of course, aluminum conductors of the same current rating or voltage-drop rating as copper conductors will be larger and may require a larger conduit. This must be considered in an economic analysis of aluminum conductor applications.

Single insulated conductors—such as TW or THHN or any of the other so-called "building wires"—must be combined with some type of raceway to make up an **NEC**-recognized "wiring method" or may be used in a cable tray or in a messenger-supported assembly (as covered in **NEC** Article 321).

Insulation must be selected according to:

1. Cost
2. Temperature rise, including both ambient temperature and temperature due to the heating effect of the load
3. Ease of installation
4. Environment of application, such as moisture, fumes, chemicals, petroleum products, acids, alkalies, etc.

Protective coverings must be selected according to:

1. Excellence of protection of the insulation against environmental factors (moisture, chemicals, acids, etc.)
2. Excellence of protection against physical damage, as by abrasion, impact, and cutting
3. Ability of metallic coverings to withstand corrosion and electrolysis

A very important consideration, indicated above, is the provision of a conductor size sufficient to handle the potential heating loads of short-circuit currents. With the expanded use of circuit-breaker overcurrent protection, coordination of protection from loads back to the source has introduced time delays in the operation of overcurrent devices. Cables in such systems must be able to withstand any impressed short-circuit currents for the duration of the overcurrent delay. For example, a motor circuit to a 100-hp motor might be required to carry as much as 15,000 A for a number of seconds. To limit damage to the cable due to heating, a much larger size conductor than is necessary for the load current alone may be required.

That concern for the ability of conductors to withstand the heating of short circuits has been expressed often by design engineers with respect to the smaller sizes of wires, as shown in the following example.

Example. A panelboard has 20-A breakers rated at 10,000 A interrupting capacity and No. 12 copper branch-circuit wiring. The available fault current at the point of breaker application is 8000 A. The short-circuit withstandability of a No. 12 copper conductor with plastic or polyethylene insulation rated 60°C is approximately 3000 A of fault or short-circuit current for one cycle.

Question. Assuming that the CB will take at least one cycle to operate, would the use of the No. 12 conductor where it is exposed to 8000 A be at odds with the

Equivalent aluminum and copper conductors based on voltage drop on three single conductors enclosed in a magnetic conduit, 80% power factor, 75°C insulation. Wire temperature and corresponding current-carrying capacities at 30°C ambient.

Copper conductors			Aluminum conductors		
Size	Voltage drop phase to phase per amp per 1000 ft	Amps	Size	Voltage drop phase to phase per amp per 100 ft	Amps
1/0	.232	150	3/0	.228	155
2/0	.193	175	4/0	.190	180
3/0	.163	200	250 kcmil	.169	205
4/0	.138	230	350 kcmil	.134	250
250 kcmil	.126	255	400 kcmil	.124	270
350 kcmil	.104	310	500 kcmil	.108	310
400 kcmil	.097	335	600 kcmil	.098	340
500 kcmil	.088	380	750 kcmil	.089	385

Copper size AWG kcmil	Compact size AWG kcmil	CONDUIT TRADE SIZE IN INCHES														
		1		1¼		1½		2		2½		3		3½		4
		Cu THW	Al XHHW	Cu THW	Al XHHW	Cu THW	Al XHHW	Cu THW	Al XHHW	Cu THW	Al XHHW	Cu THW	Al XHHW	Cu THW	Al XHHW	Cu THW	Al XHHW
4	2	3	3	5	5	7	7										
2	1/0	2	1	4	3	5	5										
1	2/0			3	3	4	4	6	7	9	9						
1/0	3/0			2	2	3	3	5	6	8	8						
2/0	4/0					3	3	5	4	7	6	10	10				
3/0	250					2	2	4	4	6	5	9	8				
4/0	300							3	3	5	5	7	7	10	10		
250	350							2	3	4	4	6	6	8	9		
350	500									3	3	4	5	6	6	8	8
400	600									2	2	4	4	5	5	7	7
500	700											3	3	4	4	6	6

FIGURE 8.14 Feeder makeups of equivalent ratings can be developed for copper or aluminum (bottom table shows number of conductors in each size of conduit).

intent expressed in the FPN following Section 240-1. This FPN states that overcurrent protection for conductors and equipment is provided for the purpose of opening the electric circuit if the current reaches a value which will cause an excessive or dangerous temperature in the conductor or conductor insulation. The 8000 A of available fault current would seem to call for conductors with that rating of short-circuit withstandability. This could mean that the branch-circuit wiring from all 20-A CBs in this panelboard must be at least No. 6 copper (the next larger size suitable for an 8000-A fault current).

Answer. As noted in UL Standard 489, a CB is required to operate safely in a circuit where the available fault current is up to the short-circuit current value for which the breaker is rated. The CB must clear the fault without damage to the insulation of conductors of proper size for the rating of the CB. A UL-listed 20-A breaker is, therefore, tested and rated to be used with wire rated for 20 A (say, No. 12 THW) and will protect the wire when applied at a point in a circuit where the available short-circuit current does not exceed the value for which the breaker is rated. This is also true of a 15-A breaker on No. 14 (15-A) wire, for a 30-A breaker on No. 10 (30-A) wire, and for all wire sizes. Underwriters Laboratories Standard 489 states:

> A circuit breaker performs successfully when operated under short-circuit conditions as described in paragraphs 21.2 and 21.3. There shall be no electrical or mechanical breakdown of the device, and the fuse that is indicated in paragraph 12.16 shall not have cleared. Cotton indicators as described in paragraphs 21.4 and 21.6 shall not be ignited. There shall be no damage to the insulation on the conductors used to wire the device. After the final operation, the circuit breaker shall have continuity in the closed position at rated voltage.

Any conductor used in a "wet location" (refer to the definition under *location* in **NEC** Article 100) must be one of the designated types, each of which has the letter "W" in its marking to indicate suitability for wet locations. Any conduit run underground is assumed to be subject to water infiltration and is, therefore, a wet location, requiring the use of only the listed conductor types within the raceway.

Although Type RHW conductor is suitable for wet locations, it is not approved for direct burial. If, however, the conductor is marked "RHW-USE"—that is, it is listed and recognized as both a single-conductor RHW and a single-conductor Type USE (underground service entrance) cable—then it is suited for direct-burial application as Type USE. "Conductors" for direct burial must be "listed"—which means certified by some kind of testing lab as suitable for direct burial in the earth.

Type UF cable is acceptable for direct earth burial (Fig. 8.15). The UL *Green Book* notes that listed USE cable is recognized for "burial directly in the earth."

All cables that are directly buried must be installed at a depth of at least 24 in below grade—as specified in **NEC** Section 300-5, but subject to reduced burial depths as covered there.

Raceways. For branch circuits, feeders, and services, conduit may be rigid metal conduit, intermediate metal conduit, rigid nonmetallic conduit, or electrical metallic tubing (for 4-in or smaller conduit). Rigid metal conduit and intermediate metal conduit of steel or aluminum provide excellent mechanical protection for conductors, enclose possible faults, and provide a low-resistance path to ground to assure quick operation of circuit protective devices in the event of fault currents. EMT offers light weight with ease of handling, cutting, and bending.

A big advantage in the use of conduit for feeders instead of a single high-capacity busway feeder is that a conduit installation reduces the distance to which short-circuit problems extend into a building from the service. Breaking the load into the smaller

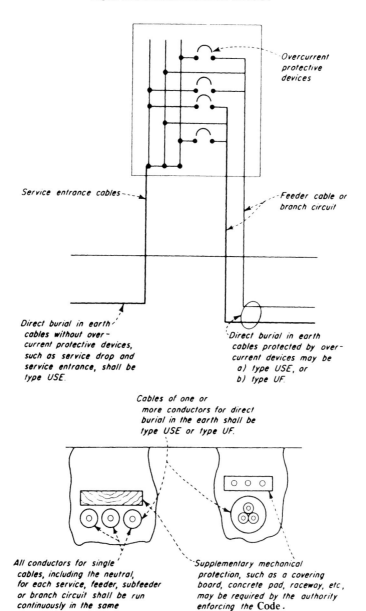

FIGURE 8.15 Type UF and Type USE cables ("U" for "underground") are recognized for direct burial in the earth.

feeders inherent in a conduit-and-cable distribution system introduces more impedance between distribution point and final branch-circuit overcurrent protective devices, thereby reducing the size of the short circuits which these final overcurrent protective devices must handle, while total impedance and consequent distribution-system losses introduced by breakup into small feeders are not appreciably increased for the overall project. And, in a conduit installation, electrical faults are localized to relatively small portions of the distribution system: Repairs can be made quickly, with little interruption of service.

The lower reactance of circuits in aluminum conduit can produce an important reduction in voltage drop. For example, in the initial design of an office building, the sizing of feeders for the given loads was based on the voltage-drop characteristics of circuits in steel conduit. In the original plan, the available kilovoltamperes at each lighting panel conformed to anticipated initial and future demands of tenants. When, however, a decision was made to install the feeders in aluminum conduit, the available tenant kilovoltampere capacity had to be recalculated on the basis of the lower reactance—and, consequently, voltage drop—of circuits in nonmagnetic enclosures. As a result, it was found that the load could be increased by 10 to 15 percent without exceeding the conductor current rating or the limit of 2 percent voltage drop in each feeder.

Galvanized rigid steel conduit and galvanized intermediate steel conduit do not generally require supplementary corrosion protection when they are directly buried in soil. The use of the word "generally" in the UL instructions indicates that it is the responsibility of the designer and/or installer to use supplementary protection where specific soils are known to corrode such conduits. Where the corrosion of underground galvanized conduit is known to be a problem, a protective jacketing or a field-applied coating of asphalt paint or equivalent material must be used on the conduit. But, UL notes on "supplementary nonmetallic coatings" must be observed with regard to resistance to corrosion.

Aluminum conduit that is directly buried in soil is said by UL to require supplementary corrosion protection. But, again, it is completely the task and responsibility of the designer and/or installer to select an effective protective coating for the aluminum conduit, because UL says "supplementary nonmetallic coatings presently used have not been recognized for resistance to corrosion." That could also be interpreted as prohibiting the use of directly buried aluminum conduit.

Galvanized rigid steel conduit and galvanized intermediate steel conduit installed in concrete do not require supplementary corrosion protection. Aluminum conduit installed in concrete definitely requires supplementary corrosion protection, but the supplementary protective coatings presently used "have not been recognized for resistance to corrosion." Watch out for this! The UL warns, "Wherever ferrous metal conduit runs directly from concrete encasement to soil burial, severe corrosive effects are likely to occur on the metal in contact with the soil." Supplementary protective coating on conduit at the crossing line can eliminate this condition.

Intermediate metal conduit (IMC) is a conduit whose wall thickness is smaller than that of rigid metal conduit but greater than that of EMT. IMC has the same threading and standard fittings, and the same general application rules, as rigid metal conduit. It actually is a light-weight rigid steel conduit which requires about 25 percent less steel than heavy-wall rigid metal conduit.

IMC may be used in any application for which rigid metal conduit is recognized, including use in all classes and divisions of hazardous locations as covered in **NEC** Sections 501-4, 502-4, and 503-3. Its thinner wall makes it lighter and less expensive than standard rigid metal conduit, but it has outstanding strength. Its lighter weight facilitates handling and installation at lower labor costs than for rigid metal conduit.

Because it has the same outside diameter as rigid metal conduit of the same trade size, it has greater interior cross-sectional area. Although this extra space is not recognized by the **NEC** as permitting the use of more conductors than can be used in the same size of rigid metal conduit, it does make wire pulling easier.

All approved rigid nonmetallic conduits are suitable for underground installations. Some types are approved for direct burial in the earth, while other types must be encased in concrete for underground applications. The nonmetallic conduits include fiber, asbestos-cement, soapstone, rigid polyvinyl chloride (PVC), fiberglass epoxy, polyethylene, and styrene conduit. Of these, medium-density polyethylene conduit and styrene conduit are not UL-listed. High-density polyethylene conduit and the others are UL-listed. The listed and labeled conduits differ widely in weight, cost, and physical characteristics, but each has certain application advantages.

The only nonmetallic conduit approved for use above ground at the present time is rigid polyvinyl chloride (PVC Schedule 40 or Schedule 80). Since not all PVC conduits are suitable for use above ground, the UL label on each conduit length will indicate if it is suitable for such use. UL applications data are detailed and divide "rigid nonmetallic conduit" into three categories, with specific instructions concerning the use of each category.

Rigid nonmetallic conduit is a general-purpose raceway for interior and exterior wiring, concealed or exposed in wood or masonry construction *under the stated conditions*. Only PVC is acceptable as a rigid nonmetallic conduit for in-building use (above ground).

For above-ground applications, rigid nonmetallic conduit must be Schedule 40 or Schedule 80 PVC conduit. Rigid nonmetallic conduit may be used above ground to carry high-voltage circuits without the need for encasing the conduit in concrete. Above-ground use is permitted indoors and outdoors.

Other rules on rigid nonmetallic conduit are:

1. Rigid nonmetallic conduit may not be used to support boxes or other equipment (Fig. 8.16).
2. When equipment grounding is required for metal enclosures of equipment used with rigid nonmetallic conduit, an equipment grounding conductor must be provided. Such a conductor must be installed in the circuit along with the circuit conductors.
3. In using nonmetallic conduit, care must be taken to ensure temperature compatibility between the conduit and the conductors in it. For instance, a conduit that has a 75°C temperature rating (the temperature at which it might melt and/or deform) must not be used with conductors which have a 90°C temperature rating and which will be loaded so they are operating at or near their temperature limit. Some available PVC rigid conduit is listed by UL and marked to indicate that it is suitable for use with conductors rated at 90°C. The UL data give the acceptable ambient temperatures and conductor temperature ratings. Conductors with 90°C insulation may be used at the higher ampacities of that temperature rating only when the conduit is encased in concrete.

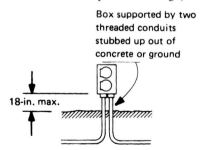

FIGURE 8.16 With threaded connections (not locknut and bushing), rigid metal conduit or IMC may support boxes, but rigid nonmetallic conduit may not support equipment or boxes.

EMT is a general-purpose raceway of the same nature as rigid metal conduit and IMC. Although rigid metal conduit and IMC afford maximum protection for conductors under all installation conditions, in many instances it is permissible, feasible, and more economical to use EMT to enclose circuit wiring rated 600 V or less. Because EMT is lighter than conduit, however, and is less rugged in construction and connection details, its use is restricted to locations (either exposed or concealed) where it will not be subjected to severe physical damage or (unless suitably protected) to corrosive agents. EMT distribution systems are constructed by combining wide assortments of related fittings and boxes. Connection is simplified by threadless components that include compression, indentation, and setscrew types.

Some questions have been raised about the acceptability of EMT directly buried in soil. The **NEC** gives EMT exactly the same recognition for direct burial that it gives to rigid steel conduit. It certainly seems clear that if galvanizing is enough corrosion protection for rigid steel conduit, it must provide equivalent protection for EMT in direct burial. In the UL listing "Electrical Metallic Tubing," a note says that "galvanized steel electrical metallic tubing in a concrete slab below grade level may require supplementary corrosion protection." (The word "may" leaves the decision up to the designer and/or installer.) But note that the rule carefully refers to "galvanized steel" EMT and not just to steel EMT.

The next note says, "In general, steel electrical metallic tubing in contact with soil requires supplementary corrosion protection." That sentence certainly admits that there are locations where soil conditions are such that supplementary corrosion protection is not required. Past experience and local soil conditions should be considered in determining the acceptability of direct burial of EMT as well as IMC and rigid conduit, with appropriate attention given to additional protection against corrosion if necessary. Of course, the ruling of the local electrical inspector should be sought and followed.

EMT up to 2 in in size has the same interior cross-sectional area as corresponding sizes of rigid metal conduit. EMT sizes of 2½ in and larger have the same outside diameter as rigid metal conduits of corresponding sizes. Accordingly, the interior cross-sectional areas (in square inches) are proportionally larger, and this provides more wiring space for greater ease of installation of conductors in these larger sizes of EMT. It is significant that the greater internal diameters of EMT provide cross-sectional areas that are from 16 to 22 percent greater than those of corresponding rigid metal conduit. But it should be stressed that these greater internal cross-sectional areas do not offer the capability of filling EMT runs with more conductors than could be used in corresponding sizes of rigid conduit. The number of conductors permitted in "conduit or tubing" is the same for rigid, IMC, and EMT, according to Tables 1 through 8 of **NEC** Chapter 9.

NEC Table 300-5 sets the basic rules on burial depths for underground applications of raceway. Rigid metal conduit and intermediate metal conduit must be buried at least 6 in below grade. EMT and plastic raceways have to be buried at least 18 in. Figure 8.17 shows some of the requirements as given in Section 300-5.

Conductors in Raceway. Circuits of different voltages up to 600 V may occupy the same wiring enclosure (cabinet, box, housing), cable, or raceway, provided all the conductors are insulated for the maximum voltage of any circuit in the enclosure, cable, or raceway. For instance, motor power conductors and motor control conductors are permitted in the same conduit. In the past, there has been a long-standing controversy about the use of control-circuit conductors in the same conduit with power leads to motors.

As shown in Fig. 8.18, a common raceway may be used for both motor power conductors and control-circuit conductors only where two or more motors are required

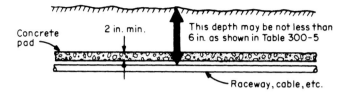

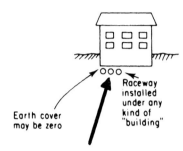

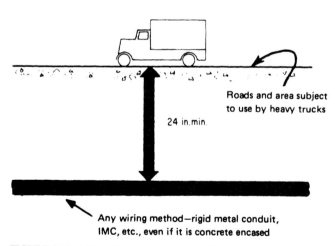

FIGURE 8.17 Some of the rules on burial depths of raceways and cables.

to be operated together to serve their load function. Many industrial and commercial installations contain machines, manufacturing operations, or processes in which a number of motors perform various parts or stages of a task. In such cases, either all the motors operate or none do; the placement of all control wires and power wires in the same raceway cannot produce a situation in which a fault in one motor circuit disables another circuit to a motor that might otherwise be kept operating.

1. This common raceway (conduit, EMT, wireway, or etc.) may contain **all** power conductors and **all** control conductors for two or more motors . . .

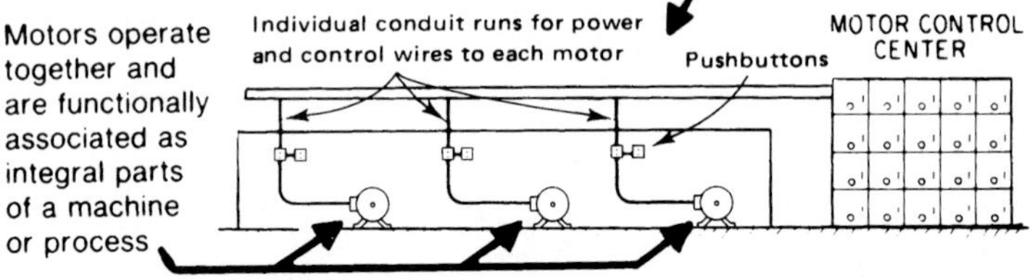

Motors operate together and are functionally associated as integral parts of a machine or process

Individual conduit runs for power and control wires to each motor

Pushbuttons

MOTOR CONTROL CENTER

2 . . . but, the intent of this **Code** rule, along with that of Section 725-15, permits a common raceway **only** where the power and control conductors are for a number of motors that operate integrally — such as a number of motors powering different stages or sections of a multi-motor process or production machine. Such usage complies with Section 725-15 (last sentence), which permits power and control wires in the same raceway, cable, or other enclosure when the equipment powered is "functionally associated" — that is, the motors have to run together to perform their task.

FIGURE 8.18 Mixing of power and control wires in a common raceway is limited.

But when a common raceway is used for power and control wires to supply separate, independent motors, a fault in one circuit could knock out others that do not have to be shut down when one goes out. With motor circuits so closely associated with vital equipment such as elevators, fans, and pumps in modern buildings, it is a matter of safety to separate such circuits and minimize the outage due to a fault in a single circuit. For safety's sake, "do not put all your eggs in one basket." But the objectionable loss of more than one motor on a single fault does not apply where all motors must be shut down when any one is stopped—as in multimotor machines and processes.

For those cases where each of several motors serves a separate, independent load with no interconnection of control circuits and no mechanical interlocking of driven loads, the use of a separate raceway for each motor is required—but only when control wires are carried in the raceways, as shown in Fig. 8.19. For the three motors shown, it would be acceptable to run the power conductors for all the motors in a single raceway and all the control circuit wires in another raceway. Of course, deratings would have to be made, and there is the definite chance of losing more than one motor on a fault in only one of the circuits in either the power raceway or the control raceway.

Conduit Fill. For the makeup of all circuits, the **NEC** regulates the maximum number of conductors that may be pulled into rigid metal conduit, rigid nonmetallic conduit, intermediate metal conduit, electrical metallic tubing, flexible metal conduit, and liquid-tight flexible metal conduit. The number of conductors permitted in a particular size of conduit or tubing is covered in **NEC** Chapter 9: Tables 1 and 3 cover conductors all of the same size used for either new work or rewiring; Tables 4 to 8 cover combinations of conductors of different sizes used for new work or rewiring. For non-lead-covered conductors, three or more to a conduit, the sum of the cross-sectional areas of the individual conductors must not exceed 40 percent of the interior cross-sectional area of the conduit or tubing, both for new work and for rewiring in existing conduit or tubing, as shown in Fig. 8.20. Note 3, which precedes the tables of **NEC** Chapter 9, permits a 60 percent fill of conduit nipples not over 24 in long, without the need for derating of ampacities.

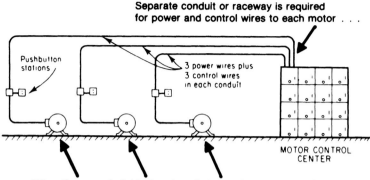

FIGURE 8.19 Separate conduit must be used to carry power and control wires to independent motors.

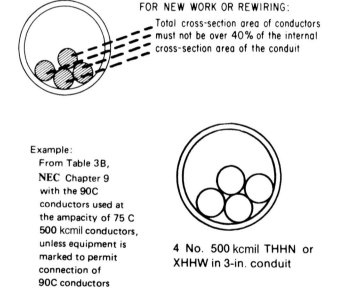

FIGURE 8.20 For three or more conductors in a conduit, the sum of their cross-sectional areas must not exceed 40 percent of the conduit cross-sectional area.

Tables 3A, 3B, and 3C of Chapter 9 give the maximum allowable fill for conduit or tubing in which all conductors are the same size up to 750 kcmil, and for conduit sizes from ½ to 6 in.

Question. What is the minimum size of conduit required for six No. 10 THHN wires?
Answer. Table 3B of Chapter 9 shows that six No. 10 THHN wires may be pulled into a ½-in conduit.

Question. What is the minimum conduit size for use with four No. 6 RHH conductors with outer covering?

Answer. Table 3C of Chapter 9 shows that a 1¼-in minimum conduit size must be used for three to five No. 6 RHH conductors.

Question. What is the minimum conduit size required for four 500-kcmil XHHW conductors?

Answer. Table 3B shows that 3-in conduit may contain four 500-kcmil XHHW (or THHN) conductors.

When all the conductors in a conduit or tubing are not the same size, the minimum required size of conduit or tubing must be calculated. Table 1 of Chapter 9 says that conduit containing three or more conductors of any type except lead-covered, for new work or rewiring, may be filled to 40 percent of the conduit cross-sectional area. Note 2 to this table refers to Tables 4 through 8 of Chapter 9 for the dimensions of conductors, conduit, and tubing to be used in calculating conduit fill for combinations of conductors of different sizes.

Example. What is the minimum size of conduit required to enclose six No. 10 THHN, three No. 4 RHH (without outer covering), and two No. 12 TW conductors (Fig. 8.21)?

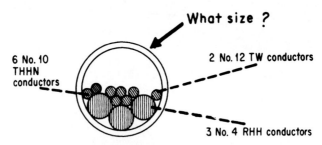

FIGURE 8.21 Minimum permitted conduit size must be calculated when conductors of different sizes are used.

Solution. The cross-sectional areas of the conductors are found in Table 5 of Chapter 9:

No. 10 THHN: 0.0184 in^2

No. 4 RHH: 0.1087 in^2

No. 12 TW: 0.0172 in^2

(Note that RHH without outer covering has the same dimensions as THW.) The total area occupied by conductors is found as follows:

Six No. 10 THHN: 6 × 0.0184	0.1104 in^2
Three No. 4 RHH: 3 × 0.1087	0.3261 in^2
Two No. 12 TW: 2 × 0.0172	0.0344 in^2
Total	0.4709 in^2

In Table 4 of Chapter 9, the fifth column from the left gives 40 percent of the cross-sectional area of each of the conduit sizes listed in the first column at the left. The 40 percent column shows that 0.34 in^2 is 40 percent fill for a 1-in conduit, and 0.60 in^2 is 40 percent fill for a 1¼-in conduit. Therefore, a 1-in conduit would be too small, and a 1¼-in conduit is the smallest that may be used for the 11 conductors.

Example. What is the minimum size of conduit for four No. 4/0 TW and four No. 4/0 XHHW conductors?

Solution. From Table 5, a No. 4/0 TW has a cross-sectional area of 0.3904 in^2. Four of these come to 4×0.3904, or 1.5616 in^2. From column 11 of Table 5, four No. 4/0 XHHW have a cross-sectional area of 4×0.3278, or 1.3112 in^2. Then the total cross-sectional area is

$$1.5616 + 1.3112 = 2.8728 \text{ in}^2$$

The fifth column of Table 4 shows that 2½-in conduit would be too small. A 3-in conduit, with a 40 percent fill of 2.9500 in^2, must be used.

Filling conduit to the **NEC** maximum is a minimal design practice; it is also frequently difficult or impossible from the practical standpoint of pulling the conductors into the conduit, owing to the usual twisting and bending of the conductors within the conduit. Bigger-than-minimum conduit should generally be used to provide some measure of spare capacity for load growth, and in many cases the conduit should be upsized considerably to allow for future installation of anticipated larger sizes of conductors.

In the sizing of conduit, neutral conductors are included in the total number of conductors because they occupy space. A completely separate consideration, however, is the relation of the neutral conductors to the number of "current-carrying" conductors in a conduit, which determines whether a derating factor must be applied:

Neutral conductors which carry only unbalanced current from phase conductors (as in the case of normally balanced 3-wire, single-phase or 4-wire, 3-phase circuits) are not counted in determining the current derating of conductors on the basis of the number in a conduit. Of course, a neutral conductor used with two phase legs of a 4-wire, 3-phase system to make up a 3-wire feeder is not a true neutral in the sense of carrying only current unbalance. Since the neutral carries the same current as the other two conductors under balanced load conditions, it must be counted as a phase conductor in derating for more than three conductors in a conduit.

Figure 8.22 shows four basic conditions of neutral loading relative to the need for counting the neutral conductor in derating a circuit to fluorescent or mercury ballasts:

Case 1. With balanced loads of equal power factor, there is no neutral current, and consequently no heating due to the neutral. For purposes of heat derating according to the **NEC,** this circuit produces the heating effect of three conductors.

Case 2. With two phases loaded and the third unloaded, the neutral carries the same current as the phases, but there is still the heating effect of only three conductors.

Case 3. With two phases fully loaded and the third phase partially loaded, the neutral carries the difference in current between the full phase value and the partial phase value, so again there is the heating effect of only three full-load phases.

Case 4. With a balanced load of fluorescent ballasts, third-harmonic generation causes a neutral current approximating the phase current, and there is the heating effect of four conductors. The neutral conductor should be counted with the phase conductors in determining conductor derating due to conduit occupancy.

618 CHAPTER EIGHT

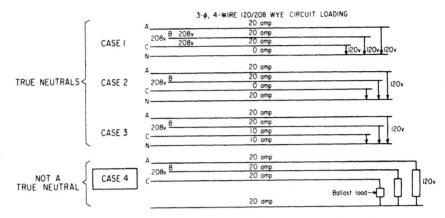

FIGURE 8.22 All neutrals must be counted for conduit fill, but only "true" neutrals do not count for load-current derating.

Because the neutral of a 3-phase, 4-wire wye feeder to a load of fluorescent or mercury ballasts will carry current even under balanced loading on the phases, such a neutral is not a true non-current-carrying conductor and should be counted as a phase wire in determining the number of conductors to arrive at a derating factor for more than three conductors in a conduit. As a result, the conductors of a 3-phase, 4-wire feeder to a fluorescent load should have a load-current rating of only 80 percent of their normal ampacity from **NEC** Table 310-16.

Ampacity Derating. Keep in mind the following:

IMPORTANT: Refer to the discussion "Note 8: ampacity derating" in Section 2 of this handbook. Note 8 to Table 310-16 on ampacity requires that the ampacity of a conductor be reduced to a value lower than the ampacity value given in the table when a raceway contains more than three current-carrying conductors. As a result, the ampacity of the conductors is changed and the conductors must be protected in accordance with the derated ampacity value and not in accordance with the tabulated value.

Because conductor "ampacity" is reduced when more than three conductors are used in a conduit, the overcurrent protection for each phase leg of a parallel makeup in a single conduit may be equal to the sum of the derated ampacities of the number of conductors used per phase leg. That would satisfy Section 240-3, which requires conductors to be protected at their ampacities. Because ampacity is reduced in accordance with the percentage factors given in Note 8 for more than three conductors in a single conduit, that derating dictates the use of multiple conduits for parallel-makeup circuits to avoid the penalty of loss of ampacity.

Figures 8.23 and 8.24 show examples of circuit makeups based on the unsafe concept of load limitation instead of ampacity derating, as applied to overcurrent rating and conductor ampacity for continuous loading [Section 220-10(b)].

Bunched or bundled cables must also have their ampacities reduced from the ampacity values shown in Table 310-16.

Note 8 does not, however, apply to conductors in wireways and auxiliary gutters. Wireways or auxiliary gutters may contain up to 30 conductors at any cross section (excluding signal circuits and control conductors used for starting duty only between a motor and its starter). The total cross-sectional area of the group of conductors must not be greater than 20 percent of the interior cross-sectional area of the wire-

THIS IS DANGEROUS
AND MUST NOT BE DONE...

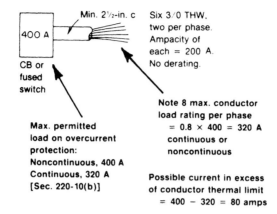

Min. 2½-in. c

Six 3/0 THW,
two per phase.
Ampacity of
each = 200 A.
No derating.

400 A

CB or
fused
switch

Max. permitted
load on overcurrent
protection:
Noncontinuous, 400 A
Continuous, 320 A
[Sec. 220-10(b)]

Note 8 max. conductor
load rating per phase
= 0.8 × 400 = 320 A
continuous or
noncontinuous

Possible current in excess
of conductor thermal limit
= 400 − 320 = 80 amps

... BUT THIS IS O.K.

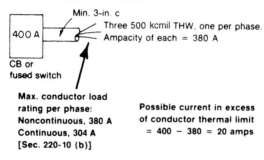

Min. 3-in. c

Three 500 kcmil THW, one per phase.
Ampacity of each = 380 A

400 A

CB or
fused switch

Max. conductor load
rating per phase:
Noncontinuous, 380 A
Continuous, 304 A
[Sec. 220-10 (b)]

Possible current in excess
of conductor thermal limit
= 400 − 380 = 20 amps

.. OR THIS

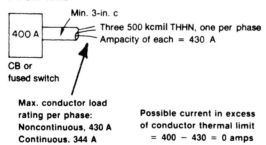

Min. 3-in. c

Three 500 kcmil THHN, one per phase
Ampacity of each = 430 A

400 A

CB or
fused switch

Max. conductor load
rating per phase:
Noncontinuous, 430 A
Continuous, 344 A

Possible current in excess
of conductor thermal limit
= 400 − 430 = 0 amps

FIGURE 8.23 Ampacity reduction is required for more than three current-carrying conductors in a conduit.

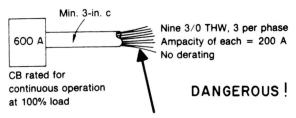

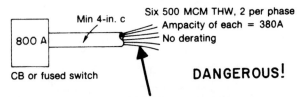

FIGURE 8.24 Parallel-conductor feeder makeup must not be used in a single conduit without ampacity reduction.

way or gutter. And ampacity derating factors for more than three conductors do not apply to wireway the way they do to wires in conduit. However, if the ampacity derating factors from Note 8 of Table 310-16 are used, there is no limit to the number of wires permitted in a wireway or an auxiliary gutter. But, the sum of the cross-sectional areas of all conductors contained at any cross section of the wireway must not exceed 20 percent of the cross-sectional area of the wireway or auxiliary gutter. More than 30 conductors may be used under those conditions.

Conductor Temperature Ratings. The following is important:

IMPORTANT: Refer to the discussion of "Conductor temperature rating" in Chapter 2 of this handbook. As discussed there, a general UL rule says that any electrical equipment rated over 100 A must be used with either 60 or 75°C conductors operating at the respective ampacities of those conductor insulation-temperature ratings unless the equipment is marked for use with conductors rated at higher temperatures. Equipment rated up to 100 A is generally limited to use with 60°C conductors or higher-temperature conductors operating at the 60°C ampacities for the particular conductor sizes, unless equipment is marked otherwise.

The UL temperature limitation is applied to the selection of feeder conductors in the steps that follow. The procedure is generally similar to the procedure described in Chapter 2 for selecting branch-circuit conductors. It can also become very involved and complex.

For the feeder in Fig. 8.25:

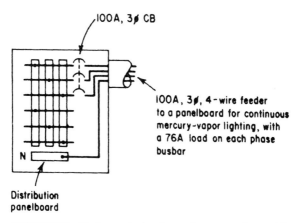

FIGURE 8.25 Feeder conductors for equipment rated up to 100 A must not exceed their 60°C ampacity.

Step 1. Because the load on the feeder is continuous, the 100-A, 3-pole CB must have its load current limited to 80 A [80 percent of its rating, per **NEC** Section 220-10(b)]. This is sufficient for the 76-A load. (A CB or fused switch may be loaded continuously to 100 percent of the CB or fuse ampere rating only when the assembly is UL-listed for such use.)

Step 2. The CB load terminals are recognized by UL for use with 60°C conductors or higher-temperature conductors loaded not over the 60°C ampacity of the given size of conductor.

Step 3. The feeder phase conductors must have an ampacity of not less than 125 percent of the continuous load of 76 A, which means not less than 95 A, because that is also required by **NEC** Section 220-10(b). The feeder neutral conductor is not subject to this limitation because it is not connected to the terminals of a heat-producing device (such as a CB or fusible switch). **NEC** Section 220-22 is the basic rule that covers the sizing of a feeder neutral. Section 220-10(b) covers the sizing of feeder phase conductors.

Step 4. If 60°C copper conductors are used for this feeder, reference must be made to the second column of Table 310-16. This feeder supplies electric-discharge lighting; therefore, Note 10(c) to the table requires that the feeder neutral be counted as a current-carrying conductor because of the harmonic currents present in the neutral. Then, because there are four current-carrying conductors in the conduit, the ampacity of each conductor must be derated to 80 percent of its value in the second column of Table 310-16. After the conductor is derated to 80 percent, it must have an ampacity of at least 95 A, as required by Section 220-10(b). From Table 310-16, a No. 1/0 TW conductor is rated at 125 A when only three conductors are used in a conduit. With four conductors in a conduit, the 125-A rating is reduced to 80 percent of that value (0.8 × 125 A), or 100 A, which is properly protected by the 100-A CB.

Note that this derating of ampacity to 80 percent, based on Note 8 of Tables 310-16 through 310-19, is in addition to the 80 percent load limitation (not a conductor derating) of Section 220-10(b). This is, in effect, a "double derating." **NEC**

Sections 210-22(c) and 220-2(a) used to contain exceptions that seemed to make such a double derating unnecessary for branch circuits. *But,* Section 220-10(b) does not contain an exception that eliminates the need for double derating of feeder conductors. It is certainly true that such a double derating provides a valuable and important amount of reserve capacity that is needed in feeder circuits to effectively minimize damage due to careless and/or "temporary" overloading during the operating life of the system.

The feeder circuit of four No. 1/0 TW conductors, rated at 100 A, would require a minimum of 2-in conduit.

NOTE: A reduced size of neutral could be used, because Section 220-10(b) does not apply to the neutral.

Step 5. If 75°C conductors were used for this feeder circuit instead of 60°C conductors, the calculations would be different. For THW copper conductors, Table 310-16 shows that No. 1 conductors, rated at 130 A for not more than three current-carrying conductors in a conduit, would have an ampacity of 0.8 × 130 A, or 104 A, when four are used in the conduit and derated. Because 125 percent of 76 A is 95 A, the 104-A conductor ampacity would satisfy Section 220-10(b) as to the ampacity of the feeder conductors for a continuous load.

Although UL listing and testing of the CB is based on the use of 60°C conductors, the use of No. 1 75°C THW conductors is acceptable because the terminals of the breaker in this case would not be loaded to more than the ampere rating of a 60°C conductor of the same size. A No. 1 60°C TW conductor is rated at 110 A when not more than three current-carrying conductors are used in a conduit. When four conductors are in one conduit, the 60°C No. 1 wires are derated to 80 percent of 110 A, or 88 A. Because that value is greater than the load of 76 A on each CB terminal, the CB terminals are not loaded in excess of the 88-A allowable ampacity of 60°C No. 1 conductors, and the UL limitation is satisfied.

Four No. 1 THW conductors, rated at 104 A, would require a minimum of 1½-in conduit. Or, four No. 1 XHHW conductors could be used in 1½-in conduit.

Step 6. If 90°C THHN conductors were to be used for this feeder (in a dry location), No. 2 copper conductors could be used. From Table 310-16, No. 2 THHN with a basic ampacity of 130 A would be derated to 80 percent of 130 A, or 104 A—which satisfies Section 220-10(b) as in step 3 above. The ampacity of a 60°C No. 2 conductor is 95 A normally; derated to 80 percent, it is 0.8 × 95, or 76 A, which gives the conductors the same rating as the load. Under such a condition the load current is not in excess of the 76-A allowable ampacity of a 60°C No. 2 conductor, and the UL limitation is satisfied.

Four No. 2 THHN, rated exactly at the required minimum rating of 96 A, would require a minimum of 1¼-in conduit, or four No. 2 XHHW could be used in 1¼-in conduit (in dry locations only).

In all the cases of conductor makeup described above, the 100-A CB would constitute proper protection.

NOTE: Of course, the voltage drop in the feeder will vary with the size of the conductors and must be accounted for.

Figure 8.26 shows an example in which feeder conductors could be used at up to their 75°C ampacity, as follows:

Step 1. Because the load on this feeder is continuous, Section 220-10(b) limits the load current to not more than 80 percent of the rating of the fuses and not more than the ampacity of the conductors. It is, then, immediately obvious that the feeder load may not exceed 80 percent of the 400-A fuse rating, or 0.8 × 400 A = 320 A.

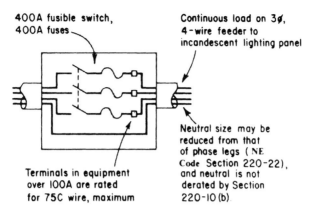

FIGURE 8.26 Equipment rated over 100 A may make use of conductors up to their 75°C ampacity.

Step 2. If 75°C conductors are to be used, because they are permitted by UL test conditions, Table 310-16 shows that 700-kcmil THW aluminum conductors rated at 375 A could be used and suitably protected by the 400-A fuses in accordance with Section 240-3(b) because 400 A is the next higher standard fuse rating (from Section 240-6) above the 375-A rating of the conductors. In that case the maximum continuous feeder load could be no more than 375 A × 0.8, or 300 A, as required by Section 220-10(b).

Step 3. If the maximum capacity of the fusible switch (0.8 × 400 A = 320 A) is desired from 75°C conductors, 800-kcmil THW aluminum conductors rated at 395 A could be used and loaded up to 80 percent of 385 A, or 316 A. If 900-kcmil THW aluminum is used, 80 percent of its 425-A ampacity is 340 A; but the load may not exceed 80 percent of the 400-A fuse rating—that is, 320 A.

Step 4. If 90°C conductors are to be used, they must be used at no more than the ampacity of a 75°C conductor of the same size as the 90°C conductor. Table 310-16 shows that 600-kcmil XHHW aluminum conductors (in a dry location) have a 385-A rating, and the 400-A fuses constitute acceptable protection for those conductors in accordance with Exception No. 1 of Section 240-3. The continuous load on the feeder phase legs would have to be limited to 80 percent of 385 A, or 308 A [Section 220-10(b)]. To check the suitability of the 90°C conductors, note that the ampere rating of a 75°C 600-kcmil aluminum conductor is given in Table 310-16 as 340 A. Because the load of 308 A is not in excess of the 340-A rating of a 75°C conductor, the UL limitation on the maximum rating of conductor termination is satisfied.

Step 5. The smaller conduit size required for the reduced size of higher-temperature conductors is a labor and material advantage.

Conductor Terminations. An extremely important element of the task of selecting switches, breakers, panelboards, switchboards, and other electrical-system equipment is the determination of the suitability of the conductor terminals for use with the circuit conductors. When copper conductors are used, there are no special problems with terminations—assuming normal care and skillful workmanship are applied. But the increasing use of aluminum conductors in feeder sizes (No. 4 and larger) demands diligent attention to the terminal arrangements to avoid undesirable effects due to the creep and cold-flow characteristics of aluminum. Effective hookup of switches, breakers, panels, motor starters, and other distribution equipment can be assured if aluminum terminations are made in prescribed ways.

Most distribution equipment has always come from the manufacturer with mechanical set-screw-type lugs for connecting circuit conductors to the equipment terminals. And with copper conductors, such terminals have been highly effective. Lugs on such equipment are commonly marked "AL-CU" or "CU-AL," indicating that the set-screw terminals are suitable for use with either copper or aluminum conductors. But, such marking on the lug itself is not sufficient evidence of suitability for use with aluminum conductors. UL requires that equipment whose terminals are suitable for use with either copper or aluminum conductors must be marked to indicate such suitability on the *label* or *wiring diagram* of the equipment—completely independently of any marking on the lugs themselves. A typical safety switch, for instance, would have lugs marked "AL-CU," but it also must have a notation on its label or nameplate that reads "Lugs suitable for copper or aluminum conductors."

Although equipment of the type described above may satisfy UL regulations and **NEC** rules [Section 110-14(a)] when aluminum conductors are terminated directly in mechanical set-screw lugs, some design engineers specify the use of only "compression-type" lugs for the terminations of aluminum conductors at all equipment terminals. Work quality and application of the correct torque are critical in eliminating the creep and cold flow of aluminum conductors in set-screw terminations.

When aluminum conductors are used with distribution equipment that comes with mechanical set-screw terminals:

- A termination device or copper pigtail may be put on the end of each aluminum conductor to provide an "end" that is suitable, tested, and proved for effective use in a set-screw-type lug. A number of manufacturers make "adapters" which are readily crimped onto the end of an aluminum conductor to "convert" the end to copper or an alloy that will not exhibit the creep and cold-flow disadvantages of aluminum in a setscrew termination.

- Or, the set-screw lug may be removed from the switch or breaker or panel and replaced with a crimp-type lug designed to accept an aluminum conductor. But, because unauthorized alteration or modification of equipment in the field voids the UL listing and can lead to very dangerous conditions, the arbitrary or unspecified changing of terminal lugs on equipment is not acceptable unless such field modification is recognized by UL and spelled out very carefully in the manufacturer's literature and on the label of the equipment itself.

Busways. One of the most popular of modern methods of electrical distribution is the busway system. Busways consist of metal enclosures containing insulator-supported busbars, constituting a complete prefabricated conductor wiring method with their own racewaylike enclosure. Although the use of a busway for industrial feeders, plug-in subfeeders, and branch-circuit systems has been developing for many years, the use of a feeder and plug-in busway in large commercial and institutional buildings has gained widespread acceptance over recent years, stimulated by developments in busway construction.

Busways are available for either indoor or outdoor use as point-to-point feeders or as plug-in takeoff routes for power. Progressive improvements in busway designs have enhanced their electrical and mechanical characteristics, reduced their physical size, and simplified the methods used to connect and support them. These developments have, in turn, reduced installation labor to the extent that busways are most favorably considered when large blocks of power must be moved to loadcenters (via low-impedance feeder busway), current must be distributed to closely spaced power-utilization points (via a plug-in busway), or rows of lighting fixtures or power tools must be energized and supported (via a lighting busway or trolley busway).

Figure 8.27 shows how busway applications are advantageous in many types of structures. For example, the upper sketch shows low-impedance feeder busways (a) used to bring power from an outside source to the main interior switchgear of a tall commercial building and then carrying power to upper-floor control centers (b and c). The lower sketch shows weatherproof feeder busway (d) connecting an outdoor transformer with the interior switchgear of an industrial plant, feeder busway (e) then connecting with the center tap of a plug-in system, and plug-in busways (f and g) carrying power to various equipment.

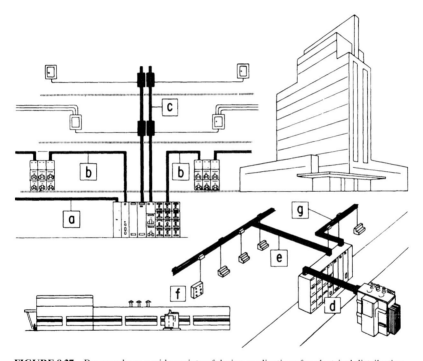

FIGURE 8.27 Busways have a wide variety of design applications for electrical distribution.

Busway conductors may be of copper or aluminum, hollow or solid, rectangular, oval, or beam-shaped. They may be insulated or bare, with phase bars grouped or interleaved, and with silver-plating applied either for entire lengths or only at connection joints and power-tap points. Busways are factory assembled and supplied in standard 10-ft sections.

Busways have proved to provide a versatile, flexible, and economical method of distribution. Offering safety, reliability, ease of layout, and efficient operation, busway systems require only basic engineering in their design. Layout possibilities are extensive; elbows, tees, offsets, and crosses permit either edgewise or flatwise orientation. Flanges permit busways to connect with switchboards or transformer tap boxes or to pass through walls, partitions, floors, and roofs in accordance with **NEC**-prescribed requisites. Power feeds or takeoffs may be located at busway ends or at center points. And they are capable of carrying large and small blocks of power from main switchboards to loadcenters to loads. For point-to-point power distribution, the overall economy of busways—in both labor and material costs—makes them particularly effective for heavy loads.

Either rigid or flexible connection may be made between adjacent straight busway sections or to power-supply sources or load takeoff devices. Utilizing metallic braid or lengths of armored cable and suitable busway attachments for flexible connections not only isolates vibration and accommodates expansion and movement; it also permits physical obstructions to be bypassed without requiring the use of special offsets or angle sections.

Busways are available specifically designed to minimize voltage drop and reactance, to serve high-amperage intermittent loads such as those created by resistance welders, or to carry high-cycle current for special lighting, laboratory, or induction-heating purposes. They also are available as slotted assemblies to accommodate moving trolleys.

Low-reactance feeder busways provide low-voltage-drop characteristics through close spacing and special construction of the busbars, which minimize reactance. This type of busway is used for all types of high-capacity feeders and risers. Because of its low reactance, it has also found application in high-frequency distribution systems.

The ampere rating of a busway must be taken as the nameplate current value and is a continuous load rating. A busway is also given a short-circuit rating in rms symmetrical amperes and must be applied only where the available short-circuit current at its supply end does not exceed that rating.

Typical ratings on a feeder busway range from 600 to 6000 A, for single-phase (two or three poles) or 3-phase (three or four poles) systems, for 120/240- or 480/277-V 3-phase, 4-wire systems, or for 3-phase, 3-wire systems—all up to 600 V. Weatherproof feeder busways are also available for exposed outdoor use. And busways are available with an extra busbar to provide equipment grounding throughout a system.

Plug-in distribution busways, with easily accessible plug-in openings for tapping to loads directly or through switches and protective devices, are available in ratings from 225 to 1500 A.

Plug-in and clamp-on devices include fused and nonfused switches and plug-in circuit breakers rated up to 800 A. Other plug-in devices include ground detectors, temperature indicators, capacitors, and transformers designed to mount directly on the busway.

Figure 8.28 is an isometric and floor plan of the complete riser and subfeed distribution system, which uses a variety of busway, in a modern commercial-industrial building housing research and development laboratories for telephone equipment. The low-impedance 1000-A busway risers feed 225-A plug-in runs on the floors, with taps made to 100-A plug-in busway along workbenches for tapping single-phase or 3-phase power to load devices as required by lab work.

Although busways have broad application potential, actual applications are limited. According to **NEC** rules, busways may be used only for exposed work. Some exception has, however, been made in cases where concealment is not permanent—such as above lift-out types of suspended ceilings. As long as it can be reached without removing any permanent part of a building, the busway is considered to be exposed. But, use above a suspended ceiling is limited to totally enclosed, nonventilated busways, without plug-in switches or CBs on the busway and only in ceiling space that is not used for air handling.

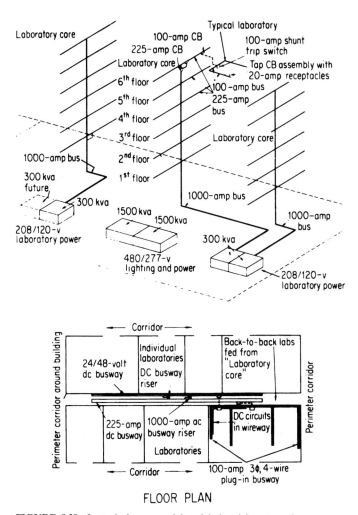

FIGURE 8.28 In typical commercial and industrial systems, busway may serve as feeders, subfeeders, and branch circuits.

Busways carry various markings to indicate the use for which the busway was investigated and listed. Busways intended to supply and support industrial and commercial lighting fixtures are marked "Lighting busway." Busways with sliding or other continuously movable means for tapping off current to load circuits are marked "Trolley busway." And, if the same busway is also acceptable for supporting and feeding lighting fixtures, it is also marked "Lighting busway." If the busway is designed to accept plug-in devices at any point along its length and is intended for general use, it is marked "Continuous plug-in busway." A busway marked "Lighting busway" and protected by overcurrent devices rated in excess of 20 A is intended for use only with fixtures having heavy-duty lampholders—unless each fixture is equipped with additional overcurrent protection to protect the lampholders.

Busway risers (vertical runs) must be supported at least every 16 ft and may be supported by a variety of spring-loaded hangers, wall brackets, or channel arrangements where they pierce floor slabs or are supported on masonry walls or columns (Fig. 8.29). Spring mounts for vertical busways may be located at successive floor-slab levels or supported by wall brackets located at intermediate elevations. Springs provide floating cradles for absorbing transient vibrations or physical shocks. Fire-resistant material must be packed into spaces between the busway and the edges of slab-piercing throat. Where it passes through a floor slab, the busway must be totally enclosed within the floor slab and for 6 ft above it. The **NEC** prohibits the use of ventilated busway in such places.

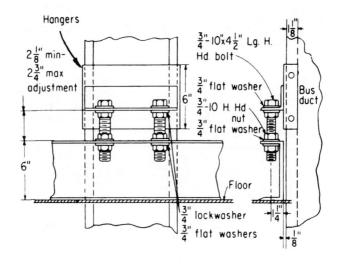

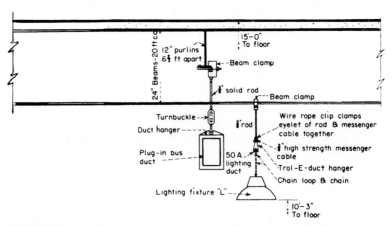

FIGURE 8.29 Specific design details should be provided to ensure safe, effective support for busway risers and horizontal runs.

Overcurrent protection—either a fused switch or a CB—is usually required in each busway subfeeder tapping power from a busway feeder of higher ampacity, where the larger busway is protected at the higher ampacity; this is necessary to protect the lower current-carrying capacity of the subfeeder. The protective device should be placed at the point at which the subfeeder connects into the feeder. However, the **NEC** permits this overcurrent protection to be omitted where busways are reduced in size if the smaller busway does not extend more than 50 ft and has a current rating at least equal to one-third the rating or setting of the overcurrent device protecting the main busway feeder, as shown in Fig. 8.30.

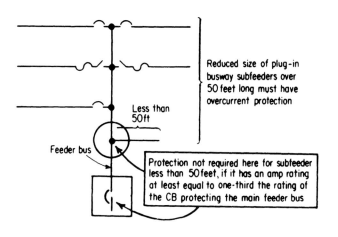

Example:

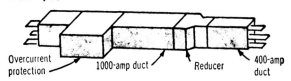

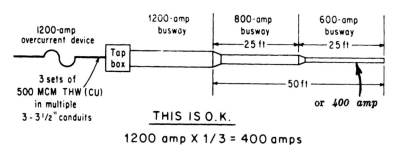

1200 amp X 1/3 = 400 amps

FIGURE 8.30 Busway subfeeder may sometimes be used without overcurrent protection at its point of supply but *only* at industrial facilities.

Branch circuits or subfeeders tapped from a busway must have overcurrent protection on the busway at the point of tap. And if they are out of reach from the floor, all fused switches and CBs must be provided with some means for a person to operate the handle of the device from the floor (hookstick, chain operator, rope-pull operator, etc.). A switch or CB is "out of reach" if the center of its operating handle, when in its highest position, is more than 6½ ft above the floor or platform on which personnel would be standing.

Figure 8.31 shows an extensive application of many busway runs as the main feeders from a loadcenter unit substation in a large industrial plant. The runs of plug-in busway supply motors and power panels directly at 480/277 V, and lighting panels at both 480/277 and 208/120 V, through local dry-type transformers.

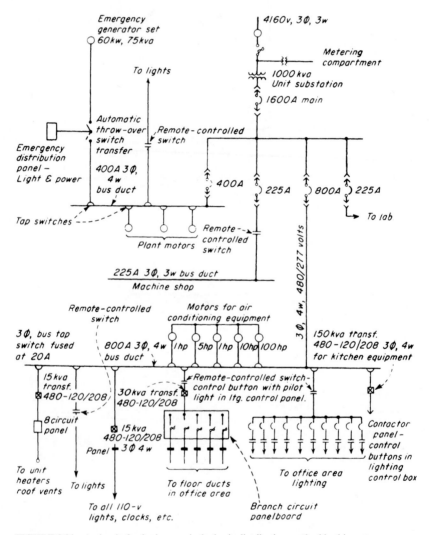

FIGURE 8.31 A plug-in feeder busway is the basic distribution method in this system.

Armored Cable. Armored cable has been used in electrical work for many years and still finds widespread application. **NEC** Article 334, entitled "Metalclad Cable," identifies so-called "interlocked armor" cable as "Type MC" and distinguishes its construction, application, and installation from standard BX-type armored cable, which is designated "Type AC."

Type AC cable is the cable assembly long used and known as BX cable. All regulations on the use of Type AC cable are given in **NEC** Article 333. Type AC armored cable is distinguished from Type MC metal-clad cables in the **NEC,** and the latter is covered in Article 334.

Type AC cable (BX) is listed and labeled by UL as "armored cable" in the *Electrical Construction Materials Directory.* The cable assembly consists of the conductors and a bonding strip (an equipment grounding conductor) within a jacket made of a spiral wrap of steel strip, with the edges of the strip interlocked, as shown in Fig. 8.32.

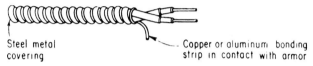

FIGURE 8.32 Basic assembly of Type AC armored cable, known as "BX."

Type AC armored-cable assemblies of two, three, or four conductors in sizes No. 14 AWG to No. 1 AWG conform to UL standards. These standards cover multiple-conductor armored cables for use in wiring systems of 600 V or less, at temperatures of 60, 75, or 90°C, depending upon the conductor insulation. Armored cables of other types which do not come under these UL standards are listed by UL as "metal-clad cable, Type MC" and covered by **NEC** Article 334. One type of MC cable is commonly called "interlocked armor cable."

Type AC cable is recognized for branch circuits and feeders but not for service entrance conductors, which must be as specified in **NEC** Section 230-43. Type MC (metal-clad) cable, such as interlocked armor cable and the other cables covered in **NEC** Article 334, is recognized for use as service entrance conductors.

Because the armor of Type AC cable is recognized as an equipment grounding conductor, its effectiveness must be ensured with an "internal bonding strip," or conductor, placed within the armor and shorting the turns of the steel jacket. The ohmic resistance of finished armor, including the bonding conductor that is required to be furnished as a part of all except lead-covered armored cable, must be within values specified by UL and checked during manufacturing. The bonding conductor that is run within the armor of the cable assembly is required by the UL standard. The function of the bonding conductor in Type AC cable is simply to short adjacent turns of the spiral-wrapped armor.

Typical assemblies of Type MC (metal-clad) cable are shown in Fig. 8.33. The top sketch shows the most common type of Type MC cable, used for many years under the name "interlocked armor cable." It is the heavy-duty industrial-feeder type of armored cable that is similar in appearance to but really different from standard BX armored cable. MC cable is a heavier-duty assembly than BX (Type AC), and great care must be taken to carefully distinguish between the two.

Type MC cable includes Type ALS and Type CS cables, in addition to the interlocked-armor and corrugated-tubing types of armor. Aluminum-sheathed (ALS) cable has insulated conductors with color-coded coverings, cable fillers, and an over-

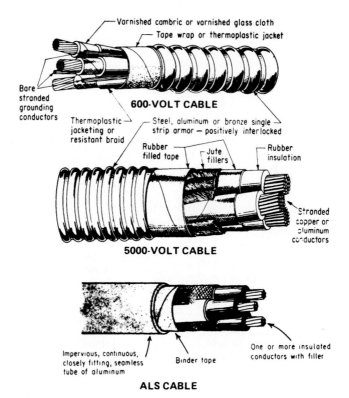

FIGURE 8.33 Typical cable assemblies that are UL-listed as Type MC cable.

all wrap of Mylar tape—all in an impervious, continuous, close-fitting, seamless tube of aluminum. It may be used for both exposed and concealed work in dry or wet locations, with approved fittings. CS cable is very similar but has a copper exterior sheath, instead of aluminum.

Type MC cable is rated by UL for use up to 5000 V, although cable for use up to 15,000 V has been available and used for many years. Type MC cable is recognized in three basic armor designs: (1) interlocked metal tape, (2) corrugated tube, and (3) smooth metallic sheath.

Because there are different forms of Type MC cable, care must be exercised to distinguish between the different constructions, owing to their varying suitability for design layouts and applications. For a long time, the interlocked-armor Type MC and the corrugated-sheath Type MC have been designated by UL as "intended for above-ground use." But **NEC** Section 334-3 recognizes Type MC cable as suitable for direct burial in the earth. However, such use must be checked with local inspection authorities.

Interlocked armor (IA) cable is available for use at 600, 5000, and 15,000 V. The high-voltage types of IA cable are very effective as primary feeders in loadcenter distribution systems. The 600-V class of IA cable is rapidly gaining in popularity for feeder applications in commercial and industrial systems.

EQUIPMENT SELECTION AND LAYOUT 633

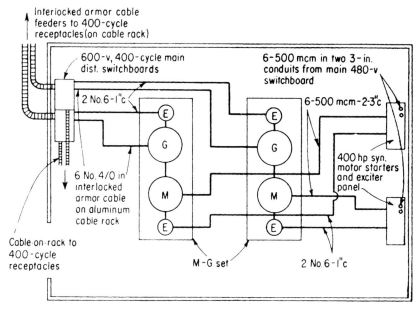

FIGURE 8.34 Cables with interlocked aluminum armor provide minimum inductive reactance in high-frequency circuits for airplane servicing.

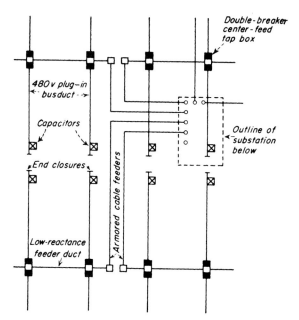

FIGURE 8.35 Armored-cable feeders were readily routed around many obstructions to supply busway runs.

Interlocked armor cable for secondary feeders is a completely flexible and protected cable assembly, available in sizes from No. 1/0 to 750 kcmil, with three or four conductors, and available in assemblies of conductors from No. 14 up for control purposes. The IA cable consists of a galvanized steel, interlocked, spiraled armor wrap around an insulated assembly of individually insulated conductors. The conductors are stranded and may be copper or aluminum. The armor may be of steel, aluminum, or bronze. Figure 8.34 shows how aluminum-armored cables were run in an aluminum cable tray to minimize inductive reactance by eliminating magnetic material (steel) from 400-H circuits in a large airplane hangar.

There are some general criteria for choosing between IA cable and busway. Where the routing is devious and there are many obstructions, IA cable often can be used to greater advantage. For straight runs, particularly long runs of high current capacity, busway is usually better. Above 1000-A capacity, busway for a feeder generally costs less. Below 500 A, a feeder may more economically consist of interlocked armor cable. However, where the advantages of either armored cable or busway are required, the cost of materials and installation is not of primary importance. As shown in Fig. 8.35, the electrical distribution system in one industrial plant made extremely effective use of both armored cable and busway. The IA-cable feeders were used to provide flexible runs around various overhead obstructions, from the 480-V switchgear to cable junction boxes on the busway runs.

CHAPTER 9
DESIGN REFERENCE DATA

GENERAL DATA FOR CALCULATIONS

DC CIRCUIT CHARACTERISTICS

Ohm's Law:

$$E = IR \quad I = \frac{E}{R} \quad R = \frac{E}{I}$$

E = voltage impressed on circuit (volts)
I = current flowing in circuit (amperes)
R = circuit resistance (ohms)

In direct current circuits, electrical power is equal to the product of the voltage and current:

$$P = EI = I^2R = \frac{E^2}{R}$$

P = power (watts)
E = voltage (volts)
I = current (amperes)
R = resistance (ohms)

AC CIRCUIT CHARACTERISTICS

The instantaneous values of an alternating current or voltage vary from zero to a maximum value each half cycle. In the practical formulae which follow, the "effective value" of current and voltage is used, defined as follows:

Effective value = 0.707 × maximum instantaneous value

Impedance:

Impedance is the total opposition to the flow of alternating current. It is a function of resistance, capacitive reactance and inductive reactance. The following formulae relate these circuit properties:

$$X_L = 2\pi f L \quad X_C = \frac{1}{2\pi f C} \quad Z = \sqrt{R^2 + (X_L - X_C)^2}$$

X_L = inductive reactance (ohms)
X_C = capacitive reactance (ohms)
Z = impedance (ohms)
f = frequency (cyles per second)
C = capacitance (farads)
L = inductance (henrys)
R = resistance (ohms)
π = 3.14

Ohm's Law for AC Circuits:

$$E = I \times Z \quad I = \frac{E}{Z} \quad Z = \frac{E}{I}$$

POWER FACTOR

Power factor of a circuit or system is the ratio of actual power (watts) to apparent power (volt-amperes), and is equal to the cosine of the phase angle of the circuit:

$$PF = \frac{\text{actual power}}{\text{apparent power}} = \frac{\text{watts}}{\text{volts} \times \text{amperes}} = \frac{KW}{KVA} = \frac{R}{Z}$$

KW = kilowatts
KVA = kilovolt-amperes = volt-amperes × 1,000
PF = power factor (expressed as decimal)

SINGLE-PHASE CIRCUITS

$$KVA = \frac{EI}{1,000} = \frac{KW}{PF} \quad KW = KVA \times PF$$

$$I = \frac{P}{E \times PF} \quad E = \frac{P}{I \times PF} \quad PF = \frac{P}{E \times I}$$

$P = E \times I \times PF$
P = power (watts)

THREE-PHASE CIRCUITS, BALANCED STAR OR WYE

$$I_N = 0 \quad I = I_p \quad E = \sqrt{3} E_p = 1.73 E_p$$

$$E_p = \frac{E}{\sqrt{3}} = \frac{E}{1.73} = 0.577 E$$

I_N = current in neutral (amperes)
I = line current per phase (amperes)
I_p = current in each phase winding (amperes)
E = voltage, phase to phase (volts)
E_p = voltage, phase to neutral (volts)

THREE-PHASE CIRCUITS, BALANCED DELTA

$$I = 1.732 \times I_p \quad I_p = \frac{I}{\sqrt{3}} = 0.577 \times I$$

$E = E_p$

POWER: BALANCED 3-WIRE, 3-PHASE CIRCUIT, DELTA OR WYE

For unity power factor (PF = 1.0):

$$P = 1.732 \times E \times I$$

$$I = \frac{P}{\sqrt{3} E} = \frac{0.577 P}{E} \quad E = \frac{P}{\sqrt{3} \times I} = \frac{0.577 P}{I}$$

P = total power (watts)

For any load:

$$P = 1.732 \times E \times I \times PF \quad VA = 1.732 \times E \times I$$

$$E = \frac{P}{PF \times 1.73 \times I} = \frac{0.577 \times P}{PF \times I}$$

$$I = \frac{P}{PF \times 1.73 \times E} = \frac{0.577 \times P}{PF \times E}$$

$$PF = \frac{P}{1.73 \times I \times E} = \frac{0.577 \times P}{I \times E}$$

VA = apparent power (volt-amperes)
P = actual power (watts)
E = line voltage (volts) phase to phase
I = line current (amperes)

POWER LOSS: ANY AC OR DC CIRCUIT

$$P = I^2 R \quad I = \sqrt{\frac{P}{R}} \quad R = \frac{P}{I^2}$$

P = power heat loss in circuit (watts)
I = effective current in conductor (amperes)
R = conductor resistance (ohms)

STANDARD LOADS FOR LIGHTING IN COMMERCIAL BUILDINGS

Watts/Sq Ft

Armories
Drill Sheds and Exhibition Halls — 5
 This does not include lighting circuits for demonstration booths, special exhibit spaces, etc.

Art Galleries
a. General — 3
b. On Paintings—50 watts per running foot of usable wall area

Auditoriums — 4

Automobile Show Rooms — 6

Banks
a. Lobby — 4
b. Counters—50 watts per running foot including service for signs and small motor applications, etc.
c. Offices and Cages — 5

Barber Shop and Beauty Parlors — 5
 This does not include circuits for special equipment.

Billiards
a. General — 3
b. Tables—450 watts per table.

Bowling
a. Alley Runway and Seats — 5
b. Pins—300 watts per set of pins.

Churches
a. Auditoriums — 2
b. Sunday School Rooms — 5
c. Pulpit or Rostrum — 5

Club Rooms
a. Lounge — 2
b. Reading Rooms — 5
 The above two uses are so often combined that the higher figure is advisable. It includes provision for convenience outlets.

Court Rooms — 5

Dance Halls — 2
 No allowance has been included for spectacular lighting, spots, etc.

Drafting Rooms — 7

Fire Engine Houses — 2

Gymnasiums
a. Main Floor — 5
b. Shower Rooms — 2
c. Locker Rooms — 2
d. Fencing, Boxing, etc. — 5
e. Handball, Squash, etc. — 5

Halls and Interior Passageways —20 watts per running foot.

Hospitals
a. Lobby, Reception Room — 3
b. Corridors—20 watts per running foot.
c. Wards — 3
 Including allowance for convenience outlets for local illumination.

d. Private Rooms — 5
 Including allowance for convenience outlets for local illumination.
e. Operating Room — 5
f. Operating Tables or Chairs
 Major Surgeries—3000 watts per area.
 Minor Surgeries—1500 watts per area.
 This and the above figure include allowance for directional control. Special wiring for emergency systems must also be considered.
g. Laboratories — 5

Hotels
a. Lobby — 5
 Not including provision for conventions, exhibits.
b. Dining Room — 4
c. Kitchen — 5
d. Bed Rooms — 3
 Including allowance for convenience outlets.
e. Corridors—20 watts per running foot.
f. Writing Room — 5
 Including allowance for convenience outlets.

Library
a. Reading Rooms — 6
 This includes allowance for convenience outlets.
b. Stack Room—12 watts per running foot of facing stacks.

Motion Picture Houses and Theatres
a. Auditoriums — 2
b. Foyer — 3
c. Lobby — 5

Museums
a. General — 3
b. Special exhibits—supplementary lighting — 5

Office Buildings
a. Private Offices, no close work — 4
b. Private Offices, with close work — 5
c. General Offices, no close work — 4
d. General Offices, with close work — 5
e. File Room, Vault, etc. — 3
f. Reception Room — 2

Post Office
a. Lobby — 3
b. Sorting, Mailing, etc. — 5
c. Storage, File Room, etc — 3

Professional Offices
a. Waiting Rooms — 3
b. Consultation Rooms — 5
c. Operating Offices — 7
d. Dental Chairs—600 watts per chair.

Railway
a. Depot—Waiting Room — 3
b. Ticket Offices—General — 5
 On Counters-50 watts per running foot.
c. Rest Room, Smoking Room — 3
d. Baggage, Checking Office — 3
e. Baggage Storage — 2
f. Concourse — 2
g. Train Platform — 2

Restaurants, Lunch Rooms and Cafeterias
a. Dining Areas — 3
b. Food Displays—50 watts per running foot of counter (including service aisle).

Schools
a. Auditoriums — 3
 If to be used as a study hall —5 watts per sq. ft.
b. Class and Study Rooms — 5
c. Drawing Room — 7
d. Laboratories — 5
e. Manual Training — 5
f. Sewing Room — 7
g. Sight Saving Classes — 7

Show Cases—25 watts per running foot.

Show Windows
a. *Large Cities
 Brightly Lighted District—700 watts per running foot of glass.
 Secondary Business Locations—500 watts per running foot of glass.
 Neighborhood Stores—250 watts per running foot of glass.
b. *Medium Cities
 Brightly Lighted District—500 watts per running foot of glass.
 Neighborhood Stores—250 watts per running foot of glass frontage.
c. *Small Cities and Towns—300 watts per running foot of glass frontage.
d. Lighting to Reduce Daylight Window Reflections—1000 watts per running foot of glass.

*Wattages shown are for white light with incandescent filament lamps. Where color is to be used, wattages should be doubled.

Stores, Large Department and Specialty
a. Main Floor — 6
b. Other Floors — 6

Stores in Outlying Districts — 5

Wall Cases—25 watts per running foot.

NOTE: Figures based on use of fluorescent equipment for large-area application, incandescent for local or supplementary lighting.

STANDARD LOADS FOR GENERAL LIGHTING IN INDUSTRIAL OCCUPANCIES

Watts/Sq Ft

AISLES, STAIRWAYS, PASSAGEWAYS 10 watts per running foot.		**DAIRY PRODUCTS**	4.0	**INSPECTION**	
		ENGRAVING	*4.5	a. Rough	3.0
ASSEMBLY		**FORGE SHOPS**		b. Medium	4.5
a. Rough	3.0	a. Welding	2.0	c. Fine	*4.5
b. Medium	4.5	**FOUNDRIES**		d. Extra Fine	*4.5
c. Fine	*4.5	a. Charging Floor, Tumbling, Cleaning, Pouring, Shaking Out	2.0	**JEWELRY AND WATCH MANUFACTURING**	
d. Extra Fine	*4.5	b. Rough Molding and Core Making	2.0	**LAUNDRIES AND DRY CLEANING**	
AUTOMOBILE MANUFACTURING		c. Fine Molding and Core Making	4.0	**LEATHER MANUFACTURING**	
a. Assembly Line	*4.5	**GARAGES**		a. Vats	2.0
b. Frame Assembly	3.0	a. Storage	2.0	b. Cleaning, Tanning and Stretching	2.0
c. Body Assembly	4.5	b. Repair and Washing	*3.0	c. Cutting, Fleshing and Stuffing	3.0
d. Body Finishing and Inspecting	*4.5	**GLASS WORKS**		d. Finishing and Scarfing	4.5
BAKERIES	4.0	a. Mixing and Furnace Rooms, Pressing and Lehr Glass Blowing Machines	3.0	**LEATHER WORKING**	
BOOK BINDING		b. Grinding, Cutting Glass to Size, Silvering	4.5	a. Pressing, Winding and Glazing	
a. Folding, Assembling, Pasting	3.0	c. Fine Grinding, Polishing, Beveling, Etching, Inspecting, etc.	*4.5	(1) Light	2.0
b. Cutting, Punching, Stitching, Embossing	4.0			(2) Dark	4.5
BREWERIES				b. Grading, Matching, Cutting, Scarfing, Sewing	
a. Brew House	3.0			(1) Light	4.5
b. Boiling, Keg Washing, etc	3.0			(2) Dark	*4.5
c. Bottling	4.0	**GLOVE MANUFACTURING**		**LOCKER ROOMS**	2.0
CANDY MAKING	4.0	a. Light Goods		**MACHINE SHOPS**	
CANNING AND PRESERVING	4.0	(1) Cutting, Pressing, Knitting, Sorting	4.5	a. Rough Bench and Machine Work	3.0
CHEMICAL WORKS		(2) Stitching, Trimming, Inspecting	4.5	b. Medium Bench and Machine Work, Ordinary Automatic Machines, Rough Grinding, Medium Buffing and Polishing	4.5
a. Hand Furnaces, Stationary Driers and Crystallizers	2.0	b. Dark Goods			
b. Mechanical Driers and Crystallizers, Filtrations, Evaporators, Bleaching	2.0	(1) Cutting, Pressing, etc.	*4.5	c. Fine Bench and Machine Work, Fine Automatic Machines, Medium Grinding, Fine Buffing and Polishing	*4.5
c. Tanks for Cooking, Extractors, Percolators, Nitrators, Electrolytic Cells	3.0	(2) Stitching, Trimming, etc.	*4.5	d. Extra Fine Bench and Machine Work, Grinding	
CLAY PRODUCTS AND CEMENTS		**HANGARS—AEROPLANE**		(1) Fine Work	*4.5
a. Grinding, Filter Presses, Kiln Rooms	2.0	a. Storage—Live	2.0	**MEAT PACKING**	
b. Moldings, Pressing, Cleaning, Trimming	2.0	b. Repair Department	*3.0	a. Slaughtering	2.0
c. Enameling	3.0	**HAT MANUFACTURING**		b. Cleaning, Cutting, Cooking, Grinding, Canning, Packing	4.5
d. Glazing	4.0	a. Dyeing, Stiffening, Braiding, Cleaning and Refining		**MILLING—GRAIN FOODS**	
CLOTH PRODUCTS		(1) Light	2.0	a. Cleaning, Grinding, Rolling	2.0
a. Cutting, Inspecting, Sewing		(2) Dark	4.5	b. Baking or Roasting	4.5
(1) Light Goods	4.5	b. Forming, Sizing, Pouncing, Flanging, Finishing and Ironing		c. Flour Grading	4.5
(2) Dark Goods	*4.5	(1) Light	3.0	**OFFICES**	
b. Pressing, Cloth Treating (Oil Cloth, etc.)		(2) Dark	6.0	a. Private and General	
(1) Light Goods	3.0	c. Sewing		(1) No close work	3.0
(2) Dark Goods	6.0	(1) Light	4.5	(2) Close work	4.5
COAL BREAKING, WASHING, SCREENING	2.0	(2) Dark	*4.5	b. Drafting Rooms	7.0
		ICE MAKING			
		a. Engine and Compressor Room	2.0		

The figures given in this table are average design loads for general lighting. In those cases marked with an asterisk (*), the load values provide only for large area lighting applications. Local lighting must then be provided as an additional load.

The figures given are based on the use of fluorescent and mercury vapor equipment of standard design. Adjustments may be made for use of higher or lower efficiency equipment. For equal lighting intensities from incandescent units, the figures must be at least doubled.

STANDARD LOADS FOR GENERAL LIGHTING IN INDUSTRIAL OCCUPANCIES

Watts/Sq Ft

PACKING AND BOXING	3.0	
PAINT MANUFACTURING	3.0	
PAINT SHOPS		
a. Dipping, Spraying, Firing, Rubbing, Ordinary Hand Painting and Finishing	3.0	
b. Fine Hand Painting and Finishing	*3.0	
c. Extra Fine Hand Painting and Finishing (Automobile Bodies, Piano Cases, etc.)	*3.0	
PAPER BOX MANUFACTURING		
a. Light	3.0	
b. Dark	4.0	
c. Storage of Stock	2.0	
PAPER MANUFACTURING		
a. Beaters, Grinding, Calendering	2.0	
b. Finishing, Cutting, Trimming	4.5	
PLATING	2.0	
POLISHING AND BURNISHING	3.0	
POWER PLANTS, ENGINE ROOMS, BOILERS		
a. Boilers, Coal and Ash Handling, Storage Battery Rooms	2.0	
b. Auxiliary Equipment, Oil Switches and Transformers	2.0	
c. Switchboards, Engines, Generators, Blowers, Compressors	3.0	
PRINTING INDUSTRIES		
a. Matrixing and Casting	2.0	
b. Miscellaneous Machines	3.0	
c. Presses and Electrotyping	4.5	
d. Lithographing	*4.5	
e. Linotype, Monotype, Typesetting, Imposing Stone, Engraving	*4.5	
f. Proof Reading	*4.5	
RECEIVING AND SHIPPING	2.0	
RUBBER MANUFACTURING AND PRODUCTS		
a. Calendars, Compounding Mills, Fabric Preparation, Stock Cutting, Tubing Machines, Solid Tire Operations, Mechanical Goods Building, Vulcanizing	3.0	
b. Bead Building, Pneumatic Tire Building and Finishing, Inner Tube Operation, Mechanical Goods Trimming, Treading	4.5	
SHEET METAL WORKS		
a. Miscellaneous Machines, Ordinary Bench Work	3.0	
b. Punches, Presses, Shears, Stamps, Welders, Spinning, Medium Bench Work	4.5	
c. Tin Plate Inspection	*4.5	
SHOE MANUFACTURING		
a. Hand Turning, Miscellaneous Bench and Machine Work	2.0	
b. Inspecting and Sorting Raw Material, Cutting and Stitching		
(1) Light	4.5	
(2) Dark	*4.5	
c. Lasting and Welting	4.5	
SOAP MANUFACTURING		
a. Kettle Houses, Cutting, Soap Chip and Powder	3.0	
b. Stamping, Wrapping and Packing, Filling and Packing Soap Powder	4.5	
STEEL AND IRON MILLS, BAR, SHEET AND WIRE PRODUCTS		
a. Soaking Pits and Reheating Furnaces	2.0	
b. Charging and Casting Floors	2.0	
c. Muck and Heavy Rolling, Shearing (Rough by Gauge), Pickling and Cleaning	2.0	
d. Plate Inspection, Chipping	*4.5	
e. Automatic Machines, Light and Cold Rolling, Wire Drawing, Shearing (fine by line)	4.5	
STONE CRUSHING AND SCREENING		
a. Belt Conveyor Tubes, Main Line Shafting Spaces, Chute Rooms, Inside of Bins	2.0	
b. Primary Breaker Room, Auxiliary Breakers under Bins	2.0	
c. Screens	3.0	
STORAGE BATTERY MANUFACTURING		
a. Molding of Grids	3.0	
STORE AND STOCK ROOMS		
a. Rough Bulky Material	2.0	
b. Medium or Fine Material requiring care	3.0	
STRUCTURAL STEEL FABRICATION	3.0	
SUGAR GRADING	5.0	
TESTING		
a. Rough	3.0	
b. Fine	4.5	
c. Extra Fine Instruments, Scales, etc.	*4.5	
TEXTILE MILLS		
a. Cotton		
(1) Opening and Lapping, Carding, Drawing, Roving, Dyeing	3.0	
(2) Spooling, Spinning, Drawing, Warping, Weaving, Quilling, Inspecting, Knitting, Slashing (over beam end)	4.5	
b. Silk		
(1) Winding, Throwing, Dyeing	4.5	
(2) Quilling, Warping, Weaving, Finishing		
Light Goods	4.5	
Dark Goods	6.0	
c. Woolen		
(1) Carding, Picking, Washing, Combing	3.0	
(2) Twisting, Dyeing	3.0	
(3) Drawing-in, Warping		
Light Goods	4.5	
Dark Goods	6.0	
(4) Weaving		
Light Goods	4.5	
Dark Goods	6.0	
(5) Knitting Machines	4.5	
TOBACCO PRODUCTS		
a. Drying, Stripping, General	3.0	
b. Grading and Sorting	*4.5	
TOILETS AND WASH ROOMS	2.0	
UPHOLSTERING		
a. Automobile, Coach, Furniture	4.5	
WAREHOUSE	2.0	
WOODWORKING		
a. Rough Sawing and Bench Work	2.0	
b. Sizing, Planing, Rough Sanding, Medium Machine and Bench Work, Gluing, Veneering, Cooperage	4.5	
c. Fine Bench and Machine Work, Fine Sanding and Finishing	6.0	

Use of these figures in computing number and loading of circuits and feeders should always be checked against the conditions and requirements of the particular area. The figures are not substitutes for lighting design.

In any case, need for particular color quality of light or special intensities or control of light must be determined as part of the lighting design, and wattages and circuit requirements must be provided accordingly.

POWER LOAD DATA

Appliance, Device or Machine	Domestic Watts From-To	Domestic Horsepower From-To	Commercial — Industrial Watts From-To	Commercial — Industrial Horsepower From-To
LIGHTING EQUIPMENT				
Airport Floods			240-3000	
Airport Landing Lights			to 1 Kw.	
Blue Printing			3-10 Kw.	
Borderlights, Prof. Stage, per ft.			200-2000	
Borderlights, Schools, per ft.			100-500	
Cove, Strips, per ft.	10-200		20-500	
Exit Signs			40-150	
Floodlights, Outdoor	75-750		200-2000	
Floodlights, Window			100-1000	
Footlights, Prof. Stage, per ft.			100-1000	
Footlights, Schools, per ft.			100-300	
Infrared lamps, per lamp	100-250		250-1000	
Luminaires (Commercial Lighting Fixtures)			100 up	
Luminous Tubing (Cold Cathode) per ft.			5-10	
Operating Rooms (Hospital)			1-10 Kw.	
Photostat Machines			1-5 Kw.	
Projectors, Amateur Movie	300-750			
Projectors, Amateur Movie and Sound	650-1250			
Projectors, Prof. Movie			1500-3500	¼-1
Projectors, Visual Lecturing			400-1000	
Reflectors, Show Case, per ft.			10-150	
Reflectors, Show Window, per ft.			100-750	
Spotlights, Ball Room	100-500		100-2000	
Spotlights, Projection Booth			2750-3300	
Spotlights, Stage or Balcony Rail			200-1500	
Spotlights, Show Windows			100-1000	
Spotlights, Statuary (Residence)	25-300			
Sterilamps, per ft.	10		10	
Vapor, Mercury, High Intensity			250-3000	
ELECTRICALLY HEATED EQUIPMENT				
Blankets	50-100			
Broilers	1000-1500			
Casseroles	100-1000			
Cookers, Food	1200			
Dishes, Chafing	160-660			
Driers, Clothes	1500-5000	¼	1-10 Kw.	½
Driers, Hair	200-550		300-1200	
Friers, Deep Fat	1300		1-2 Kw.	
Frying Pans	1200			
Heaters, Air	.4-9 Kw.		.4-9 Kw.	
Heaters, Immersion Type	150-1000		200-2500	
Heaters, Organ Chamber	1-3 Kw.		2-10 Kw.	
Heaters, Permanent Wave Mach			2-4 Kw.	
Heaters, panel per sq. ft.	17-50		17-50	
Heaters, Soil per 60-ft. & 120-ft. Lengths	400-800			
Heaters, Space Elements	1-3 Kw.		1-3 Kw.	
Heat Pumps		1-5		1 up
Heaters, Tank Type Water	1-5 Kw.		1-5 Kw.	
Ironers, Clothes	1200-3300	1/20-¼		

DESIGN REFERENCE DATA

Appliance, Device or Machine	Domestic		Commercial — Industrial	
	Watts From-To	Horsepower From-To	Watts From-To	Horsepower From-To
Irons, Flat	500-1200		500-2500	
Irons, Soldering	60-500		200-400	
Irons, Waffle	300-1320		1-3 Kw.	
Lamps, Health and Sun	30-1500		250-1500	
Machines, Vending			100-1000	
Machines, Popcorn			3-6 Kw.	
Makers, Coffee	450-1200		1-6 Kw.	
Ovens, Baking & Roasting	660-4000		5-15 Kw.	
Ovens, Bread & Pie			12-55 Kw.	
Ovens, Industrial Annealing			5-30 Kw.	
Ovens, Industrial Enamelling			10-100 Kw.	
Plates, Hot, Grills, Griddle Tops	480-2000		1-6 Kw.	
Pots, Glue			100-1500	
Ranges	5-17 Kw.		8-25 Kw.	
Roaster	1-2 Kw.			
Sterilizers, Dental & Doctor			1000-3000	
Toasters, Bread & Sandwich	420-1400		2-5.5 Kw.	
Waffle Iron	1000-1500		2-6 Kw.	
Warmers, Bottle	300-600			
Warmers, Cafeteria Food			4-6 Kw.	
Warmers, Plate	110-500		1-2 Kw.	
Warmers, Soup & Seafood			450-1000	
MOTOR-OPERATED EQUIPMENT				
Air Conditioning Systems		1-5		1 up
Automatic Heating Equipment		$1/8$-$1/2$		$1/4$-10
Blowers, Organ		1-3		2-$7 1/2$
Blowers, Pneumatic Tube Systems				10-30
Blowers, Portable Cleaning				$1/6$-3
Cash Registers				$1/20$-$1/8$
Cleaners, Vacuum Built-in		$1/2$-5		1-$7 1/2$
Cleaners, Vacuum Portable		$1/30$-$3/4$		
Clippers, Hedge		$1/30$-$1/4$		
Compressors, Air (Gasoline Station)				1-5
Compressors, Air (Temp. Regul. System)		$1/4$-2		$1/2$-3
Compressors, Refrigeration				5-5000
Conditioners, Air (Room Type)		$1/3$-$1 1/2$		$1/3$-5
Coolers, Water		$1/8$-$1/3$		$1/8$-$1/3$
Cranes, Travelling Lift				5-30
Cranes, Travelling Bridge				1
Dehumidifiers				$1/8$-$1/3$
Dental Chair Units			500-1500	
Disposal Units (Garbage)		$1/3$-		
Door Openers, Private Garage		$1/8$-1		
Door Openers, Commercial				$1/4$-5
Drills, Portable $1/8$ to $1/2$ in.		$1/8$-$1/4$		$1/6$-$1/4$
Drills, Portable $5/8$ & Larger				$1/4$-1
Dumbwaiters				$1/2$-5
Elevators, 1-Ton Freight				3-10
Elevators, 5-Ton Freight				$7 1/2$-20
Elevators, 10 Pass.				$7 1/2$-20
Elevators, 25 Pass.				10-50

CHAPTER NINE

Appliance, Device or Machine	Domestic		Commercial — Industrial	
	Watts	Horsepower	Watts	Horsepower
	From-To	From-To	From-To	From-To
Escalators	....	....	....	10-40
Extractors Steam Laundry	....	....	....	5-20
Fans, Bracket & Desk	30-100	....	30-100	....
Fans, Ceiling	80-125	....	80-125	....
Fans, Pedestal	125-300	....	125-300	....
Fans, Ventilating 10-in.	35-45	....	35-45	....
Fans, Ventilating 12-24 in.	....	1/40-1/4	....	1/40-1/4
Fans, Ventilating 30-in. & up	....	....	....	3/8-3
Fans, Attic	....	1/4-1/2	....	....
Freezers, Food	....	1/8-1/2	....	1/8 up
Freezers, Ice Cream	....	1/20-1/4	....	1/8 up
Grinders, Coffee	1/8-1/4	....	....	1/6-3/4
Grinders, Meat	....	1/20-1/4	....	1/4-1
Hoists, Ash & Cinder	....	....	....	1-5
Hoists, Tramrail 1-ton	....	....	....	1 1/2-3
Hoists, Tramrail 5-ton	....	....	....	6-10
Hoists, Warehouse Loading	....	....	....	1-3
Lathes, Home Shop	....	1/8-1	....	....
Machines (Floor) Sanding	....	1/8-1	....	1-3
Machines (Floor) Terrazzo	....	1/8-1	....	1-5
Machines (Floor) Waxing	....	1/8-1	....	1/4-1
Machines Sewing	....	1/50-1/20	....	1/4-2
Machines Office, Adding	....	....	....	1/20-1/10
Machines Office, Addressing	....	....	....	1/8-1/3
Machines Office, Billing	....	....	....	1/10-1/2
Machines Office, Bookkeeping	....	....	....	1/10-1/2
Machines Office, Computing	....	....	1-5 Kw.	....
Machines Office, Dictation	....	....	....	-1/30
Machines Office, Record Shaving	....	....	....	-1/20
Machines Office, Sealing & Stamping .	....	....	....	1/10-1/6
Machines Office, Typewriters	....	....	....	1/30-1/10
Mangles, Laundry	....	....	....	7 1/2-20
Mixers, Beverage	30-100	....	30-100	....
Mixers, Dough	....	1/10-1/4	....	5-20
Mixers, Food	....	1/10-1/4	....	1/6-2
Mowers, Lawn	....	1/3-1/2	....	1/3-1
Pumps, Boiler Feed	....	....	....	1-5
Pumps, Brine	....	....	....	2-20
Pumps, Drinking Water Circ.	....	....	....	1/2-5
Pumps, Fire Protection	....	....	....	20-150
Pumps, Fuel	....	1/8-1	....	1/4-3
Pumps, Household Water	....	1/6-1	....	....
Pumps, Milking Machines	....	....	....	1/2-2
Pumps, Pool & Illum. Fountain	....	....	....	1/4-5
Pumps, Roof Storage Tank	....	....	....	7 1/2-25
Pumps, Sump	....	1/8-1/2	....	1/4-3
Pumps, Vacuum	....	....	....	2-5

DESIGN REFERENCE DATA

Appliance, Device or Machine	Domestic		Commercial — Industrial	
	Watts From-To	Horsepower From-To	Watts From-To	Horsepower From-To
Refrigerators		1/8 - 1/3		1/2 - 2
Saws, Band (Home Work Shop)		1/4 - 1/2		
Sprayers, Paint & Insecticide		1/20 - 3/4		
Stage, Curtain Control Motor				1/4 - 1
Stage, Orchestra Lift				15-25
Stage, Organ Lift				3-7 1/2
Stokers, Coal		1/6 - 1/3		1/2 - 5
Tumblers, Laundry Drying				3-10
Washers, Clothes	300-1800	1/8 - 1/2		
Washers, Dish	500-1800	1/8 - 1/2		
Washers, Steam Laundry				1/2 - 5
				3-15
MAGNETS, RECTIFIERS, TRANSFORMERS				
Chargers, Battery	600-750		1-15 Kw.	
Diathermy, Therapeutic			250-750	
Electroplating			5-20 Kw.	
Furnaces, Induction			5-500 Kw.	
Magnets, Lifting Metal			4-15 Kw.	
Magnets, Metal Extracting			1 1/2 - 5 Kw.	
School Laboratory Panel			5-50 Kw.	
Transformers, Bell Ringing	25-50			
Transformers, Signal Systems, Relay	25-50		50-500	
Valves, Gas & Liquids, 1-in. & Less			200-1000	
Valves, Above 1 in.			500-3000	
Welders, Light Duty Spot & Arc			3-20 Kw.	
Welders, Heavy Duty & Arc			20-100 Kw.	
X-Ray — Dental & Doctor			2-25 Kw.	
X-Ray Hospital			10-40 Kw.	
COMMUNICATIONS AND SIGNALLING EQUIPMENT				
Airport Communications			500-5000	
Alarms, Burglar	10-60		100-1000	
Alarms, Fire	10-60		100-1000	
Amplifiers, Radio Distribution			200-1000	
Annunciators, Home 5/8- to 2 1/2-in. Lamps, Each	1.8-2.4			
Annunciators, Large Systems—(110-Volt Lamps, Each)	-10		-10	
Bells, 2 1/2 in. to 4 in.	5-10		5-10	
Bells, Larger			10-30	
Buzzers	5-6		5-6	
Chimes, Door Single and Multiple-Tone	15-25		15-25	
Chimes, Church Systems				1/2 - 2
Clocks, Master Impulse	1-2		1-2	
Clocks, Secondary Type	.5-1		.5-1	
Gongs, Horns, Howlers	10-30		10-30	
High Fidelity Systems	75-500		100-750	
Radio, Amateur Transmitting	100-2000			
Radio, Home Receivers	50-500			
Sirens, Small & Heavy-duty			65-250	1-7 1/2
Television receivers	300-1000		300-1000	
Whistles, Motor Compressor				1/20 - 1

CHAPTER NINE

MOTOR SELECTION CHART

Basic Motor Types	HP Ratings	Motor Operating Classifications	Application Data
Direct-current motors 115, 230 volts d.c.	Fractional to several hundred hp.	Series-wound	For high starting torque, with speed control depending upon load—cranes, hoists, elevators, electric railway cars, locomotives and trucks. In fractional hp. sizes—sewing machines, vacuum cleaners, electric fans, and hair driers (usually the universal type motor).
	Fractional to several thousand hp	Shunt-wound	For constant-speed or adjustable-speed, with good speed regulation—drives for milling machines, centrifugal pumps, lathes, conveyors, grinders, blowers, shapers and elevators.
		Compound-wound	Offers constant-speed and variable-speed action, with a definite zero load speed and high starting torque. Used with fly wheels on punch presses, power shears, hoists, rolling mill drives and conveyors.
Universal motor 115, 230 volts d.c. or 1 phase a.c.	Fractional and several hundred hp.	Series-wound or compensated series-wound	A special adaptation of the direct-current series motor. Applications are same as for fractional hp. d.c. series type. In larger sizes, for electric railways.
Single-phase a.c. motors 110-220-440-550 volts	Fractional hp.	Shaded-pole induction motor	For small appliances and devices requiring low starting torque—small fans, motion picture projectors and similar small constant-speed devices.
	Fractional to 5 hp.	Split-phase start induction motor	For moderate starting torque and constant-speed—small machine tools, oil burner motors and small appliances.
		Reactor-start split-phase induction motor	Same as split-phase start induction motor, but requires less starting current.
	Fractional to 15 hp.	Capacitor-start split-phase induction motor	For high starting torque, constant speed and low noise—air conditioning equipment, large fans and commercial refrigeration equipment.
		Capacitor-start and run induction motor	Similar to capacitor-start split-phase induction motor, but in larger sizes, with increased power factor and smoother, quieter operation.
	Fractional to 40 hp.	Repulsion motor	For varying-speed, high starting torque and low starting current—pumps, stokers, conveyors, compressors and similar applications when polyphase current is not available.
		Compensated repulsion motor	Similar to repulsion motor, but with improved power factor and constant- or varying speed.
		Repulsion-induction motor	Similar to repulsion motor, but with constant- or varying-speed.
		Repulsion-start induction motor	Similar to repulsion motor, with high starting torque and fairly constant speed.
3-phase a.c. motors 208-220-440-550-2200-2300-4000 volts	½ to 400 hp.	Squirrel-cage induction motor	For high reliability and efficiency at essentially constant-speed, requiring little maintenance. Depending upon construction, classifications are as follows: normal-torque, normal starting current; normal-torque, low starting current; high-torque, low starting current; high-slip; low starting torque, normal starting current; low starting torque, low starting current. For rotary compressors, machine tools, large fans, light conveyors, milling machines, agitators, elevators, hoists, punch presses, centrifugal pumps and blowers.
	½ to several thousand hp.	Wound-rotor (slip-ring) induction motor	For limited speed control and speed adjustments under fluctuating load, with low starting current—conveyors, fans, lift bridges, cranes, hoists and drives for metal-rolling mills.
	20 to several thousand hp.	Synchronous	For power factor correction and for exact slow-speed drives and maximum efficiency on continuous loads above 75 hp.

DESIGN REFERENCE DATA

Occupancy	Decibel Range
Apartments and hotels	35–45
Average factory	70–75
Classrooms and lecture rooms	35–40
Hospitals, auditoriums, churches	35–40
Private offices, conference rooms	40–45
Offices—small	53
—medium (3 to 10 desks)	58
—large	64
—factory	61
Stores—average	45–55
—large (5 or more clerks)	61
Residence—without radio	53
—with radio, conversation	60
Radio, recording, television	25–30
Theaters, music rooms	30–35
Street—average	80

Note: Manufacturers now sound rate dry-type transformers to meet or exceed NEMA Audible Sound Level standards. Select a transformer with a decibel rating lower than the ambient sound level of the area in which it is to be installed.

NEMA AUDIBLE SOUND LEVELS
For dry-type general-purpose specialty transformers
600-volt or less, single or 3-phase

TRANSFORMER RATING (KVA)	AVERAGE SOUND LEVEL (DECIBELS)
0-9	40
10-50	45
51-150	50
151-300	55
301-500	60

NEMA AUDIBLE SOUND LEVELS
Oil-immersed and dry-type self-cooled transformers
15,000-volt insulation class and below

KVA	OIL IMMERSED (DECIBELS)	DRY TYPE (DECIBELS)	
		Ventilated	Sealed
0-300	55 DB	58 DB	57 DB
301-500	56 DB	60 DB	59 DB
501-700	57 DB	62 DB	61 DB
701-1000	58 DB	64 DB	63 DB
1001-1500	60 DB	65 DB	64 DB
1501-2000	61 DB	66 DB	65 DB
2001-3000	63 DB	68 DB	66 DB

Calculation Of Short-Circuit Currents—Point-To-Point Method.

Adequate interrupting capacity and protection of electrical components are two essential aspects required by the —— National Electrical Code in Sections 110-9, 110-10, 230-98, and 240-1. The first step to assure that system protective devices have the proper interrupting rating and provide component protection is to determine the available short-circuit currents. The application of the point-to-point method permits the determination of available short-circuit currents with a reasonable degree of accuracy at various points for either 3φ or 1φ electrical distribution systems. This method assumes unlimited primary short-circuit current (infinite bus).

Basic Short-Circuit Calculation Procedure.

Procedure		Formulae	
Step 1	Determine transf. full-load amperes from either: a) Name plate b) Table SC-2 c) Formula	3φ transf.	$I_{FLA} = \dfrac{KVA \times 1000}{E_{L-L} \times 1.73}$
		1φ transf.	$I_{FLA} = \dfrac{KVA \times 1000}{E_{L-L}}$
Step 2	Find transf. multiplier.	—	Multiplier $= \dfrac{100}{\text{Transf.\%Z}}$
Step 3	Determine transf. let-thru short-circuit current (Table SC-4 or formula).	—	†I_{SCA} = Transf.$_{FLA}$ × multiplier
Step 4	Calculate "f" factor.	3φ faults	$f = \dfrac{1.73 \times L \times I}{C \times E_{L-L}}$
		1φ line-to-line (L-L) faults on 1φ, center-tapped transformers	$f = \dfrac{2 \times L \times I}{C \times E_{L-L}}$
		1φ line-to-neutral (L-N) faults on 1φ, center-tapped transformers	$f = \dfrac{2 \times L \times I^*}{C \times E_{L-N}}$
		L = length (feet) of circuit to the fault. C = constant from Table SC-1. For parallel runs, multiply C values by the number of conductors per phase. I = available short-circuit current in amperes at beginning of circuit.	
Step 5	Calculate "M" (multiplier) or take from Table SC-3.	$M = \dfrac{1}{1 + f}$	
Step 6	Compute the available short-circuit current (symmetrical) at the fault.	I_{SCA} at fault $= I_{SCA}$ at beginning of crk. $\times M$	

†**Note 1.** Motor short-circuit contribution, if significant, may be added to the transformer secondary short-circuit current value as determined in Step 3. Proceed with this adjusted figure through Steps 4, 5, and 6. A practical estimate of motor short-circuit contribution is to multiply the total load current in amperes by 4.

Reprinted, with permission, from the "Electrical Protection Handbook" of the Bussman Division, McGraw-Edison Company, St. Louis, Missouri.

DESIGN REFERENCE DATA

Example Of Short-Circuit Calculation.

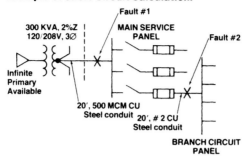

FAULT #1

Step 1 $I_{FLA} = \dfrac{KVA \times 1000}{E_{L-L} \times 1.73} = \dfrac{300 \times 1000}{208 \times 1.73} = 834A$

Step 2 $\text{Multiplier} = \dfrac{100}{\text{Trans. \%Z}} = \dfrac{100}{2} = 50$

Step 3 $I_{SCA} = 834 \times 50 = 41,700A$
At Transformer Secondary

Step 4 $f = \dfrac{1.73 \times L \times I}{C \times E_{L-L}} = \dfrac{1.73 \times 20 \times 41,700}{18,100 \times 208} = .383$

Step 5 $M = \dfrac{1}{1+f} = \dfrac{1}{1 + 0.383} = .723$ (See Table SC-3)

Step 6 $I_{SCA} = 41,700 \times .723 = 30,150A$
Fault #1

FAULT #2

Step 4 Use I_{SCA} @ Fault #1 to calculate

$f = \dfrac{1.73 \times 20 \times 30,150}{4760 \times 208} = 1.05$

Step 5 $M = \dfrac{1}{1+f} = \dfrac{1}{1 + 1.05} = 0.49$ (See Table SC-3)

Step 6 $I_{SCA} = 30,150 \times 0.49 = 14,770A$
Fault #2

Note: For simplicity, the motor contribution was not included.

*Note 2. The L-N fault current is higher than the L-L fault current at the secondary terminals of a single-phase center-tapped transformer. The short-circuit current available (I) for this case in Step 4 should be adjusted at the transformer terminals as follows:
At L-N center tapped transformer terminals
I = 1.5 × L-L Short-Circuit Amperes at Transformer Terminals
At some distance from the terminals, depending upon wire size, the L-N fault current is lower than the L-L fault current. The 1.5 multiplier is an approximation and will theoretically vary from 1.33 to 1.67. These figures are based on change in turns ratio between primary and secondary, infinite source available, zero feet from terminals of transformer, and 1.2 × %X and 1.5 × %R for L-N vs. L-L resistance and reactance values. Begin L-N calculations at transformer secondary terminals, then proceed point-to-point.

Calculation Of Short-Circuit Currents At Second Transformer In System.

Use the following procedure to calculate the level of fault current at the secondary of a second, downstream transformer in a system when the level of fault current at the transformer primary is known.

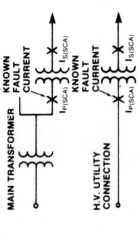

Procedure For Second Transformer In System.

Procedure	Formulae	
Step 1 Calculate "f" ($I_{P(SCA)}$, known).	3φ transformer ($I_{P(SCA)}$ and $I_{S(SCA)}$ are 3φ fault values.)	$f = \dfrac{I_{P(SCA)} \times V_P \times 1.73 \, (\%Z)}{100,000 \times KVA_{TRANS.}}$
	1φ transformer ($I_{P(SCA)}$ and $I_{S(SCA)}$ are 1φ fault values; $I_{S(SCA)}$ is L-L.)	$f = \dfrac{I_{P(SCA)} \times V_P \times (\%Z)}{100,000 \times KVA_{TRANS.}}$
Step 2 Calculate "M" (multiplier) or take from Table SC-3.	—	$M = \dfrac{1}{1+f}$
Step 3 Calculate short-circuit current at secondary of transformer. (See Note 1 under "Basic Procedure")	—	$I_{S(SCA)} = \dfrac{V_P}{V_S} \times M \times I_{P(SCA)}$

$††I_{P(SCA)}$ = Available fault current at transformer primary
$I_{S(SCA)}$ = Available fault current at transformer secondary
V_p = Primary voltage L-L.
V_s = Secondary voltage L-L.

KVA_{TRANS} = KVA rating of transformer.
%Z = Percent empedance of transformer.
Note — To calculate fault level at the end of a conductor run, follow Steps 4, 5, and 6 of Basic Procedure.

Table SC-1A. "C" Values For Conductors and *Busway.

AWG Or MCM	Copper Three Single Conductors				Copper Three Conductor Cable				Aluminum-Three Single Conductors Or Three Conductor Cables			
	Magnetic Duct		Nonmagnetic Duct		Magnetic Duct		Nonmagnetic Duct		Magnetic		Nonmagnetic Duct	
	600V And 5KV Nonshielded	5KV Shielded And 15KV	600V And 5KV Nonshielded	5KV Shielded And 15KV	600V And 5KV Nonshielded	600V And 5KV Nonshielded	600V And 5KV Nonshielded	600V And 5KV Nonshielded	600V And 5KV Nonshielded	600V And 5KV Nonshielded	600V And 5 Nonshielded	
12	617	—	—	—	—	—	—	—	—	—	—	
10	982	—	—	—	—	—	—	—	—	—	—	
8	1230	1230	1230	1230	1230	1230	1230	1230	—	—	—	
6	1940	1940	1950	1940	1950	1950	1950	1950	1180	1180		
4	3060	3040	3080	3070	3080	3080	3090	3090	1870	1870		
3	3860	3830	3880	3870	3880	3880	3900	3900	2360	2360		
2	4760	4670	4830	4780	4830	4830	4850	4850	2960	2970		
1	5880	5750	6020	5920	6020	6020	6100	6100	3720	3750		
1/0	7190	6990	7460	7250	7410	7410	7580	7580	4670	4690		
2/0	8700	8260	9090	8770	9090	9090	9350	9350	5800	5880		
3/0	10400	9900	11500	10700	11100	11100	11900	11900	7190	7300		
4/0	12300	10800	13400	12600	13400	13400	14000	14000	8850	9170		
250	13500	12500	14900	14000	14900	14900	15800	15800	10300	10600		
300	14800	13600	16700	15500	16700	16700	17900	17900	11900	12400		
350	16200	14700	18700	17000	18600	18600	20300	20300	13500	14200		
400	16500	15200	19200	17900	19500	19500	21100	21100	14800	15800		
450	17300	15900	20400	18800	20700	20700	22700	22700	—	—		
500	18100	16500	21500	19700	21900	21900	24000	24000	17200	18700		
600	18900	17200	22700	20900	23300	23300	25700	25700	18900	21000		
700	—	—	—	—	—	—	—	—	20500	23100		
750	20200	18300	24700	22500	25600	25600	28200	28200	21500	24300		
1000	—	—	—	—	—	—	—	—	23600	27600		

*Note—See next page for Busway

Table SC-1B. "C" Values For Busway.

Ampacity	Plug-In Busway		Feeder Busway		High Imped. Busway
	Copper	Aluminum	Copper	Aluminum	Copper
225	28700	23000	18700	12000	—
400	38900	34700	23900	21300	—
600	41000	38300	36500	31300	—
800	46100	57500	49300	44100	—
1000	69400	89300	62900	56200	15600
1200	94300	97100	76900	69900	16100
1350	119000	104200	90100	84000	17500
1600	129900	120500	101000	90900	19200
2000	142900	135100	134200	125000	20400
2500	143800	156300	180500	166700	21700
3000	144900	175400	204100	188700	23800
4000	—	—	277800	256400	—

Table SC-2A. Three-Phase Transformer—Full-Load Current Rating (In Amperes).

Voltage (Line-To-Line)	Transformer KVA Rating								
	150	167	225	300	500	750	1000	1500	2000
208	417	464	625	834	1388	2080	2776	4164	5552
220	394	439	592	788	1315	1970	2630	3940	5260
240	362	402	542	722	1203	1804	2406	3609	4812
440	197	219	296	394	657	985	1315	1970	2630
460	189	209	284	378	630	945	1260	1890	2520
480	181	201	271	361	601	902	1203	1804	2406
600	144	161	216	289	481	722	962	1444	1924

Table SC-2B. Single-Phase Transformer—Full-Load Current Rating (In Amperes).

Voltage	Transformer KVA Rating									
	25	50	75	100	150	167	200	250	333	500
115/230	109	217	326	435	652	726	870	1087	1448	2174
120/240	104	208	313	416	625	696	833	1042	1388	2083
230/460	54	109	163	217	326	363	435	544	724	1087
240/480	52	104	156	208	313	348	416	521	694	1042

Table SC-3. "M" (Multiplier).*

f	M	f	M
0.01	0.99	1.50	0.40
0.02	0.98	1.75	0.36
0.03	0.97	2.00	0.33
0.04	0.96	2.50	0.29
0.05	0.95	3.00	0.25
0.06	0.94	3.50	0.22
0.07	0.93	4.00	0.20
0.08	0.93	5.00	0.17
0.09	0.92	6.00	0.14
0.10	0.91	7.00	0.13
0.15	0.87	8.00	0.11
0.20	0.83	9.00	0.10
0.25	0.80	10.00	0.09
0.30	0.77	15.00	0.06
0.35	0.74	20.00	0.05
0.40	0.71	30.00	0.03
0.50	0.67	40.00	0.02
0.60	0.63	50.00	0.02
0.70	0.59	60.00	0.02
0.80	0.55	70.00	0.01
0.90	0.53	80.00	0.01
1.00	0.50	90.00	0.01
1.20	0.45	100.00	0.01

* $M = \dfrac{1}{1 + F}$

Table SC-4. Short-Circuit Currents Available from Various Size Transformers.

Voltage And Phase	KVA	Full Load Amps	% Impedance	†Short Circuit Amps
120/240 1 ph.	25	104	1.6	10,300
	37½	156	1.6	15,280
	50	209	1.7	19,050
	75	313	1.6	29,540
	100	417	1.6	38,540
	167	695	1.8	54,900
120/208 3 ph.	150	417	2.0	20,850
	225	625	2.0	31,250
	300	834	2.0	41,700
	500	1388	2.0	69,400
	750	2080	5.0	41,600
	1000	2776	5.0	55,520
	1500	4164	5.0	83,280
	2000	5552	5.0	111,040
	2500	6950	5.0	139,000
277/480 3 ph.	150	181	2.0	9,050
	225	271	2.0	13,550
	300	361	2.0	18,050
	500	601	2.0	30,059
	750	902	5.0	18,040
	1000	1203	5.0	24,060
	1500	1804	5.0	36,060
	2000	2406	5.0	48,120
	2500	3007	5.0	60,140

†Three-phase short-circuit currents based on "infinite" primary. Single-phase short-circuit currents on 100,000 KVA primary.

ESTIMATED SHORT CIRCUIT AT END OF LOW-VOLTAGE FEEDER

Power-system maximum estimated short-circuit currents, as functions of distance along feeder conductors fed from standard 3-phase radial secondary unit substations, can be read directly in rms symmetrical amperes from a series of curves (Figs. 9.1 through 9.30). The one-line diagram shows the typical radial circuit investigated. The conditions on which the curves are based were as follows:

1. The fault was a bolted 3-phase short circuit.
2. The primary 3-phase short-circuit duty was 500 MVA (60 cycles) for all curves. A typical supply-system X/R at the low-voltage bus was used in calculating the curves for each case.
3. Motor contributions through the bus to the point of short circuit were included in the calculations on the basis of 100 percent contribution for the 240-, 480-, and 600-V systems and 50 percent contributions for the 208-V systems.
4. The feeder-conductor impedance values used in the calculations are indicated for various conductor sizes.

These curves can also be used to select feeder conductor sizes and lengths needed to reduce short-circuit duties to desired smaller values. Note that conductors thus selected must be further checked to assure adequate load and short-circuit capabilities and acceptable voltage drop.

Coordinated ratings are based on two protective devices operating in series with all short-circuit current flowing through the upstream device. If any current bypasses the upstream device (such as motor contribution fed in on load side of upstream device), a fully rated system, not a coordinated rated system, should be used.

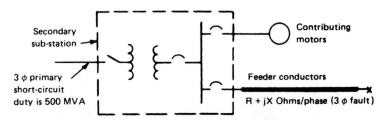

Typical circuit investigated to show effect on short-circuit duty as point of fault is moved away from the low-voltage bus along the feeder conductors.

Feeder Impedance Values Used in Investigation

Feeder Conductor Size/Phase	Resistance (R) Ohms/Phase/1000 Ft	60-cycle Inductive Reactance (X) Ohms/Phase/1000 Ft
#4	0.3114	0.0492
#1/0	.1231	.057
250 MCM	.0534	.0428
2-500 MCM	.0144	.0201
4-750 MCM	.0053	.0099

Reprinted, with permission, from "Short-circuit Current Calculations," publication EESG II-AP-1, of the Apparatus Distribution Sales Division, General Electric Company.

GENERAL DATA FOR CALCULATIONS

DC CIRCUIT CHARACTERISTICS

Ohm's Law:

$$E = IR \qquad I = \frac{E}{R} \qquad R = \frac{E}{I}$$

E = voltage impressed on circuit (volts)
I = current flowing in circuit (amperes)
R = circuit resistance (ohms)

In direct current circuits, electrical power is equal to the product of the voltage and current:

$$P = EI = I^2 R = \frac{E^2}{R}$$

P = power (watts)
E = voltage (volts)
I = current (amperes)
R = resistance (ohms)

AC CIRCUIT CHARACTERISTICS

The instantaneous values of an alternating current or voltage

SINGLE-PHASE CIRCUITS

$$KVA = \frac{EI}{1,000} = \frac{KW}{PF} \qquad KW = KVA \times$$

$$I = \frac{P}{E \times PF} \qquad E = \frac{P}{I \times PF} \qquad PF$$

$P = E \times I \times PF$
P = power (watts)

THREE-PHASE CIRCUITS, BALAN WYE

$$I_N = 0 \qquad I = I_p \qquad E = \sqrt{3} E_p = 1.$$

$$E_p = \frac{E}{\sqrt{3}} = \frac{E}{1.73} = 0.577E$$

I_N = current in neutral (amperes)
I = line current per phase (amperes)
I_p = current in each phase winding (or
E = voltage, phase to phase (volts)
E_p = voltage, phase to neutral (volts)

FIGURE 9.1 Transformer: 150 kVA, 208 V, 2.0 percent Z.

DESIGN REFERENCE DATA

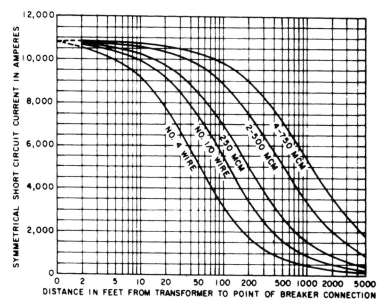

FIGURE 9.2 Transformer: 150 kVA, 208 V, 4.5 percent Z.

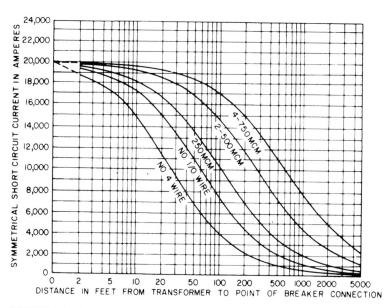

FIGURE 9.3 Transformer: 150 kVA, 240 V, 2.0 percent Z.

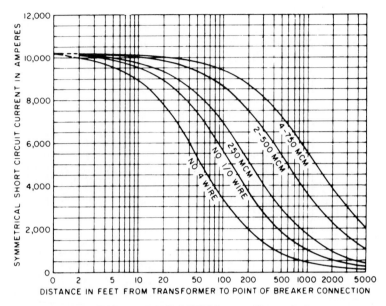

FIGURE 9.4 Transformer: 150 kVA, 240 V, 4.5 percent Z.

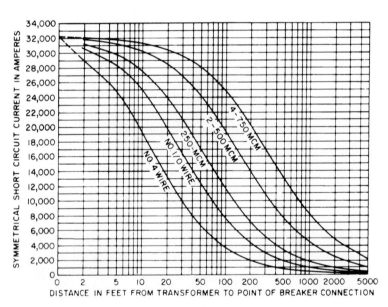

FIGURE 9.5 Transformer: 225 kVA, 208 V, 2.0 percent Z.

DESIGN REFERENCE DATA

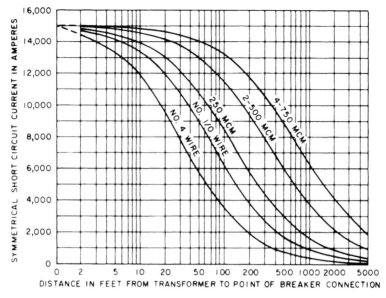

FIGURE 9.6 Transformer: 225 kVA, 208 V, 4.5 percent Z.

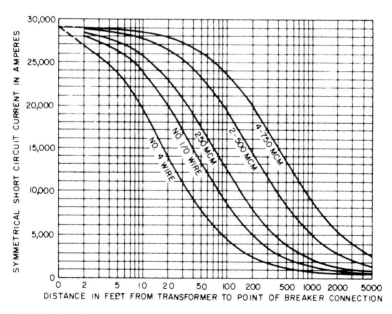

FIGURE 9.7 Transformer: 225 kVA, 240 V, 2.0 percent Z.

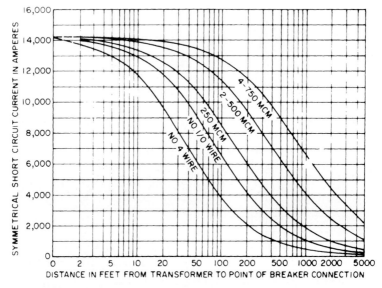

FIGURE 9.8 Transformer: 225 kVA, 240 V, 4.5 percent Z.

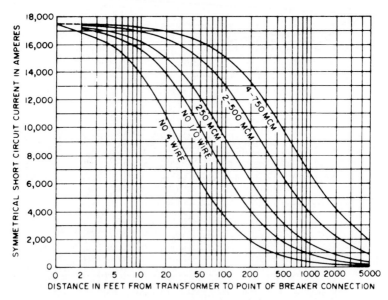

FIGURE 9.9 Transformer: 300 kVA, 208 V, 4.5 percent Z.

DESIGN REFERENCE DATA 657

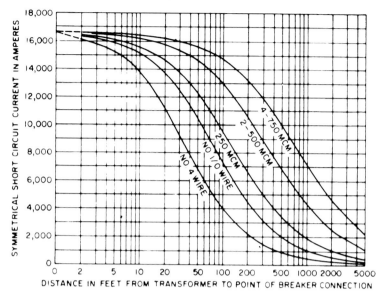

FIGURE 9.10 Transformer: 300 kVA, 240 V, 4.5 percent Z.

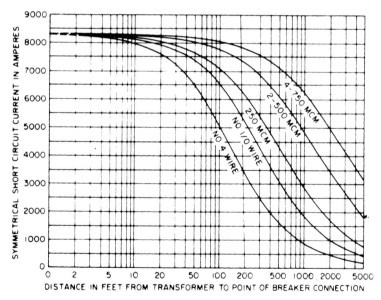

FIGURE 9.11 Transformer: 300 kVA, 480 V, 4.5 percent Z.

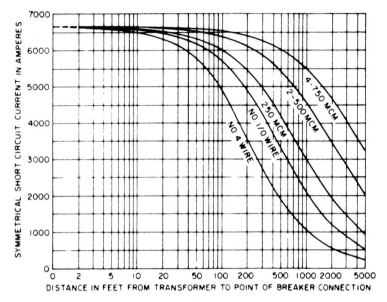

FIGURE 9.12 Transformer: 300 kVA, 600 V, 4.5 percent Z.

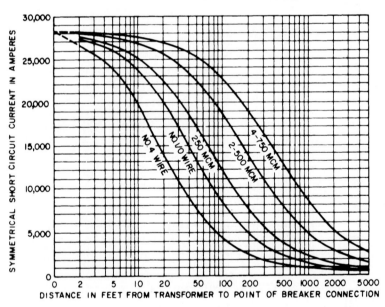

FIGURE 9.13 Transformer: 500 kVA, 208 V, 4.5 percent Z.

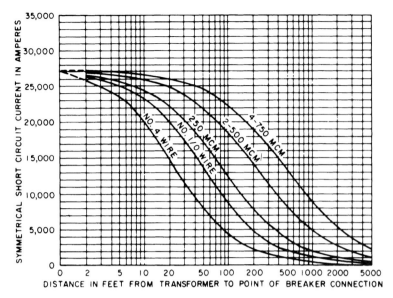

FIGURE 9.14 Transformer: 500 kVA, 240 V, 4.5 percent Z.

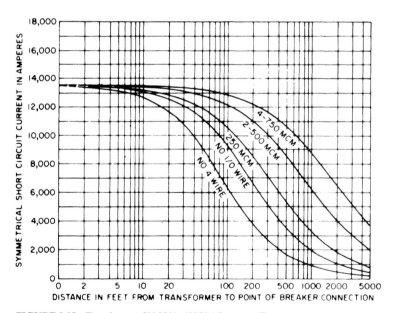

FIGURE 9.15 Transformer: 500 kVA, 480 V, 4.5 percent Z.

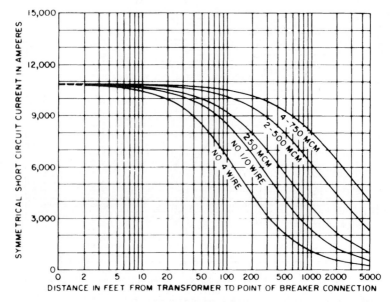

FIGURE 9.16 Transformer: 500 kVA, 600 V, 4.5 percent Z.

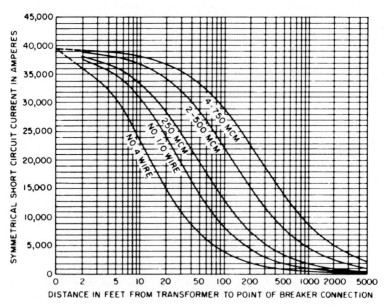

FIGURE 9.17 Transformer: 750 kVA, 208 V, 5.75 percent Z.

DESIGN REFERENCE DATA

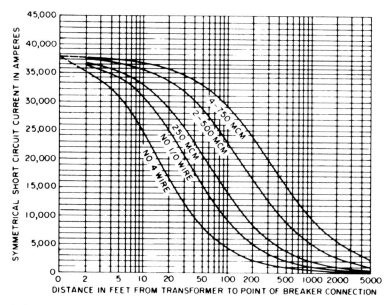

FIGURE 9.18 Transformer: 750 kVA, 240 V, 5.75 percent Z.

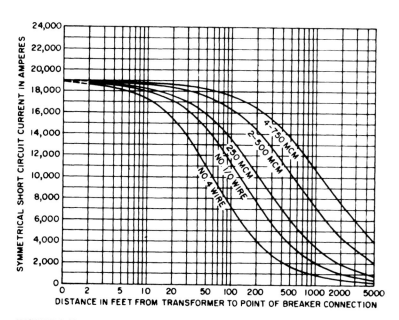

FIGURE 9.19 Transformer: 750 kVA, 480 V, 5.75 percent Z.

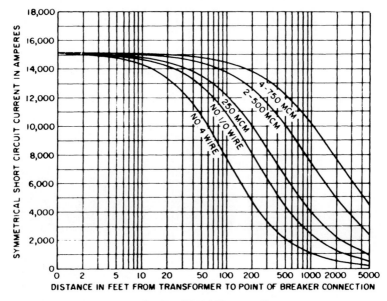

FIGURE 9.20 Transformer: 750 kVA, 600 V, 5.75 percent Z.

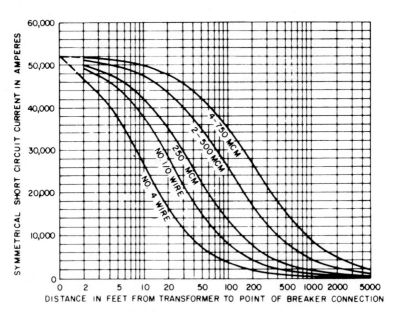

FIGURE 9.21 Transformer: 1000 kVA, 208 V, 5.75 percent Z.

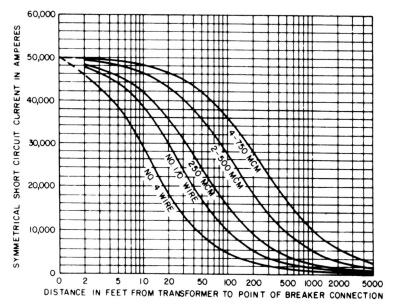

FIGURE 9.22   Transformer: 1000 kVA, 240 V, 5.75 percent Z.

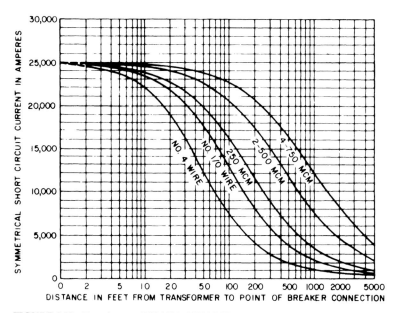

FIGURE 9.23   Transformer: 1000 kVA, 480 V, 5.75 percent Z.

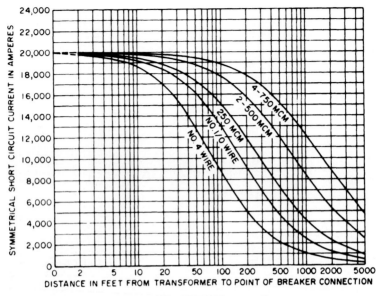

FIGURE 9.24 Transformer: 1000 kVA, 600 V, 5.75 percent Z.

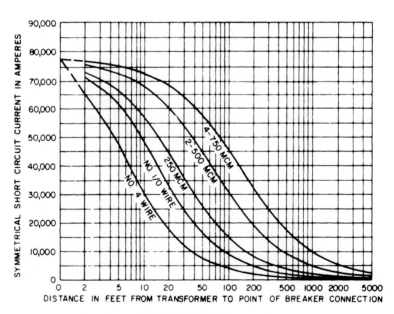

FIGURE 9.25 Transformer: 1500 kVA, 208 V, 5.75 percent Z.

DESIGN REFERENCE DATA **665**

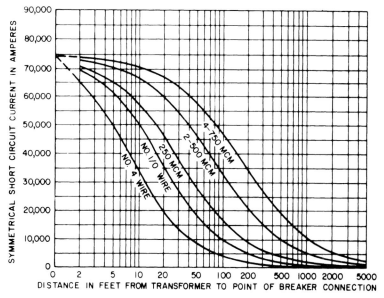

FIGURE 9.26 Transformer: 1500 kVA, 240 V, 5.75 percent Z.

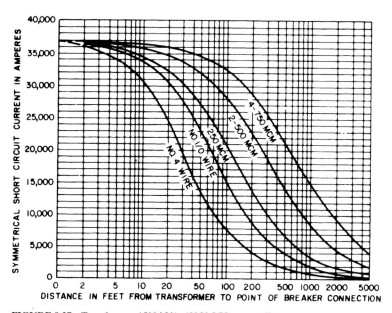

FIGURE 9.27 Transformer: 1500 kVA, 480 V, 5.75 percent Z.

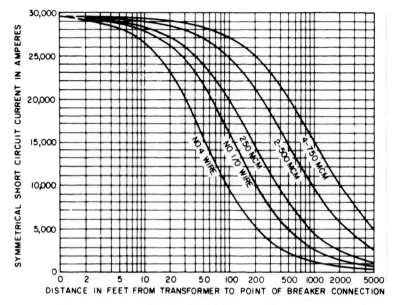

FIGURE 9.28   Transformer: 1500 kVA, 600 V, 5.75 percent Z.

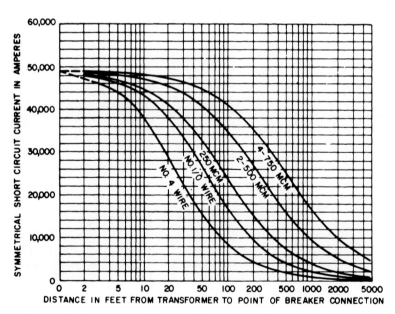

FIGURE 9.29   Transformer: 2000 kVA, 480 V, 5.75 percent Z.

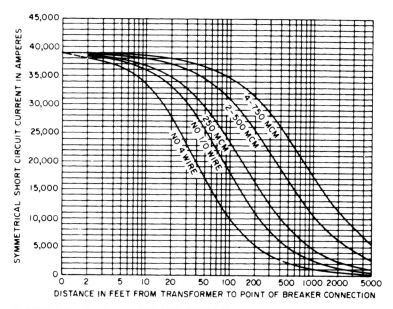

FIGURE 9.30 Transformer: 2000 kVA, 600 V, 5.75 percent Z.

NOMOGRAPHS SHOW IC RATINGS OF FUSES AND CBS

Given the available rms fault current at a service or transformer output, just use a straight edge to determine short-circuit fault current at the end of any given length of cables from No. 8 up to 750-kcmil copper, 277 and 120 V ac, in either steel or aluminum conduit (Figs. 9.31 through 9.67). Curves are also given for determining short-circuit current on the secondary side of 480-208Y/120 V ac transformers.

In this age of computers, many people might wonder about the value of nomographs for determining reduction of 3-phase short-circuit currents in cables and in 480-208Y/120-V ac transformers. Certainly nomographs should never be viewed as an ideal substitute or replacement for basic short-circuit calculations, as the engineering profession has always made them.

But for the very many instances when a complete set of electrical drawings for a given project have to be checked fairly quickly in a limited time frame, nomographs provide an extremely valuable tool for engineers, contractors, electricians, and inspectors. When you're confronted with checking schedules on panelboards, motor controllers and control centers, and switchboard/switchgear, and you find out that short-circuit ratings are not shown (and very often were never calculated), then the nomographs shown in this chapter can help to reduce the panic and provide a reading on the safety characteristics of the overall systems.

Just how good are these nomographs? In making short-circuit calculations, the first step is to construct a one-line diagram of the system. From the one-line diagram, the impedances of sources (service, generators, motors), feeders, transformers, etc., are then developed. From this data an impedance diagram is constructed. Then the impedances are added vectorially, then reduced and combined until there is only one impedance.

When the voltage of the source is divided by the impedance that has been determined—following Ohm's Law—the 3-phase fault current is calculated. Of course, this is all a relatively simple explanation of a procedure that can be rather complex and involved. But the explanation given here is intended to make a point: The cable short-circuit diagrams were developed from an equation reading as follows:

$$\frac{1}{A} = \frac{1}{B} + \frac{1}{C}$$

Although that would represent an arithmetic addition of impedances, when the calculations were made, they were actually made vectorially. As a result, when these nomographs are used, the actual value of short-circuit current indicated on the nomograph will be correct or slightly higher than the real calculated values. Thus, the short-circuit values shown will be conservative and on the safe side.

When plotting the reduction of short-circuit currents through transformers, the values of impedance used were the lowest manufacturers' impedances that could be found. So using these nomographs will again yield results that are always on the safe, conservative side of actual values.

Let's Take an Example

Take a look at the nomograph for 500-kcmil cables at 277 V (Fig. 9.65). Assume this cable terminates at a panelboard, and the fault current rating of the panelboard is to be determined.

Assume a fault current of 150,000 rms amperes is available at the switchboard that feeds the panel. Say the length of the 500-kcmil feeder run is 150 ft. Put one end of your straight edge at 150 on the left vertical scale of available fault current. Put the other end of the straight edge at 150 ft on the right vertical scale of cable length in feet. Then read the number where the straight edge crosses the two center scales which show the available fault current at the end of the 150-ft feeder run (Fig. 9.33).

By interpolation, you will read: 32,000 A for the circuit in aluminum conduit or, 29,000 A for the circuit if it is in steel conduit.

Conclusion: The panelboard and CBs or fuses at the load end of the 150-ft feeder run must have a short-circuit rating of *either* 29,000 or 32,000 A, depending on the material of the conduit or raceway enclosing the circuit.

Transformer Plot

For using the transformer curve shown in Fig. 9.32, the procedure is even simpler.

Take a 75-kVA transformer with a primary available fault current of 80,000 rms amperes. Run a line up vertically from the value of 80,000 on the horizontal coordinate at bottom (primary-side fault current available) to its intersection with the curve for the 75-kVA transformer. Then from that intersection point, run a line horizontally to the left. Note that the horizontal line intersects the left side coordinate (fault current on transformer's secondary terminals) at 7.6 kA, or 7600 A on the 208Y/120-V ac secondary side. Note that above 100,000 A, the curves are nearly flat. Therefore, the 100,000-A figure can be used for all fault currents exceeding 100,000 A on the primary side of the transformer (Fig. 9.34).

As can be seen, these nomographs do not solve all problems. But for low-voltage projects, they can be a great help.

DESIGN REFERENCE DATA 669

The nomographs here provide a handy and easily used technique for everyone's use in assuring compliance with the strict and very important rule of **NEC** Sec. 110-9, which says:

110-9. Interrupting Rating. Equipment intended to break current at fault levels shall have an interrupting rating sufficient for the system voltage and the current which is available at the line terminals of the equipment.

That **Code** rule requires that each and every fuse and circuit breaker must have a short-circuit interrupting rating in symmetrical rms amperes that is at least equal to the fault current the circuit can deliver into a bolted phase-to-phase short on the ungrounded conductors of a circuit. The nomographs indicate the minimum short-circuit duty rating required of a protective device for the available current at the line (not load) terminals of the fuse or CB.

FIGURE 9.31

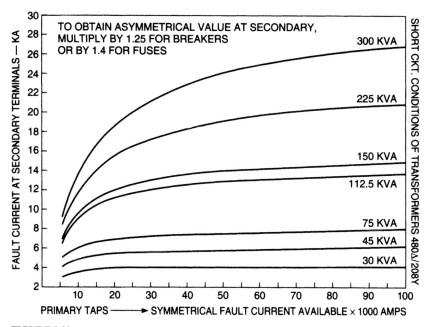

FIGURE 9.32

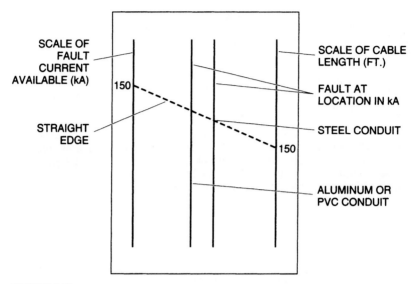

FIGURE 9.33

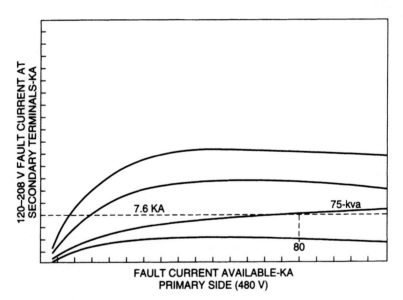

FIGURE 9.34

DESIGN REFERENCE DATA 671

#8 Cu. CABLE – 120 VOLTS

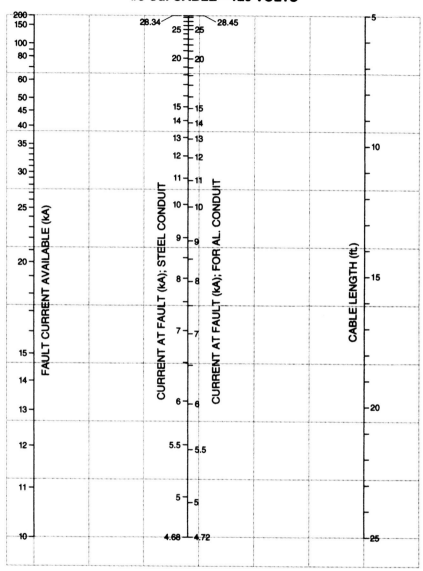

FIGURE 9.35

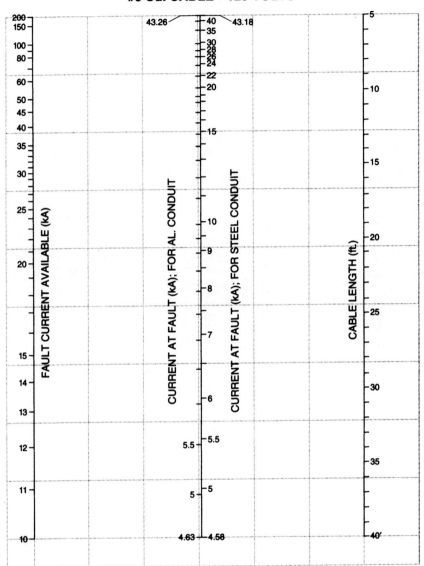

FIGURE 9.36

DESIGN REFERENCE DATA 673

#4 Cu. CABLE – 120 VOLTS

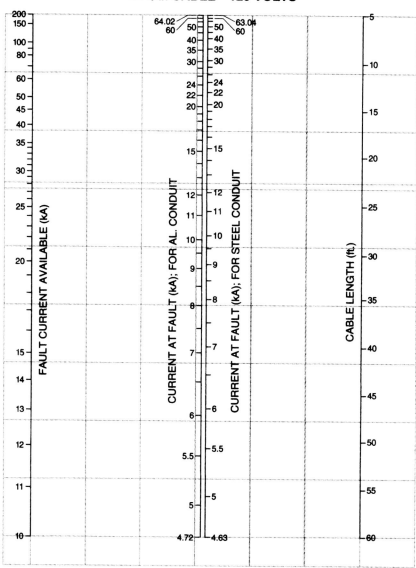

FIGURE 9.37

#2 Cu. CABLE – 120 VOLTS

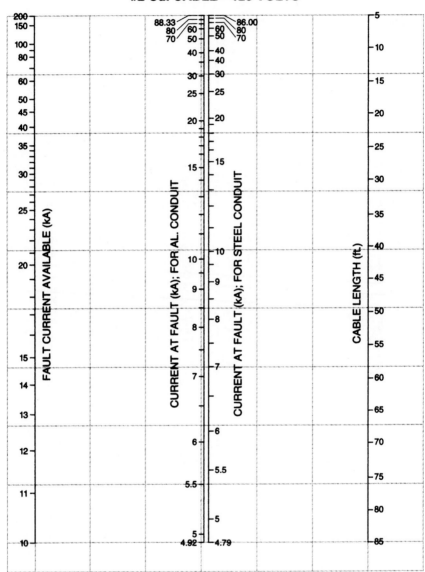

FIGURE 9.38

DESIGN REFERENCE DATA

#1 Cu. CABLE – 120 VOLTS

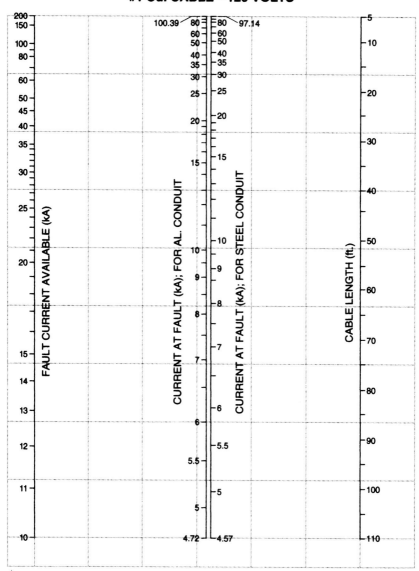

FIGURE 9.39

#1/0 Cu. CABLE – 120 VOLTS

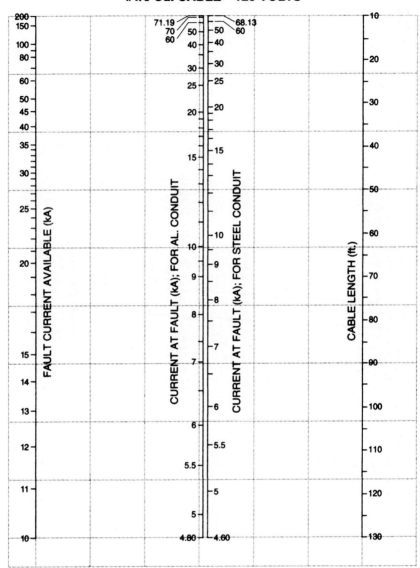

FIGURE 9.40

DESIGN REFERENCE DATA 677

#2/0 Cu. CABLE – 120 VOLTS

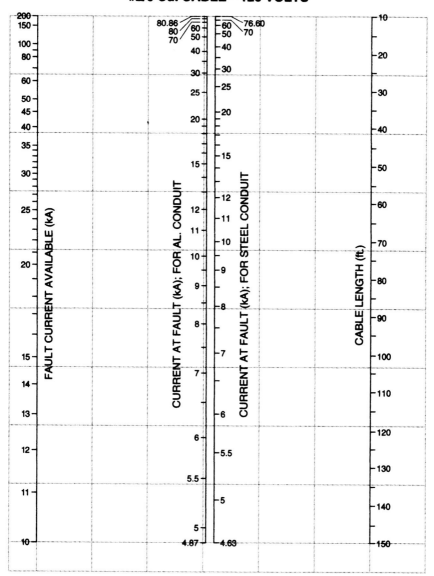

FIGURE 9.41

#3/0 Cu. CABLE – 120 VOLTS

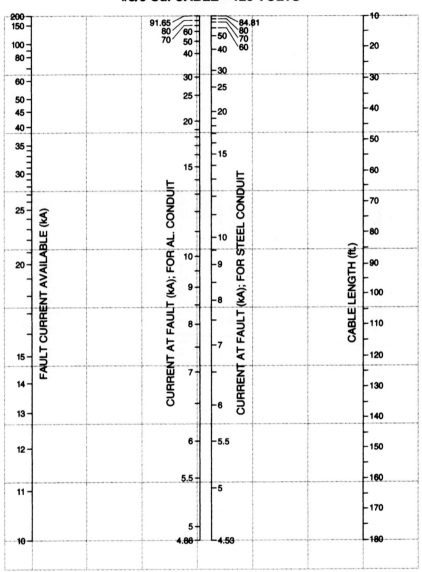

FIGURE 9.42

#4/0 Cu. CABLE – 120 VOLTS

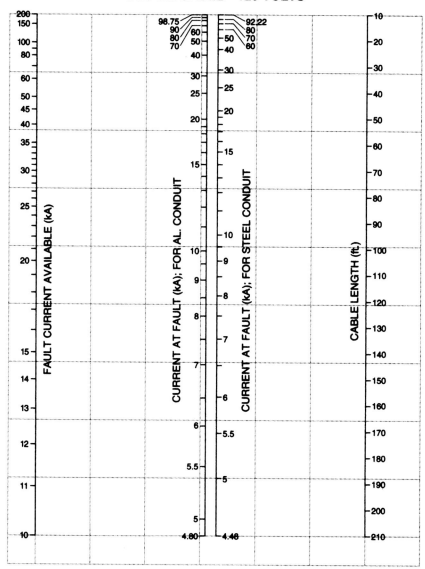

FIGURE 9.43

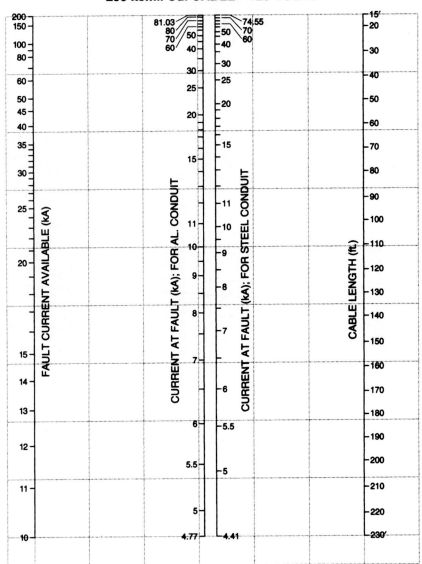

FIGURE 9.44

DESIGN REFERENCE DATA

300 kcmil Cu. CABLE – 120 VOLTS

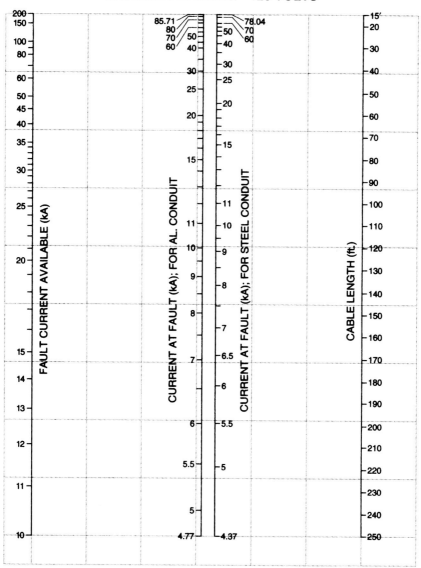

FIGURE 9.45

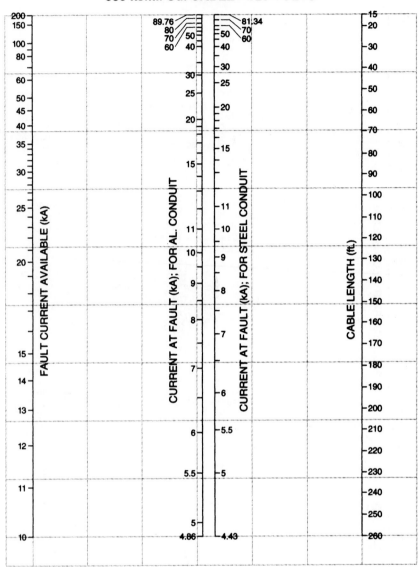

FIGURE 9.46

DESIGN REFERENCE DATA

400 kcmil Cu. CABLE – 120 VOLTS

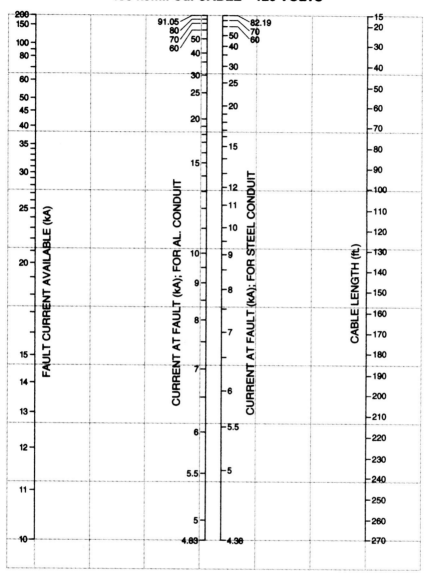

FIGURE 9.47

500 kcmil Cu. CABLE – 120 VOLTS

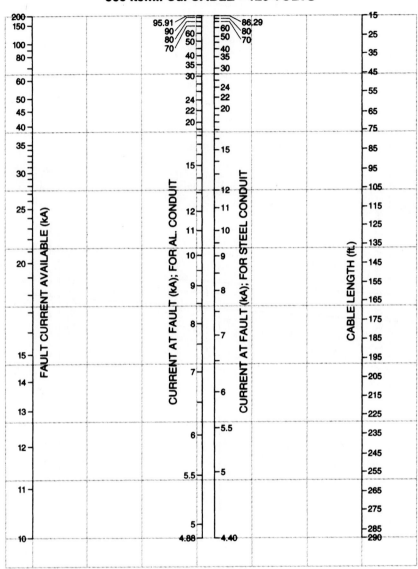

FIGURE 9.48

600 kcmil Cu. CABLE – 120 VOLTS

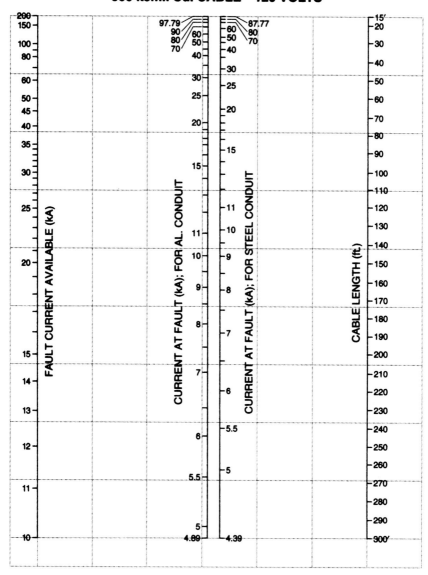

FIGURE 9.49

750 kcmil Cu. CABLE – 120 VOLTS

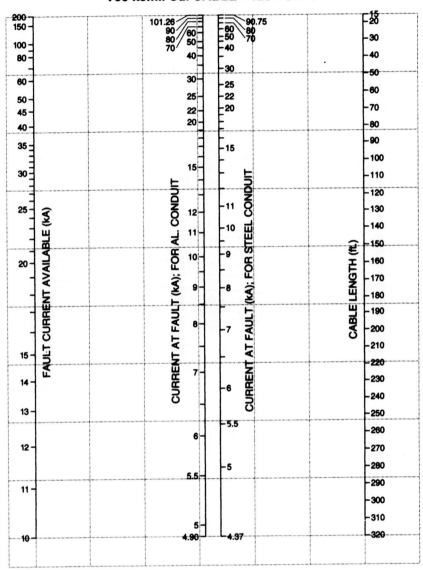

FIGURE 9.50

#8 Cu. CABLE – 277 VOLTS

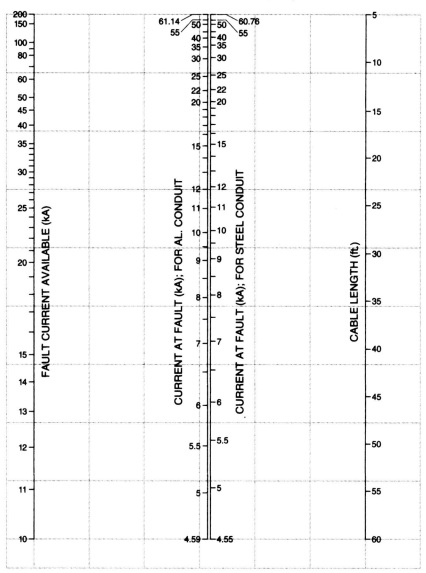

FIGURE 9.51

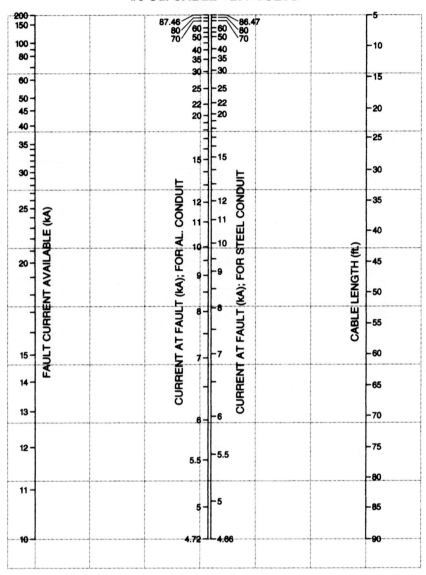

FIGURE 9.52

DESIGN REFERENCE DATA

#3 CU. CABLE – 277 VOLTS

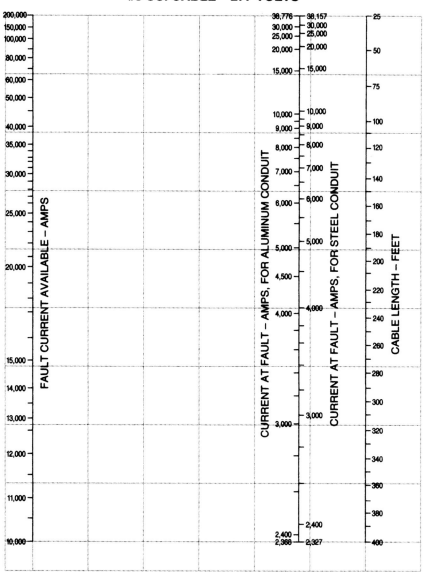

FIGURE 9.53

#4 CU. CABLE – 277 VOLTS

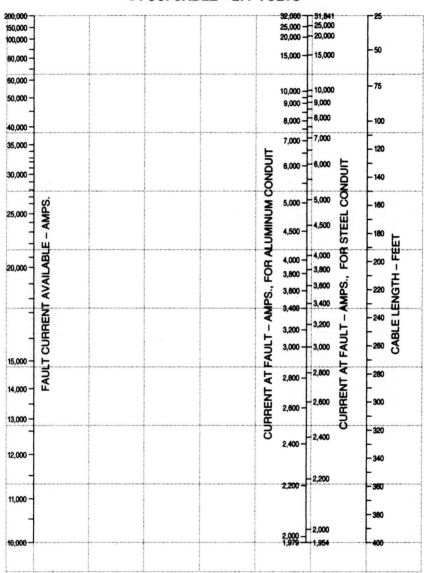

FIGURE 9.54

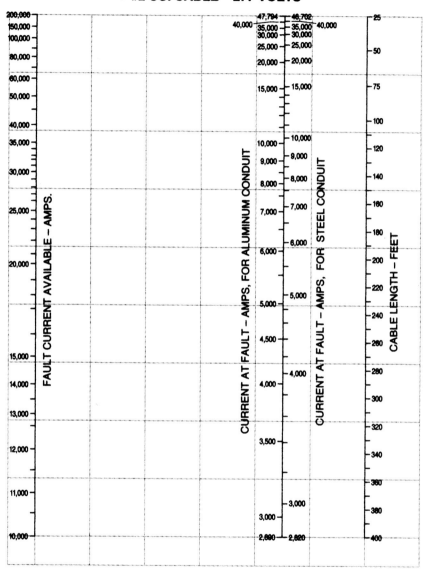

FIGURE 9.55

#1 Cu. CABLE – 277 VOLTS

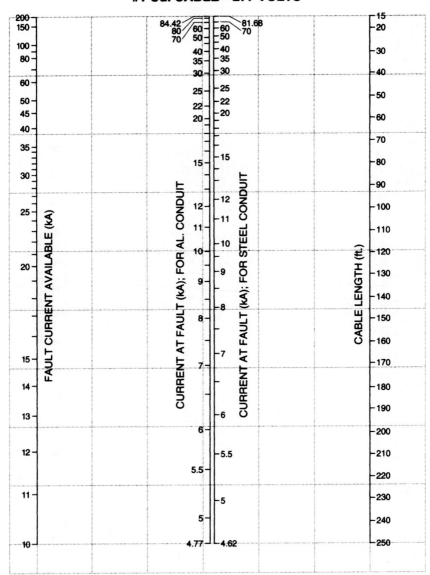

FIGURE 9.56

DESIGN REFERENCE DATA

#1/0 CU. CABLE – 277 VOLTS

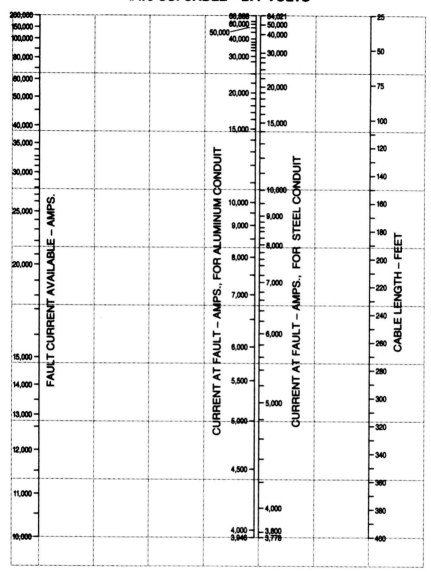

FIGURE 9.57

#2/0 Cu. CABLE – 277 VOLTS

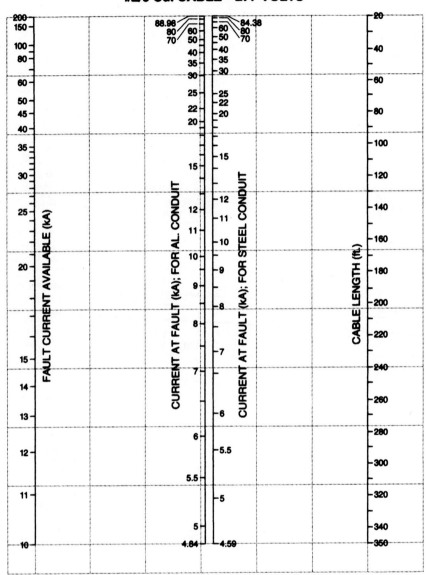

FIGURE 9.58

DESIGN REFERENCE DATA

3/0 COPPER CABLE – 277 V

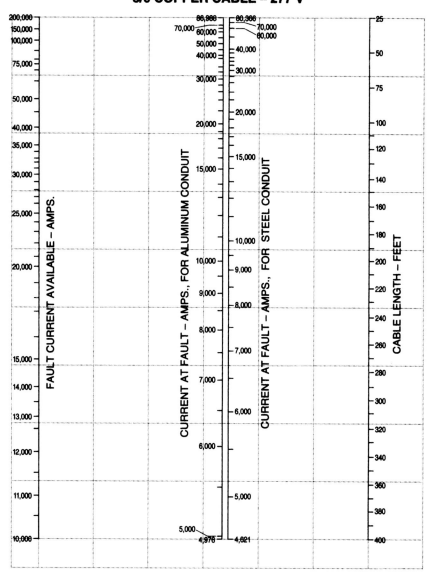

FIGURE 9.59

4/0 COPPER CABLE – 277 V

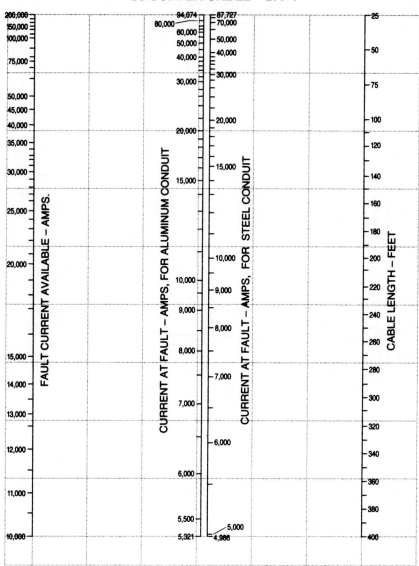

FIGURE 9.60

250 kcmil CU. CABLE – 277 V

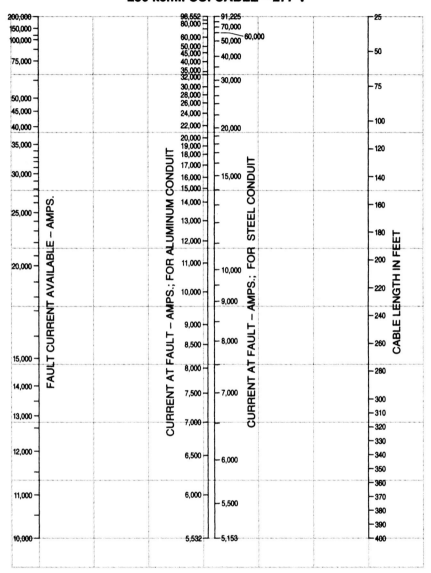

FIGURE 9.61

300 kcmil Cu. CABLE – 277 VOLTS

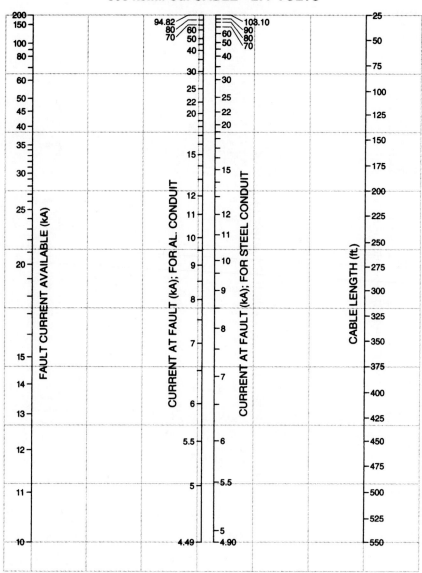

FIGURE 9.62

DESIGN REFERENCE DATA

350 kcmil CU. CABLE – 277 V

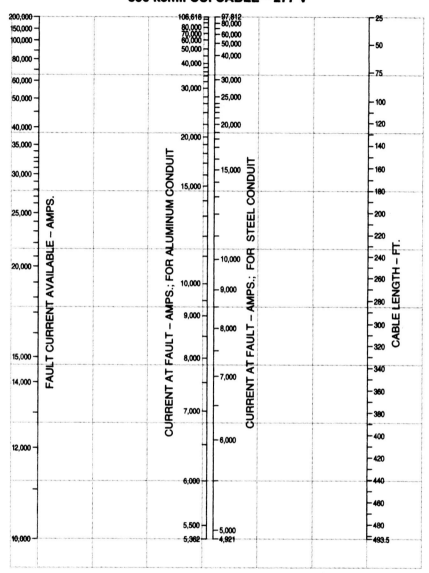

FIGURE 9.63

400 kcmil Cu. CABLE – 277 VOLTS

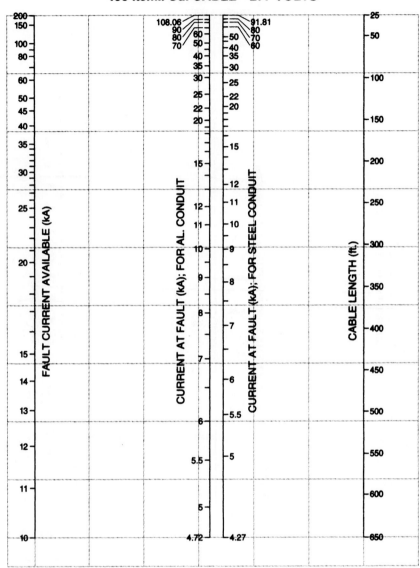

FIGURE 9.64

500 kcmil CU. CABLE – 277 V

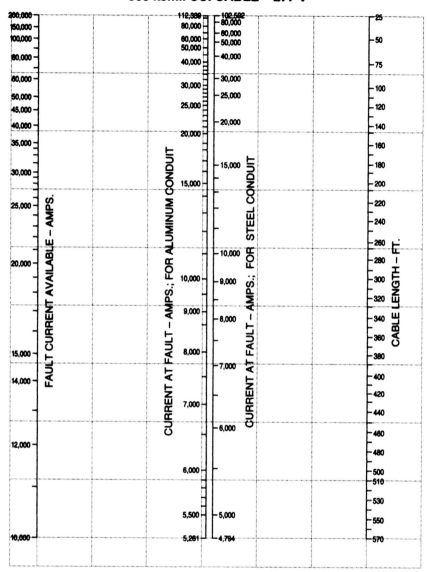

FIGURE 9.65

600 kcmil Cu. CABLE – 277 VOLTS

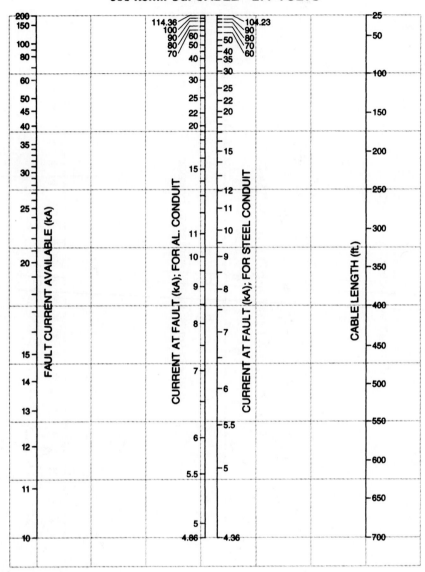

FIGURE 9.66

DESIGN REFERENCE DATA 703

750 kcmil Cu. CABLE – 277 VOLTS

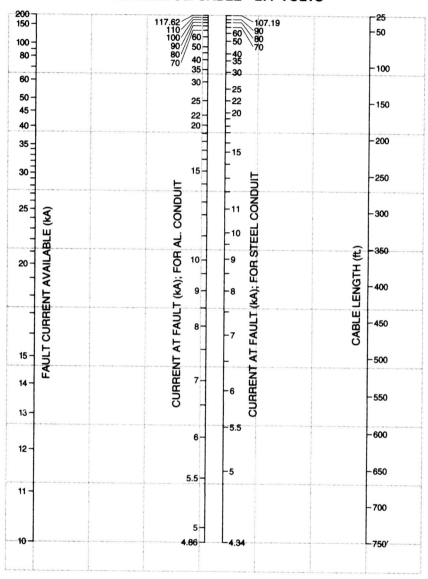

FIGURE 9.67

INDEX

Index note: The *f.* after a page number refers to a figure.

AC circuit characteristics calculations, 636
Acceptable, 4
Accepted, 4
Accidental starting of control circuits, 230–233
  hot-leg switching, 232, 234*f.*
  transformers, 231, 232*f.*
    inside starter, 232, 235*f.*
    outside starter, 232, 236*f.*
  two-pole START button, 231–233
  two-pole STOP button, 232, 233*f.*
  ungrounded system, 230–231
Agricultural buildings, 374, 375*f.*
Air conditioning equipment, 173–174
  branch circuits, 57–59, 190–192
  grounding, 382, 383*f.*
  lighting and power feeders, 270, 271*f.*
  utilization voltages, 25
Air-handling ceilings, 68–69
  wiring methods, 65–68
Airplane hangar, 101, 103*f.*
American National Standards Institute (ANSI):
  ANSI/IEEE publication C57.110 1986, 421, 423–427
  curve shift, 433–443
    current flow, 435, 436*f.*
    examples, 436–443
    recommendations, 436, 437*f.*
  Standard 37.17, 354
  Standard C-841 ["Voltage Ratings for Electric Power Systems and Equipment (60 Hz)"], 24–26
Ampacity:
  control circuits, 212

Ampacity (*Cont.*):
  lighting and appliance branch circuits, 86–93
    circuit makeup and loading impact, 90–91
    conductor insulation, 90–93
    correction factors, 86
    derating, 88–91, 618
    electric-discharge load, 92*f.*, 93
    90-degree C wires, 87–89, 91–93
    resistive load, 93
ANSI (*see* American National Standards Institute)
Apartments (*see* Dwelling units)
Appliance branch circuits (*see* Lighting and appliance branch circuits)
Approval, 4, 6
Approved, 4, 6
Armored cable, 631–634
Artificial internal environments, 308–309
ATS (automatic transfer switches), 601–604
Attic, lighting and appliance branch circuits for, 50–52, 59, 122
Audible sound levels, 645
Automatic protective devices, 2
Automatic transfer switches (ATS), 601–604
Autotransformers, 484–489
  applications, 486–487
  buck transformer, 487
  grounding, 487–489
  hookup, 484, 485*f.*
Auxiliary Protective Signaling Systems, 3

Banks, lighting and appliance branch circuits for, 101

Basement, lighting and appliance branch circuits for, 48–50, 59, 119, 122
Bathroom, lighting and appliance branch circuits for, 38f., 45, 48f., 59, 60f., 111, 114–116, 123
Boathouse, 119
Bonding:
  services, 564–567
    piping, 570–574
    techniques, 565–568, 570f., 571f.
  transformers for control circuits, 228
Bonding jumpers:
  lighting and power feeder grounding, 390–391
  services, 568–570
Boost (buck) transformer, 487
Branch circuits, 15
  accessibility, 16–17
  capacity and current rating, 15
  flexibility, 16–17
  layout, 15
  load sizing, 16–17
  multiwire, 19–20
    neutral conductor, 20
  overcurrent protection, 20–22
  receptacle, 20, 22f.
  types of, 19
  (*See also* Lighting and appliance branch circuits; Motor branch circuits)
Buck (boost) transformer, 487
Building ground, 370, 372–379, 382–386, 542, 553, 555
Busway wiring methods, 624–630

Cable, armored, 631–634
Capacitor kilovars (reactive kilovoltamperes), 395
Capacity for lighting and power feeders, 265–306
  branch-circuit load, 265–267, 271–273
  common neutral versus separate neutral, 287–292
    equal nonlinear load, 288–289
    harmonics, 288
    heating effects, 291–292
    phase and neutral loading, 287, 288f., 290
  conductor size selection, 279–281
    example, 281
    power loss calculations, 281, 282f.
  continuous-duty motor loads, 268–269
  demand factors, 267, 273–274

Capacity for lighting and power feeders (*Cont.*):
  for growth, 274–276
    sizing, 276, 277f.
  multiple conductors, 299, 303–306
    ampacity derating, 304–305
    common neutral, 306, 308f.
    versus single conductor, 306, 307f.
  neutral sizing, 281–286
    electric-discharge lighting reduction, 284, 287f.
    feeder to ballast load, 284, 286f.
    lighting load type, 284, 285f.
    neutral-connected loads, 282, 283f.
    phase-to-phase loads, 284, 286f.
    resistance load reduction, 282–284
  sizing versus panelboard rating, 276–280
    circuit breaker rating, 277–279
  voltage drop minimization, 292–299
    circuit makeup, 294–295, 299–303
    formulas, 293, 294f.
    limitations, 293, 296
    loadcenter distribution, 296, 297f.
    voltage regulators, 296–299
CB (*see* Circuit breaker)
CBEMA (Computer and Business Equipment Manufacturer Association), 421–427
Ceiling, lighting and appliance branch circuit controls for, 121, 122f.
Central Station Protective Signaling Devices, 3
Certified, 4–6
CF (crest factor), 421–427
Circuit breaker (CB):
  air (ac power), 320–322
  automatic transfer switches, 601, 602f.
  encased (insulated-case), 322
  instantaneous-trip, 180–181, 321
  lighting and appliance branch circuits, 77–82, 84–86, 124, 125f., 142
  lighting and power feeders, 278, 312–318, 320–322
    installation location, 335, 338f.
    short-circuit protection, 322–331
    versus fuse, 331–333
  molded-case, 320–322, 330, 331f.
  motor branch circuit controller, 192, 193f.
  nomograph IC ratings, 667–703
    example, 653f., 668, 701f.
    transformer plot, 653f., 654f., 668
  tripping mechanisms, 321–322

Clock motors, 192
Co-op units (*see* Dwelling units)
Color coding:
   conductors, 381
   lighting and appliance branch circuits, 93–97
      alternatives, 97
      grounded conductor, 94–95
      grounding conductor, 94, 95, 97, 98*f*.
      hot conductor, 94–97
      neutrals, 94, 96*f*.
      safety precautions, 97
Commercial buildings, lighting and appliance branch circuits for, 16–17, 19, 31, 33*f*., 82, 123–124, 637
Computer and Business Equipment Manufacturer Association (CBEMA), 421–427
Computer equipment:
   branch circuits, 66, 89, 102, 104*f*., 169, 170*f*.
   lighting and power feeders, 287
Condominium (*see* Dwelling units)
Conductors:
   color coding, 381
   lighting and appliance branch circuits, 63–69
      air-handling ceilings, 68–69
         wiring systems, 65–68
      calculations, 64
      color coding, 93–98
      data processing systems, 66
      environmental air spaces, 65–66
      overcurrent protection, 88
      suspended ceiling, 68
      temperature rating, 83–86, 620
         ampacity, 86
         circuit breaker, 84–86
         distribution and control equipment, 84
         terminals, 84, 86
   lighting and power feeders:
      grounding, 379, 381–382, 384–390
      multiple conductors, 299, 303–306
      size selection, 279–281
   motor branch circuits, 174, 176–178
      current-carrying capacities determination, 177
      example, 177–178
      sizing, 174, 176
      wound-rotor motors, 176, 177*f*.
   motor feeders, 393–394
   transformers, 479–480

Conductors (*Cont.*):
   wiring methods, 605–608
      in raceways, 612–623
      terminations, 623–624
Contactor control, 126–133
   magnetic, 126–127, 129*f*.
   mechanically held, 127–128, 130–133
      coil-circuit wires, 131, 133, 134*f*.
      heating equipment, 131, 133*f*.
      office buildings, 133, 135*f*.
      remote-control switch, 130, 131*f*.
      split-bus panelboard, 130
      widespread lighting, 130–132
Control circuits, 174, 212–245
   accidental starting protection, 230–233
      hot-leg switching, 232, 234*f*.
      transformers, 231, 232*f*.
         inside starter, 232, 235*f*.
         outside starter, 232, 236*f*.
      two-pole START button, 231–233
      two-pole STOP button, 232, 233*f*.
      ungrounded system, 230–231
   ampacity, 212
   Class 1, 214, 215*f*.
   Class 2, 213, 214
   Class 3, 213, 214
   control wires in raceways, 215–218
      high-voltage starters, 218
      multimotors, 216, 217*f*.
      non-associated motors, 216–218
      single motor, 216, 277, 278
   disconnects, 228–230
      remote, 229–230
   elements, 212
   versus load circuits, 212
   overcurrent protection, 212–214, 218–225
      magnetic starters, 219
      remote-control device, 220–222
      secondary wires, 220, 223*f*.
      transformers, 221–225
   plan layout, 239–245
      diagrams, 242, 243*f*., 245
      needs assessment, 240–242
      power panel schedule, 244*f*., 245
   transformers, 225–228
      accidental starting protection, 231, 232*f*.
      bonding, 228
      grounding, 226–228
      overcurrent protection, 221–225
      requirement exceptions, 226–227
   voltage regulation, 234–239
      example, 235–236, 239

Control circuits
  voltage regulation (*Cont.*):
    power factor, 235–237
    power-factor capacitors, 236–239
      kilovar multipliers, 237–240
Control equipment, 2, 84
Controller, 191, 193*f.*
Controls:
  lighting and appliance branch circuits, 119–133
    alternating-current snap switches, 119
    circuit breaker, 124, 125*f.*
    contactor control, 126–133
      coil-circuit wires, 131, 133, 134*f.*
      heating equipment, 131, 133*f.*
      magnetic, 126–127, 129*f.*
      mechanically held, 127–128, 130–133
      office buildings, 133, 135*f.*
      remote-control switch, 130, 131*f.*
      split-bus panelboard, 130
      widespread lighting, 130–132
    in dwelling units, 120–122
    full-circuit control, 123–124
    in hotels and motels, 123
    in industrial buildings, 123
    local control, 119
    low-voltage relay switching, 125–128
    mercury switches, 119, 121*f.*
    panelboards, 146, 149*f.*
    receptacles, 119
    remote and automatic control, 124–125
    screwless terminal switches, 123
    snap switches, 123, 124*f.*
    split-wire duplex receptacle, 119–121
  motor branch circuits, 191–201
    across-the-line controller, 199–201
    circuit breaker, 192, 193*f.*
    controller, 191, 193*f.*
    example, 195
    full-voltage controller, 199–200
    general-use switch, 192, 193*f.*
    hermetic motor, 194
    installed location, 195, 197–199
    layout plans, 197–199
    low-voltage protection, 200–201
    manual starters, 200, 209
    plug-and-receptacle connection, 192, 193*f.*
    protective device, 192, 193*f.*
    reduced-voltage controller, 199
    short-circuit protection, 179–180
    starter poles, 195, 196*f.*

Controls
  motor branch circuits (*Cont.*):
    3-wire control circuit, 201
    2-wire control circuit, 201
    undervoltage protection, 200–201
Crawl space, lighting and appliance branch circuits for, 50–52, 119
Crest factor (CF), 421–427
Current rating of service feeders, 499–508
  residential service entrance conductors, 506–508
  optional calculation, 508–513
  sizing, 499, 500*f.*
    example, 501–506
Custom-made equipment, 5

Damp location, 64
Data-processing equipment, 66, 89, 287, 583–584
DC circuit characteristics calculations, 636
De-icing installations, 36
Demand factor, 273–274
Design planning (*see* Planning)
Direct grade level access, 117
Disconnects:
  control circuits, 228–230
    remote, 229–230
  motor branch circuits, 206–211
    example, 207–209
    in-sight, 210
    out-of-sight, 210–211
    requirements, 206, 207*f.*
    selection based on ampere rating, 207, 208*f.*
  services, 517–531
    disconnect grouping, 521–523
      emergency disconnect, 522, 577, 579*f.*
      identification marking, 521
    disconnect rating, 529–531
      multiple services, 530, 531*f.*
    disconnect types, 529
      manual operation, 529, 530*f.*
    installation location, 523, 525–529
      equipment grouping, 526, 527*f.*, 529
      in multi-occupancy buildings, 523, 526*f.*, 529
      number of sets, 527, 528*f.*
      readily accessible, 523, 525*f.*
    layout arrangements, 517, 519*f.*
    multibuilding premises, 531, 532*f.*, 535
    multiple service entrance grounding, 561–563

Disconnects
  services (*Cont.*):
    number of disconnects, 519, 520
    switchboards, 589, 590*f.*
Distribution equipment, branch circuits for, 84
Distribution system, 247, 585
  circuit configurations and voltages, 247, 249*f.*
  design guidelines, 264–265
  elements, 247, 248*f.*
  feeder, 247, 264
  lighting feeder, 247
  loadcenter, 250
  main, 247, 264
  panelboard, 250
  power feeder, 247
  subfeeder, 250, 264
  switchboard, 250
  voltage levels, 250–263
    480-V, 3-phase, 3-wire system, 253–254
    480Y/277-V (460Y/265-V), 3-phase, 4-wire system, 253–256
    4160Y/2400-V, 5-phase, 4-wire system, 254, 256–259
    4800-V system, 256
    layouts, 250, 251*f.*
    low-voltage service, 258, 260*f.*
    120/208-V, 3-phase, 4-wire system, 250–252
    120/240-V, 3-wire, single-phase system, 250
    7200-V, 3-phase system, 256
    600-V system, 258, 262*f.*
    special load equipment, 258–259, 263*f.*
    13.2Y/7.2-kV (13.8-kV), 3-phase, 4-wire system, 256, 259*f.*
    13,200-V system, 259
    240-V, 3-phase, 3-wire system, 253
    2400-V, 3-phase delta system, 254
    variation from nominal, 259, 263*f.*
Diversity factor, 273–274
Dormitories, lighting and appliance branch circuits for, 33, 59, 105, 117, 118*f.*
Dry location, 64
Duct system, 66
Dwelling units, 3
  lighting and appliance branch circuits, 16, 42–48
    calculations, 42–43
    controls, 120–122
    fixed room dividers, 43–45

Dwelling units
  lighting and appliance branch circuits (*Cont.*):
    GFCI protection, 111, 114–119
    loads, 16, 36, 38*f.*
    outdoor receptacle, 46–49, 59
    outlets, 59, 61*f.*
    panelboards, 144–145, 147*f.*, 148*f.*
    receptacles, 43–45, 50, 51*f.*, 102, 105–106
    sliding room dividers, 43*f.*, 44
    voltage limits, 17, 29–30, 33, 34
  lighting and power feeders, 267–270
  services, 491–492, 506–513

Electric railway, 33
Electrical Appliance and Utilization Equipment Directory (Orange Book), 6, 69
Electrical Code for One- and Two-Family Dwellings, 3
Electrical Construction Materials Directory (Green Book), 6, 83, 194, 315–317, 444, 511, 550–551, 576, 603, 608, 631
Electrical design planning (*see* Planning)
Electrical Metalworking Machine Tools, 3
Electrical plans, 1–2
  for control circuits, 242, 243*f.*
  for lighting and appliance branch circuits, 162–164, 166–169
Electrical symbols, 162, 163*f.*
Electrical Testing Laboratories, Inc., 4
Electrodes, made, 550–552
Emergency supply to services, 573–584
  emergency equipment, 575–581
    automatic transfer, 578, 580*f.*
    battery-pack light units, 577, 579*f.*
    disconnects, 577, 579*f.*
    hookups, 580, 581*f.*
    lighting, 578, 580*f.*
    services from independent sources, 577, 578*f.*
    water-cooled generator, 577
    wiring, 580–582
  standby systems, 581–584
    manual transfer switch, 582–583
    uninterruptible power supply, 583–584
Environmental air spaces, 65–66
Equipment ground, 365–367, 379–381, 542
  conductors, 386–390
Equipment selection and layout, 3, 585–634
  custom-made equipment, 5
  emergency equipment, 575–581

710 INDEX

Equipment selection and layout (*Cont.*):
  motor centers, 604
  switchboards, 250, 585–590
    circuit makeup, 589–590
    disconnect identification, 589, 590*f.*
    headroom and illumination, 588, 589*f.*
    installation location, 587–588
    minimum clearances, 586–587
  switching devices, 593–604
    automatic transfer switches, 601–604
      circuit breaker, 601, 602*f.*
      operating coils, 601, 602*f.*
    fault-interruption service, 594–595
    high-capacity switches, 600–601
    selection considerations, 599–600
    short-circuit withstandability, 595–597, 599
    sizing, 593, 594*f.*
      motor and lighting load in power panel, 593
      motor load only in power panel, 593
      transformer inrush, 597–599
  (*See also* Panelboards; Wiring methods)
Essential Electrical Systems for Health Care Facilities, 3

Factory Mutual Engineering Corp., 4
Fan, branch circuit for, 97–98
Fault-interruption service, 594–595
FCC (flat-conductor cable) wiring, 110–111, 113*f.*
Feeder, 247, 264
  (*See also* Current rating of services feeders; Lighting and power feeders; Motor feeders)
Feeder circuits, 71
Feeder distribution panels, 591, 592*f.*
Feeder taps (*see* Taps)
Fire safety equipment, 3, 180, 220, 225, 523, 524*f.*, 535, 538–539
Flammable Combustible Liquids Code, 3
Flat-conductor cable (FCC) wiring, 110–111, 113*f.*
480-V, 3-phase, 3-wire system, 253–254
480/277-V (460/265-V) wye-connected system, 17–18, 160, 161*f.*, 355–356, 415–416
480Y/277-V (460Y/265-V), 3-phase, 4-wire system, 253–256
4160Y/2400-V, 5-phase, 4-wire system, 254, 256–259
4800-V system, 256

Fuse:
  classifications, 318
    Class G, 318–320
    Class H, 318–320
    Class J, 318–320
    Class K, 318–320
    Class L, 318–320
    Class R, 318–320
    Class T, 318–320
    current-limiting, 318, 319, 329–333
    time-delay, 319–320
  lighting and power feeders, 312–313, 318–320
    versus circuit breaker, 331–333
    short-circuit protection, 328–333
    nomograph IC ratings, 667–703
      example, 653*f.*, 668, 701*f.*
      transformer plot, 653*f.*, 654*f.*, 668
  operating characteristics, 320

Garage, lighting and appliance branch circuits for, 50, 59, 111, 114*f.*, 115, 117, 120–122
Gasoline dispensing pumps, 147, 149*f.*
GFCI protection:
  in dwelling units, 111, 114–119
  in hotels and motels, 111, 115, 117
GFP (*see* Ground-fault protection)
Green Book (Electrical Construction Materials Directory), 6, 83, 194, 315–317, 444, 511, 550–551, 576, 603, 608, 631
Ground-fault protection (GFP):
  lighting and power feeders, 347–357, 536
    building disconnect, 356–357
    480Y/277-V system, 355–356
    low-level arcing, 348, 349*f.*
    residual method, 354
    sensing, 350–351
    time-delay, 352–355
    zero-sequence sensing, 351–352
    zone-selective instantaneous system, 353–354
  motor feeders, 406–413
  services, 536–542
    alternative fault-current paths, 539, 540*f.*
    disconnect subdividing, 538
    generator, 539, 541
    rating, 537
    transfer switch, 539, 541*f.*
Grounding, 358
  autotransformers, 487–489
  control circuit transformers, 226, 228

Grounding (*Cont.*):
  lighting and power feeders, 358–391
    bonding jumpers, 390–391
      arrangements, 391
    building ground, 370, 372–379, 382–386, 542, 553, 555
      conductors, 382, 384–386
      outbuildings, 370, 372–376
      over 250 V circuits, 382, 383*f.*
      protective device, 382, 384*f.*
      at supply side, 370, 372*f.*
    conductors, 379, 381–382, 384–390
      color coding, 381
    equipment ground, 365–367, 379–381, 542
      alternating-current equipment, 380–381
      conductors, 379, 386–390
    ground connections, 367–370, 552, 553
      grounded neutral system, 367–369
      grounded service conductor, 367, 368*f.*
      grounding-electrode conductor, 367
      subpanel bonding, 370, 371*f.*
    Occupational Safety and Health Administration regulations, 364–365
    wiring-system, 358–363
      grounded conductor, 359, 360*f.*
      grounded electrical system, 358, 359*f.*
      grounded neutral system, 363
      120-V control circuit, 360, 361*f.*
      under 50-V, 362
      ungrounded systems, 363
  services, 542–573
    bonding, 564–567
      burn-clear, 563, 564
      bushings and fittings, 568, 571*f.*
      conduit sleeve, 567, 569*f.*
      flex, 565, 566*f.*
      low-impedance path, 564, 565*f.*
      piping, 570–574
      skin effect, 567, 568*f.*
      techniques, 565–568, 570*f.*, 571*f.*
      wiring methods, 568, 570*f.*
    bonding jumpers, 568–570
      parallel conduits, 570, 573*f.*
      sizing, 569–570, 572*f.*, 573*f.*
    building metal frame, 546, 547*f.*
    connections, 552–554
      bonding, 553–555
      bonding jumper, 553, 557*f.*

Grounding
  services
    connections (*Cont.*):
      common grounding-electrode conductor, 553, 556*f.*
      electrode system, 474, 544–545
      ground rod, 546, 548*f.*
      made electrodes, 550–552
        ground-rod installation, 550, 551*f.*
        resistance to earth measurement, 551, 552*f.*
      multiple service entrance disconnects, 561–563
        bonding, 563, 564
        layout, 561, 563*f.*
      neutral, 554–557
        circuit conductor, 555, 557*f.*
        to load side, 555–556, 558*f.*
      pole-type, 560–562
      Ufer system, 549
      unused-neutral, 557–560
      water-pipe electrode, 545, 546
    transformers, 445, 469–477, 542
      bonding jumper, 470–472
      examples, 480–481
      exemptions, 476, 478*f.*
      grounding electrode, 471*f.*, 473–474
      grounding electrode conductor, 471–473, 475–476
      secondary neutral, 470, 474, 475*f.*
      unacceptable methods, 476, 477*f.*
      zig-zag, 487–489
Grounding Electrical Distribution Systems for Safety (Soares), 387, 389
Group installation of motor branch circuits, 188, 189*f.*

Hallway, lighting and appliance branch circuits for, 43, 50, 59, 120, 123
Harmonics of transformers, 421–427
  CBEMA crest factor, 421–427
  effects on K-factor, 428
  IEEE method, 421, 423–427
  sizing, 421–422
Hazardous Location Equipment Directory (Red Book), 6
Hazardous locations safety provisions, 2
Health care facilities, 3, 603
  distribution system voltage levels, 258, 261*f.*
  electrical design planning, 13
  ground-fault protection, 357, 358*f.*

712 INDEX

Health care facilities (*Cont.*):
  lighting and appliance branch circuits, 28
  lighting and power feeders, 268, 360–361
Heating equipment, 174
  baseboard heater, 45, 46*f.*
  electric-resistance duct heaters supplementary overcurrent protection, 15
  lighting and appliance branch circuits, 59, 98, 131, 133*f.*
  lighting and power feeders, 269–271
  utilization voltages, 25
High-capacity switches, 600–601
Hospitals (*see* Health care facilities)
Hotels and motels:
  lighting and appliance branch circuits:
    calculations, 43
    controls, 123
    GFCI protection, 111, 115, 117
    outlets, 59
    receptacles, 45, 50, 105
    voltage limits, 29, 33
  lighting and power feeders, 267
Household Fire Warning Equipment, 3
Houses (*see* Dwelling units)

I^2R loss in motor feeders, 394–396
IBEW, 42
ICEA, 387*f.*, 389
Identification marks, 97
  (*See also* Color coding)
IEEE/ANSI Standard C57, 433
IEEE harmonics methods, 421, 423–427
In plug connection fuses, 15
Industrial buildings:
  busway wiring methods, 626, 627*f.*, 629
  control circuit layouts, 240
  electrical design planning, 9
  lighting and appliance branch circuits, 16–17, 82, 101, 103*f.*
    controls, 123
    layouts, 19
    loads, 16–17, 28, 638–639
    voltage limits, 31, 33*f.*
  lighting and power feeders, 310, 311*f.*, 347, 348*f.*, 362–363
  motor feeders, 395–397
Institutional buildings, lighting and appliance branch circuits in, 19, 31, 33*f.*, 82
Insurance company regulations, 3

K-factor of transformers, 421–428
  CBEMA crest factor, 421–427

K-factor of transformers (*Cont.*):
  computation, 423, 424*f.*, 427, 428
  harmonic orders effects, 428
  IEEE method, 421, 423–427
  sizing, 421–422
Kilovars, 395
Kilovoltampere rating of transformers, 420–421
Kitchen, lighting and appliance branch circuits in, 43–45, 47*f.*, 52–62, 119, 120*f.*, 123

Labeled, 4–6
Laundry facilities:
  lighting and appliance branch circuits, 48, 49*f.*, 119
  lighting and power feeders, 267, 270
Life Safety Code (NFPA No. 101), 3, 574
Lighting and appliance branch circuits, 15–171
  ampacity, 86–93, 618
    circuit makeup and loading impact, 90–91
    conductor insulation, 90–93
    correction factors, 86
    derating, 88–91
    electric-discharge load, 92*f.*, 93
    90-degree C wires, 87–89, 91–93
    resistive load, 93
  basic concepts, 15–17, 19–22
  color coding, 93–97
    alternatives, 97
    grounded conductor, 94–95
    grounding conductor, 94, 95, 97, 98*f.*
    hot conductor, 94–97
    neutrals, 94, 96*f.*
    safety precautions, 97
  conductors, 63–69
    air-handling ceilings, 68–69
      wiring systems, 65–68
    calculations, 64
    color coding, 93–98
    environmental air spaces, 65–66
    overcurrent protection, 88
    suspended ceiling, 68
    temperature rating, 83–86, 620
      ampacity, 86
      circuit breaker, 84–86
      distribution and control equipment, 84
      terminals, 84, 86
  controls, 119–133
    alternating-current snap switches, 119

Lighting and appliance branch circuits
controls (*Cont.*):
  circuit breaker, 124, 125*f*.
  contactor, 126–133
    coil-circuit wires, 131, 133, 134*f*.
    heating equipment, 131, 133*f*.
    magnetic, 126–127, 129*f*.
    mechanically held, 127–128, 130–133
    remote-control switch, 130, 131*f*.
    split-bus panelboard, 130
    widespread lighting, 130–132
  full-circuit, 123–124
  local, 119
  low-voltage relay switching, 125–128
  mercury switches, 119, 121*f*.
  receptacles, 119
  remote and automatic, 124–125
  screwless terminal switches, 123
  snap switches, 123, 124*f*.
  split-wire duplex receptacle, 119–121
installation requirements, 38–41
  cord connection, 40, 41*f*.
  flex as grounding, 40
  flexible raceways and fixture wires, 38, 39*f*.
  lampholder, 40
  wire ratings, 38, 39*f*.
loads, 34–37
  continuous, 28–30
  general purpose, 35
  multioutlet 30-A, 35, 36*f*.
  multioutlet 40-A and 50-A, 36, 37*f*.
modular wiring, 42
non-general outlet loads, 59–63
  convenience receptacle, 60
  cord-and-plug-connected, 62–63
  heavy-duty lampholder, 60
  motor-operated devices, 60
  recessed lighting, 60
number required, 22–25
  actual load, 23–24
  capacity, 25–26
  connected load, 23
  example, 26–27
  total computed load, 22–24
  utilization voltages, 24–28
  volt-ampere-per-square foot basis, 23
overcurrent protection, 69–82
  cartridge fuse, 75, 76, 78*f*.
  circuit breaker, 77–82
  Class R fuse, 75–77
  grounded conductor, 71, 73*f*.

Lighting and appliance branch circuits
overcurrent protection (*Cont.*):
  multiwire circuits, 78–80
  panelboards, 145–146
  plug fuses, 75, 76*f*.
  rating, 69, 70*f*.
  readily accessible, 74–75
  shock hazard, 81
  600 V fuse, 76
  split-wired receptacles, 81–82
  tap wire sizes, 74
  two-pole disconnect, 81
  type S fuse, 75, 76*f*.
  ungrounded conductor, 71, 72*f*.
  wire size, 71, 72*f*.
panelboards, 133–134, 136–147, 265, 278–279, 591
  capacity supplement, 146
  circuit disconnect, 147, 150*f*.
  controls, 146, 149*f*.
  example, 134, 136–138
  layouts, 133, 136*f*.
  loadcenter, 133
  overcurrent protection, 145–146
  panel main protection, 138–142
    circuit breaker, 142
    exemptions, 139, 141*f*., 142, 144*f*.
    fuse, 142
    location, 142, 143*f*.
    snap switches, 142–143, 145*f*.
    from transformers, 143, 146*f*.
  rating, 134, 137*f*.
plan layouts, 160–171
  components embedded in concrete, 169
  computer receptacles, 169, 170*f*.
  electrical plans, 162–164
    drawings, 168*f*., 169
    isometric drawings, 166–168
  electrical symbols, 162, 163*f*.
  480Y/277-V system, 160, 161*f*.
  panelboards, 160, 162*f*., 164–166
  60- and 400-Hz system, 169, 171*f*.
  special hookups, 166
receptacles, 97–106
  amperage loading schedule, 102, 105*f*.
  application, 106–111
    aluminum conductors, 108, 110*f*.
    ampere rating, 108, 111*f*.
    FCC wiring, 110–111, 113*f*.
    grounded metallic system, 106–109
    poke-through receptacle, 108, 110, 112*f*., 113*f*.

Lighting and appliance branch circuits
  receptacles
    application (*Cont.*):
      rating, 106
      layouts, 101
      load, 98, 501
      multioutlet assemblies, 101
      nonresidential occupancies, 99–102
      number required, 98–100
      plug-in devices, 98
      pole receptacle, 102, 104*f*.
      small-appliance, 106
      spacing, 106
    summary, 159–160
    voltage drop, 148, 150–159
      copper conductors, 150, 154*f*.
      copper loss calculations, 150, 152*f*.
      dc motor, 153, 157*f*.
      formulas, 150, 151*f*.
      homerun, 150, 153*f*.
      skin effect, 158
      wire resistances, 152, 156*f*.
      wire size selection, 150, 155*f*.
    voltage limits, 29–34
      120-V loads, 30, 31*f*.
      over 120-V loads, 30, 32*f*.
      split wiring, 34, 35*f*.
      split-wired receptacles, 33–35
      277-V loads, 30–32
      ungrounded circuits, 32–33
    voltage selection, 17–18
      480/277-V (460/265-V) wye-connected system, 17–18
      loadcenter layout, 17
      120/280-V, 3-wire system, 18
      open-delta 3-phase system, 18
      single-phase, 3-wire system, 17
      three-phase, 3-wire system, 18
      three-phase, 4-wire (red-leg) delta system, 18
      three-phase, 4-wire wye system, 17
    (*See also* Branch circuits; Dwelling units)
Lighting and power feeders, 247–391
  capacity, 265–306
    branch-circuit load, 265–267, 271–273
    common neutral versus separate neutral, 287–292
      equal nonlinear load, 288–289
      harmonics, 288
      heating effects, 291–292
      phase and neutral loading, 287, 288*f*., 290

Lighting and power feeders
  capacity (*Cont.*):
    conductor size selection, 279–281
      example, 281
      power loss calculations, 281, 282*f*.
    continuous-duty motor loads, 268–269
    demand factors, 267, 273–274
    for growth, 274–276
    sizing, 276, 277*f*.
    multiple conductors, 299, 303–306
      ampacity derating, 304–305
      common neutral, 306, 308*f*.
      versus single conductor, 306, 307*f*.
    neutral sizing, 281–286
      electric-discharge lighting reduction, 284, 287*f*.
      feeder to ballast load, 284, 286*f*.
      lighting load type, 284, 285*f*.
      neutral-connected loads, 282, 283*f*.
      phase-to-phase loads, 284, 286*f*.
      resistance load reduction, 282–284
    sizing versus panelboard rating, 276–280, 314
      circuit breaker rating, 277–279
    voltage drop minimization, 292–299
      circuit makeup, 294–295, 299–303
      formulas, 293, 294*f*.
      limitations, 293, 296
      loadcenter distribution, 296, 297*f*.
      voltage regulators, 296–299
  grounding, 358–391
    bonding jumpers, 390–391
      arrangements, 391
    building ground, 370, 372–379, 382–386, 542, 553, 555
      conductors, 382, 384–386
      outbuildings, 370, 372–376
      over 250 V circuits, 382, 383*f*.
      protective device, 382, 384*f*.
      at supply side, 370, 372*f*.
    conductors, 379, 381–382, 384–390
      color coding, 381
    equipment ground, 365–367, 379–381, 542
      alternating-current equipment, 380–381
      conductors, 379, 386–390
    ground connections, 367–370, 552, 553
      grounded neutral system, 367–369
      grounded service conductor, 367, 368*f*.
      grounding-electrode conductor, 367

Lighting and power feeders
  grounding
    ground connections (*Cont.*):
      subpanel bonding, 370, 371*f*.
      Occupational Safety and Health Administration regulations, 364–365
    wiring-system, 358–363
      grounded conductor, 359, 360*f*.
      grounded electrical system, 358, 359*f*.
      grounded neutral system, 363
      120-V control circuit, 360, 361*f*.
      under 50-V, 362
      ungrounded systems, 363
  motor feeders and, 400–406
    examples, 401–406
  overcurrent protection, 306, 308–357
    automatic response, 309, 310*f*.
    with circuit breaker, 312–318, 320–333, 335, 338*f*.
    continuous current rating, 312
    with fuse, 312–313, 318–320, 328–333
    ground-fault protection, 347–357, 536
      building disconnect, 356–357
      480Y/277-V system, 355–356
      low-level arcing, 348, 349*f*.
      residual method, 354
      sensing, 350–351
      time-delay, 352–355
      zero-sequence sensing, 351–352
      zone-selective instantaneous system, 353–354
    installation location, 333–335
      busway, 334–337
      readily accessible, 333–335, 338*f*.
    short-circuit protection, 322–331, 390*f*.
      calculations, 324, 325
      device selection coordination, 325, 326*f*.
      elements, 323
      flow patterns, 326–328
      interrupting capacity, 325
      operating speed, 325–327
    sizing, 311–312, 314–322
      continuous loading, 314, 317–318
      switch assembly, 317
      at supply, 311, 313*f*.
    taps, 311, 335–347, 454, 455, 461
      examples, 338–347
        panelboard with unprotected conductors, 339–341
        single-motor circuit, 338–340
        transformers, 341–345

Lighting and power feeders
  taps
    examples (*Cont.*):
      unprotected tap, 343–347
      under ten feet, 336–337, 339*f*.
      under twenty-five feet, 337, 340*f*.
    (*See also* Distribution system)
Lightning Protection Code, 3
Listed, 4–6
Load circuits, 212
Load factor, 273
Loadcenter, 250
Loads:
  branch-circuit sizing, 16–17
  capacity growth allowance, 6–7
  distribution system voltage levels, 258–259, 263*f*.
  lighting and appliance branch circuits, 34–37
  continuous, 28–30
  in dwelling units, 16, 36, 38*f*.
  general purpose, 35
  in industrial buildings, 16–17, 28
  multioutlet 30-A, 35, 36*f*.
  multioutlet 40-A and 50-A, 36, 37*f*.
  non-general outlets, 59–63
  receptacle, 98, 102, 105*f*.
  lighting and power feeders, 271–273
Local codes, 3
  lighting and appliance branch circuits voltage selection, 18
  transformers, 444
Local Protective Signaling Devices, 3
Luminary fuses, 15

Made electrodes, 550–552
Main, 247, 264
Manually switch together, 77–78
Metalworking machine tools, 3
Mobile Homes, 3
Model State Electrical Law, Inspection and Electrical Installations, 3
Modular wiring, 42
Motels (*see* Hotels and motels)
Motor branch circuits, 173–211
  air conditioning and refrigeration, 190–191
    rating, 191, 192*f*.
    short circuit fault, 191, 192*f*.
  conductors, 174, 176–178
    current-carrying capacities determination, 177
    example, 177–178

716 INDEX

Motor branch circuits
 conductors (*Cont.*):
  sizing, 174, 176
  wound-rotor motors, 176, 177*f*.
 controls, 191–201
  across-the-line controller, 199–201
  branch-circuit circuit breaker, 192, 193*f*.
  branch-circuit protective device, 192, 193*f*.
  controller, 191, 193*f*.
  example, 195
  full-voltage controller, 199–200
  general-use switch, 192, 193*f*.
  hermetic motor, 194
  installed location, 195, 197–199
  layout plans, 197–199
  low-voltage protection, 200–201
  manual starters, 200, 209
  plug-and-receptacle connection, 192, 193*f*.
  reduced-voltage controller, 199
  starter poles, 195, 196*f*.
  3-wire control circuit, 201
  2-wire control circuit, 201
  undervoltage protection, 200–201
 design elements, 174, 175*f*.
 disconnects, 206–211
  example, 207–209
  in-sight, 210
  out-of-sight, 210–211
  requirements, 206, 207*f*.
  selection based on ampere rating, 207, 208*f*.
 overcurrent protection, 201–206
  exceptions, 203–204, 206
  requirements, 201–203
  setting adjustments, 204, 205*f*.
  three-phase starter, 204
 short-circuit protection, 178–190
  any rating motor, 183–189
  example, 181–182, 188
  with fuses, 185–186
  group installation, 188, 189*f*.
  instantaneous-trip CB, 180–181
  magnetic, 180–181
  motor controllers (NJOT), 179–180
  motor short-circuit protector, 180, 182
  multimotors, 182–190
  multiple fractional-horsepower, 186, 188*f*.
  rating, 178, 179
  sizing, 176*f*., 179

Motor branch circuits
 short-circuit protection (*Cont.*):
  small motors, 182–183
  tap conductors, 188–190
  voltage selection, 18
Motor centers, 604
Motor feeders, 393–413
 circuit protection, 406–413
 conductor sizing, 393–394
 lighting and power feeders and, 400–406
  examples, 401–406
 overcurrent protection, 396, 399–400
  examples, 396, 398–400
  sizing, 396, 398*f*.
 voltage drop and power factor, 394–396
  capacitor kilovars, 395
  capacitor power-factor correction, 395–397
Motor overload devices, 15
Motor selection chart, 644
Motor short-circuit protector (MSCP), 180, 182
Motor-starter control circuits (*see* Control circuits)
MSCP (motor short-circuit protector), 180, 182
Multifamily dwellings (*see* Dwelling units)

National Electric Code (NEC), 3
 Article 100, definitions:
  branch circuits, 19
  dwelling units, 111
  location, 608
  one-family dwelling, 46, 529
  receptacle, 20, 22*f*.
  services, 492
 Article 200, lighting and appliance branch circuits, 19, 276
 Article 210, lighting and appliance branch circuits, 19–20, 99, 532
 Article 215, conductor voltage, 99, 532
 Article 220, Branch-circuit, Feeder, and Service Calcs, 19, 99, 268*f*., 530, 533–535
 Article 230, services, 494, 533
 Article 240, equipment grounding conductors, 389, 532
 Article 310, conductors, 83
 Article 321, wiring conductors, 606
 Article 328, FCC wiring, 111
 Article 333, armored cable, 631
 Article 334, Metalclad Cable, 631
 Article 422, appliances, 174

INDEX 717

National Electric Code (NEC) (*Cont.*):
Article 424, space-heating equipment, 174
Article 426, Fixed Outdoor Electric De-icing and Snow Melting Equipment, 36
Article 430, motor-operated devices, 19, 60, 173–174, 195, 232–233, 268, 533
Article 430(e), motor feeders, 406
Article 430(g), remote-control circuit, 213
Article 440, Air-conditioning and Refrigeration Equipment, 60, 173–174
Article 440(g), room air-conditioner, 174
Article 514, Gasoline Dispensing and Service Stations, 147
Article 517, Health Care Facilities, 603
Article 517(c), emergency supply, 574
Article 604, modular wiring, 42
Article 645, data processing systems, 66
Article 685, lighting and power feeder overcurrent protection, 313
Article 700, Emergency Systems, 574–576, 603
Article 701, Legally Required Standby Systems, 581, 603
Article 702, Optional Standby Systems, 582, 603
Article 725, remote control wiring, 126, 131, 212, 213, 215
branch circuit, 15
branch circuit load limitations, 35
health care facilities requirements, 13
load capacity growth allowance, 6–7
Occupational Safety and Health Administration and, 4
safety provisions, 2–3
Section 90-1(c), safety precautions, 454
Section 110-1, equipment grounding conductors, 389
Section 110-2, electrical products and equipment, 5–6, 83, 422, 444
Section 110-3, equipment examination, identification, installation and use, 6, 422, 444
Section 110-3(b), listed or labeled equipment, 6, 83, 385
Section 110-10, lighting and power feeder short-circuit protection, 326
Section 110-14(a), conductor terminations, 624
Section 200-6, conductor identification, 97
Exceptions, grounded conductor color coding, 94

National Electric Code (NEC) (*Cont.*):
Section 210-1, lighting and appliance loads, 19
Section 210-4, multiwire branch circuits, 78
Exception No. 1, multiwire branch circuits, 78
Exception No. 2, two-pole switch, 82
Section 210-4(b), multiwire branch circuits two-pole disconnect, 54*f*., 81
Section 210-4(d), hot conductor color coding, 95
Section 210-5, conductor color coding, 93, 96*f*.
Section 210-5(b), grounding conductor color coding, 95
Exception No. 1, grounding conductor color coding, 95
Section 210-5(b)(1), lighting and power feeders grounding, 364
Section 210-6, lighting and appliance branch circuits voltage limits, 29–30, 470
Section 210-6(a), split-wired receptacles, 33–35, 364, 365
Exception No. 1, lighting and power feeders grounding, 365
Exception No. 2, lighting and power feeders grounding, 365
Section 210-6(b), lighting and power feeders grounding, 364, 365
Section 210-6(b)(1), medium-based screw-shell lampholders, 31
Section 210-6(c), permanently connected utilization equipment, 31, 34
Section 210-6(c)(1), medium-based screw-shell lampholders, 31
Section 210-6(c)(3), electric-discharge fixtures, 30
Section 210-6(c)(5), lighting and appliance branch circuits voltage limits for appliances, 30
Section 210-6(d), outdoor branch circuits, 31–33
Section 210-8, dwelling unit bathroom GFCI protection, 36
Section 210-8(a), Dwelling Units, 111
Section 210-8(a)(1), bathroom GFCI protection, 111
Section 210-8(a)(2), garage receptacle, 50, 115, 117
Exception No. 2, GFCI protection for dedicated appliances, 115, 117

National Electric Code (NEC) (*Cont.*):
- Section 210-8(a)(3), outdoor receptacle GFCI protection, 117
- Section 210-8(a)(4), basement and crawl space receptacles, 50, 119
  - Exception No. 1, refrigerator or freezer GFCI protection, 119
  - Exception No. 2, GFCI protection, 50, 119
  - Exception No. 3, sump pump GFCI protection, 119
- Section 210-8(a)(5), kitchen sink GFCI protection, 119
- Section 210-8(a)(6), boathouse GFCI protection, 119
- Section 210-8(b), bathroom receptacle, 38*f.*, 111, 115
- Section 210-8(b)(2), rooftop receptacle, 51
- Section 210-9, Circuits Derived from Autotransformers, 32, 485
  - Exception No. 1, in existing installations, 486–487
- Section 210-10(b), lighting and appliance branch circuits overcurrent protection, 77
- Section 210-19, branch-circuit wire ampacity, 74, 91, 93
- Section 210-19(a), lighting and appliance branch circuits overcurrent protection rating, 70*f.*, 150
- Section 210-19(c):
  - Exception No. 1, flexible raceways and fixture wires for taps, 38
  - Exception No. 1b, flexible raceways and fixture wires, 38
  - Exception No. 2, flexible raceways and fixture wires, 38
- Section 210-20, lighting and appliance branch circuits overcurrent protection tap wire sizes, 74
- Section 210-21(a), fluorescent fixtures, 36, 40
- Section 210-21(b), receptacle ampere rating, 108, 111*f.*
- Section 210-22(c), branch circuit total load, 28, 29*f.*, 91, 622
- Section 210-23(b) panelboard overcurrent protection, 146
- Section 210-50(c), appliance receptacles, 43–44
- Section 210-52, garage receptacles, 99, 115
- Section 210-52(a), receptacle spacing, 43–47, 50, 105, 106

National Electric Code (NEC) (*Cont.*):
- Section 210-52(b)(1), Exception No. 3, plug-in lamps, 57
- Section 210-52(c), dwelling unit receptacles, 45, 105
- Section 210-52(d), kitchen small appliance spacing, 56
- Section 210-52(f), basement receptacle, 50
- Section 210-52(g), basement and garage receptacles, 48, 117, 119
- Section 210-52(h), hallway receptacle, 43
- Section 210-60, bathroom wash basin receptacle, 45
- Section 210-62, store window receptacles, 99
- Section 210-70, dwelling unit lighting and appliance branch circuits, 59, 120–121
  - Exception No. 1, wall-switch-controlled receptacle outlet, 122–123
  - Exception No. 2, remote, central and automatic control, 123
- Section 210-70(a), stairway outlet, 59
  - Exception No. 1, wall-switched receptacle, 55, 56, 59
  - Exception No. 2, remote, central and automatic lighting control, 59
- Section 210-70(b), hotels and motels lighting and appliance branch circuits, 59, 123
- Section 210-70(c), wall-switched controlled lighting outlet, 59
- Section 215-2, feeder conductor voltage drop, 150
- Section 215-4, lighting and power feeder common neutral, 306
- Section 215-8, high-leg conductor color coding, 95, 97*f.*
- Section 215-10, ground fault protection, 356, 357
- Section 220, Branch-Circuit and Feeder Calculations, 24
- Section 220-2, voltage values, 24–27
- Section 220-2(a), branch circuit total load, 28, 622
- Section 220-2(c), receptacle circuits, 57
- Section 220-2(c)(4), receptacles required, 99, 100*f.*
- Section 220-3, computed load, 23, 24, 134, 267, 271
- Section 220-3(a), conductor load, 91, 137

INDEX

National Electric Code (NEC) (*Cont.*):
Section 220-3(b), total computed load, 22, 105, 265, 266*f*.
Section 220-3(c), lighting and power feeder branch-circuit loads, 272
   Exception No. 1, multioutlet assemblies, 101
Section 220-3(c)(5), receptacle voltage values, 26
Section 220-4, Branch Circuits Required, 23, 24, 105, 265, 266
Section 220-4(a), branch circuits minimum number required, 23, 105, 106, 267
Section 220-4(b), kitchen receptacles, 52–53
Section 220-4(b)(1), service feeders, 509
   Exception No. 3, wall switch-controlled kitchen small appliance receptacles, 55–57
Section 220-4(b)(2), kitchen countertop receptacles, 52–53
Section 220-4(c), feeder capacity for laundry facilities, 267
Section 220-4(d), branch circuit panelboard, 23, 24, 134, 137–138, 265–267, 271, 278
Section 220-10(a), feeder sizing, 23, 266–267
Section 220-10(b), panelboard main protection, 276–279, 281, 314, 384, 401, 454, 462, 464, 468, 482, 502–505, 618, 621–623
   Exception, service feeders, 505
Section 220-11, lighting and power feeder branch-circuit loads, 273, 464
Section 220-13, feeder capacity, 268, 271
Section 220-15, Exception No. 1, service feeders, 499
Section 220-17, dwelling units lighting and power feeder capacity, 268
   Exception No. 1, service feeders, 499
Section 220-19, service feeders, 506
Section 220-20, service feeders, 268, 499
Section 220-21, air-conditioning equipment feeder load, 270, 508
Section 220-22, lighting and power feeder neutral sizing, 281, 284, 290, 510, 511, 621
Section 220-30 to Section 220-32, dwelling units heating load, 269
Section 220-30, service feeders, 508, 509, 511, 513

National Electric Code (NEC) (*Cont.*):
Section 220-52(b)(1), kitchen small appliance receptacles, 53–56
Section 221-10, panelboard rating, 276
Section 225-8, service disconnects, 531
Section 225-8(b), outbuilding service disconnects, 531
   Exception No. 1, building ground, 377, 535–536
Section 230-2, services, 496–498, 518, 519, 525
   Exception No. 7, multiservices, 496–499, 563
Section 230-6, service disconnects, 529
Section 230-10 to Section 230-21, service feeders, 506–508
Section 230-24(b), Exception, service disconnects, 529
Section 230-30 to Section 220-41, service feeders, 500
Section 230-31, services, 492
Section 230-40, Exception No. 2, service entrance conductors, 496–499
Section 230-41(b)(2), service feeders, 507
Section 230-42, service feeders, 500
Section 230-42(b)(2), service feeder conductor rating, 510
Section 230-43, armored cable, 631
Section 230-45, multiservices, 496
Section 230-63, service bonding, 565
Section 230-71, building ground disconnect, 377
Section 230-71(a), service disconnects, 519
Section 230-72, service disconnects, 517, 521, 527
Section 230-72(a), service disconnect layouts, 517–519
Section 230-72(b), emergency service disconnects, 521, 523
Section 230-72(d), multi-occupancy building service disconnects, 523, 525, 526, 528
   Exception, multiple service disconnects, 525
Section 230-80, service disconnects, 530, 534
Section 230-82, Exception No. 5, service disconnects, 523
Section 230-90:
   Exception No. 1, service overcurrent protection for motors, 532–533
   Exception No. 2, service overcurrent protection fuse rating, 533–534

National Electric Code (NEC)
  Section 230-90 (*Cont.*):
    Exception No. 3, service overcurrent protection fuse and circuit breaker sets, 534–535
    Exception No. 4, multi-occupancy building service disconnects, 523, 525, 538
  Section 230-90(a), service overcurrent protection, 534
    Exception No. 3, panelboards in dwelling units, 144, 530
  Section 230-95, ground fault protection, 354–357, 536–538, 541
    Exception No. 2, services, 539
  Section 230-95(a), ground fault protection time delay, 351, 537, 538
  Section 230-95(c), service ground-fault protection, 542
  Section 238-23, services, 492
  Section 240-1, FPN, conductor overcurrent protection, 88, 389, 606, 608
  Section 240-2(b) to Section 240-2(c), tap conductors, 462
  Section 240-3, conductor overcurrent protection, 88, 212, 313, 385, 446, 597, 618
    Exception No. 1, lighting and power feeder, 313, 623
    Exception No. 5, transformers, 220
  Section 240-3(a), lighting and appliance branch circuits overcurrent protection, 69, 71
  Section 240-3(b), conductor overcurrent protection, 90, 276, 466, 535
  Section 240-3(c), lighting and power feeder overcurrent protection, 313, 385
  Section 240-3(f), motor circuit conductors, 156
  Section 240-3(i), transformer overcurrent protection, 447, 462
  Section 240-4, flexible raceways and fixture wires for taps, 38
  Section 240-6, circuit breaker rating, 183, 185, 186, 312, 466, 533
  Section 240-10, readily accessible overcurrent protection, 40, 74–75
  Section 240-12, Electrical System Coordination, 204, 310
  Section 240-13, ground fault protection, 356

National Electric Code (NEC) (*Cont.*):
  Section 240-20(b), lighting and appliance branch circuits overcurrent protection, 77
  Section 240-21, lighting and appliance branch circuits overcurrent protection for ungrounded conductor, 71, 457, 460, 462
  Section 240-21(b), ten-foot feeder taps, 335, 337, 344, 455, 480
  Section 240-21(c), feeder taps under twenty-five feet, 337–340, 343–345, 347
  Section 240-21(d), twenty-five-foot feeder taps, 335, 341, 343–345, 347, 455, 458, 460
  Section 240-21(e), unprotected feeder tap lengths, 343–347
  Section 240-21(i), unprotected feeder taps, 343
  Section 240-21(j), industrial feeder taps, 347
  Section 240-21(m), one hundred-foot feeder taps, 335
  Section 240-24, suspended ceiling, 68, 333, 334, 339, 340
  Section 240-40, cartridge fuse, 186
  Section 240-60, 300-volt type cartridge fuse, 75
  Section 240-60(b), low-voltage cartridge fuse, 75–76, 319
  Section 240-83(d), circuit breaker for fluorescent lighting, 124
  Section 241-21(d)(2), feeder tap conductors, 459
  Section 250-5, lighting and power feeders grounding, 359, 364
  Section 250-5(b), grounding of transformers for control circuits, 226, 359, 365
    Exception No. 3, ungrounded operation, 226–227, 360, 364
  Section 250-6(a), grounding and bonding of transformers for control circuits, 228
  Section 250-23, building ground, 372, 560*f.*, 561
    Exception No. 2, neutral conductor bonding, 370
    Exception No. 5, lighting and power feeder, 369
  Section 250-23(a), panel neutral block bonding, 376, 560

National Electric Code (NEC)
Section 250-23(a), panel neutral block bonding (*Cont.*):
Exception No. 2, on load side, 376
Section 250-24, branch circuits, 20, 370, 372, 374
Section 250-24(a), panel neutral block bonding, 376
Exception No. 1, building ground, 370
Exception No. 2, grounded circuit conductor bonding, 370, 372, 374, 376–377, 379
Section 250-24(b), grounding electrode connection, 374
Exception No. 1, building ground, 370
Exception No. 2, ungrounded system, 374, 376
Section 250-24(c), remote disconnect, 376, 377
Section 250-24(c)(2), bonding conductor, 379
Section 250-24(d), bonding conductor, 379
Section 250-26, transformer grounding, 445, 469, 474, 480, 481, 486
Section 250-26(a), transformer grounding connections, 472, 477, 553
Section 250-26(b), transformer bonding connections, 472
Exception, small control and signaling, 476
Section 250-51, lighting and power feeder grounding, 366, 389
Section 250-51(2), equipment grounding conductors, 389
Section 250-54, building ground, 372
Section 250-57, grounding conductor with insulation color coding, 95, 228, 381
Section 250-57(b), grounding conductor color coding, 95, 474
Section 250-60, lighting and power feeder grounding, 369, 370
Section 250-72, bonding jumpers, 569
Section 250-74:
Exception No. 1, grounded metallic system with surface-mounted box, 106, 108
Exception No. 2, grounded metallic system with wire springs and machine screws, 108, 109*f*.
Exception No. 3, non-self-grounded metallic system, 108

National Electric Code (NEC) (*Cont.*):
Section 250-79(c), transformer bonding jumpers, 472, 477, 480, 484, 570, 572
Section 250-79(d), bonding jumpers, 391
Section 250-80, piping bonding, 572
Section 250-81, lighting and power feeder grounding, 367, 372, 77, 474, 476, 542, 545, 546
Section 250-81(a) to Section 250-81(d), service grounding, 543, 545, 549
Section 250-81(a), transformer grounding, 473, 542, 545, 546
Section 250-83, transformer grounding, 476, 542, 543, 546
Section 250-91(a), Exception No. 2, multiple service entrance disconnect grounding, 562
Section 250-91(b), building ground, 370, 372
Exceptions No. 1 and No. 2, flex as grounding, 40
Section 250-94, Exception No. 1(a), service grounding, 545, 546
Section 250-95, Conductors in Parallel Sizing, 304, 386–389, 391
Section 250-112, transformer grounding, 473
Section 255-8, building ground, 377
Exceptions No. 1 and No. 2, remote disconnect, 376
Section 255-8(b), building ground disconnect, 377
Section 300-3, control wires in raceways, 215
Section 300-3(c), control wires in raceways for high-voltage motor starters, 218
Section 300-3(c)(1), control wires in raceways for multimotors, 216
Section 300-5, underground circuits installation, 65, 514, 608, 612
Section 300-21, FCC wiring, 108, 111
Section 300-22, air-handling ceilings, 65, 68
Section 300-22(a), air-handling ceilings, 65
Section 300-22(b), air-handling ceilings within ducts and plenums, 65, 66, 68
Section 300-22(c), air-handling ceilings in other than ducts and plenums, 65, 66, 68, 69
Exception No. 1, liquidtight flexible metal conduit, 66
Exception No. 2, integral fan systems, 66, 69

National Electric Code (NEC)
  Section 300-22(c), air-handling ceilings in other than ducts and plenums (*Cont.*):
    Exception No. 3, non-habitable areas, 66, 69
    Exception No. 4, prefabricated cable, 66, 69
    Exception No. 5, joist or stud spaces, 66, 69
  Section 310-4, Conductors in Parallel, 299, 303–304
    Exception No. 1, in elevators, 306
    Exception No. 2, smaller than No.1/0, 306
    Exception No. 3, smaller than No.1/0, 306
    FPN, unused-neutral service, 559–560
  Section 310-8, underground conductors, 65
  Section 310-15, conductor ampacity, 88
  Section 330-3(9), MI cable, 65
  Section 334-3, ALS cable, 65, 632
  Section 334-4, ALS cable, 65
  Section 350-5, equipment ground, 379
  Section 350-58(a), equipment ground, 381
  Section 350-79(e), bonding jumper, 381
  Section 351-7, equipment ground, 379
  Section 364-11, lighting and power feeder installation location, 334
  Section 364-12, lighting and power feeder installation location, 334, 339, 340
  Section 380-8, screwless terminal switches, 68, 123, 333, 334
  Section 380-8(b), snap switches, 123
  Section 384-3(f), phase legs identification, 94
  Section 384-13, panelboards, 133, 134, 137, 276, 277, 279, 464
  Section 384-14, panelboards, 138–140, 278–279, 519
  Section 384-16, panelboards, 138, 139, 142, 339–340, 480
  Section 384-16(a), panelboards, 142, 144f., 276, 521
    Exception No. 2, dwelling units, 144
  Section 384-16(b), panelboards with snap switches, 142–143
  Section 384-16(c), branch circuit total load, 28–30, 90, 91, 93, 145, 146, 482
  Section 384-16(d), panelboards from transformers, 143, 447
  Section 402-5, Class 1 control circuits, 214

National Electric Code (NEC) (*Cont.*):
  Section 410-30, cord connection, 40, 41f.
  Section 410-31, branch circuits, 20
  Section 410-56(b), aluminum conductors, 108, 110f.
  Section 410-78, autotransformer-type ballasts, 32, 33, 364, 365
  Section 422-15(c), infrared lamp industrial heating, 33
  Section 424-22(b), lighting and appliance branch circuits overcurrent protection in suspended ceilings, 74
  Section 424-22(c) to (e), lighting and appliance branch circuits overcurrent protection in suspended ceilings, 16, 74
  Section 426-4, snow-melting and deicing installations, 36
  Section 430-3(b), transformer overcurrent protection, 446
  Section 430-6, nameplate current rating, 177
  Section 430-6(a), full-load current for motors, 183
  Section 430-11, motor branch circuit disconnects, 209
  Section 430-22, motor circuit, 153, 156
  Section 430-22(a), branch-circuit wire sizing, 176
  Section 430-24 to Section 430-26, continuous-duty motor loads, 183, 185, 268–269
  Section 430-24, motor feeders, 393, 400, 501
  Section 430-26, service feeders, 501
  Section 430-31, service ground-fault protection, 538–539
  Section 430-32, overcurrent protection, 203, 204
  Section 430-33, service feeders, 501
  Section 430-34, overcurrent protection, 204
  Section 430-36, overcurrent protection, 71, 310
  Section 430-37, lighting and appliance branch circuits overcurrent protection for grounded conductor, 71
  Section 430-40, motor short-circuit protector, 182, 188
  Section 430-44, overcurrent protection, 204
  Section 430-51 to Section 430-58, short-circuit protection, 178

INDEX 723

National Electric Code (NEC) (*Cont.*):
  Section 430-52, circuit breaker, 181–183, 185, 186, 188, 385, 399, 407, 413, 483, 532
    Exception No. 2b, motor feeders, 400
    FNP, short-circuit protection overload relay table, 188
  Section 430-52(a), motor feeders, 408–410, 412
    Exception, instantaneous-trip circuit protection, 409, 412
  Section 430-53, Several Motors or Loads on One Branch Circuit, 191
  Section 430-53(a), short-circuit protection for small motors, 182–183, 186, 199
  Section 430-53(b), short-circuit protection for motors of any rating, 183, 185, 186, 188
  Section 430-53(c), group installation motors, 188
  Section 430-58, instantaneous-trip circuit breaker, 181
  Section 430-61 to Section 430-63, motor feeders, 406
  Section 430-62, feeder taps, 347, 399, 400, 403, 405, 406, 534
  Section 430-62(a), motor feeders, 406–409
  Section 430-63, feeder taps, 347, 406, 534
  Section 430-71 to Section 430-74, control circuits, 212, 213
  Section 430-72, magnetic motor starter control circuits, 133, 212–213, 218–220
    Exception No. 4, service ground-fault protection, 539
  Section 430-72(a), transformer overcurrent protection, 220, 223*f*.
  Section 430-72(b), Overcurrent Protection, 215, 219, 227
    Exception No. 1, magnetic starter, 219–220, 227
    Exception No. 2, control-circuit, 212, 220–222
    Exception No. 3, transformers, 220, 222, 223*f*.
    Exception No. 4, safety equipment, 220, 225
  Section 430-72(b)(1), Class 1 control circuit, 214
  Section 430-73, accidental starting of control circuit, 232

National Electric Code (NEC) (*Cont.*):
  Section 430-73(c), transformer overcurrent protection, 221, 223*f*.
    Exception No. 1, rated less than 50 VA, 221
    Exception No. 2, rated less than 2 A, 221, 224*f*.
  Section 430-74, accidental starting of control circuit, 232
  Section 430-74(a), control circuit disconnects, 228–229, 232
    Exception No. 1, remote disconnects, 229
    Exception No. 2, remote disconnects in hazardous location, 229
  Section 430-81, motor controllers, 191
  Section 430-83, motor controllers, 191, 194
    Exception No. 1, enclosed switches, 194
    Exception No. 3, torque motors, 192
  Section 430-89, motor controller speed-limiting devices, 199
  Section 430-102, suspended ceiling, 68, 210
    Exception No. 1, high-voltage motor disconnect, 210
    Exception No. 2, industrial machinery disconnect, 210–211
  Section 430-102(b), motor controllers in suspended ceiling, 68
  Section 430-107, suspended ceiling, 68
  Section 430-109, Exception Nos. 2, 3 and 4, enclosed switches, 194, 206
  Section 430-110(a), instantaneous-trip circuit breaker, 181
  Section 430-111, enclosed switches, 194
  Section 430-112, circuit breaker as disconnect, 186, 210, 211
  Section 430-113, control circuit remote disconnects, 230
  Section 440-6, nameplate current rating, 177
  Section 440-6(a), nameplate current rating for sealed motor compressor, 178
  Section 440-12, refrigeration compressors, 209
  Section 440-22, fuse rating, 178
  Section 440-22(a), sealed hermetic motor-compressor, 190
  Section 440-22(b), branch circuit short-circuit and ground-fault protection sizing, 190–191
  Section 440-32, continuous operating branch-circuit conductor rating, 178

National Electric Code (NEC) (*Cont.*):
  Section 440-41, motor-compressor controls, 195
  Section 440-52, overcurrent protection, 203
  Section 440-52(a), overload relay, 178
  Section 440-60 to Section 440-64, room air-conditioner, 174, 191
  Section 440-62(a), room air conditioner as single motor, 191
  Section 445-5, feeder taps, 343
  Section 450-1, Exception No. 2, transformer overcurrent protection, 227
  Section 450-3, transformer overcurrent protection, 220, 221, 223*f.*, 455, 460
  Section 450-3(a)(1), transformer inrush, 597
  Section 450-3(b), transformer overcurrent protection in primary current, 227, 449, 453–454, 457, 481, 482
    Exception No. 1, protective device ratings, 467
  Section 450-3(b)(1), transformer overcurrent protection, 223, 225, 435, 450, 460, 462, 466, 481, 597
    Exception No. 1, transformer coordination, 450, 462, 466
  Section 450-3(b)(2), feeder taps, 341, 343, 451, 455, 460, 597
  Section 450-4, lighting and power feeders grounding, 364, 487, 488
  Section 450-13, suspended ceiling, 68
    Exception No. 1, dry-type transformers, 68
  Section 460-7, control circuit voltage regulation, 237
  Section 460-9, overcurrent protection, 204
  Section 501-4, raceway wiring methods, 610
  Section 501-16(b), service bonding, 568
  Section 502-4, raceway wiring methods, 610
  Section 503-3, raceway wiring methods, 610
  Section 514-5, circuit disconnect, 147
  Section 517-104, distribution system, 261*f.*, 364
  Section 517-160, lighting and power feeders grounding, 361
  Section 645-2(c), raised floor for data-processing circuits, 66
  Section 685-2, lighting and power feeder overcurrent protection, 313

National Electric Code (NEC) (*Cont.*):
  Section 700-26, service ground-fault protection, 541
  Section 701-7, emergency equipment, 576
  Section 725-2(e), remote-control circuit, 213
  Section 725-11 to Section 725-20, Class 1 control circuits, 314
  Section 725-12, control circuit overcurrent protection, 131, 213, 314*f.*, 218
    Exception No. 2, remote-control conductors, 222–225
    Exception No. 3, remote-control conductors, 131, 133, 218–219
    Exception No. 4, coil-circuit remote control, 133
  Section 725-14, control wires in raceways, 215
  Section 725-15, Class 1 control circuit, 214–216, 218
  Section 725-16(b), Class 1 control circuits, 214
  Section 725-17, Class 1 control circuit, 214, 215
  Section 725-17(a), control wires in raceways, 215
  Section 725-35, control-circuit overcurrent protection, 213
  Section 800-53(a), air-handling ceilings, 68
  Table 220-3(b), watts per square foot, 23, 99, 101, 265
  Table 220-11, lighting and power feeder branch-circuit loads, 271, 273, 499, 501, 530, 534
  Table 220-13, Demand Factors for Nondwelling Receptacle Loads, 268*f.*, 499
  Table 220-18, laundry facilities feeder load, 270, 499
  Table 220-19, cooking appliances, 60–61, 268, 272, 499, 506, 507
    Note 1, 18-kW range, 62
  Table 220-30, service feeders, 508–510, 530
  Table 221-3(b), dwelling unit receptacles, 105
  Table 250-94, bonding jumpers, 470–472, 477, 480, 481, 543–545, 559, 560, 562–563, 568–572
  Table 250-95, building ground, 379, 382, 384–389, 391, 572
  Table 300-5, raceway wiring methods, 612

INDEX   725

National Electric Code (NEC) (*Cont.*):
  Table 310-16 to Table 310-19, ampacity, 64, 83, 84, 86–91, 93, 156, 174, 219, 220, 276, 281, 304, 385, 459, 462, 479, 504, 511, 606, 618, 621–623
    Note 8, ampacity derating, 29, 30*f.*, 64, 87–91, 93, 214, 215, 276, 284, 304–306, 401, 461, 467, 482, 502–504, 618, 620, 621
    Note 10(c), resistive load, 89, 93, 276, 284, 504, 621
  Table 350-3, flexible raceways and fixture wires, 38
  Table 373-6, panelboards, 592
  Table 430-22, overcurrent protection, 204
  Table 430-22(a), nameplate current rating, 176
  Table 430-23(c), ampacity, 176
  Table 430-37, overcurrent protection, 204
  Table 430-72(b), overcurrent protection, 219, 220
  Table 430-147 to Table 430-151, rated motor voltages, 27–28, 174, 176, 182, 183, 195, 209, 272, 399, 400, 403, 409
  Table 430-152, protective device ratings, 176, 178–180, 182, 183, 185, 208, 394, 396, 400, 406–409, 411–413
  wire resistance table, 152, 156*f.*, 157
National Fire Protection Association (NFPA), 3
  Life Safety Code (NFPA No. 101), 3, 574
  Standard 90A, 66
NEC (*see* National Electric Code)
NEMA, 133, 315–316
  audible sound levels, 645
  fuse, 318
  kilovoltampere rating of transformers, 421
  panelboards, 592
  Switchboards, 586
  switching devices, 599, 600
NFPA (*see* National Fire Protection Association)
Noise levels, 645

Occupational Safety and Health Administration (OSHA), 3
  custom-made equipment, 5
  emergency supply, 574–575
  lighting and power feeders, 364–365
  motor branch circuit controls, 199
  National Electric Code and, 4
  poke-through receptacle, 110

Occupational Safety and Health Administration (OSHA) (*Cont.*):
  Subpart S, 4–6
  switchboards, 589
  third-party certification, 4
  transformers, 225, 444
Office buildings:
  electrical design planning, 10–11
  lighting and appliance branch circuits, 22–24, 28, 99, 101, 102, 133, 135*f.*
    contactor control, 133, 135*f.*
  lighting and power feeders, 268, 269*f.*
Ohm's Law, 443–444, 636
One-family houses (*see* Dwelling units)
115/230-V (120/240-V), 3-wire, single-phase system, 250, 415
120/208-V, 3-phase, 4-wire system, 250–252, 415, 416
120/240-V (115/230-V), 3-wire, single-phase system, 250, 415
120/280-V, 3-wire system, 18
Open-delta 3-phase system, 18
Orange Book (Electrical Appliance and Utilization Equipment Directory), 6, 69
OSHA (*see* Occupational Safety and Health Administration)
Outdoor receptacles, branch circuits for, 46–49, 59, 117, 120, 123, 130–131
Outlets:
  lighting and appliance branch circuits, 59–63
  lighting and power feeders, 267–268
Overcurrent protection:
  branch circuits, 20–22
  busway wiring methods, 629
  control circuits, 212–214, 218–225
    magnetic starters, 219
    remote-control device, 220–222
    secondary wires, 220, 223*f.*
    transformers, 221–225
  lighting and appliance branch circuits, 69–82
    cartridge fuse, 75, 76, 78*f.*
    circuit breaker, 77–82
    Class R fuse, 75–77
    conductors, 88
    grounded conductor, 71, 73*f.*
    multiwire circuits, 78–80
    panelboards, 145–146
    plug fuses, 75, 76*f.*
    rating, 69, 70*f.*
    readily accessible, 74–75
    shock hazard, 81
    600 V fuse, 76

Overcurrent protection
  lighting and appliance branch circuits (*Cont.*):
    split-wired receptacles, 81–82
    in suspended ceilings, 74
    tap wire sizes, 74
    two-pole disconnect, 81
    type S fuse, 75, 76*f.*
    ungrounded conductor, 71, 72*f.*
    wire size, 71, 72*f.*
  lighting and power feeders, 306, 308–357
    automatic response, 309, 310*f.*
    with circuit breaker, 312–313
    continuous current rating, 312
    with fuse, 312–313
    ground-fault protection, 347–357
      building disconnect, 356–357
      480Y/277-V system, 355–356
      low-level arcing, 348, 349*f.*
      residual method, 354
      sensing, 350–351
      time-delay, 352–355
      zero-sequence sensing, 351–352
      zone-selective instantaneous system, 353–354
    installation location, 333–335
      busway, 334–337
      readily accessible, 333–335, 338*f.*
    short-circuit protection, 322–331, 390*f.*
      calculations, 324, 325
      device selection coordination, 325, 326*f.*
      elements, 323
      flow patterns, 326–328
      interrupting capacity, 325
      operating speed, 325–327
    sizing, 311–312, 314–322
      continuous loading, 314, 317–318
      switch assembly, 317
    at supply, 311, 313*f.*
  motor branch circuits, 201–206
    exceptions, 203–204, 206
    requirements, 201–203
    setting adjustments, 204, 205*f.*
    three-phase starter, 204
  motor feeders, 396, 399–400
    examples, 396, 398–400
    sizing, 396, 398*f.*
  services, 531–536
    conductor ampacity, 532, 533*f.*
    conductor overload, 534
  supplementary, 15–16

Overcurrent protection (*Cont.*):
  transformers, 446–449
    examples, 481–484
    layout, 446, 447*f.*
    on primary, 446, 448*f.*
    on secondary, 447, 449*f.*

Panelboards, 250, 591–593
  feeder distribution panels, 591, 592*f.*
  installation guidelines, 592–593
  lighting and appliance branch circuits, 133–134, 136–147, 265, 278–279, 591
    capacity supplement, 146
    circuit disconnect, 147, 150*f.*
    controls, 146, 149*f.*
    in dwelling units, 144–145, 147*f.*, 148*f.*
    example, 134, 136–138
    layouts, 133, 136*f.*
    loadcenter, 133
    overcurrent protection, 145–146
    panel main protection, 138–142
      circuit breaker, 142
      exemptions, 139, 141*f.*, 142, 144*f.*
      fuse, 142
      location, 142, 143*f.*
      snap switches, 142–143, 145*f.*
      from transformers, 143, 146*f.*
    plan layouts, 160, 162*f.*, 164–166
    rating, 134, 137*f.*, 136
  lighting and power feeders, 274–276
    sizing versus rating, 276–280, 314
  power distribution panels, 591
  service-equipment, 591
  special, 591
Paper and pulp mill power supply source, 8
Piping, bonding of, 570–574
Planning, 1–13
  accessibility, 7
  background information, 2
  budget, 1
  building's electrical requirements, 1
  checklists, 9–13
    health care facilities, 13
    industrial building, 9
    office building, 10–11
    school, 12–13
    shopping center, 11–12
  codes and standards conformity, 1, 2
  electrical circuiting concepts implementation, 1
  electrical plans, 1–2
  electrical system installation, 1

Planning (*Cont.*):
  energy supply characteristics, 1
  flexibility, 7
  insurance company regulations, 3
  load capacity growth allowance, 6–7
  local codes compliance, 3
  National Electric Code compliance, 3, 5–6
  Occupational Safety and Health Administration compliance, 3–6
  reliability, 7
  safety provisions, 2–3
  system analysis, 7–9
    building type, 7, 8
    equipment standardization, 8
    power supply source, 7–9
    third-party certification, 4–6
    wiring concepts and configurations selection, 1
Plug receptacles (*see* Receptacles)
Poke-through receptacle, 108, 110, 112*f.*, 113*f.*
Power distribution panels, 591
Power factor, 636
  of motor feeders, 394–396
Power feeders (*see* Lighting and power feeders)
Power load data, 640–643
Power loss calculations, 636
Power supply source, 7–9
Process control equipment, 3
Proprietary Protective Signaling Devices, 3
Protective devices, automatic, 2
Protective signaling devices, 3
Public Fire Service Communications, 3
Purged Enclosures for Electrical Equipment, 3

Raceways, 90
  control circuit wires in, 215–218, 277, 278
  wiring methods, 608, 610–612
  conductors in, 612–623
Reactive kilovoltamperes (capacitor kilovars), 395
Readily accessible, 74–75
Receptacles:
  lighting and appliance branch circuits, 97–106
    amperage loading schedule, 102, 105*f.*
    application, 106–111
      aluminum conductors, 108, 110*f.*
      ampere rating, 106, 108, 111*f.*
      FCC wiring, 110–111, 113*f.*
      grounded metallic system, 106–109

Receptacles
  lighting and appliance branch circuits
    application (*Cont.*):
      poke-through receptacle, 108, 110, 112*f.*, 113*f.*
    controls, 119–121
    in dormitories, 105
    in dwelling units, 43–45, 50, 51*f.*, 102, 105–106
    for fan, 97–98
    for heating equipment, 98
    in hotels and motels, 45, 105
    layouts, 101
    load, 98, 501
    multioutlet assemblies, 101
    for nonresidential occupancies, 99–102
    number required, 98
    plan layouts, 169, 170*f.*
    for plug-in devices, 98
    pole, 102, 104*f.*
    for small-appliances, 97–98, 102, 106
    spacing requirements, 106
  lighting and power feeders, 267–268
Recommended Practices for Electrical Equipment Maintenance, 3
Red Book (Hazardous Location Equipment Directory), 6
Red-leg (three-phase, 4-wire) delta system, 18, 636
Refrigeration equipment, 173–174
  branch circuits, 59, 119, 190–192, 209
  utilization voltages, 25
Remote Station Protective Signaling Systems, 3
Residential buildings (*see* Dwelling units)
Rooftop, lighting and appliance branch circuits for, 50–52

Safety precautions:
  color coding, 97
  in planning stage, 2–3
  shock hazard, 81
Schools:
  electrical design planning, 12–13
  lighting and appliance branch circuits, 28, 101, 102*f.*
  lighting and power feeders, 268, 297–299
Service-equipment panelboards, 591
Services, 491–584
  design procedure, 493–499
    conductors, 496, 497*f.*, 499
    multiservices, 494, 496*f.*

Services
  design procedure (*Cont.*):
    service lateral, 497, 498*f.*
    utility regulations, 494, 495*f.*
  disconnects, 517–531
    grouping, 521–523
      emergency disconnect, 521–523, 577, 579*f.*
      identification marking, 521
    installation location, 523, 525–529
      equipment grouping, 526, 527*f.*, 529
      in multi-occupancy buildings, 523, 526*f.*, 529
      number of sets, 527, 528*f.*
      readily accessible, 523, 525*f.*
    layout arrangements, 517, 519*f.*
    multibuilding premises, 531, 532*f.*, 535
    number of, 519, 520
    rating, 529–531
      multiple services, 530, 531*f.*
    types of, 529
      manual operation, 529, 530*f.*
  in dwelling units, 491–492, 506–513
  emergency supply, 573–584
    emergency equipment, 575–581
      automatic transfer, 578, 580*f.*
      battery-pack light units, 577, 579*f.*
      disconnects, 577, 579*f.*
      hookups, 580, 581*f.*
      lighting, 578, 580*f.*
      services from independent sources, 577, 578*f.*
      water-cooled generator, 577
      wiring, 580–582
    standby systems, 581–584
      manual transfer switch, 582–583
      uninterruptible power supply, 583–584
  feeder current rating, 499–508
    residential service entrance conductors, 506–508
      optional calculation, 508–513
    sizing, 499, 500*f.*
      example, 501–506
  ground-fault protection, 536–542
    alternative fault-current paths, 539, 540*f.*
    disconnect subdividing, 538
    generator, 539, 541
    rating, 537
    transfer switch, 539, 541*f.*
  grounding, 542–573
    bonding, 564–567

Services
  grounding
    bonding (*Cont.*):
      burn-clear, 563, 564
      bushings and fittings, 568, 571*f.*
      conduit sleeve, 567, 569*f.*
      flex, 565, 566*f.*
      low-impedance path, 564, 565*f.*
      piping, 570–574
      skin effect, 567, 568*f.*
      techniques, 565–568, 570*f.*, 571*f.*
      wiring methods, 568, 570*f.*
    bonding jumpers, 568–570
      parallel conduits, 570, 573*f.*
      sizing, 569–570, 572*f.*, 573*f.*
    building metal frame, 546, 547*f.*
    connections, 552–554
      bonding, 553–555
      bonding jumper, 553, 557*f.*
      common grounding-electrode conductor, 553, 556*f.*
    electrode system, 474, 544–545
    ground rod, 546, 548*f.*
    made electrodes, 550–552
      ground-rod installation, 550, 551*f.*
      resistance to earth measurement, 551, 552*f.*
    multiple service entrance disconnects, 561–563
      bonding, 563, 564
      layout, 561, 563*f.*
    neutral, 554–557
      circuit conductor, 555, 557*f.*
      to load side, 555–556, 558*f.*
    pole-type, 560–562
    Ufer system, 549
    unused-neutral, 557–560
    water-pipe electrode, 545, 546
  overcurrent protection, 531–536
    conductor ampacity, 532, 533*f.*
    conductor overload, 534
  overhead, 492, 493
  service drop, 492, 493
  service entrance, 493
  service lateral, 492, 493, 497, 498*f.*
  service point, 492
  underground, 492, 493*f.*
  underground residential distribution, 492
  wiring methods, 513–516
    concrete encasement, 514, 515*f.*
    flexible conduit, 514
    head position, 515–516

Services
  wiring methods (*Cont.*):
    underground, 514–518
7200-V, 3-phase system, 256
Shock hazard, 81
Shopping center electrical design planning, 11–12
Short-circuit protection:
  calculations:
    point-to-point method, 646–647
    transformer in system, 648–651
  lighting and power feeders, 322–331, 390*f.*
    calculations, 324, 325
    device selection coordination, 325, 326*f.*
    elements, 323
    flow patterns, 326–328
    interrupting capacity, 325
    operating speed, 325–327
  low-voltage feeder, 651–667
  motor branch circuits, 178–190
    group installation, 188, 189*f.*
    instantaneous-trip CB, 180–181
      example, 181–182
      magnetic, 180–181
    motor controllers (NJOT), 179–180
    motor short-circuit protector, 180, 182
    multimotors, 182–190
      any rating, 183–189
      example, 188
      with fuses, 185–186
      multiple fractional-horsepower, 186, 188*f.*
      small motors, 182–183
    rating, 178, 179
    sizing, 176*f.*, 179
    tap conductors, 188–190
  motor feeders, 406–413
  short-circuit withstandability, 595–597, 599
Show window:
  lighting and appliance branch circuits, 50, 52*f.*, 62, 63*f.*, 99
  lighting and power feeders, 267, 272
Single-family houses (*see* Dwelling units)
Single-phase, 3-wire system, 17, 636
600-V system, 258, 262*f.*
Skin effect, 158, 291, 567, 568*f.*
Snow-melting and deicing installations, 36
Soares, Eustace, 387, 389
Sound levels, 645
Spray Application, 3
Stairway, lighting and appliance branch circuits for, 59, 120, 123

Standard on Intrinsically Safe Process Control Equipment, 3
Standby systems, 581–584
Starting, accidental, of control circuits, 230–236
Stores:
  lighting and appliance branch circuits for, 28, 50, 52*f.*, 62, 63*f.*, 99, 101
  lighting and power feeders for, 267, 268, 272
Subfeeder, 250, 264
Sump pump, 119
Supplementary overcurrent protection, 15–16
Suspended ceilings, lighting and appliance branch circuits in, 68–69, 74
Switchboards, 250, 585–590
  circuit makeup, 589–590
  disconnect identification, 589, 590*f.*
  headroom and illumination, 588, 589*f.*
  installation location, 587–588
  minimum clearances, 586–587
Switchboards (NEMA), 586
Switching devices, 593–604
  automatic transfer switches, 601–604
    circuit breaker, 601, 602*f.*
    operating coils, 601, 602*f.*
  fault-interruption service, 594–595
  high-capacity switches, 600–601
  selection considerations, 599–600
  short-circuit withstandability, 595–597, 599
  sizing, 593, 594*f.*
    motor and lighting load in power panel, 593
    motor load only in power panel, 593
  transformer inrush, 597–599
System grounding (*see* Equipment ground)

Taps:
  lighting and power feeders, 311, 335–347, 454, 455, 461
    examples, 338–347
      panelboard with unprotected conductors, 339–341
      single-motor circuit, 338–340
      transformers, 341–345
      unprotected tap, 343–347
      under ten feet, 336–337, 339*f.*
      under twenty-five feet, 337, 340*f.*
  tap conductor, 188–190
  transformer, 416–417
TCD Comment No. 11-30 Substantiation, 406–407

730 INDEX

Temperature ratings for lighting and appliance branch circuits, 83–86, 620
Thermal cutouts, 15
Third-party certification, 4–6
13.2Y/7.2-kV (13.8-kV), 3-phase, 4-wire system, 256, 259*f.*
13,200-V system, 259
Three-phase, 3-wire system, 18, 636
Three-phase, 4-wire (red-leg) delta system, 18, 636
Three-phase, 4-wire wye system, 17, 636
Torque motors, 192
Town-house (*see* Dwelling units)
Transformers, 415–489
  autotransformers, 484–489
    applications, 486–487
    grounding, 487–489
    hookup, 484, 485*f.*
  circuit breaker and fuse nomograph IC ratings, 653*f.*, 654*f.*, 668
  conductors, 479–480
  control circuits, 225–228
    accidental starting protection, 231, 232*f.*
      inside starter, 232, 235*f.*
      outside starter, 232, 236*f.*
    bonding, 228
    grounding, 226–228
    overcurrent protection, 221–225
    requirement exceptions, 226–227
  coordination, 449–461
    built-in overload protection, 460
    feeder and transformer tap layout, 457–460
    inrush current, 449
    primary protection, 456–457
    secondary-circuit conductor protection, 454–456
    time-current trip-curves, 450–453
  destruct curves, 432–446
    ANSI curve shift, 433–443
      current flow, 435, 436*f.*
      examples, 436–443
      recommendations, 436, 437*f.*
    enclosure, 444
    grounding provisions, 445
    mounting, 444
    per unit system, 433, 443–444
    transformer category, 433–435
    UL listing, 444
    winding connections, 445–446
  efficiency, 430–432

Transformers
  efficiency (*Cont.*):
    Class 220, 80-degree C rise transformer, 432
    Class 220, 115-degree C rise transformer, 431–432
    Class 220, 150-degree C rise transformer, 431, 432
  examples, 461–469, 477–484
  frequency rating, 429
  grounding, 469–477, 542
    bonding jumper, 470–472
    examples, 480–481
    exemptions, 476, 478*f.*
    grounding electrode, 471*f.*, 473–474
    grounding electrode conductor, 471–473, 475–476
    secondary neutral, 470, 474, 475*f.*
    unacceptable methods, 476, 477*f.*
  harmonics, 421–427
    CBEMA crest factor, 421–427
    effects on K-factor, 428
    IEEE method, 421, 423–427
    sizing, 421–422
  impedance, 323–324
  K-factor, 421–428
    CBEMA crest factor, 421–427
    computation, 423, 424*f.*, 427, 428
    harmonic orders effects on, 428
    IEEE method, 421, 423–427
    sizing, 421–422
  kilovoltampere rating, 420–421
  lighting and power feeders, 341–345
  noise rating, 429
  number of phases, 429
  overcurrent protection, 446–449
    examples, 481–484
    layout, 446, 447*f.*
    on primary, 446, 448*f.*
    on secondary, 447, 449*f.*, 648–651
  switching devices, 597–599
  temperature rating, 429–430
    ambient temperature, 430
    Class 105, 429
    Class 150, 429–430
    Class 185, 430
    Class 220, 430
    duty cycle, 430
    relative loading, 430
  total demand load, 479
  voltage rating, 416–420
    circuit design, 418

INDEX 731

Transformers
  voltage rating (*Cont.*):
    examples, 418–420
    rated secondary current, 416
    rated secondary voltage, 416
    taps, 416–417
240-V, 3-phase, 3-wire system, 253
2400-V, 3-phase delta system, 254

Ufer grounding system, 549
UL (*see* Underwriters Laboratories Inc.)
Underfloor space, lighting and appliance branch circuits in, 59, 122
Underground residential distribution (URD), 492
Underwriters Laboratories Inc. (UL), 4, 5
  appliances and utilization equipment, 83–84
  baseboard heater, 45
  circuit breaker, 82, 84–86, 278, 315, 316f., 333
  conductors, 347, 385
  distribution and control equipment, 84
  electric-discharge lighting fixture, 30
  Electrical Appliance and Utilization Equipment Directory (Orange Book), 6, 69
  Electrical Construction Materials Directory (Green Book), 6, 83, 194, 315–317, 444, 511, 550–551, 576, 603, 608, 631
  emergency equipment, 575–576
  equipment terminals, 90
  equipment terminations temperature ratings, 83
  FCC wiring, 111
  fuse, 76, 318, 319, 332
  Hazardous Location Equipment Directory (Red Book), 6
  made electrodes, 550
  modular wiring, 42
  motor branch circuits, 179, 182, 206
  overcurrent protection, 314, 482
  panelboards, 133, 145–146, 250
  poke-through receptacle, 108, 112f., 113f.
  services, 504, 505, 517
  Standard 489, 608
  Standard UL-1008, 603
  switchboard, 315
  switching devices, 593, 599, 600, 603
  transformers, 444, 482
  wiring methods, 608, 610–612, 620–624, 631, 632

Uninterruptible power supply (UPS), 583–584
URD (underground residential distribution), 492
Utility regulations, 494, 495f.
Utility room, lighting and appliance branch circuits for, 59, 122
Utilization equipment, 4, 83–84
Utilization voltages (*see* Voltage)

Voltage:
  capacitor kilovars, 395
  control circuit regulation, 234–239
    example, 235–236, 239
    power factor, 235–240
  distribution system levels, 250–263
    480-V, 3-phase, 3-wire system, 253–254
    480Y/277-V (460Y/265-V), 3-phase, 4-wire system, 253–256
    4160Y/2400-V, 5-phase, 4-wire system, 254, 256–259
    4800-V system, 256
    layouts, 250, 251f.
    low-voltage service, 258, 260f.
    120/208-V, 3-phase, 4-wire system, 250–252
    120/240-V, 3-wire, single-phase system, 250
    7200-V, 3-phase system, 256
    600-V system, 258, 262f.
    special load equipment, 258–259, 263f.
    13.2Y/7.2-kV (13.8-kV), 3-phase, 4-wire system, 256, 259f.
    13,200-V system, 259
    240-V, 3-phase, 3-wire system, 253
    2400-V, 3-phase delta system, 254
    variation from nominal, 259, 263f.
  lighting and appliance branch circuits, 17–18
  utilization voltage, 24–28
  voltage drop, 148, 150–159
    copper conductors, 150, 154f.
    copper loss calculations, 150, 152f.
    dc motor, 153, 157f.
    formulas, 150, 151f.
    homerun, 150, 153f.
    skin effect, 158
    wire resistances, 152, 156f.
    wire size selection, 150, 155f.
  voltage limits, 29–34
    in dwelling units, 17, 29–30, 33, 34
    in hotels and motels, 29, 33

Voltage
  lighting and appliance branch circuits
    voltage limits (*Cont.*):
      in institutional buildings, 31, 33*f*.
      120-V loads, 30, 31*f*.
      over 120-V loads, 30, 32*f*.
      split wiring, 34, 35*f*.
      split-wired receptacles, 33–35
      277-V loads, 30–32
      ungrounded circuits, 32–33
    voltage selection:
      480/277-V (460/265-V) wye-
        connected system, 17–18
      loadcenter layout, 17
      120/280-V, 3-wire system, 18
      open-delta 3-phase system, 18
      single-phase, 3-wire system, 17
      three-phase, 3-wire system, 18
      three-phase, 4-wire (red-leg) delta
        system, 18
      three-phase, 4-wire wye system, 17
  lighting and power feeders, 292–299
  motor feeders, 394–396
    capacitor power-factor correction,
      395–397
  transformers, 416–420
    circuit design, 418
    examples, 418–420
    rated secondary current, 416
    rated secondary voltage, 416
    taps, 416–417
Voltamperes, 279

Watts, 279
Wet location, 65
Winding connections of transformers,
  445–446
Wiring methods, 604–634
  armored cable, 631–634
    interlocked armor (IA), 632–634
    Type AC (BX), 631
    Type MC, 631–632
  busways, 624–630

Wiring methods
  busways (*Cont.*):
    applications, 625
    isometric and floor plan, 626, 627*f*.
    overcurrent protection, 629
    plug-in feeder, 630
    risers and horizontal runs, 628
  conductor terminations, 623–624
  conductors, 605–608
    conductor size, 605
    example, 606, 608
    feeder makeup, 605–607
    insulation, 606
    protective coverings, 606
    underground installation, 608, 609*f*.
  conductors in raceways, 612–623
    ampacity derating, 618–620
      circuit makeup, 618–620
    conductor temperature ratings, 620–623
    conduit fill, 614–618
      cross-sectional areas, 614, 615*f*.
      examples, 616–617
      neutrals, 617–618
    independent motors, 614, 615*f*.
    power and control wire intermix,
      612–614
  control circuits, 215–218
  emergency supply, 580–582
  FCC, 110–111, 113*f*.
  grounding with, 358–363
  lighting and appliance branch circuits,
    17–18, 42
  in air-handling ceilings, 63–68
  raceways, 608, 610–612
    burial depths, 612, 613*f*.
    electrical metallic tubing, 608, 610, 612
    intermediate metal conduit (IMC), 610,
      612
    rigid nonmetallic conduit, 611
  services, 513–516
  underground installation, 605

Zig-zag grounding, 487–489

ABOUT THE AUTHORS

Joseph F. McPartland, "Mr. Electrical Construction," is the nation's foremost expert on electrical design and contruction. He is the author of 26 books on electrical design and construction.

Brian J. McPartland is an electrical consultant and editor of *Electrical Contractors: Design and Installation Update,* a bimonthly technical report. He has held positions in both product engineering and sales with various electrical equipment manufacturers and was chief editor at "edi" (*Electrical Design and Installation*) magazine. Both he and Joseph F. McPartland are co-authors of *McGraw-Hill's National Electrical Code® Handbook,* 21st Edition and *McGraw-Hill's Handbook of Electrical Construction Calculations.*